Algorithms for Continuous Optimization

NATO ASI Series

Advanced Science Institutes Series

A Series presenting the results of activities sponsored by the NATO Science Committee, which aims at the dissemination of advanced scientific and technological knowledge, with a view to strengthening links between scientific communities.

The Series is published by an international board of publishers in conjunction with the NATO Scientific Affairs Division

A **Life Sciences**	Plenum Publishing Corporation
B **Physics**	London and New York
C **Mathematical**	Kluwer Academic Publishers
and Physical Sciences	Dordrecht, Boston and London
D **Behavioural and Social Sciences**	
E **Applied Sciences**	
F **Computer and Systems Sciences**	Springer-Verlag
G **Ecological Sciences**	Berlin, Heidelberg, New York, London,
H **Cell Biology**	Paris and Tokyo
I **Global Environmental Change**	

NATO-PCO-DATA BASE

The electronic index to the NATO ASI Series provides full bibliographical references (with keywords and/or abstracts) to more than 30000 contributions from international scientists published in all sections of the NATO ASI Series.
Access to the NATO-PCO-DATA BASE is possible in two ways:

– via online FILE 128 (NATO-PCO-DATA BASE) hosted by ESRIN,
Via Galileo Galilei, I-00044 Frascati, Italy.

– via CD-ROM "NATO-PCO-DATA BASE" with user-friendly retrieval software in English, French and German (© WTV GmbH and DATAWARE Technologies Inc. 1989).

The CD-ROM can be ordered through any member of the Board of Publishers or through NATO-PCO, Overijse, Belgium.

Algorithms for Continuous Optimization
The State of the Art

edited by

Emilio Spedicato

**Department of Mathematics,
University of Bergamo,
Bergamo, Italy**

Kluwer Academic Publishers

Dordrecht / Boston / London

Published in cooperation with NATO Scientific Affairs Division

Proceedings of the NATO Advanced Study Institute on
Algorithms for Continuous Optimization: The State of the Art
Il Ciocco, Barga, Italy
September 5–18, 1993

A C.I.P. Catalogue record for this book is available from the Library of Congress.

ISBN-13: 978-94-010-6652-5   e-ISBN-13: 978-94-009-0369-2
DOI: 10.1007/ 978-94-009-0369-2

Published by Kluwer Academic Publishers,
P.O. Box 17, 3300 AA Dordrecht, The Netherlands.

Kluwer Academic Publishers incorporates the publishing programmes of
D. Reidel, Martinus Nijhoff, Dr W. Junk and MTP Press.

Sold and distributed in the U.S.A. and Canada
by Kluwer Academic Publishers,
101 Philip Drive, Norwell, MA 02061, U.S.A.

In all other countries, sold and distributed
by Kluwer Academic Publishers Group,
P.O. Box 322, 3300 AH Dordrecht, The Netherlands.

Table of Contents

Preface

The NATO Advanced Study Institute on "Algorithms for continuous optimization: the state of the art" was held September 5-18, 1993, at Il Ciocco, Barga, Italy. It was attended by 75 students (among them many well known specialists in optimization) from the following countries: Belgium, Brasil, Canada, China, Czech Republic, France, Germany, Greece, Hungary, Italy, Poland, Portugal, Rumania, Spain, Turkey, UK, USA, Venezuela. The lectures were given by 17 well known specialists in the field, from Brasil, China, Germany, Italy, Portugal, Russia, Sweden, UK, USA.

Solving continuous optimization problems is a fundamental task in computational mathematics for applications in areas of engineering, economics, chemistry, biology and so on. Most real problems are nonlinear and can be of quite large size. Developing efficient algorithms for continuous optimization has been an important field of research in the last 30 years, with much additional impetus provided in the last decade by the availability of very fast and parallel computers. Techniques, like the simplex method, that were already considered fully developed thirty years ago have been thoroughly revised and enormously improved.

The aim of this ASI was to present the state of the art in this field. While not all important aspects could be covered in the fifty hours of lectures (for instance multiobjective optimization had to be skipped), we believe that most important topics were presented, many of them by scientists who greatly contributed to their development.

Of the 20 contributions presented in this volume, 19 are based on the lectures given by the official lecturers. The extra contribution, by professor Sargent, one of the founders of continuous optimization, deals with an interesting topic which was the subject of a lively special session. In addition to the lectures more than 30 contributed papers were presented during the ASI. A selection of these will appear in a special issue of the journal Optimization Methods and Software.

The first paper, by F. Giannessi, gives a general approach to optimality conditions via a separation scheme. The following three papers, by C.G. Broyden, Å. Björck and J.M. Martínez, present algorithms for linear equations, least squares and nonlinear equations, which are subproblems appearing in the context of continuous optimization. The papers by R. Fletcher and N. Deng, Z. Li, deal with the unconstrained nonlinear optimization, the second paper specializing on the issue of nonquadratic models (many of the results presented here are new and were obtained while preparing the lecture). Algorithms for constrained nonlinear optimization are considered in the chapters by M.C. Bartholomew-Biggs, G. Di Pillo, Y. Evtushenko and V.G. Zhadan, and A. Conn, N. Gould and P.C. Toint. The paper of Evtushenko and Zhadan presents in particular several important results obtained in the last twenty years in the former Soviet Union and previously practically unknown in the West. The new technique called ABS methods is described by E. Spedicato and Z. Xia, who show how it leads to a unifying framework for many optimization methods. The subject of nonsmooth optimization is treated by C. Lemaréchal and J. Zowe, who discuss also some important applications. The classic simplex method, in the modern efficient formulations, and the interior point methods for solving linear programming problems are discussed by D. F. Shanno, with a thorough consideration of computational issues. J. Stoer gives theoretical results on the newly proposed infeasible interior point methods. The important linear complementarity problem is considered in detail by J. Judice. R. Sargent discusses then the "big-M" problem, solving an apparent paradox. Global optimization via deterministic techniques is then considered by Y.G. Evtushenko and M.A. Potapov, for the case of small or moderate number of variables. Some other techniques for this problem are briefly considered by R. Schnabel, who addresses the opportunities offered by parallel computers. L.C. W. Dixon discusses in two contributions the new technique of automatic differentiation, which is likely to give in the future great improvement in the performance of optimization algorithms, and the approximation of function via neural networks, another recently developed subject.

While some papers indicate two authors, the lectures were actually given by the following lecturers: Deng, Evtushenko, Conn, Spedicato, Zowe.

The ASI was organized in the framework of the activities of NATO Scientific Affairs Division. NATO provided the main financial support. Additional support and collaboration are acknowledged from the Department of Mathematics, University of Bergamo, CNR (Comitato 1, Comitato 11, Progetto Finalizzato Trasporti), Banca Popolare di Bergamo and Unione Industriali di Bergamo.

An important event during the ASI was the celebration of the 60th birthday of Professor C.G. Broyden, a lecturer and one of the great creative scientists in optimization (and in other fields). He is a coauthor of the most famous and used in practice algorithm in nonlinear optimization, the BFGS methods, and we were fortunate that also two other coauthors of the method, Professor R. Fletcher and D.F. Shanno, were present at the event and as lecturers.

This ASI could not have been organized without the scientific collaboration of the codirectors professors L.C.W. Dixon and D.F. Shanno and the organizational help of Dr. M.T. Vespucci, R. Wang and particularly Z. Huang, to whom I express my sincere thanks.

This volume is dedicated to professors C.G. Broyden, L.C.W. Dixon and D.F. Shanno, whose beautiful work in the field of Quasi-Newton methods at the beginning of the seventies was for me inspiration for working in optimization.

Bergamo University, February 1994

Emilio Spedicato
ASI Director

List of Contributors

Mike C. BARTHOLOMEW-BIGGS
Department of Information Science
University of Hertfordshire
Hatfield Campus, College Lane
Hatfield Herts, AL10 9AB
UK

Åke BJÖRCK
Department of Mathematics
Linköping University
S-581 83 Linköping
SWEDEN

Charles G. BROYDEN
Facoltá di Scienze dell'Informazione
Universitá di Bologna
Via Sacchi 3, 47023 Cesena
ITALY

Andy CONN
Mathematical Science Department
IBM T.J. Watson Research Center
P.O.Box 218
Yorktown Heights, NY 10598-0218
USA

Naiyang DENG
Division of Basic Science
Beijing Agricultural Engineering University
Beijing 100083
P.R.CHINA

Gianni DI PILLO
Dipartimento di Informatica e Sistemistica
Universitá di Roma "La Sapienza"
Via Buonarroti, 12
00185 Roma
ITALY

Laurence DIXON
Department of Information Science
University of Hertfordshire
Hatfield Campus, College Lane
Hatfield Herts, AL10 9AB
UK

Yuri EVTUSHENKO
Computer Center
Russian Academy of Sciences
40, Vavilov Street
117967 Moscow GSP-1
RUSSIA

Roger FLETCHER
Department of Mathematical Sciences
University of Dundee
Dundee DD1 4HN
UK

Franco GIANNESSI
Dipartimento di Matematica
Universitá di Pisa
Via F. Buonarroti 2
56100 Pisa
ITALY

Nicholas GOULD
Central Computing Department
Rutherford Appleton Laboratory
Chilton, Oxon, OX11 0AL
Oxfordshire

Joaquim JUDICE
Departamento de Matemática
Universidade de Coimbra
3000 Coimbra
PORTUGAL

Claude LEMARÉCHAL
INRIA, Rocquencourt
B. P. 105
78153 LE CHESNAY
FRANCE

Zhengfeng LI
Division of Basic Science
Beijing Agricultural Engineering University
Beijing 100083
P.R.CHINA

José Mario MARTÍNEZ
Department of Applied Mathematics
IMECC-UNICAMP, CP 6065
University of Campinas
13081 Campinas SP
BRAZIL

Michael A. POTAPOV
Computer Center
Russian Academy of Sciences
40, Vavilov Street
117967 Moscow GSP-1
RUSSIA

Roger SARGENT
Center of Process Systems Engineering
Imperial College
London, SW7 2BY
UK

Robert SCHNABEL
Department of Computer Science
University of Colorado at Boulder
ECOT 7-3 Engineering Center
Campus Box 430
Boulder, Colorado 80309-0430
USA

David SHANNO
RUTCOR
Rutgers Center for Operations Research
Rutgers University
New Brunswick
New Jersey 08903-5062
USA

Emilio SPEDICATO
Dipartimento di Matematica
Universitá di Bergamo
Piazza Rosate 2
24129 Bergamo
ITALY

Josef STOER
Institut für Angewandte Mathematik
Universität Würzburg
Am Hubland
W-8700 Würzburg
GERMANY

Philippe TOINT
Département de Mathematiques
Facultés Universitaires N-D de la Paix
61, rue de Bruxelles
B-5000 Namur
BELGIUM

Zunquan XIA
Department of Applied Mathematics
Dalian University of Technology
Dalian 116024
P.R. China

Vitali G. ZHADAN
Computer Center
Russian Academy of Sciences
40, Vavilov Street
117967 Moscow GSP-1
RUSSIA

Jochem ZOWE
Friedrich-Schiller-Universität Jena
Institut für Angewandte Mathematik
07740 JENA
GERMANY

List of Participants

BELGIUM

Willem K. BRAUERS
University of Antwerpen
RUCA
Birontlaan 97
B-2600, Berchen-Antwerpen

Philippe TOINT
Departement de Mathématiques
Facultées Universitaires N-D de la Paix
61, rue de Bruxelles
B-5000 Namur

BRASIL

Ana FRIEDLANDER
IMECC-Unicamp
CP 6065
University of Campinas
13081 Campinas SP

BULGARIA

Kostadin IVANOV
Institute for Nuclear Research
and Nuclear Energy
Bulgarian Academy of Sciences
Blvd. Tzarigradsko Shaussee 72
Sofia 1784

CANADA

Syed M. AHSAN
Economics Department
Concordia University
1455 de Maisonneuve Blvd. W.
Montreal, PQ H3G 1M8

Bert BUCKLEY
Department of Mathematics
Royal Roads Military College
FMO Victoria, B.C. V0S 1B0

Paul H. CALAMAI
Department of Systems Design Engineering
University of Waterloo
Waterloo, Ontario, N2L 3G1

Henry WOLKOWICZ
Faculty of Mathematics
Department of Combinatorics
and Optimization
University of Waterloo
Waterloo, Ontario, N2L 3G1

P.R. CHINA

Zhi CHEN
Department of Mathematics
Beijing Polytechnic University
Beijing 100022

Meifang ZHU
Department of Mathematics
Beijing Polytechnic University
Beijing 100022

CZECH REPUBLIC

Oitka DRKOSOVA
ICIS, Academy of Sciences
Pod vodarenskou vezi 2
Prague 8

Ladislav LUKSAN
ICIS, Academy of Sciences
Pod vodarenskou vezi 2
Prague 8

Miroslav TUMA
ICIS, Academy of Sciences
Pod vodarenskou vezi 2
Prague 8

FRANCE

Susana GOMEZ
INRIA
Domaine de Rocquencourt
78153 Le Chesnay Cedex

Clovis GONZAGA
INRIA
Domaine de Rocquencourt
78153 Le Chesnay Cedex

Sonnana B. HAMMA
Departement de Mathématiques
Université de Limoges
87060 Limoges Cedex

Joao PATRICIO
CERFACS
Ave. de Coriolis 42
31057 Toulouse Cedex

Claudia SAGASTIZABAL
INRIA
Domaine de Rocquencourt
78153 Le Chesnay Cedex

GERMANY

Sonja BERNER
Fachbereich Mathematik
Universität Wuppertal
Gauss-Strasse 20
5600 Wuppertal

Volker DRYGALLA
Fachbereich Mathematik
Martin Luther Universität
4010 Halle

Eildert GRÖNEVELD
Institute of Animal Husbandry
and Animal Behaviour
Federal Agricultural Research Center (FAL)
3057 Neustadt 1

Florian JARRE
Institut für Angewandte Mathematik
Universität Würzburg
Am Hubland
8700 Würzburg

Helmut MAURER
Institut für Numerische Mathematik
Westfälische Wilhelms-Universität Münster
Einsteinstr. 62
4400 Münster

Arnold NEUMAIER
Institut für Angewandte Mathematik
Universtät Freiburg
Hermann-Herder-Str. 10
7800 Freiburg

GREECE

Dimitri MANOLAS
Laboratory of Fluid Mechanics and Energy
Chemical Engineering Department
University of Patras
26110, Patras

HUNGARY

Sandor FEGYVERNEKI
Institute of Mathematics
University of Miskolc
3515 Miskolc

Aurel GALANTAI
Institute of Mathematics
University of Miskolc
3515 Miskolc

ITALY

Carla ANTONI
Dipartimento di Matematica
Universitá di Pisa
V. Buonarroti 2
56100 Pisa

Francisco FACCHINEI
Dipartimento di Informatica e Sistemistica
Universitá di Roma "La Sapienza"
Via Buonarroti, 12
00185 Roma

Zhijian HUANG
Dipartimento di Matematica
Universitá di Bergamo
Piazza Rosate 2
24129 Bergamo

Marco LOCATELLI
Dipartimento di Scienza e Informatica
Universitá of Milano
Via Comelico 39/41
20135 Milano

Giandomenico MASTROENI
Dipartimento di Matematica
Universitá di Pisa
V. Buonarroti 2
56100 Pisa

Roberto MUSMANNO
Dipartimento di Elettronica,
Informatica e Sistemistica
Universitá della Calabria
87036 Arcavacata di Rende

Maddalena NONATO
Dipartimento di Informatica
Universitá di Pisa
Corso Italia 40
56125 Pisa

Laura PALAGI
Dipartimento di Informatica e Sistemistica
Universitá di Roma "La Sapienza"
Via Buonarroti, 12
00185 Roma

Luisa EQUI PIERAZZINI
Dipartimento di Informatica
Universitá di Pisa
Corso Italia 40
56125 Pisa

Massimo ROMA
Dipartimento di Informatica e Sistemistica
Universitá di Roma "La Sapienza"
Via Buonarroti, 12
00185 Roma

Cinzia SIDOTI
Centro Ricerche FIAT
Strada Torino 50
Orbassano

Gerardo TORALDO
Dipartimento di Ingegneria
e Fisica dell'Ambiente
Universitá della Basilicata
Via Della Tecnica
85100 Potenza

Maria Teresa VESPUCCI
Dipartimento di Matematica
Universitá di Bergamo
Piazza Rosate 2
24129 Bergamo

Rong WANG
Dipartimento di Matematica
Universitá di Bergamo
Piazza Rosate 2
24129 Bergamo

Giovanni ZILLI
Dipartimento di Metodi e Modelli
Universitá
Via Belzoni 7
35131 Padova

POLAND

Slawomir FILIPEK
Department of Chemistry
Warsaw University
1 Pasteur St.
02093 Warsaw

Katarzyna GRECKA
Department of Technology
and Management
Politechnika Gdanska
80952 Gdansk
Majakowskiego 11/12

PORTUGAL

Maria L. ALCANTARA DA CUNHA
FCT/UNL
Department of Materials Science
Quinta da Torre
2825 Monte de Caprarica

Luis MERCA FERNANDES
Department of Mathematics
Escola Superior de Tecnologia de Tomar
2300 Tomar

Maria MIRANDA GUEDES
Grupo de Matematica Aplicada
Universidade de Porto
Rua Das Taipas
135-4000 Porto

Luis F. PORTUGAL
Departamento de Ciencias da Terra
Universidade de Coimbra
3000 Coimbra

ROMANIA

Daniela TEODORESCU
METROREX
Dimitru Colescu 38
Bucharest

SPAIN

Manuel GALAN
Departamento de Matematicas
Universidad de las Palmas
35017 Las Palmas de Gran Canarias

Cecilia POLA
Departamento de Matematicas,
Estadistica y Computacion
Facultad de Ciencias
Universidad de Cantabria
39005 Santander, Cantabria

SWITZERLAND

Karl FRAUENDORFER
Institute of Operations Research
University of St. Gallen
Bodenstrasse 6
CH-9000 ST Gallen

TURKEY

Kuban ALTINEL
Department of Industrial Engineering
Bogazici University
80815, Bebek-Istanbul

Elif Fahriye KÖKSAL
Department of Chemistry
Gukurova University
Adana

Muhammet KÖKSAL
Engineering Faculty
Inonu University
44069 Malatya

Recai KOLAY
Dokuz Eylul Universitesi
Muhendislik Fakultesi
Elektronik Bolumu
35100 Bornova-Izmir

Kenan TAS
Department of Mathematics
Faculty of Science
Hacettepe University
06532 Beytepe-Ankara

Gulay TOHUMOGLU
Electrical and Electronics
Engineering Department
Gaziantep University
27310 Gaziantep

UK

Walter FORSTER
Department of Mathematics
University of Southampton
Southampton

Nicholas GOULD
Central Computing Department
Rutherford Appleton Laboratory
Chilton, Oxon, OX11 0AL
Oxfordshire

Marli HERNANDEZ-FIGUEROA
Department of Information Science
University of Herfordshire
Hatfield Campus, College Lane
Hatfield Herts, AL10 9AB

Hugo Enrique HERNANDEZ
Department of Electronic
& Electrical Engineering
University College London
London WC1E 7JE

Sven LEYFFER
Department of Mathematics
University of Dundee
Dundee, DD1 4HN

Martin REED
Department of Mathematics & Statistics
Brunel University
Uxbridge UB8 3PH

Roger SARGENT
Center of Process Systems Engineering
Imperial College
London, SW7 2BY

Shicai ZHU
Center for Process Systems Engineering
Imperial College
London, SW7 2BY

USA

Miguel F. ANJOS
SCCM, Building 460
Stanford University
Stanford, CA 94305-2140

Robert FOURER
Department of Industrial Engineering
Northwestern University
Evanston IL 60208-3119

Stefano HERZEL
Department of Operations Research
Cornell University
Ithaca, NY 14853

Roy P. JOHANSON
Department of Mechanical Engineering
and Applied Mechanics
University of Michigan
ANN ARBOR, MI 48109-2125

Panos PAPALAMBROS
Department of Mechanical Engineering
and Applied Mechanics
University of Michigan
ANN ARBOR, MI 48109-2125

Christopher SIKORSKI
Computer Science Department
University of Utah
Salt Lake City, UT 84112

Valeria SIMONCINI
Center for Supercomputing Research
and Development
University of Illinois at Urbana-Champaign
465 C&SRL, 1308 West Main Street
Urbana, IL 61801-2307

VENEZUELA

Angel Ramon SANCHEZ DELGADO
Department of Mathematics
University Simon Bolivar
Caracas 89000

Bernardo FEIJOO
Department of Mathematics
University Simon Bolivar
Caracas 89000

General Optimality Conditions
via a Separation Scheme

Franco Giannessi

Department of Mathematics
University of Pisa
Via F. Buonarroti 2
56100 Pisa
ITALY

Abstract: The paper develops an approach to optimality conditions based on a separation scheme in the image space. The key point of the approach is a necessary condition for a class of nonsmooth problems with unilateral constraints. This approach can be extended to several other topics, including optimization with bilateral constraints, sufficient optimality conditions, duality, variational inequalities.

1. INTRODUCTION

Nonsmooth Optimization has become an important field both for theoretical analysis and for the applications; see for instance [4,9]. This has stimulated the extension of both the theory and the algorithms, which have been developed – along more than two centuries – for the smooth case.

The generalization of optimality conditions has been one of the first purposes; it has been done by adopting various and different approaches. The present paper aims to point out one of them, namely that which exploits a separation scheme in the image space defined as the space where the images of the given functions run. Such an approach has been conceived by a few authors, independently each other, on seventies and eighties. Here it will be described by analyzing a particular case, i.e. a necessary condition for problems with unilateral constraints. The same approach can be used to investigate other topics, as necessary conditions in presence of bilateral constraints, sufficient conditions, duality, penalty methods. Moreover, it will be shown that the same approach can be exploited to study fields different from Optimization, like that of Variational Inequalites.

Assume that we are given the positive integer m, a nonempty subset X of a Hilbert space H whose norm is denoted by $\| \cdot \|$, and the real–valued functions

1

E. Spedicato (ed.), Algorithms for Continuous Optimization, 1–23.
© 1994 *Kluwer Academic Publishers.*

$f : X \to \mathbb{R}$ and $g : X \to \mathbb{R}^m$. Consider the following problem:

$$\min f(x) \quad , \quad x \in R := \{x \in X : g(x) \geq 0\}. \tag{1}$$

In the sequel we will set $I = \{1, \ldots, m\}$, $g(x) = (g_i(x), \; i \in I)$, and, given $\bar{x} \in H$, $N_\rho(x) := \{x \in H : \|x - \bar{x}\| < \rho\}$ with $\rho > 0$ will denote a neighbourhood of $\bar{x}$. Besides the finite dimensional case where $X \subseteq \mathbb{R}^n$, we will consider another particular case, namely, the one where $X = C^1[a, b]$, with $a, b \in \mathbb{R}$, and X is equipped with the norm $\|x\|_\infty := \max_{t \in [a,b]} |x(t)|$, and where

$$f(x) := \int_a^b \psi_0(t, x(t), \; x'(t))dt \; ; \; g_i(t) := \int_a^b \psi_i(t, x(t), \; x'(t))dt \;, \; i \in I, \tag{2}$$

the functions $\psi_i : \mathbb{R}^3 \to \mathbb{R}$, $i \in \{0\} \cup I$ being given. Fixed endpoint conditions can be included in the definition of X.

A crucial aspect of the present analysis is based on the concept of image of (1), as has been shown in [9], which now will be briefly recalled. Obviously, $\bar{x} \in R$ is a global minimum point of (1) iff the system

$$f(\bar{x}) - f(x) > 0 \;, \quad g(x) \geq 0 \;, \quad x \in X \tag{3}$$

is impossible. Hence, in order to state optimality for (1) we are led to prove the disjunction of the sets $\mathcal{H}$ and $\mathcal{K}$, where

$$\mathcal{H} := \{(u, v) \in \mathbb{R}^{1+m} : u > 0 \;, \; v \geq 0\} \;,$$
$$F(x) := (f(\bar{x}) - f(x), g(x)) \;, \quad \mathcal{K} := F(X).$$

$\mathcal{K}$ is called the image of (1). To prove directly whether or not $\mathcal{H} \cap \mathcal{K} = \emptyset$ is generally impracticable. Then, it is natural to try to show such a disjunction indirectly by proving, for instance, that the two sets lie in two disjoint halfspaces or, more generally, in two disjoint level sets, respectively. This separation approach is exactly equivalent to looking for a system that is in alternative with (3). In this order of ideas, the set $\mathcal{K}$ plays a key role. See [9] for details about sets and concepts that will be used here.

If a local (instead of global) minimum point is the goal, we have merely to replace X by its intersection with N. Note that $\mathcal{K}$ depends on $\bar{x}$ (a change of $\bar{x}$ turns out in a translation of $\mathcal{K}$ in the direction of u); such a dependence is taken for granted.

The problems that can be reduced to the scheme (1) share the characteristic of having a finite dimensional image. Hence, certain problems, for instance of geodesic type, are not covered by (1).

2. $\mathcal{C}$–DIFFERENTIABILITY

Now, we will introduce a class of functions within which it is possible to establish a necessary optimality condition for (1).

In the following definition f denotes any of those functions appearing in (1). Set $Z := X - \bar{x}$ and assume that Z be a convex cone with apex at the origin. Denote by $\mathcal{G}$ the set of all functions $\mathcal{D}f : X \times Z \to \mathbb{R}$ which are positively homogeneous of degree one with respect to the 2–nd argument, i.e.

$$\mathcal{D}f(\bar{x}; \alpha z) = \alpha \mathcal{D}f(\bar{x}; z) \ , \ \forall \alpha \in \mathbb{R}_+; \tag{4}$$

$\mathcal{L}$ and $\mathcal{C}$ will denote the subsets of $\mathcal{G}$, whose elements are continuous linear and convex (and hence sublinear), respectively; G will denote any subset of $\mathcal{G}$.

Definition 1. Let $G \subseteq \mathcal{G}$. A function $f : X \to \mathbb{R}$ is said G–*differentiable* at $\bar{x}$, iff there exists $\mathcal{D}_G f \in G$, such that

$$\lim_{z \to 0} \frac{1}{\|z\|}[\epsilon(\bar{x}; z) := f(x) - f(\bar{x}) - \mathcal{D}_G f(\bar{x}; z)] = 0. \tag{5}$$

$\mathcal{D}_G f$ is said the G–*derivative* of f. The function f is said *directionally derivable* at $\bar{x}$ in the direction z, iff the limit

$$f'(\bar{x}; z) := \lim_{\alpha \downarrow 0} \frac{1}{\alpha}[f(\bar{x} + \alpha z) - f(\bar{x})] \tag{5'}$$

exists.

It is easy to see that a $\mathcal{L}$–differentiable function is a differentiable one, and that a G–differentiable function is directionally derivable – the G–derivative being the directional derivative – but not conversely as Examples 4 and 5 show.

The class of $\mathcal{C}$–differentiable functions has been introduced in [10] (under the term generalized differentiable functions, and as a subclass of that of semidifferentiable ones) and developed in [11]. Independently of this in the same period the above class (under the term of B–differentiable functions) has been introduced in [18].

In the following a necessary optimality condition for (1) will be established within the class of $\mathcal{C}$–differentiable functions. To understand this class let us consider some examples.

Example 1. Let us set $H = X = R = \mathbb{R}$, $f : \mathbb{R} \to \mathbb{R}_+$ with $f(x) = |x| + x^2$ if $x \in \mathbb{Q}$ and $f(x) = |x| + 2x^2$ if $x \notin \mathbb{Q}$, $\mathbb{Q}$ being the set of rational numbers. It is easy to check that f is $\mathcal{C}$–differentiable at $\bar{x} = 0$ with $\mathcal{C}$–derivative $\mathcal{D}f(0; z) = |z|$. Note that f is continuous at $\bar{x} = 0$ only.

Example 2. Let us set $H = X = R = \mathbb{R}$, $f : [-1, 1] \to \mathbb{R}_+$, with

$$f(x) := \begin{cases} -x & \text{, if } -1 \leq x \leq 0, \\ f_n(x) := -x^2 + (3 \cdot 2^{-n} + 1)x - 2^{1-2n} & \text{, if } 2^{-n} < x \leq 2^{1-n}, \\ & \quad n = 1, 2, \ldots. \end{cases}$$

By setting

$$\mathcal{D}f(\bar{x}; x - \bar{x}) :=$$

$$:= \begin{cases} -(x - \bar{x}) & \text{, if } -1 \leq \bar{x} \leq 0, \\ (1 - 2^{-n-1}(x - \bar{x}) & \text{, if } x \leq 2^{-n}, \ \bar{x} = 2^{-n}, \ n = 1, 2, \ldots \\ (1 + 2^{-n})(x - \bar{x}) & \text{, if } x > 2^{-n}, \ \bar{x} = 2^{-n}, \ n = 1, 2, \ldots \\ (-2\bar{x} + 3 \cdot 2^{-n} + 1)(x - \bar{x}) & \text{, if } 2^{-n} < \bar{x} < 2^{1-n}, \quad n = 1, 2, \ldots \end{cases}$$

we see that (4) is fulfilled at any $\bar{x} \in [-1, 1]$. This is easily seen by noting that 1 ± 2^{-n} are right and left derivatives of $f_n(x)$. Note that f is C–differentiable at any $\bar{x} \in [-1, 1]$, but is not convex on $[-1, 1]$; however f is locally Lipschitz at any $\bar{x} \in [-1, 1]$.

Example 3. Let us set $H = X = R = \mathbb{R}$, $f : [-1, 1] \to \mathbb{R}_+$, with

$$f(x) := \begin{cases} -x & \text{, if } -1 \leq x \leq 0, \\ 2^{-n}(-4x^2 + 12 \cdot 2^{-n}x - 8 \cdot 2^{-2n})^{1/2} & \text{, if } 2^{-n} < x < 2^{1-n}, \\ & \quad n = 1, 2, \ldots \end{cases}$$

This case is quite analogous to the preceding one with the sole exception that $\mathcal{D}f = \infty$ at $\bar{x} = 2^{-n}$; indeed f is not Lipschitz at $\bar{x} = 2^{-n}$.

Example 4. Let us set $H = X = R = \mathbb{R}^2$, $\bar{x} = 0$, $\|z\| = \|z\|_2$,

$$\varphi(\alpha) := \begin{cases} 1 & \text{, if } 0 < \alpha \leq 1 \text{ and } \alpha \geq 3, \\ 3 - 2\alpha & \text{, if } 1 < \alpha \leq 2, \\ 2\alpha - 5 & \text{, if } 2 < \alpha < 3, \end{cases}$$

$$f(x) := \begin{cases} (x_1^2 + x_2^2)^{1/2} & \text{, if } x \not> 0, \\ \varphi\left(\frac{x_2}{x_1}\right)(x_1^2 + x_2^2)^{1/2} & \text{, if } x > 0. \end{cases}$$

We find that (5)' is fulfilled with $f'(\bar{x}; z) = (z_1^2 + z_2^2)^{1/2}$, while it is not possible to verify (5). Hence, a directionally derivable function is not necessarily G–differentiable in the sense that $G \subseteq \mathcal{G}$ and $\mathcal{D}_G f \in G$ such that (5) is satisfied do not necessarily exist. Note that f is continuous (but not locally Lipschitz) and $f'(0; z) > 0$, $\forall z \in \mathbb{R}^2 \setminus \{0\}$; notwithstanding this, $\bar{x}$ is not a minimum point; hence

f generalizes the so–called Peano function showed by Peano to detect the famous Lagrange mistake.

Example 5. Let us set $H = X = R = \mathbb{R}^2$, $\bar{x} = 0$, $\|z\| = \|z\|_2$,

$$f(x) := \begin{cases} |x_1| + |x_2| & \text{, if } x_2 \leq x_1^2, \\ (5 - 4\alpha) \cdot |x_1|^{2\alpha - 1} & \text{, if } x_2 = \alpha x_1^2, \ 1 \leq \alpha \leq \frac{3}{2} \\ (6\alpha - 10) \cdot x_1^2 & \text{, if } x_2 = \alpha x_1^2, \ \frac{3}{2} < \alpha < 2 \\ x_2 & \text{, if } x_2 \geq 2x_1^2. \end{cases}$$

f is directionally derivable at $x = \bar{x}$ in any direction:

$$f'(0; z) = \lim_{\alpha \downarrow 0} \frac{1}{\alpha} f(\alpha z_1, \alpha z_2) = \left\{ \begin{array}{ll} z_2 & \text{, if } z_2 \geq z_1^2 \\ |z_1| + |z_2| & \text{, if } z_2 < z_1^2 \end{array} \right\} > 0,$$

$\forall z \in \mathbb{R}^2 \setminus \{0\}.$

The same remarks made for Example 4 hold here.

Definition 2. Let X be convex, $G \subseteq \mathcal{C}$ and $f : X \to \mathbb{R}$. The *generalized subdifferential* of a G–differentiable function f at $\bar{x} \in X$ is given by

$$\underline{\partial}_G f(\bar{x}) := \{ t \in H' : \mathcal{D}_G(\bar{x}; z) \geq \langle t, z \rangle , \ \forall z \in Z \}, \tag{6}$$

H' being the continuous dual of H; t is called the *generalized subgradient* of f at $\bar{x}$. If (6) holds with $G \subseteq -\mathcal{C}$ ($-\mathcal{C}$ is the set of superlinear functions) and with reverse inequality, then the RHS of (6) is denoted by $\bar{\partial}_G f(\bar{x})$ and is called the *generalized superdifferential* of f at $\bar{x}$. The pair $(\underline{\partial}_G g(\bar{x}), \bar{\partial}_G f(\bar{x}))$ is called the *generalized differential* of f at $\bar{x}$.

Note that (6) is nothing more than the classic subdifferential of Convex Analysis (denoted by ∂) of $\mathcal{D}_G f$, namely $\underline{\partial}_G f(\bar{x}) = \partial \mathcal{D} f(\bar{x}; 0)$. Note also that (6), as well as ∂f, do not collapse – from a formal viewpoint – to the classic differential when f is differentiable; this would happen defining $\underline{\partial}_G f(\bar{x}) = \{\langle t, z \rangle : t \in H', \ \mathcal{D}_G(\bar{x}; z) \geq \langle t, z \rangle, \ \forall z \in Z \}$; however, in this case $\underline{\partial}_G f$ is a singleton and its unique element is the gradient of f.

In Examples 1–3 we have $\underline{\partial}_\mathcal{C} f(0) = [-1, 1]$. In Example 2 we have $\underline{\partial}_\mathcal{C} f(2^{-n}) = [1 - 2^{-n-1}, 1 + 2^{-n}]$, $n = 1, 2, \ldots$.

Example 6. $H = \mathbb{R}$, $f : \mathbb{R} \to \mathbb{R}$ with $f(x) = x^2 \operatorname{sen} \frac{1}{x}$ if $x \neq 0$ and $f(0) = 0$. We find $\mathcal{D}_\mathcal{C} f(0; z) = 0 \ \forall z \in \mathbb{R}$ (indeed f is differentiable). Thus $\underline{\partial}_\mathcal{C} f(0) = \{0\}$, while the Clarke subdifferential (see [11]) is [-1,1].

Now we will consider some properties of $\mathcal{C}$–differentiable functions, which will be useful in the sequel (see [11]).

Proposition 1. Let X be open and convex.
(i) If $f : X \to \mathbb{R}$ is $\mathcal{C}$–differentiable, then

$$\underline{\partial}_{\mathcal{C}}(\alpha f)(x) = \alpha \underline{\partial}_{\mathcal{C}} f(x) \quad , \quad \forall \alpha > 0 \, , \, \forall x \in X; \tag{7}$$

at $\alpha = 0$ the thesis is trivial, provided that $\underline{\partial}_{\mathcal{C}} f(x) \neq \emptyset$.

(ii) If $f_i : X \to \mathbb{R}$, $i = 1, \ldots, r$ are $\mathcal{C}$–differentiable, with proper $\mathcal{C}$–derivatives, and $\bigcap\limits_{i=1}^{r} ri(\operatorname{dom} \mathcal{D}_{\mathcal{C}} f_i) \neq \emptyset$, then

$$\underline{\partial}_{\mathcal{C}} \left[\sum_{i=1}^{r} f_i(x) \right] = \sum_{i=1}^{r} \underline{\partial}_{\mathcal{C}} f_i(x) \, . \tag{8}$$

(iii) $\underline{\partial}_{\mathcal{C}} f(x)$ is closed and convex and, if $\mathcal{D}_{\mathcal{C}} f$ is proper, bounded.

(iv) If f is convex, then $\underline{\partial}_{\mathcal{C}} f(x) = \partial f(x)$.

Proof. (i) Obvious consequence of the definition. (ii)–(iv). Apply Th.23.8 of [11].

A proposition quite analogous to Proposition 1 holds for $\bar{\partial}$. The following propositions are specializations of Propositions 4.5, 4.6 and 4.7 of [11].

Proposition 2. Let $X \subseteq \mathbb{R}^n$ be open and convex. A convex function $f : X \to \mathbb{R}$, having X as effective domain, is $\mathcal{C}$–differentiable at any $\bar{x} \in X$, and its unique directional $\mathcal{C}$–derivative coincides with the directional derivative of f.

Proposition 3. Let X be open and convex. If $f_1 : X \to \mathbb{R}$ is convex, with X as effective domain, and $f_2 : X \to \mathbb{R}$ is differentiable, then $f_1 + f_2$ is $\mathcal{C}$–differentiable at any $\bar{x} \in X$, and its $\mathcal{C}$–derivative is the sum of the directional derivative and f_1 and of the derivative of f_2, and its generalized subdifferential is

$$\underline{\partial}_{\mathcal{C}}[f_1(\bar{x}) + f_2(\bar{x})] = \partial f_1(\bar{x}) + \{f_2'(\bar{x})\}.$$

Proposition 4. Let $X \subseteq \mathbb{R}^n$ be open and convex; let the functions $f_i : X \to \mathbb{R}$, $i = 1, \ldots, r$ be $\mathcal{C}$–differentiable at $\bar{x} \in X$. Then the sup–function

$$f(x) := \sup\{f_1(x), \ldots, f_r(x)\}$$

is $\mathcal{C}$–differentiable at $\bar{x}$, having as $\mathcal{C}$–derivative the supremum of the $\mathcal{C}$–derivatives of those f_i such that $f_i(\bar{x}) = f(\bar{x})$.

Propositions quite analogous to Propositions 2–4 hold at $G = -\mathcal{C}$.

3. HOMOGENEIZATION AND SEMISTATIONARITY.

In Sect. 5 a necessary condition will be stated for problem (1), where f, $-g_i$, $i \in I$ are $\mathcal{C}$–differentiable. To this aim we will heavily use the image of (1) to establish some preliminary propositions.

Lemma 1 (Homogeneization)
Let f, $-g_i$, $i \in I$ be $\mathcal{C}$–differentiable at $\bar{x} \in X$. If $\bar{x}$ is a minimum point of (1), then the system:

$$\mathcal{D}_{\mathcal{C}} f(\bar{x}; z) < 0 \; ; \; \mathcal{D}_{\mathcal{C}} g_i(\bar{x}; z) > 0 \, , \, i \in I^0 \; ;$$
$$g_i(\bar{x}) + \mathcal{D}_{\mathcal{C}} g_i(\bar{x}; z) \geq 0 \, , \, i \in I \backslash I^0; \; z \in Z \tag{9}$$

is impossible, where $I^0 := \{ i \in I : g_i(\bar{x}) = 0, \; \epsilon_i(\bar{x}; z) \not\equiv 0 \}$ and ϵ_i is the remainder (5) associated to g_i.

Proof. Ab absurdo, suppose now that (9) is possible; let $z = \hat{z}$ be a solution. Then $\alpha \hat{z}$ is a solution of (9) $\forall \alpha \in]0,1]$, since $g_i(\bar{x}) \geq 0$ and $\mathcal{D}_{\mathcal{C}} f$, $\mathcal{D}_{\mathcal{C}} g_i$ satisfy (4). The assumption implies that the remainders:

$$\epsilon_0(\bar{x}; z) := f(x) - f(\bar{x}) - \mathcal{D}_{\mathcal{C}} f(\bar{x}; z) \; ;$$
$$\epsilon_i(\bar{x}; z) := g_i(x) - g_i(\bar{x}) - \mathcal{D}_{\mathcal{C}} g_i(\bar{x}; z), \, i \in I$$

fulfil (5), so that $\exists \hat{\alpha} \in]0,1]$ such that:

$$\frac{\epsilon_0(\bar{x}; \hat{\alpha}\hat{z})}{\|\hat{\alpha}\hat{z}\|} < \frac{-\mathcal{D}_{\mathcal{C}} f(\bar{x}; \hat{z})}{\|\hat{z}\|} \; ; \; \frac{\epsilon_i(\bar{x}; \hat{\alpha}\hat{z})}{\|\hat{\alpha}\hat{z}\|} > -\frac{\mathcal{D}_{\mathcal{C}} g_i(\bar{x}; \hat{z})}{\|\hat{z}\|} \, , \, i \in I^0.$$

From these inequalities, by setting $z^* = \hat{\alpha}\hat{z}$ and noting that $g_i(\bar{x}) = 0 \; \forall i \in I^0$, we have:

$$\mathcal{D}_{\mathcal{C}} f(\bar{x}; z^*) + \epsilon_0(\bar{x}; z^*) < 0 \; ; \; g_i(\bar{x}) + \mathcal{D}_{\mathcal{C}} g_i(\bar{x}; z^*) + \epsilon_i(\bar{x}; z^*) > 0 \, , \, i \in I^0. \tag{10a}$$

$\forall i \in I \backslash I^0$ either $g_i(\bar{x}) = 0$ and $\epsilon_i \equiv 0$ or $g_i(\bar{x}) > 0$. In the former case, by setting $z^* := \hat{z}$, we obviously have:

$$g_i(\bar{x}) + \mathcal{D}_{\mathcal{C}} g_i(\bar{x}; z^*) + \epsilon_i(\bar{x}; z^*) \geq 0. \tag{10b}$$

In the latter case $\exists \alpha^0 \in]0,1]$ such that:

$$g_i(\bar{x}) + \mathcal{D}_{\mathcal{C}} g_i(\bar{x}; \alpha \hat{z}) > 0 \, , \, \forall \alpha \in]0, \alpha^0] \, ,$$

and thus $\exists \tilde{\alpha} \in]0, \alpha^0]$ such that:

$$\frac{\epsilon_i(\bar{x}; \tilde{\alpha}\hat{z})}{\|\tilde{\alpha}\hat{z}\|} \geq -\frac{g_i(\bar{x}) + \alpha^0 \mathcal{D}_{\mathcal{C}} g_i(\bar{x}; \hat{z})}{\alpha^0 \|\hat{z}\|} \geq$$
$$\geq -\frac{g_i(\bar{x}) + \alpha \mathcal{D}_{\mathcal{C}} g_i(\bar{x}; \hat{z})}{\alpha \|\hat{z}\|} \, , \, \forall \alpha \in]0, \alpha^0],$$

where the 1–st inequality holds since again the remainder ϵ_i fulfils (5) and the 2–nd side is fixed and negative, the 2–nd inequality holds since the 2–nd side is obviously the maximum of the 3–rd on $]0, \alpha^0]$. By setting $z^* := \tilde{\alpha}\hat{z}$ it follows that:

$$g_i(\bar{x}) + \mathcal{D}_{\mathcal{C}} g_i(\bar{x}; z^*) + \epsilon_i(\bar{x}; z^*) \geq 0. \tag{10c}$$

Collecting all (10), recalling that $g_i(\bar{x}) = 0$, $i \in I^0$, and using the definition of the remainders ϵ_i, we obtain the possibility of system (3), which contradicts the minimality of $\bar{x}$. □

The impossibility of system (9) can be expressed as the disjunction of two sets of the image space associated to (1) in Sect.1. To this end introduce the sets:

$$\mathcal{H}_h := \{(u, v) \in \mathbb{R} \times \mathbb{R}^m : u > 0 \, ; \, v_i > 0 \, , \, i \in I^0 \, ; \, v_i \geq 0 \, , \, i \in I \backslash I^0\};$$

$$\mathcal{K}_h := \{(u, v) \in \mathbb{R} \times \mathbb{R}^m : u = -\mathcal{D}_{\mathcal{C}} f(\bar{x}; z) \, ; \, v_i = g_i(\bar{x}) + \mathcal{D}_{\mathcal{C}} g_i(\bar{x}; z) \, ,$$
$$i \in I \, ; \, z \in Z\}.$$

It is easily seen that the impossibility of system (9) holds iff

$$\mathcal{H}_h \cap \mathcal{K}_h = \emptyset. \tag{9$'$}$$

Note that system (9) is set up with the homogeneous parts of the functions of (3) and hence represents the homogenization of (3); thus $\mathcal{K}_h$ is the image of the homogenization of (1); $\mathcal{H}_h$ simply follows the changes in the types of inequalities in going from (3) to (9).

When f, g_i, $i \in I$ are differentiable, then Lemma 1 collapses to the well known Linearization Lemma [1], system (9) being replaced by

$$\langle f'(\bar{x}), z \rangle < 0 \, ; \, \langle g_i'(\bar{x}), z \rangle > 0 \, , \, i \in I^0 \, ; \, \langle g_i'(\bar{x}), z \rangle \geq 0 \, , \, i \in I \backslash I^0. \tag{9$''$}$$

Note that this lemma, as well as Proposition 5, can be slightly sharpened by requiring differentiability or $\mathcal{C}$–differentiability only for those g_i such that $g_i(\bar{x}) = 0$, and continuity for the remaining ones. Lemma 1 can be generalized to semidifferentiable functions [11].

The generalization of the concept of stationary point, which is associated with that of necessary conditions, has received much attention. The crucial part is the kind of convergence that is required. The following definition seems to be quite general, even if it is clear that it is not possible to handle every problem with a single kind of convergence.

Definition 3. $\bar{x} \in X$ will be called a *lower semistationary point* of a problem of type $\min\limits_{x \in X} f(x)$, iff

$$\liminf_{x \to \bar{x}} \frac{f(x) - f(\bar{x})}{\|x - \bar{x}\|} \geq 0. \tag{11}$$

Proposition 5. (i) If $\bar{x}$ is a local minimum point of f on X, then (11) holds. (ii) If X and f are convex, then a lower semistationary point of f on X is a global minimum point. (iii) If f is differentiable, then (11) becomes

$$\langle f'(\bar{x}), x - \bar{x} \rangle \geq 0 \quad , \quad \forall x \in X, \tag{11}'$$

which, if $\bar{x} \in \text{int } X$, collapses to

$$f'(\bar{x}) = 0. \tag{11}''$$

(iv) If X and f are convex, then (11) becomes

$$f'(\bar{x}; x - \bar{x}) \geq 0 \quad , \quad \forall x \in X. \tag{11}'''$$

Proof. (i) Since $\bar{x}$ is a local minimum, $\exists N_\rho(\bar{x})$ such that $f(x) \geq f(\bar{x})$, $\forall x \in N_\rho(\bar{x})$, so that $[f(x) - f(\bar{x})/\|x - \bar{x}\| \geq 0$, $\forall x \in N_\rho(\bar{x})$, and therefore (11) holds. (ii) Ab absurdo, suppose that $\exists \hat{x} \in X$ such that $f(\hat{x}) < f(\bar{x})$. This fact and the convexity of f imply

$$\frac{f(\bar{x} + \alpha(\hat{x} - \bar{x})) - f(\bar{x})}{\|\alpha(\hat{x} - \bar{x})\|} \leq \frac{f(\hat{x}) - f(\bar{x})}{\|\hat{x} - \bar{x}\|} < 0 \ , \ \forall \alpha \in]0, 1].$$

Hence, we are led to the inequalities

$$\liminf_{x \to \bar{x}} \frac{f(x) - f(\bar{x})}{\|x - \bar{x}\|} \leq \liminf_{\alpha \downarrow 0} \frac{f(\bar{x} + \alpha(\hat{x} - \bar{x})) - f(\bar{x})}{\|\alpha(\hat{x} - \bar{x})\|} \leq$$

$$\leq \frac{f(\hat{x}) - f(\bar{x})}{\|\hat{x} - \bar{x}\|} < 0 \ ,$$

which contradict (11). (iii) Let $f(x) = f(\bar{x}) + \langle f'(\bar{x}), x - \bar{x} \rangle + \epsilon(\bar{x}; x - \bar{x})$ be the expansion of f at $\bar{x}$ with $\epsilon/\|x - \bar{x}\| \to 0$ as $x \to \bar{x}$; set $x - \bar{x} = tz$ with $t \in \mathbb{R}_+$, and $S := \{z : \|z\| = 1\}$. (11) becomes

$$\liminf_{x \to \bar{x}} \left[\langle f'(\bar{x}), \frac{x - \bar{x}}{\|x - \bar{x}\|} \rangle + \frac{\epsilon(\bar{x}; x - \bar{x})}{\|x - \bar{x}\|} \right] =$$

$$= \sup_{\tau > 0} \inf_{\substack{z \in S \\ t \in [0, \tau]}} \left[\langle f'(\bar{x}), z \rangle + \frac{1}{t} \epsilon(\bar{x}; tz) \right] \geq 0. \tag{12}$$

(11)$'$ follows, if we show that the latest inequality implies

$$\ell := \inf_{z \in S} \langle f'(\bar{x}), z \rangle \geq 0.$$

Ab absurdo, suppose that $\ell < 0$. Then, $\forall \eta > 0 \ \exists z_\eta \in S$ such that $\langle f'(\bar{x}), z_\eta \rangle < \ell + \eta$. Without any loss of generality assume that $\ell + \eta < 0$. Select $\delta \in]0, -(\ell+\eta)[$. If τ is small enough, because of the differentiability, we have $\frac{1}{t}\epsilon(\bar{x}; tz) < \delta$, $\forall z \in S$, $\forall t \in]0, \tau]$, and in particular at $z = z_\eta$. Therefore

$$\langle f'(\bar{x}), z_\eta \rangle + \frac{1}{t}\epsilon(\bar{x}; tz_\eta) < \ell + \eta + \delta < 0 , \ \forall t \in]0, \tau],$$

which shows that the infimum of the form in the 1–st square brackets of (12) is less than a negative constant. This contradicts the assumption. If $\bar{x} \in \text{int } X$, $\exists x', x'' \in X$ with $\bar{x} = \frac{1}{2}(x' + x'')$; then (11)' implies $\langle f'(\bar{x}), x' - x'' \rangle \geq 0$, $\langle f'(\bar{x}), x'' - x' \rangle \geq 0$ and hence (11)''.

(iv) It is enough to repeat the proof of (iii) with the following changes: the expansion of f is now $f(x) = f(\bar{x}) + f'(\bar{x}; x - \bar{x}) + \epsilon(\bar{x}; x - \bar{x})$; the scalar product which appears in (12) and consequently in the sequel must be replaced with $f'(\bar{x}; x - \bar{x})$ and then $f'(\bar{x}; z)$ or $f'(\bar{x}; z_\eta)$. □

Note that (11)', which is equivalent to require the non–negativity of the directional derivative like (11)''', is a particular case of Variational Inequality.

Denote by

$$L(x; \theta, \lambda) := \theta f(x) - \langle \lambda, g(x) \rangle$$

the Lagrangian function associated to (1); in the particular case (2) it becomes

$$L(x; \theta, \lambda) := \int_a^b [\theta \psi_0(t, x(t), x'(t)) - \sum_{i=1}^m \lambda_i g_i(t, x(t), x'(t))] dt.$$

The positive polar of $\mathcal{K}_h - \bar{k}$ (where $\bar{k} := (\bar{u}, \bar{v}) := (0, g(\bar{x}))$ is the image of $x = \bar{x}$ or $z = 0$) is given by

$$(\mathcal{K}_h - \bar{k})^* := \{k^* \in \mathbb{R}^{1+m} : \langle k^*, k - \bar{k} \rangle \geq 0, \ \forall k \in \mathcal{K}_k\} =$$
$$= \{(u^*, v^*) \in \mathbb{R} \times \mathbb{R}^m : u^* u + \langle v^*, v - \bar{v} \rangle \geq 0 , \ \forall (u, v) \in \mathcal{K}_h\};$$

when $\mathcal{K}_h - \bar{k}$ is a linear variety its polar collapses to the orthogonal complement and is denoted by $(\mathcal{K}_k - \bar{k})^\perp$. In the sequel a $*$ as apex of a set will denote its positive polar.

Lemma 2 (Semistationarity)
Let f and $-g$ be G–differentiable with $G \subseteq \mathcal{G}$. (i) If

$$-(\theta, \lambda) \in (\mathcal{K}_h - \bar{k})^*, \tag{13}$$

then

$$\liminf_{x \to \bar{x}} \frac{L(x; \theta, \lambda) - L(\bar{x}; \theta, \lambda)}{\|x - \bar{x}\|} \geq 0. \tag{14}$$

If $\lim_{z \to 0} \mathcal{D}_G f\left(\bar{x}; \frac{z}{\|z\|}\right)$ and $\lim_{z \to 0} \mathcal{D}_G g_i\left(\bar{x}; \frac{z}{\|z\|}\right)$, $i \in I$ exist, then the lower limit of (14) collapses to the ordinary limit. (ii) If $\bar{x} \in \text{int } X$ and f, g_i, $i \in I$ are differentiable (so that $G = \mathcal{L}$), then (i) becomes: if

$$-(\theta, \lambda) \in (\mathcal{K}_h - \bar{k})^\perp, \tag{13}'$$

then

$$L_x'(\bar{x}; \theta, \lambda) = 0. \tag{14}'$$

Proof. (13) is equivalent to

$$-(\theta, \lambda) \in \{(u^*, v^*) \in \mathbb{R} \times \mathbb{R}^m : \langle (u^*, v^*), (u - \bar{u}, v - \bar{v}) \rangle \geq 0 ,$$
$$\forall (u, v) \in \mathcal{K}_h\},$$

or

$$\mathcal{D}_G L(x; z; \theta, \lambda) - \mathcal{D}_G L(\bar{x}; z; \theta, \lambda) \geq 0 , \quad \forall x \in X, \tag{15}$$

where

$$\mathcal{D}_G L(x; z; \theta, \lambda) := \theta \mathcal{D}_G f(\bar{x}; z) - \sum_{i \in I} \lambda_i \mathcal{D}_G g_i(\bar{x}; z).$$

If G is closed under addition – this happens if $G = \mathcal{L}$ or $G = \mathcal{C}$ – then $\mathcal{D}_G L$ is the G–derivative of L. Divide both sides of (15) by $\|z\|$ and add to them

$$\frac{1}{\|z\|} \bar{\epsilon}(\bar{x}; z; \theta, \lambda) := \frac{1}{\|z\|} [\theta \epsilon_0(\bar{x}; z) - \sum_{i \in I} \lambda_i \epsilon_i(\bar{x}; z)];$$

then (15) becomes

$$\frac{1}{\|z\|} [L(x; \theta, \lambda) - L(\bar{x}; \theta, \lambda)] \geq \frac{1}{\|z\|} \bar{\epsilon}(\bar{x}; z; \theta, \lambda), \quad \forall x \in X \backslash \{\bar{x}\}.$$

Now (14) follows, since $\bar{\epsilon}/\|z\| \to 0$ as $x \to \bar{x}$. The remaining part is obvious. (ii) Since $\mathcal{K}_h$ is now affine, the polar becomes the orthogonal complement and hence liminf collapses to lim and this is zero since both $\geq$ and $\leq$ must hold. $\quad\square$

Lemma 2 holds for a wider class of semidifferentiable functions.

4. Conic extension

Definition 4. Given a set S of the image space $\mathbb{R}^{1+m}$, the set $\mathcal{E}(S) := S - \text{cl}\,\mathcal{H}$ is called the *conic extension* of S with respect to the cone $\text{cl}\,\mathcal{H}$.

Lemma 3 (Conic extension)
(i) $\mathcal{H} \cap \mathcal{K} = \emptyset \iff \mathcal{H} \cap \mathcal{E}(\mathcal{K}) = \emptyset$.
(ii) $\mathcal{H}_h \cap \mathcal{K}_h = \emptyset \iff \mathcal{H}_h \cap \mathcal{E}(\mathcal{K}_h) = \emptyset \iff \mathcal{H}_h \cap \mathcal{E}(\mathcal{K}_h - \bar{k}) = \emptyset \Rightarrow \mathcal{E}(\mathcal{K}_h) \cap \text{int}\mathcal{H} = \emptyset$.
(iii) If f is C–differentiable and $g_i, i \in I$ are $(-C)$–differentiable, then $\mathcal{E}(\mathcal{K}_h)$ and $\mathcal{E}(\mathcal{K}_h - \bar{k})$ are convex.
(iv) If $f, g_i,\ i \in I$ are differentiable, then $\mathcal{E}(\mathcal{K}_h)$ is the sum of an affine variety and $\mathbb{R}_-^{1+m}$.

Proof.
(i) $\Leftarrow$ is obvious consequence of $\mathcal{K} \subseteq \mathcal{E}(\mathcal{K})$. $\Rightarrow$ Ab absurdo, if $a \in \mathcal{K}$, $b \in \text{cl}\,\mathcal{H}$, $a - b \in \mathcal{H}$, then $a = (a - b) + b \in \mathcal{H} + \text{cl}\,\mathcal{H} = \mathcal{H}$, so that $\mathcal{H} \cap \mathcal{K} = \emptyset$ is contradicted.
(ii) The 1–st statement is proved as (i). To prove the 2–nd one it is enough to note that $g(\bar{x}) \geq 0$ so that $\bar{k} \in \text{cl}\,\mathcal{H}$ and

$$\mathcal{E}(\mathcal{K}_h - \bar{k}) := \{(a - \bar{k}) - b : a \in \mathcal{K}_h \; ; \; b \in \text{cl}\,\mathcal{H}\} =$$
$$= \{a - (\bar{k} + b) : a \in \mathcal{K}_h \; ; \; \bar{k} + b \in \text{cl}\,\mathcal{H} + \text{cl}\,\mathcal{H} = \text{cl}\,\mathcal{H}\} = \mathcal{E}(\mathcal{K}_h).$$

The last implication is obvious consequence of $\text{int}\,\mathcal{H} \subset \mathcal{H}_h$. (iii) It is enough to prove the convexity of $\mathcal{E}(\mathcal{K}_h - \bar{k})$, since $\mathcal{E}(\mathcal{K}_h)$ is a translation of it. If $\mathcal{K}_h = \{\bar{k}\}$, then the thesis is trivial since $\mathcal{E}(\mathcal{K}_h) = \bar{k} - \text{cl}\,\mathcal{H}$. Otherwise, consider $\Phi(z) := (-\mathcal{D}_C f(\bar{x}; z)$, $\mathcal{D}_{-C} g_i(\bar{x}; z)$, $i \in I^0)$ and any two distint elements of $\mathcal{E}(\mathcal{K}_h)$, which can be written as $\Phi(z^i) - h^i$, with $h^i \in \text{cl}\,\mathcal{H}$, $z^i \in Z$, $i = 1, 2$; $\forall \alpha \in [0,1]$ their convex combination, say $\Psi(\alpha)$, is such that

$$\Psi(\alpha) = (1 - \alpha)\Phi(z^1) + \alpha\Phi(z^2) - h(\alpha) \leq \Phi(z(\alpha)) - h(\alpha),$$

where $h(\alpha) := (1 - \alpha)h^1 + \alpha h^2 \in \mathcal{H}$, and $z(\alpha) := (1 - \alpha)z^1 + \alpha z^2 \in Z$. Hence we have $\Psi(\alpha) = \Phi(z(\alpha)) - \hat{h}(\alpha)$, where $\Phi(z(\alpha)) \in \mathcal{K}_h$ and $\hat{h}(\alpha) := [\Phi(z(\alpha)) - (1 - \alpha)\Phi(z^1) - \alpha\Phi(z^2)] + h(\alpha) \in \text{cl}\,\mathcal{H}$, since the square bracket is non–negative. This shows $\Psi(\alpha) \in \mathcal{E}(\mathcal{K}_h - \bar{k})$. (iv) obvious. $\qquad \square$

Note that the 1–st part of (i) of the above lemma is equivalent to the impossibility of system (3).

Lemma 4 (Sign of multipliers)

Let f and $-g$ be $\mathcal{C}$–differentiable at $\bar{x} \in X$. If $\bar{x}$ is a minimum point of (1), then

$$(\mathcal{K}_h - \bar{k})^* \cap [\mathbb{R}^{1+m}_- \setminus \{0\}] \neq \emptyset. \tag{16}$$

If φ and g are differentiable, then (16) becomes

$$(\mathcal{K}_h - \bar{k})^\perp \cap [\mathbb{R}^{1+m}_- \setminus \{0\} \neq \emptyset. \tag{16}'$$

Proof. By Lemma 1 $(9)'$ holds and hence, by Lemma 3 (ii), we have $\mathcal{H}_h \cap \mathcal{E}(\mathcal{K}_h - \bar{k}) = \emptyset$. Since $\mathcal{H}_h$ is obviously convex and, by Lemma 3 (iii), also $\mathcal{E}(\mathcal{K}_h - \bar{k})$ is convex, it follows that $\mathcal{H}_h$ and $\mathcal{E}(\mathcal{K}_h - \bar{k})$ are separable; this means that $\exists \pi \in \mathbb{R}^{1+m} \setminus \{0\}$ such that

$$\mathcal{H}_h \subseteq \{k \in \mathbb{R}^{1+m} : \langle \pi, k \rangle \leq 0\} \; ; \; \mathcal{E}(\mathcal{K}_h - \bar{k}) \subseteq$$
$$\subseteq \{k \in \mathbb{R}^{1+m} : \langle \pi, k \rangle \geq 0\} := \sigma.$$

The former of the above inclusions implies $\pi \leq 0$. The latter implies $\pi \in \sigma^* \subseteq \mathcal{E}(\mathcal{K}_h - \bar{k})$ and thus $\pi \in \mathcal{E}^*(\mathcal{K}_h - \bar{k}) \setminus \{0\}$; hence $\pi \in (\mathcal{K}_h - \bar{k})^*_h \setminus \{0\}$, since $\mathcal{K}_h - \bar{k} \subseteq \mathcal{E}(\mathcal{K}_h - \bar{k}) \Rightarrow \mathcal{E}^*(\mathcal{K}_h - \bar{k}) \subseteq (\mathcal{K}_h - \bar{k})^*$. (16) follows. $(16)'$ is now obvious. $\qquad \square$

Lemma 5. (Complementarity)

Let f and $-g$ be $\mathcal{C}$–differentiable at $\bar{x} \in X$. If $\bar{x}$ is a minimum point of (1), then $\exists (-\bar{\theta}, -\bar{\lambda}) \in (\mathcal{K}_h - \bar{k})^* \cap [\mathbb{R}^{1+m}_- \setminus \{0\}]$, such that

$$\langle \bar{\lambda}, g(\bar{x}) \rangle = 0 \; . \tag{17}$$

Proof. Ab absurdo, suppose that (17) is false. Then, $\forall \pi := (-\bar{\theta}, -\bar{\lambda}) \in (\mathcal{K}_h - \bar{k})^* \cap [\mathbb{R}^{1+m}_- \setminus \{0\}] \neq \emptyset$ (the latest inequality being consequence of Lemma 4) we have $\langle \lambda, g(\bar{x}) \rangle > 0$ and hence $\langle \pi, -\bar{k} \rangle > 0$, since $\bar{k} = (0, g(\bar{x}))$. This fact, Lemma 3(iii), and the inclusion $\mathcal{E}^*(\mathcal{K}_h - \bar{k}) \subseteq (\mathcal{K}_h - \bar{k})^*$, imply:

$$-\bar{k} \in \text{int} \{\mathcal{E}^*(\mathcal{K}_h - \bar{k}) \cap [\mathbb{R}^{1+m}_- \setminus \{0\}]\}^* = \text{int} [\mathcal{E}^*(\mathcal{K}_h - \bar{k}) \cap \mathbb{R}^{1+m}_-]^* =$$

$$\text{int cl conv} [\mathcal{E}(\mathcal{K}_h - \bar{k}) \cup \mathbb{R}^{1+m}_-] = \text{int} \, \mathcal{E}(\mathcal{K}_h - \bar{k}).$$

Denote by $S(k)$ an open sphere of $\mathbb{R}^{1+m}$ with center at k and suitably small radius. From the latest relations we deduce that $S(-\bar{k}) \subset \mathcal{E}(\mathcal{K}_h - \bar{k})$ and thus $S(0) \subset \mathcal{E}(\mathcal{K}_h)$, which implies $\mathcal{E}(\mathcal{K}_h) \cap \text{int}\, \mathcal{H} \neq \emptyset$. By Lemmas 1 and 3(ii) this contradicts the minimality of $\bar{x}$. $\qquad \square$

5. A NECESSARY CONDITION

Lemma 6 (Sign of directional derivative)
(i) Let f and $-g$ be G–differentiable at $\bar{x} \in X$. If $\bar{x} \in X$ is a lower semistationary point of $L(x; \theta, \lambda)$ with $\theta \geq 0$ and $\lambda \geq 0$, then

$$\inf_{z \in S} \mathcal{D}_G L(\bar{x}; z; \theta, \lambda) \; \geq \; 0 \; , \tag{18}$$

where $S := \{z \in Z : \|z\| = 1\}$ and $\mathcal{D}_G L$ is the G–derivative of the Lagrangian function L. (ii) If f and g are differentiable, then (18) becomes

$$V_L(\bar{x}; z; \theta, \lambda) = 0 \quad , \quad \forall z \in S, \tag{18$'$}$$

where V_L is a (continuous) linear functional (first variation of L). If, in addition, $X = H = \mathbb{R}^n$, then, instead of (18)$'$, we find

$$\langle L'_x(\bar{x}; \theta, \lambda), z \rangle = 0 \quad , \quad \forall z \in S, \tag{18$''$}$$

where L'_x denotes the gradient of L.

Proof. (i) Set $x = \bar{x} + \alpha z$ with $\alpha \in \mathbb{R}_+$. (11) holds iff

$$\sup_{\tau > 0} \inf_{\substack{z \in S \\ \alpha \in]0, \tau]}} [\mathcal{D}_G L(\bar{x}; z; \theta, \lambda) + \frac{1}{\|\alpha z\|} \epsilon_L(\bar{x}; \alpha z; \theta, \lambda)] \geq 0 \; , \tag{19}$$

where $\mathcal{D}_G L := \theta \mathcal{D}_G f(\bar{x}; z) - \sum_{i \in I} \mathcal{D}_G g_i(\bar{x}; z)$ is the G–derivative of L and $\epsilon_L := \theta \epsilon_0 - \sum_{i \in I} \lambda_i \epsilon_i$. Ab absurdo, suppose that

$$\ell := \inf_{z \in S} \mathcal{D}_G L(\bar{x}; z; \theta, \lambda) < 0.$$

Since $\epsilon_L / \|\alpha z\| \to 0$ as $\alpha \downarrow 0$, we find that the infimum in (19) is less than a negative constant, so that its supremum is still negative; this fact, which holds a priori if $\ell = -\infty$, contradicts (19) and proves (18). (ii) To achieve (18)$'$ and (18)$''$ it is enough to note that $z \in Z \Rightarrow -z \in Z$, and that the LHS of (18)$'$ and (18)$''$ are linear. $\square$

Note that, when $X = H = \mathbb{R}^n$ and $G = \mathcal{C}$, then the infimum in (18) is achieved, since, with $\theta \geq 0$ and $\lambda \geq 0$, $\mathcal{D}_{\mathcal{C}} L$ is convex and hence continuous, and S is compact.

For a generic function f, consider the following set of positively homogeneous functions of degree 1:

$$\mathcal{C}^0 := \{\mathcal{D}_{\mathcal{C}^0} f \in \mathcal{G} : \bar{\gamma} + \operatorname{epi} \mathcal{D}_{\mathcal{C}^0} f \subseteq \operatorname{cl} T_H[\bar{\gamma}; \operatorname{epi}(f - f(\bar{x}))]\},$$

where $\bar{\gamma} \in \operatorname{graph} f$ and T_H denotes the hypertangent cone $^{(*)}$ to $\operatorname{epi}(f - f(\bar{x}))$ at $\bar{\gamma}$. In [11] it is shown that $\mathcal{C}^0 \subseteq \mathcal{C}$. The $\mathcal{C}^0$–derivative is known as Clarke derivative.

Theorem 1 (Necessary condition)

Let f be $\mathcal{C}$–differentiable, let $g_i, i \in I$ be $(-\mathcal{C})$–differentiable at $\bar{x} \in X$, and $ri(\operatorname{dom} \mathcal{D}_{\mathcal{C}} f) \cap ri(\operatorname{dom}(-g_1) \cap \ldots \cap \operatorname{dom}(-g_m)) \neq \emptyset$. If $\bar{x}$ is a minimum point of (1), then there exist multipliers $\bar{\theta} \in \mathbb{R}$ and $\bar{\lambda} \in \mathbb{R}^m$, such that

$$\inf_{z \in S} \mathcal{D}_{\mathcal{C}} L(\bar{x}; z; \bar{\theta}, \bar{\lambda}) \geq 0 \ , \tag{20a}$$

$$g(\bar{x}) \geq 0 \ , \ \bar{\theta} \geq 0 \ , \ \bar{\lambda} \geq 0 \ , \ (\bar{\theta}, \bar{\lambda}) \neq 0 \ , \tag{20b}$$

$$\langle \bar{\lambda}, g(\bar{x}) \rangle = 0 \ , \tag{20c}$$

where $\mathcal{D}_{\mathcal{C}} L$ is the $\mathcal{C}$–derivative of L and $S := \{z : \|z\| = 1\}$. (20a) is equivalent to

$$0 \in \bar{\theta} \underline{\partial}_c f(\bar{x}) - \sum_{i \in I} \bar{\lambda}_i \bar{\partial}_{-c} g_i(\bar{x}) \ , \tag{20a$'$}$$

which becomes

$$0 \in \bar{\theta} \partial f(\bar{x}) - \sum_{i \in I} \bar{\lambda}_i \partial g_i(\bar{x}) \ , \tag{20a$''$}$$

if, in particular, X, f and $-g$ are convex. When f and g are differentiable on X and $\bar{x} \in \operatorname{int} X$, then (20a) collapses to $V_L = 0$ along $x = \bar{x}$, where V_L is the first variation of L, while in case (2) it becomes:

$$\Psi'_x(t, \bar{x}, \bar{x}'; \bar{\theta}, \bar{\lambda}) - \frac{d}{dt} \Psi'_{x'}(t, \bar{x}, \bar{x}'; \bar{\theta}, \bar{\lambda}) = 0, \tag{20a$'''$}$$

where $\Psi := \theta \psi_0(t, x, x') - \sum_{i \in I} \lambda_i \psi_i(t, x, x')$ is the integrand of L; if $X = H = \mathbb{R}^n$, then (20a) collapses to

$$L'_x(\bar{x}; \bar{\theta}, \bar{\lambda}) = 0 \ , \tag{20a$''''$}$$

where L'_x is the gradient of L with respect to x.

$^{(*)}$ The hypertangent cone to set S at $\bar{s} \in \operatorname{cl} S$ is defined as the set of $\bar{s} + s$ for which there exists $\bar{\alpha} > 0$ and a neighbourhood N of $\bar{s}$, both depending on s, such that, $\forall s' \in S \cap N$ and $\forall \alpha \in]0, \bar{\alpha}[$, we have $s' + \alpha s \in S$.

Proof. Lemma 5 implies the existence of $(-\bar{\theta}, -\bar{\lambda}) \in (\mathcal{K}_h - \bar{k})^*$ (so that, by Lemma 2, $\bar{x}$ is a lower semistationarity point of L and, by Lemma 6 with $G = \mathcal{C}$, (20a) holds) such that $\bar{\theta} \geq 0$, $\bar{\lambda} \geq 0$, $(\bar{\theta}, \bar{\lambda}) \neq 0$, $\langle \bar{\lambda}, g(\bar{x}) \rangle = 0$, and thus (20b, c) hold too. $\mathcal{D}_{\mathcal{C}}L$ is sublinear, so that $\underline{\partial}_{\mathcal{C}}L \neq \emptyset$. (20a) implies that $0 \in \underline{\partial}_{\mathcal{C}}L$. This relation, due to Proposition 1(i), (ii), becomes:

$$0 \in \bar{\theta}\underline{\partial}_{\mathcal{C}}f(\bar{x}) + \sum_{i=1}^{m} \bar{\lambda}_i \underline{\partial}_{\mathcal{C}}[-g_i(\bar{x})] = \bar{\theta}\underline{\partial}_{\mathcal{C}}f(\bar{x}) - \sum_{i=1}^{m} \bar{\lambda}_i \bar{\partial}_{-\mathcal{C}}g_i(\bar{x}).$$

(20a)$''$ is obvious, since in that case $\underline{\partial}_{\mathcal{C}}L = \partial L$. Let f and g be differentiable and $\bar{x} \in$ int X; since both z and $-z$ are in S, $\mathcal{D}_{\mathcal{C}}L$ collapses to V_L or to L'_x and (20a) to $V_L = 0$ or to (20a)$'''$, respectively when H is infinite or finite dimensional. In the first case it is well known that – the functionals being differentiable – the variation can be equivalently replaced with Euler variation, so that classic deductions lead to (20a)$'''$ in case (2). $\qquad \square$

6. QUASI–VARIATIONAL INEQUALITIES.

In Sect.1 it has been shown that optimality can be reduced to separation of two sets in the image space, namely $\mathcal{H}$ and $\mathcal{K}$. Now we will show that such a separation scheme can be considered as the "root" for other theories besides that of constrained extrema. Indeed, in the given space H, the separation has been used since a long time to investigate optimality; however, without introducing the image space it is difficult to conceive such a common root for different theories. Now we will briefly show how the theory of Quasi–Variational Inequalities can be reduced to the scheme adopted in the previous sections.

Let us consider a Quasi–Variational Inequality (in short, QVI) in the form: to find $y \in K(y)$, such that

$$\langle F(y), x - y \rangle \geq 0 \;, \; \forall x \in K(y) := \{x \in X(y) : g(y; x) \in \mathcal{C}\} \neq \emptyset, \qquad (21)$$

where $\mathcal{C}$ is a closed and convex cone with apex at the origin, $X(y)$ is a subset of a Banach space H, and $g : X^2(y) \to \mathbb{R}^m$. (21) obviously collapses to a Variational Inequality (in short, VI) when K is independent of y. In [8], Sect. 3, it has been proposed to associate an image to a VI; that proposal can be trivially extended to (21), for which it becomes: $y \in K(y)$ is a solution of (21) iff the system (in the unknown x)

$$u := \langle F(y), y - x \rangle > 0 \;, \; v := g(y; x) \in \mathcal{C} \;, \; x \in X(y), \qquad (22)$$

is infeasible. The space where (u, v) runs is the *image space* associated to (21), and the set

$$\mathcal{K}(y) := \{(u, v) \in \mathbb{R} \times \mathbb{R}^m : u = \langle F(y), y - x \rangle, \; v = g(y; x), \; x \in X(y)\}$$

is the *image* of (21). To system (22) we associate the set $\mathcal{H} := \{(u,v) \in \mathbb{R} \times \mathbb{R}^m : u > 0, \ v \in \mathcal{C}\}$, which depends on the types of inequalities only; another obvious remark is that the impossibility of (22) is equivalent to $\mathcal{H} \cap \mathcal{K}(y) = \emptyset$. To show this disjunction in [8] a separation scheme is proposed; this approach has been developed in [9]. It will now be slightly modified and applied to (21) in order to define a general class of gap functions; concepts and notations of [9] will be here understood.

Since the present purpose is merely that to show how a QVI (or VI) can be reduced to the same scheme like constrained extrema and investigated through the same analysis, then for the sake of simplicity in the remaining part of this section we will develop the following particular case:

$$I := \{1, \ldots, m\} \quad g(y;x) = (g_i(y;x), \ i \in I) \ ,$$
$$\mathcal{C} = O_p \times \mathbb{R}_+^{m-p} \quad , \quad O_p := (0, \ldots, 0) \in \mathbb{R}^p \ ,$$

where $p, m \in \mathbb{Z}_+$ with $0 \leq p \leq m$, and $\mathcal{C} = \mathbb{R}_+^m$ or $\mathcal{C} = O_m$ according to $p = 0$ or $p = m$, respectively. Consider the function

$$w(y; u, v; \lambda, \omega) := u + \langle \lambda, \gamma(y; v; \omega) \rangle \ , \tag{23}$$
$$u \in \mathbb{R}, \ v \in \mathbb{R}^m, \ \lambda \in \mathcal{C}^*, \ \omega \in \Omega,$$

where $\mathcal{C}^* = \{\lambda \in \mathbb{R}^m : \lambda_i \geq 0, \ i = p+1, \ldots, m\}$ is the positive polar of $\mathcal{C}$, and

$$\gamma(y; v; \omega) := (\gamma_i(y; v_i; \omega_i), \ i \in I) \ ,$$
$$\gamma_i : H \times \mathbb{R} \times \Omega_i \to \mathbb{R}, \quad \omega = (\omega_i, \ i \in I) \ , \quad \omega_i \in \Omega_i \ , \quad \Omega = \underset{i=1}{\overset{m}{\times}} \Omega_i;$$

γ and Ω must be such that, $\forall y \in X(y)$,

$$\text{lev}_{>0} w \supset \mathcal{H} \quad ; \quad \bigcap_{\omega \in \Omega} \text{lev}_{>0} w = \text{cl}\,\mathcal{H}. \tag{24}$$

Under these conditions (23) is a weak separation function in the sense of [9].

Each γ_i may be considered as a transformation of g_i; for this reason, $\forall y \in X(y)$, $\forall \omega_i \in \Omega_i$, γ_i must be such that

$$(\gamma_i(y; g_i(y;x); \omega_i), \ i \in I) \in \mathcal{C} \iff (g_i(y;x), \ i \in I) \in \mathcal{C}. \tag{25}$$

If $p = 0$, so that $\mathcal{C} = \mathbb{R}_+^m$, examples of γ_i (considered also in [7], p. 352) are:

$$\gamma_i(y; v_i; \omega_i) = v_i \exp(-\omega_i v_i), \quad \omega_i \in \Omega_i := \mathbb{R}_+, \tag{23$'$}$$

$$\gamma_i(y; v_i; \omega_i) = 1 - \exp(-\omega_i v_i), \quad \omega_i \in \Omega_i := \mathbb{R}_+ . \tag{23$''$}$$

Obviously, (23) is not the most general class of separation functions we can consider. For instance, (23) may be replaced with:

$$w(y; u, v; \omega) := u + \gamma(y; v; \omega), \quad \omega \in \Omega,$$

where $\gamma : H \times \mathbb{R}^m \times \Omega \to \mathbb{R}$ and Ω must be such that $\mathrm{lev}_{>0} \ w \supset \mathcal{H}$.
The dependence of γ on y is motivated by the fact that, in spite of what happens for constrained extremum problems, the change of y does not merely imply a shift of $\mathcal{K}(y)$. The above remarks lead us to consider the *transformed image* of (21):

$$\mathcal{K}(y; \omega) := \{(u, v) \in \mathbb{R} \times \mathbb{R}^m : u = \langle F(y), y - x \rangle,$$
$$v = \gamma(y; g(y; x); \omega), \ x \in X(y)\}$$

and its *conic extension:*

$$\mathcal{E}(y; \omega) := \mathcal{K}(y; \omega) - \mathrm{cl}\,\mathcal{H},$$

where cl denotes closure.

7. THE WEAK CASE.

A function $\psi : K^0 \to \mathbb{R}$ with $K^0 := \{y \in H : y \in K(y)\}$ is said a *gap function* iff $\psi(y) \geq 0 \ \forall y \in K^0$ and $\psi(y) = 0$ iff y is a solution of (21). Since we will set up a gap function as a by-product of the separation scheme in the image space, it is natural to expect to find two classes of gap functions, corresponding to weak and strong separation functions. The preceding definitions and notations correspond to the weak separation; hence the function (where the dependence on ω is understood):

$$\psi_w(y) := \min_{\lambda \in C^*} \max_{x \in X(y)} [\langle F(y), y - x \rangle + \langle \lambda, \gamma(y; g(y; x); \omega) \rangle], \qquad (26)$$

will be showed to be a gap function for (21), and will be called a *weak gap function*, since it comes from (23).

Theorem 2. Let $y \in K(y)$. Assume that the extrema in (26) exist and for each $y \in K(y)$ there exists $\omega(y) \in \Omega$, such that
(i) $\mathcal{E}(y; \omega(y))$ is convex;
(ii) $\{(u, v) \in \mathcal{H} : v = 0\} \not\subseteq T(\mathcal{E}(y; \omega(y)))$, where T denotes the Bouligand tangent cone at the origin.
Then y is a solution of (21) iff $\psi_w(y) = 0$.

Proof. Let y be a solution of (21), so that $g(y; y) \in C$. Then, the inclusion $\mathrm{lev}_{>0} w \supset \mathcal{H}$ implies

$$w(y; \langle F(y), y - y \rangle; g(y; y); \lambda, \omega(y)) \geq 0 , \quad \forall \lambda \in C^*.$$

Hence

$$\max_{x \in X(y)} \ w(y; \langle F(y), y - x \rangle, \ g(y; x); \lambda, \omega(y)) \geq 0 \ , \quad \forall \lambda \in C^*,$$

so that $\psi_w(y) \geq 0$. Ab absurdo, assume that $\psi_w(y) > 0$. Then $\exists \alpha > 0$, such that

$$\max_{x \in X(y)} \ w(y; \langle F(y), y - x \rangle, \ g(y; x); \lambda, \omega(y)) \geq \alpha > 0 \ , \quad \forall \lambda \in C^*. \tag{27}$$

Since y solves (21), i.e. (22) is impossible or $\mathcal{H} \cap \mathcal{K}(y; \omega(y)) = \emptyset$ or, equivalently, $\mathcal{H} \cap \mathcal{E}(y; \omega(y)) = \emptyset$, then (i) and the obvious convexity of $\mathcal{H}$ imply that $\mathcal{H}$ and $\mathcal{E}(y; \omega(y))$ can be linearly separated; (ii) implies that there exists disjunctive separation, namely $\mathcal{E}(y; \omega(y))$ can be included in one closed halfspace and $\mathcal{H}$ in its complement, or $\exists \bar{\lambda} \in C^*$ such that:

$$\mathcal{E}(y; \omega(y)) \subseteq \{(u, v) \in \mathbb{R} \times \mathbb{R}^m : u + \langle \bar{\lambda}, v \rangle \leq 0\},$$

and hence

$$w(y; \langle F(y), y - x \rangle, \ g(y; x); \bar{\lambda}, \omega(y)) \leq 0 \ , \quad \forall x \in X(y),$$

which contradicts (27). Now assume that $\psi_w(y) = 0$. Then $\exists \tilde{\lambda} \in C^*$ such that

$$\max_{x \in X(y)} \ w(y; \langle F(y), y - x \rangle, \ g(y; x); \tilde{\lambda}, \omega(y)) = 0,$$

so that

$$\langle F(y), y - x \rangle + \langle \tilde{\lambda}, \gamma(y; g(y; x); \omega(y)) \rangle \leq 0 \ , \ \forall x \in X(y). \tag{28}$$

Ab absurdo, assume that $\exists \tilde{x} \in K(y)$ such that

$$\langle F(y), y - \tilde{x} \rangle > 0.$$

Then, since $\langle \lambda, \gamma(y; g(\tilde{x}); \omega(y)) \rangle \geq 0 \ , \ \forall \lambda \in C^*$, we find

$$\langle F(y), y - \tilde{x} \rangle + \langle \lambda, \gamma(y; g(y; \tilde{x}); \lambda, \omega(y)) \rangle > 0 \ , \ \forall \lambda \in C^*.$$

This inequality, at $\lambda = \tilde{\lambda}$, contradicts (28). $\qquad\qquad\square$

Note that $\psi_w(y) \geq 0 \ , \ \forall y \in K(y)$, even if in the proof of Theorem 2, this property has not been used.

If QVI collapses to VI, namely $K(y)$ is independent of y and is not furtherly specified, so that g is not given, then such a VI can be obtained by the format (21) with $g(y; x) \in C$ identically verified. Hence (21) is not less general than the usual formats. It follows that in the format (21) some or all elements of g can be used for special purposes. For instance, if we set $m = 1$ and $C = \mathbb{R}_+$, take $g(y; x)$ and $X(y)$

independent of y, and set $g(y;x) = 0 \; \forall x \in X(y)$, $\gamma(y;v;\omega) = v$, $\forall v \in \mathbb{R}$, $\forall \omega \in \Omega := \mathbb{R}$ (the definition of Ω is of no importance), then (26) collapses to

$$\psi_w(y) = \max_{x \in X} \langle F(y), y - x \rangle, \tag{29}$$

which is the gap function introduced by Auslander [2].

In Theorem 2 we do not assume that $K(y)$ is convex even when $K(y)$ is independent of y, unlike what happens in [2,6,20]. However it may happen that, $K(y)$ being not convex, (i) of Theorem 2 turns out to be satisfied.

8. THE STRONG CASE

Now it will be shown that, correspondingly to what happens for the general situation of the Theory of Optimization, we have the strong case here too. This leads us to define another class of gap functions. Consider again (21), and the function:

$$s(y; u, v; \omega) := u - \delta(y; v; \omega) \, , \; u \in \mathbb{R}, \; v \in \mathbb{R}^m, \; \omega \in \Omega, \tag{30}$$

where $\delta : H \times \mathbb{R}^m \times \Omega \to \mathbb{R}$ is, for each y, such that $\mathrm{lev}_{>0}\, s \subset \mathcal{H}$, and δ and Ω must be such that the following conditions are satisfied

$$\exists \hat{\omega} \in \Omega \;\; \text{s.t.} \;\; \mathcal{K}(y) \cap \mathrm{lev}_{=0}\, s(y; u, v; \hat{\omega}) \neq \emptyset; \tag{31a}$$

$$\mathcal{H} \cap \mathcal{K}(y) \neq \emptyset \; \Rightarrow \; \exists \hat{\omega} \in \Omega \;\; \text{and} \;\; \exists (\hat{u}, \hat{v}) \in \mathcal{H} \cap \mathcal{K}(y)$$
$$\text{s.t.} \;\; s(y; \hat{u}, \hat{v}; \hat{\omega}) > 0. \tag{31b}$$

Given a kind of δ we may have several possible sets Ω. In this case we can choose Ω in order to simplify the subsequent development. Instead of (26) we consider now the function:

$$\psi_s(y) := \max_{\omega \in \Omega} \max_{x \in X(y)} [\langle F(y), y - x \rangle - \delta(y; g(y;x); \omega)], \tag{32}$$

which will be called a *strong gap function*; the term strong is motivated by the fact that (30) is a strong separation function in the sense of [11].

Theorem 3. Let $y \in K(y)$. Assume that the maxima in (32) exist, and that condition (31) holds. Then y is a solution of (21) iff $\psi_s(y) = 0$.

Proof. Because of (31a) $\exists \hat{\omega} \in \Omega$ and $\exists \hat{x} \in K(y)$ such that:

$$s(y; \langle F(y), y - \hat{x} \rangle, \; g(y; \hat{x}); \hat{\omega}) = 0;$$

the inclusion $\text{lev}_{>0}\ s \subset \mathcal{H}$ implies:

$$s(y; \langle F(y), y - x \rangle, g(y;x); \omega) \le 0 ,\ \forall x \in X(y) ,\ \forall \omega \in \Omega.$$

Then the necessity follows. Now assume that $\psi_s(y) = 0$. Then:

$$\max_{x \in X(y)} s(y; \langle F(y), y - x \rangle,\ g(y;x); \omega) \le 0 ,\ \forall \omega \in \Omega. \tag{33}$$

Ab absurdo, assume that $\exists \tilde{x} \in K(y)$ such that $\langle F(y), y - \tilde{x} \rangle > 0$. Then:

$$(\langle F(y), y - \tilde{x} \rangle,\ g(y; \tilde{x})) \in \mathcal{H} \cap \mathcal{K}(y),$$

so that $\mathcal{H} \cap \mathcal{K}(y) \ne \emptyset$. Hence, because of (31b), $\exists \hat{\omega} \in \Omega$ and $\exists \hat{x} \in K(y)$ such that:

$$s(y; \langle F(y), y - \hat{x} \rangle,\ g(y; \hat{x}); \hat{\omega}) > 0.$$

This inequality contradicts (33) at $\omega = \hat{\omega}$. $\qquad\qquad\qquad\qquad\qquad\square$

Now consider VI. Then we still consider the format (21), with $X(y)$ convex and independent of y and thus replaced by X; we set $m = 1$, $C = \mathbb{R}_+$, $g(y;x) = \langle x - y,\ G(x - y) \rangle$, with G a positive definite square matrix of order n. Since the constraint $g(y;x) \ge 0$ is identically true, so that $K(y) = X$ and $\mathcal{K}(y) \subseteq \{(u,v) \in \mathbb{R} \times \mathbb{R} : v \ge 0\}$, we have that, with the position

$$\delta(y; v; \omega) = \begin{cases} \omega v , & \text{if } v \ge 0 \\ +\infty , & \text{if } v < 0 \end{cases} ,\quad y \in \mathbb{R}^n,\ \omega \in \mathbb{R}_+ ,$$

the function s obviously satisfies the inclusion $\text{lev}_{>0}\ s \subset \mathcal{H}$. At $x = y$ we find that $(u = 0,\ v = 0) \in \mathcal{K}(y)$ and $s(y; 0, 0; \omega) = 0$, $\forall y \in X$, $\forall \omega \ge 0$; hence (31a) is verified at any $\omega \ge 0$. In order to discuss (31b), note that, $\forall y \in X$, we have

$$\mathcal{K}(y) = \{(u,v) \in \mathbb{R} \times \mathbb{R} : u = \langle F(y), y - x \rangle ,$$
$$v = \langle x - y, G(x - y) \rangle ,\ x \in X\}$$
$$\subset \{(u,v) \in \mathbb{R} \times \mathbb{R} : v \ge 0\};$$

$$\tilde{x} \in X;\ \langle F(y), y - \tilde{x} \rangle > 0 \ \Rightarrow\ \exists \hat{x} \in X \text{ s.t.}$$
$$\langle F(y), y - \hat{x} \rangle > \langle \hat{x} - y, G(\hat{x} - y) \rangle.$$

The 1-st part of condition (31b) implies the fulfilment of the 1-st part of the above condition, whose 2-nd part is equivalent to the 2-nd part of (31b) with $\hat{\omega} = 1$. Hence (31b) holds with $\Omega = \{1\}$. Thus (32) becomes

$$\psi_s(y) = \max_{x \in X}[\langle F(y), y - x \rangle - \langle x - y,\ G(x - y) \rangle],$$

which is the gap function considered by Fukushima [6]. Note that we could have chosen other sets Ω. We have made the simplest choice: a singleton. This is not unique. We might have chosen $\Omega = \{\hat{\omega} > 0\}$ finding the same result, or $\Omega = \{0\}$ finding again (29). The Auslander gap function is so particular that the two approaches become the same.

In a quite analogous way, by setting $g(y; x) = G(x, y)$, G being non–negative, continuously differentiable on $X \times X$, strongly convex on X with respect to $x, \forall y \in X$, and such that $G(x, x) = 0$, and

$$\nabla_x G(y, y) = 0 \quad \forall y \in X,$$

then we recover the gap function considered by Zhu and Marcotte [20].

Note that we have $\psi_s(y) \geq 0$, $\forall y \in K(y)$, as shown by the proof of Theorem 3.

REFERENCES

[1] ABADIE J., "On the Kuhn–Tucker theorem". In "Nonlinear Programming", North–Holland, Amsterdam, 1967, pp. 19–36.

[2] AUSLANDER A., "Optimization. Méthodes Numériques." Masson, Paris, 1976.

[3] BEN–TAL A. and ZOWE J., "Necessary and sufficient optimality conditions for a class of nonsmooth minimization problems." Mathem. Progr., Vol. 24, 1982, pp. 70–91.

[4] CAMBINI A., "Nonlinear separation theorems, duality, and optimality conditions", in "Optimization and Related Fields", Edited by R. Conti, E. De Giorgi, and F. Giannessi, Springer–Verlag, Berlin, 1986 pp. 57–93.

[5] CLARKE F.H., DEM'YANOV V.F. and GIANNESSI F. (eds.), "Nonsmooth Optimization and Related Topics." Plenum Press, New York, 1989.

[6] FUKUSHIMA M., "Equivalent differentiable optimization problems and descent methods for asymmetric variational inequalities problems." Mathem. Progr. Vol. 53, 1992, pp. 99–110.

[7] GAUVIN J., "Theory of nonconvex programming". École Polytecnique, Montréal, 1993, pp. 1–58.

[8] GIANNESSI F., "Theorems of the alternative, quadratic programs and complementarity problems". In "Variational Inequalities and Complementarity Problems", R.W. Cottle et al. eds., J. Wiley, New York, 1980, pp. 151–186.

[9] GIANNESSI F., "Theorems of the alternative and optimality conditions, Journal of Optimization Theory and Applications", Vol. 42, 1984, pp. 331–36.

[10] GIANNESSI F., "Theorems of the alternative for multifunctions with applications to optimization. Necessary conditions". Tech. Paper No. 131, Dept. of Mathem., Univ. of Pisa, 1986, pp. 1–127.

[11] GIANNESSI F., "Semidifferentiable functions and necessary optimality conditions", Journal of Optimization Theory and Applications, Vol. 60, N.2, 1989, pp. 191–241.

[12] GIANNESSI F. (ed.), "Nonsmooth Optimization. Methods and Applications". Gordon and Breach Science Publishers, Amsterdam, 1992.

[13] HIRIART URRUTY J.-B., "Refinements of necessary optimality conditions in nondifferentiable programming I". Appl. Mathem. and Optimization, Vol 5, 1979, pp. 63–82.

[14] IOFFE A., "A Lagrange multiplier rule with small convex–valued subdifferentials for nonsmooth problems of mathematical programming involving equality and nonfunctional constraints". Mathem. Progr., Vol. 58, 1993, pp. 137–145.

[15] OETTLI W., "Optimality conditions for programming problems involving multivalued mappings", in "Modern Applied Mathematics", Edited by B. Korte, North–Holland, Amsterdam, Part. 2, 1982, pp. 195–226.

[16] POURCIAU B.H., "Modern multiplier rules". Am. Math. Monthly, June–July 1980, pp. 433-452.

[17] REILAND T.W. ,"Optimality conditions and duality in continuous programming, I: convex programming and a theorem of the alternative, Journal of Mathematical Analysis and Applications, Vol. 77, 1980, pp. 297–325.

[18] ROBINSON S.M., "Local structure of feasible sets in nonlinear programming, part III: stability and sensitivity". Mathematical Programming Study, No. 30, 1987, pp. 45–66.

[19] TARDELLA F., "On the image of a constrained minimum problem and some applications to the existence of the minimum", Journal of Optimization Theory and Applications, Vol. 60, 1989, pp. 93–104.

[20] ZHU D.L. and MARCOTTE P., "A general descent framework for monotone variational inequalities". To appear in Journal Optimization Theory and Applications.

Linear Equations in Optimisation

C. G. BROYDEN

Facoltá di Scienze dell'Informazione
Universitá di Bologna
Via Sacchi 3, 47023 Cesena
Italy

ABSTRACT. Linear equations that arise from problems of constrained or unconstrained optimisation generally have certain specific properties, e.g. symmetry or a particular structure, that enable special-purpose methods to be developed for their solution. In this paper some of these properties are outlined and the ways in which they can be exploited are discussed. Some seven methods are considered, including the direct ones of Aasen and Biggs and the conjugate-gradient methods of Hestenes and Stiefel.

1. Introduction

The linear algebraic equations encountered in both constrained and unconstrained optimisation are considerably more specific than those met with normally. The matrix of coefficients is usually symmetric and, in the case of constrained optimisation, has a characteristic structure. Thus the algorithms used for solving linear equations in the context of optimisation are more specialised than algorithms like Gaussian elimination, LU decomposition or QU decomposition which are used to solve completely general sets of equations. Since the theme of this ASI is optimisation it seems appropriate to consider algorithms which try to make use of the particular properties of the systems being solved rather than to give a general survey of the type the author gave at a previous ASI [7] devoted to solving linear algebraic equations. We therefore assume a certain familiarity with the standard methods which include, in addition to those cited above, the method of Choleski for symmetric positive definite systems and, to a lesser extent, the methods of Aasen, Bunch and Parlett, and Bunch and Kaufman for symmetric indefinite ones. Most of these algorithms may be found in any standard text of numerical linear algebra (see, e.g., [16], [26]).

We thus consider only algorithms which set out to solve the equations

$$Ax = b \tag{1}$$

where A is always symmetric and often structured. We note that although A can be positive definite it is often not so, and never is when it has the characteristic structure of the Lagrangean Hessian. This effectively rules out the use of Choleski decomposition as a general purpose method even when, as for unconstrained optimisation, the

25

E. Spedicato (ed.), Algorithms for Continuous Optimization, 25–35.

matrix A is often the Hessian matrix of the function to be minimised. It is certainly true that in the neighbourhood of a local minimum of a suitably continuous and differentiable function (and that means a large number of the problems encountered in practice) the Hessian matrix *is* positive definite but far from the solution, unless more is known about the function, all that may be assumed is that the Hessian is symmetric. If A is an *approximation* to the Hessian then since it is desirable for this to be positive definite, even when remote from the solution, it is sometimes expressed as LL^T with only L being stored. This guarantees positive definiteness if L is real and nonsingular, properties that may easily be verified or imposed. Thus the Choleski method is sometimes used to force positive-definiteness upon matrices that are not naturally so, or which should be so but have become polluted by roundoff error.

2. Method 1 (Choleski's Method) and 2 (Aasen's Method)

Since the Choleski method is the simplest algorithm for solving linear equations we begin with a brief description. If A is real, symmetric and positive definite it is easy to show that it may be expressed as

$$A = LL^T \tag{2}$$

where L is a real and nonsingular lower triangular matrix. The proof is inductive, showing that if such a factorisation is possible for the r-th order leading principal submatrix of A then it is also true for the leading principal submatrix of order $r + 1$ provided that this submatrix is positive definite, as it must be if the original matrix is positive definite. If A is not positive definite the process comes to a halt with the computer trying to obtain the square root of a negative number, and then forcing A to be positive definite is based on replacing this number by a suitable positive one. A related problem occurs if A is nearly singular. In this case the positive number (the *pivot*) becomes very small, and as it is subsequently used as a divisor this smallness can give rise to numerical instability. In this case the existence of rounding error can also make this number negative, calling for the procedure described above for indefinite matrices to be implemented. If, though, A is positive definite and rounding error can be avoided then the Choleski method is both simple, short and stable. No pivoting is needed to guarantee stability and the method can be written in a few lines of code. It is one of the most useful methods available to those with linear equations to solve and no-one involved in optimisation, where the conditions for its successful application frequently occur, can afford to be in ignorance of it.

The need to take square roots can be removed if the matrix A is factorised according to

$$A = LDL^T \tag{3}$$

where L is now a unit lower triangular and D a diagonal matrix. It is always possible to carry out this factorisation, even for indefinite matrices, if the leading principal submatrices of A are nonsingular but in this case the factorisation is often highly numerically unstable and the method is not recommended unless A is known to be positive definite, i.e unless every diagonal element of D is positive. In this case the decomposition (2) may be marginally more stable.

Where A is known to be indefinite the method of Aasen [1] can be used. In this method A is expressed as the product of a unit lower triangular matrix L, a symmetric tridiagonal matrix T and the transpose of L, thus

$$PAP^T = LTL^T \tag{4}$$

where P is a permutation matrix introduced to express the row-column interchanges that are a feature of the method. Its success owes much to this pivoting strategy. Suppose we wish to reduce A to upper triangular form by Gaussian elimination and that, at an intermediate stage of the process, the coefficient matrix has the form

$$\begin{bmatrix} U & V \\ 0 & C \end{bmatrix},$$

where U is upper triangular. In order to maintain numerical stability, before subtracting appropriate multiples of the first row of $C = [c_{ij}]$ from subsequent rows (so that the dimension of U is increased by one) we usually exchange the first row of C by the r-th row, where $\|c_{r1}\| \geq \|c_{j1}\|$, $\forall j$. This ensures that when a multiple of the first row of C is subtracted from another row then that multiple is less than or equal to one in absolute value. The problem arises if we also wish to retain symmetry. If, in order to achieve this, we exchange the first and r-th columns of C then the element that we have so carefully placed in the top left-hand corner of C will now be replaced by some other element. The net result is that we cannot achieve the dual aims of symmetry and stability simultaneously when using Gaussian elimination. Aasen's idea was to use a similar technique to reduced A to upper Hessenberg form. At an intermediate stage of the process let the coefficient matrix have the form

$$M = \begin{bmatrix} H & K \\ E & B \end{bmatrix}$$

where H is now upper Hessenberg and E is null apart from the upper right-hand corner element. In order to increase the dimension of H by one Aasen subtracts the appropriate multiple of the *second* row of B from all subsequent rows but before doing so exchanges the *second* row of B with the r-th where $\|b_{r1}\| \geq \|b_{j1}\|$, $j \geq 2$. This is equivalent to pre-multiplying M by a permutation matrix and, since the first row of B is never involved, this operation always leaves E unchanged and hence does not alter the structure of M. If now, in order to preserve symmetry, a similar column interchange is carried out, the first column of B is unchanged so that the element placed in the b_{21} position remains in place. The reduction process can now take place with both stability and symmetry, and it is straightforward to see that the resulting matrix will be upper Hessenberg. Since this reduction is equivalent to premultiplying PAP^T by the unit lower triangular matrix L^{-1} (which is similar but not identical to the lower triangular factor obtained by Gaussian elimination) we have

$$L^{-1}PAP^T = H \tag{5}$$

where H is upper Hessenberg. Hence $L^{-1}PAP^TL^{-T} = HL^{-T}$. But the product of an upper Hessenberg matrix with an upper triangular matrix is upper Hessenberg and, since A is symmetric, so is $L^{-1}PAP^TL^{-T}$. Thus HL^{-T} is both upper Hessenberg and symmetric, hence tridiagonal, say T. Equation (4) follows. Now it is readily verified that with the particular form of L given by Aasen's method the operation of postmultiplying H by L^{-T} leaves the principal diagonal of H and all the elements to its left unchanged. It is therefore not necessary to compute HL^{-1} explicitly since T can be simply written down once H has been determined. Once T has been computed the solution of $LTL^Tx = b$ can be obtained by solving successively $Ly = b$, $Tz = y$ (using Gaussian elimination or some more specialised technique, and sacrificing symmetry) and $L^Tx = z$ for y, z and x respectively.

Other direct methods [10], [11] and [23] have been proposed for solving the symmetric indefinite problem but the comparisons of Barwell and George [3] indicate that Aasen's method is marginally the most accurate. We omit descriptions of the others in the interests of brevity.

3. Method 3 (The Conjugate Gradient Method and Variations)

If A is large and sparse, as is often the case when solving optimisation problems, direct methods require too much space to be seriously considered and we are forced to look at indirect or quasi-iterative ones. One of the more popular is the method of conjugate gradients [17], [18], [19] which is based on finding the stationary values of the quadratic function

$$\phi(x) \equiv (1/2)x^TGx - x^Th \tag{6}$$

where G is a symmetric but not necessarily definite nth order matrix and H an nth order vector, related to A and b in ways subsequently to be defined. If x_1 is an initial value of x, S_i a constant $n \times i$ matrix and if x_{i+1} denotes the value of x for which $\phi(x)$ is stationary, where x is constrained to satisfy

$$x \equiv x_1 + S_iz \tag{7}$$

and where $z \in R^i$ is a vector of independent variables, it is well-known (see e.g. Luenberger [20]) that

$$x_{i+1} = x_1 - S_i(S_i^TGS_i)^{-1}S_i^Tg_1 \tag{8}$$

where g_1 denotes the gradient of ϕ evaluated at x_1. Moreover, if

$$S_i = [p_1, p_2, \ldots, p_i] \tag{9}$$

where p_j, $1 \leq j \leq i$, are constant vectors such that $c_j = p_j^T G p_j$ are nonzero and

$$p_i^T G p_j = 0, \quad i \neq j, \tag{10}$$

then x_{i+1} may also be expressed as

$$x_{i+1} = x_i - p_i c_i^{-1} p_i^T g_i \tag{11}$$

where

$$g_i = G x_i - h \tag{12}$$

The sequence of vectors $\{p_j\}$ that satisfy equation (10) can be generated [6] by p_0 null and

$$p_{i+1} = K g_{i+1} - p_i \alpha_i, \quad i \geq 0 \tag{13}$$

where K is an arbitrary symmetric matrix, or p_0 null, p_1 arbitrary and

$$p_{i+1} = K G p_i - p_i \alpha_i - p_{i-1} \beta_{i-1} \quad i \geq 1 \tag{14}$$

The constants α_i and β_{i-1} are chosen to satisfy $p_{i-1}^T G p_{i+1} = 0$ and $p_i^T G p_{i+1} = 0$, choices that are always possible if $p_{i-1}^T G p_{i-1} \neq 0$ and $p_i^T G p_i \neq 0$. These conditions are satisfied for p_{i-1} and p_i non-null if G is positive definite, and algorithms for which G has this property have been called *b-stable*. It can be shown [6] that if K is positive definite another kind of numerical instability (*w-instability*) may be avoided, an instability where successive steps become very small and the algorithm "dies".

Three principal methods have been proposed for solving equation (1) for A symmetric which give rise to 2-term formulae for the vectors p_i:

No	G	h	K	Names	References
1	A	b	I	cg	[17] [18] [19] [25]
2	A^2	Ab	A^{-1}	cr	[17] [18] [19] [25]
3	A	b	M^{-1}	pcg	[2] [12]

The first two, the original conjugate gradient and conjugate residual methods introduced by Hestenes and Stiefel in 1952 are both *b-* and *w*-stable if A is positive definite but if A is indefinite then Method 2 is only *b*-stable and Method 1 is only *w*-stable. Their use for indefinite A is therefore not recommended. Method 3, the preconditioned conjugate gradient method, is *b-* and *w*-stable if both A and M are definite. The preconditioning matrix M is chosen to reduce the condition number of KG in order to improve the convergence of the method. See Dennis and Turner [12] or Ashby *et al* [2] for further details.

For the three-term formula with A symmetric the following possibilities arise:

No	G	h	K	Names	References
4	A	b	I	Nazareth	[13] [21] [22
5	A^2	Ab	A^{-1}		[17] [18] [19] [25]
6	I	$A^{-1}b$	A	SYMMLQ	[14] [24] [27]

Methods 4 and 5 are simply straightforward three-term versions of methods 1 and 2 but method 6 is a very sophisticated algorithm for solving large indefinite problems. For further details the appropriate references should be consulted.

4. Solving the Stationary Lagrangean Equations

The above methods, where A is symmetric but lacks structure, may usually be applied to unconstrained problems. If, however, a problem is constrained the equations to be solved (see the chapters on constrained optimisation in this volume) often have the form

$$\begin{bmatrix} B & J^T \\ J & 0 \end{bmatrix} \begin{bmatrix} p \\ z \end{bmatrix} = - \begin{bmatrix} g \\ c \end{bmatrix} \tag{15}$$

They are derived by attempting to find the stationary value of the Lagrangean function $\phi(x) + z^T c(x)$ by zeroing its first partial derivatives with respect to both x and z using Newton's method. In the equation, B is the Lagrangean Hessian with respect to x (or an approximation thereto), J is the Jacobian of $c(x)$ (the constraints), g is the gradient of $\phi(x)$, p is the *correction* to x to give it its new value but z *is* the new value of z. Since we assume that M is nonsingular it follows that J, which always has at least as many columns as rows, has full row rank.

We first prove

Theorem 1 *Let M be defined by*

$$M = \begin{bmatrix} B & J^T \\ J & 0 \end{bmatrix} \tag{16}$$

and assume that B is nonsingular. Then M is singular if and only if $JB^{-1}J^T(= K$ say) is singular.

Proof. If K is singular then $\exists\, x \neq 0$ such that $Kx = 0$. Hence

$$\begin{bmatrix} B & J^T \\ J & 0 \end{bmatrix} \begin{bmatrix} B^{-1}J^T x \\ x \end{bmatrix} = 0$$

so that M is singular.

Conversely, if M is singular, then $\exists z = \begin{bmatrix} y \\ x \end{bmatrix} \neq 0$ such that

$$\begin{bmatrix} B & J^T \\ J & 0 \end{bmatrix} \begin{bmatrix} y \\ x \end{bmatrix} = 0.$$

Hence

$$y + B^{-1}J^T x = 0, \tag{17}$$

$Jy = 0$ and $x \neq 0$ since if $x = 0$ then, from (17), $y = 0$ so that $z = 0$ and contradiction arises. But since $Jy = 0$ it follows from equation (17) that $Kx = 0$ so that K is singular. □

Method 4

This theorem gives us one possible way of solving equation (15) when B is nonsingular. From equation (15) we have

$$Bp + J^T z = -g \tag{18}$$

which becomes, on pre-multiplication by JB^{-1},

$$Jp + Kz = -JB^{-1}g \tag{19}$$

with K as previously defined. Hence, since $Jp = -c$,

$$Kz = c - JB^{-1}g \tag{20}$$

and this equation may be solved for z. Equation (18) may then be solved for p. This method does, however, break down if B is singular and would be expected to give inaccurate results if B were badly conditioned, and this must be seen as a weakness of this method.

Method 5

Another method that has been proposed to solve equations (15) and (16) using their structure relies on the use of orthogonal transformations. Let Q be an orthogonal matrix such that

$$QJ^T = \begin{bmatrix} U \\ 0 \end{bmatrix} \tag{21}$$

where U is upper triangular and is, since J is assumed to have full row rank, nonsingular. From the first row of equation (15) we have, since Q is orthogonal,

$$QBQ^T Qp + QJ^T z = -Qg \tag{22}$$

and if we define $s = Qp$, $h = Qg$ and $G = QBQ^T$ this may be written, from equation (21) and with obvious partitioning,

$$\begin{bmatrix} G_{11} & G_{12} \\ G_{21} & G_{22} \end{bmatrix} \begin{bmatrix} s_1 \\ s_2 \end{bmatrix} + \begin{bmatrix} U \\ 0 \end{bmatrix} z = - \begin{bmatrix} h_1 \\ h_2 \end{bmatrix} \tag{23}$$

or

$$G_{11}s_1 + G_{12}s_2 + Uz = -h_1 \tag{24}$$

and

$$G_{21}s_1 + G_{22}s_2 = -h_2 \tag{25}$$

From the second row of equation (15) we obtain $JQ^TQp = -c$ which becomes, from equation (21),

$$U^Ts_1 = -c \tag{26}$$

Thus s may be found by solving equation (24) for s_1 and equation (25) for s_2 (it may easily be shown that M is nonsingular if and only if G_{22} is nonsingular). Finally equation (24) then yields z if needed. The correction vector p is then obtained from s by $p = Q^Ts$. Unlike the previous method, this method works even if B is singular and is generally more numerically stable. For further details see [15].

Another problem that can crop up in constrained optimisation is the solution of the equation

$$(B + (1/r)J^TJ)p = -(g + (1/r)J^Tc) \tag{27}$$

(or some variant thereof) which arises from the attempt to solve the constrained optimisation problem by minimising a penalty function, in this case that of Courant

$$P(x) \equiv \phi(x) + (2r)^{-1}c(x)^Tc(x) \tag{28}$$

Finding the minimum of $P(x)$ by finding the value of x that zeroes its gradient by Newton's method leads precisely to equation (27). The problem is that as r is forced to zero the condition number of $M = B + (1/r)J^TJ$ becomes increasingly large so that if M and the right-hand side $g + (1/r)J^Tc$ are computed as a single matrix and single vector respectively then severe numerical instability occurs. It is necessary to keep the different parts of the matrix and vector separate in order to obtain a satisfactory algorithm, and we now describe two ways of doing this.

Method 6 (Biggs)

In this method [4], [5], equation (25) is pre-multiplied by rJB^{-1} to give

$$(rI + K)Jp = -(JB^{-1}gr + Kc) \tag{29}$$

where $K = JB^{-1}J^T$ as previously defined. If now z is defined by

$$Jp = zr - c \tag{30}$$

and $zr - c$ substituted for Jp in both equations (27) and (29) we obtain, respectively,

$$Bp + J^Tz = -g \tag{31}$$

and

$$(rI + K)z = c - JB^{-1}g, \tag{32}$$

from the second of which z may be determined provided that $rI + K$ is nonsingular. This value may then be substituted into equation (31) and p thus determined.

In order for this method to succeed not only does B need to be nonsingular but K must not have any eigenvalues equal to $-r$, and the near-occurrence of either of these conditions could lead to numerical instability. Note that equations (18) and (31) are identical, and so are (20) and (32) if $r = 0$. The method thus tends to method 4 in the limit as r is driven to zero. Despite its numerical weaknesses the method is very ingenious and deserves better than to masquerade, as it so frequently does, as an SQP method.

Method 7 (Broyden and Attia)

The second way of solving equation (27) requires the same orthogonal transformations as were used in method 5. Equation (27) may be written

$$Q(B + (1/r)J^T J)Q^T Qp = -Q(g + (1/r)J^T c) \qquad (33)$$

from which we obtain, with the same definitions as above and after some manipulation,

$$U^T s_1 = zr - c \qquad (34)$$

and

$$G_{21}s_1 + G_{22}s_2 = -h_2 \qquad (35)$$

where

$$Uz = -(h_1 + G_{11}s_1 + G_{12}s_2) \qquad (36)$$

These equations may be written

$$\begin{bmatrix} rI & U^T & 0 \\ U & G_{11} & G_{12} \\ 0 & G_{21} & G_{22} \end{bmatrix} \begin{bmatrix} z \\ s_1 \\ s_2 \end{bmatrix} = - \begin{bmatrix} c \\ h_1 \\ h_2 \end{bmatrix} \qquad (37)$$

and may be solved directly by a modification of Aasen's method. As in the method of Biggs, the above algorithm can break down for certain values of r but unlike the method of Biggs it does not require the non-singularity of B. Note also that as r tends to zero then the method tends to method 5.

Another feature is that if r is small then equations (34)–(36) may be solved by successive approximation, thus:

1. Set z null

2. Solve equation (34) for s_1

3. Solve equation (35) for s_2

4. Solve equation (36) for z

5. If not converged then repeat from (2).

See Broyden and Attia [8], [9] for further details.

References

[1] Aasen, J. O., "On the Reduction of a Symmetric Matrix to Tridiagonal Form," BIT 11, 233 − 242, 1971.

[2] Ashby, S.F., Manteuffel, T.A. and Saylor, P.E., "A Taxonomy for Conjugate Gradient Methods, SIAM J. Numer. Anal. 27, 1542-1568, 1990.

[3] Barwell, V. and George, J.A., "A Comparison of Algorithms for Solving Symmetric Indefinite Systems," ACM Trans. Math. Soft. 2, 242-251, 1976.

[4] Biggs, M.C., "Constrained Minimisation using Recursive Equality Constrained Quadratic Programming", in Numerical Methods for Optimisation, edited by F.A. Lootsma, Academic Press, London, 1972.

[5] Biggs, M.C., "Recursive Quadratic Programming Based on Penalty Functions for Constrained Optimisation," in Nonlinear Optimisation, Theory and Algorithms, edited by L.C.W. Dixon, E. Spedicato and G.P. Szego, Birkhauser, Boston Massachusetts, 1980.

[6] Broyden, C.G., "Block Conjugate Gradient Methods," Optimization Methods and Software, 2, 1-17, 1993.

[7] Broyden, C.G., "Classical Methods for Linear Equations," in Computer Algorithms for Solving Linear Algebraic Equations - The State of the Art, edited by Emilio Spedicato, NATO ASI Series F: Computer and Systems Sciences, Vol. 77, Springer, 1991.

[8] Broyden, C.G. and Attia, N.F., "A Smooth Sequential Penalty Function Method for Solving Nonlinear Programming Problems," in Systems Modelling and Optimisation, edited by P. Thoft Christensen, Springer, Berlin, 1983.

[9] Broyden, C.G. and Attia, N.F., "Penalty Functions, Newton's Method and Quadratic Programming," Journ. Opt. Theory and Applics. 58, 3, 377-385, 1988.

[10] Bunch, J.R. and Parlett, B.N., "Direct Methods for Solving Symmetric Indefinite Systems of Linear Equations," SIAM J. Numer. Anal. 8, 639-655, 1971.

[11] Bunch, J.R. and Kaufman, L., "Some Stable Methods for Calculating Inertia and Solving Symmetric Linear Systems," Math. Comp. 31, 162-179, 1977.

[12] Dennis, J.E. and Turner, K., "Generalized Conjugate Directions", Linear Algebra Applics. 88/89, 187-209, 1987.

[13] Dixon, L.C.W., "On Nazareth's Three Term Conjugate Gradient Method," Technical Report No. 133, Hatfield Polytechnic Numerical Optimisation Centre, 1983.

[14] Fletcher, R., "Conjugate Gradient Methods for Indefinite Systems," in *Proceedings of Dundee Conference on Numerical Analysis,* Lecture Notes in Mathematics 506, edited by G. A. Watson, Springer, Berlin-Heidelberg, 1976.

[15] Gill, P.E., Murray, W. and Wright, M.H., *Numerical Linear Algebra and Optimisation,* Vol 2, Addison-Wesley, 1991.

[16] Golub, G.H. and Van Loan, C., *Matrix Computations,* Johns Hopkins University Press, Baltimore, 1983.

[17] Hestenes, M.R., "The Conjugate Gradient Method for Solving Linear Systems," in *Proceedings of Symposia on Applied Mathematics, Vol VI, Numerical Analysis,* Mc.Graw-Hill, New York, 1956.

[18] Hestenes, M.R., *Conjugate Direction Methods in Optimisation,* Springer, Berlin-Heidelberg, 1980.

[19] Hestenes, M.R. and Stiefel, E., "Methods of Conjugate Gradients for Solving Linear Systems," J. Res. Nat. Bureau of Standards 49, 409-436, 1952.

[20] Luenberger, D.G., *Introduction to Linear and Nonlinear Programming,* Addison-Wesley, Reading, Mass., 1984.

[21] Nazareth, L., "A Conjugate Gradient Algorithm without Line Searches," JOTA 23, No 3, 373-387, 1977.

[22] Nocedal, J, "On the Method of Conjugate Gradients for Function Minimisation," Ph.D. Thesis, Rice University, 1978.

[23] Parlett, B.N. and Reid, J.K., "On the Solution of a System of Linear Equations whose Matrix is Symmetric but not Definite," BIT 10, 386-397, 1970.

[24] Paige, C. C. and Saunders, M.A., "Solution of Sparse Indefinite Systems of Linear Equations," SIAM J. Numer. Anal. 12, No. 4, 617-629, 1975.

[25] Reid, J.K., "On the Method of Conjugate Gradients for the Solution of Large Sparse Systems of Equations," in: *Large Sparse Sets of Linear Equations,* edited by J. K. Reid, Academic Press, London and New York, 1971.

[26] Stewart, G.W., *Introduction to Matrix Computations,* Academic Press, New York, 1973.

[27] Stoer, J. and Freund, R.W., "On the Solution of Large Indefinite Systems of Linear Equations by Conjugate Gradient Methods," in: *Computing Methods in Applied Sciences and Engineering* V, edited by R. Glowinski and J.L. Lions, 35-5, North Holland, Amsterdam, 1982.

Generalized and Sparse Least Squares Problems

ÅKE BJÖRCK
Department of Mathematics
Linköping University
S-581 83 Linköping
Sweden

ABSTRACT. Least squares problems arise frequently in optimization, e.g., in interior point methods. This paper surveys methods for solving least squares problems of non-standard form such as generalized and sparse problems. Algorithms for standard and banded problems are first studied. Methods for solving generalized least squares problems are then surveyed. The special case of weighted problems is treated in detail. Iterative refinement is discussed as a general technique for improving the accuracy of computed solutions. Least squares problems where the solution is constrained by linear equality constraints or quadratic constraints are also treated.

Graph theoretic methods for reordering rows and columns to reduce fill in when solving sparse least squares problems are surveyed. The numerical phase of sparse Cholesky and sparse QR factorization is then discussed. In particular the multifrontal method, which currently is the most efficient implementation, is described.

1. Least Squares Problems

1.1. INTRODUCTION

Let $A \in \mathbf{R}^{m \times n}$ be a rectangular matrix, and $b \in \mathbf{R}^m$ a vector. A fundamental computational problem is the **linear least squares problem**

$$\min_{x \in \mathcal{S}} \|x\|_2, \qquad \mathcal{S} = \{x \in \mathbf{R}^n \mid \|Ax - b\|_2 = \min\}. \tag{1.1}$$

Least squares problems arise frequently in applications. An important example in optimization is in interior point methods, where the Karush-Kuhn-Tucker optimality condition give rise to a generalized least squares problem, see Wright [54].

Many surveys over computational methods for problem (1.1) exist, see, e.g., Lawson and Hanson [37], Björck [9], [10]. Here we focus on methods for solving least squares problems of non-standard form. In particular we consider in detail methods for weighted and sparse problems.

In Section 1 we first briefly review the method of normal equations and methods based on the QR factorization of A for the standard least squares problem. We then

37

E. Spedicato (ed.), Algorithms for Continuous Optimization, 37–80.
© 1994 *Kluwer Academic Publishers.*

consider problems where A is a (rectangular) banded matrix, and show that these can be handled by simple modifications of methods for dense problems.

In Section 2.1 we discuss the method of Paige for generalized least squares prob lems involving two matrices A and W, where $W = B^T B$ is symmetric positive definite matrix. In Section 2.2 we consider methods which rely on factorizing the symmetric indefinite system matrix directly, and discuss its numerical stability. The important special case when W is diagonal, the weighted least squares problem, is treated in Section 2.3. Iterative refinement in fixed precision is an important technique to improve solutions computed by methods which are not backward stable. This is the topic of Section 2.4.

In the simplest constrained least squares problem the solution is required to satisfy a subsystem of equations exactly. In Section 3 we compare three different methods for solving such problems. An important technique for regularizing ill-conditioned least squares problems is to include a quadratic inequality constraint. In Section 4 we give an overview of methods for solving such constrained problems.

Sparse least squares problems are treated in Sections 5 and 6. The symbolic phase of the computation is covered in Section 5 and the numerical phase in Section 6. Graph theoretic tools are introduced in 5.2 and transformation to block triangular form in 5.3. Techniques for reordering column for sparsity are discussed in Section 5.4. The Cholesky and QR factorization both aim at computing a sparse upper triangular factor R, and share the same symbolic phase. In the numerical phase the difference is that in the Cholesky factorization one first forms numerically the matrix $A^T A$ whereas in the QR factorization the matrix A is transformed directly after the rows have been preordered. The row sequential method for QR factorization is described in Section 6.1. The currently most efficient implementations of both sparse Cholesky and QR factorizations use a multifrontal approach. This is described for the QR factorization in Section 6.2. Finally, in Section 6.3 methods are studied for updating the solution to a sparse problem when a few dense equations are added.

We have treated generalized problems separately from sparse problems. It should be stressed that in practice the interest is often in generalized *and* sparse problems. Unfortunately, in many cases efficient algorithms for such problems remain to be developed.

1.2. THE METHOD OF NORMAL EQUATIONS

The set S of all least squares solutions is characterized by $x \in S \iff A^T(b - Ax) = 0$, i.e., the residual $r = b - Ax$ is orthogonal to $\mathcal{R}(A)$. Hence any least squares solution satisfies the **normal equations**

$$A^T A x = A^T b. \qquad (1.2)$$

The solution x is unique if and only if rank$(A) = n$. In this case $A^T A$ is positive definite, and the Cholesky factorization $A^T A = R^T R$, R upper triangular, diag$(R) >$

0, exists and is unique. The least squares solution $x = (A^TA)^{-1}A^Tb$ is obtained by solving the two triangular systems of equations

$$R^Tz = A^Tb, \quad Rx = z.$$

The method of normal equations dates back to Gauss. The total number of flops to compute x is (neglecting lower order terms) $mn^2/2 + n^3/6$. (One flop is here one multiplication *and* one addition).

If rank $(A) = m$ in (1.1), then the system $Ax = b$ is consistent, and the unique solution x of minimum norm satisfies the **normal equations of the second kind**

$$AA^Tz = b, \qquad x = A^Tz. \tag{1.3}$$

Here AA^T is symmetric and positive definite and the solution $x = A^T(AA^T)^{-1}b$ can be computed from the Cholesky factorization of AA^T.

To compute the matrix A^TA in (1.2) we can partition $A = (a_{.1}, a_{.2}, \ldots, a_{.n})$ by columns, and use the inner product formulation

$$(A^TA)_{jk} = a_j^Ta_k, \quad 1 \le j \le k \le n. \tag{1.4}$$

It is only necessary to compute and store the upper triangular part of A^TA. This requires $\frac{1}{2}n(n+1)m + mn$ flops, and if $m \gg n$ can be viewed as a data compression. The inner product formulation (1.4) is not suitable for large problems since each column needs to be accessed many times. An alternative row oriented algorithm, which uses only *one pass* through the matrix A, is obtained by partitioning A by rows, $A^T = (a_{1.}, a_{2.}, \ldots, a_{m.})$, and using the outer product form

$$A^TA = \sum_{i=1}^{m} a_{i.}a_{i.}^T \quad 1 \le j \le k \le n. \tag{1.5}$$

Outer product form is also preferable to use if the matrix A is sparse. Note that in both formulas we can get A^Tb by adjoining b to A and forming $(A, b)^T(A, b)$.

Using the standard model for floating point computation it is easy to show that the computed matrix satisfies

$$\overline{A^TA} = A^TA + E, \qquad |e_{ij}| < 1.06mu \sum_{k=1}^{m} |a_{ik}||a_{jk}|.$$

where $mu < 0.1$, and u is the machine unit. However, it is *not true* that $\overline{A^TA} = (A + E)^T(A + E)$ for some error matrix E, so the rounding errors in forming the matrix A^TA are not in general equivalent to small perturbations of the initial data matrix A. Hence the method of normal equations is not backwards stable. When A^TA is ill-conditioned it might be necessary to use double precision in forming and solving the normal equations in order to avoid loss of significant information. Another problem with using the normal equations is that it essentially squares the condition number,

$$\kappa(A^T A) = \kappa^2(A), \qquad \kappa(A) = \sigma_1/\sigma_r,$$

$r = \text{rank}(A)$. Although a term proportional to $\kappa^2(A)\|r\|_2$ occurs in the perturbation analysis for the least squares problem, this is only relevant for large residual problems. Hence *the normal equations can be much worse conditioned than the least squares problem from which they originated.*

A important way to improve the accuracy of a solution $\bar{x}$ computed by the method of normal equations is by iterative refinement, see Section 2.4. This requires that the data matrix A is saved and used to compute the residual $b - A\bar{x}$. In this way information lost when $A^T A$ was formed can be recovered.

1.3. QR FACTORIZATION

We now consider methods for solving problem (1.1) which are based on orthogonal transformations, and avoid the squaring of the condition number in the method of normal equations. These methods are based on the QR factorization of A. Assuming that the matrix A has full column rank, there exists a factorization

$$A = Q \begin{pmatrix} R \\ 0 \end{pmatrix} = Q_1 R, \qquad Q = (Q_1, Q_2), \tag{1.6}$$

where Q is square orthogonal and R upper triangular. Since Q is orthogonal the singular values of R equal those of A and $\kappa(R) = \kappa(A)$. Here Q_1 is uniquely determined and $\mathcal{R}(A) = \mathcal{R}(Q_1)$. Since

$$A^T A = (R^T \ 0)Q^T Q \begin{pmatrix} R \\ 0 \end{pmatrix} = R^T R,$$

R equals the unique Cholesky factor of $A^T A$ if $\text{diag}(R) > 0$. · Note, however, that this is only true in *exact computation.*

Given the QR factorization of A the unique least squares solution can be computed from

$$Rx = c_1, \qquad Q_1^T b = c_1. \tag{1.7}$$

When the matrix A has full row rank, $\text{rank}(A) = m \leq n$, the QR factorization of A^T (which is equivalent to the LQ factorization of A) can be used to compute the minimum norm solution of the underdetermined system

$$x = Q_1 y, \qquad R^T y = b. \tag{1.8}$$

The QR factorization (1.6) can be computed by premultiplying A by a sequence of elementary orthogonal transformations. **Householder reflectors** are orthogonal matrices of the form

$$P = I - uu^T/\gamma, \quad \gamma = u^T u/2,$$

where u is called the Householder vector. It is easily verified that P is symmetric and orthogonal. The product Pa where a is a given vector can be computed without explicitly forming P itself from

$$Pa = (I - uu^T/\gamma)a = a - u(u^Ta)/\gamma. \tag{1.9}$$

It follows that $Pa \in \text{span}[a, u]$, $Pu = -u$, i.e., P reverses u and $Pa = a$, for $a \perp u$. The effect of the transformation Pa for a general vector a is to reflect a in the $(m-1)$-dimensional hyperplane characterized by the normal vector u. Given a vector a such that $\|a\|_2 = \sigma$, $\alpha_1 = a^Te_1$, we define a Householder vector by

$$u = a + \text{sign}\,(\alpha_1)\sigma e_1, \qquad \gamma = \sigma(\sigma + |\alpha_1|).$$

Then $Pa = -\text{sign}\,(\alpha_1)\sigma e_1$, i.e., the transformation zeroes all components except the first one in a. Using a sequence $P_1, \ldots, P_n$ of such transformations elements below the diagonal in a matrix A can be zeroed and the factorization (1.6) with $Q = P_1 \cdots P_n$ is obtained. It is usually advantageous to keep Q in product form and not compute Q explicitly.

Another useful class of orthogonal transformations is **Givens rotations**. In $\mathbf{R}^2$ the matrix representing a rotation clockwise through an angle θ is given by

$$G(\theta) = \begin{pmatrix} c & s \\ -s & c \end{pmatrix}, \quad c = \cos\theta, \quad s = \sin\theta.$$

Note that $G^{-1}(\theta) = G^T(\theta) = G(-\theta)$. In $\mathbf{R}^n$ the matrix $G_{ij}(\theta)$ representing a rotation in the plane spanned by the unit vectors e_i and e_j, $i < j$, is a rank two modification of the unit matrix I_m. Premultiplying a vector $a = (\alpha_1, \ldots, \alpha_m)^T$ by $G_{ij}(\theta)$ we get $\tilde{\alpha}_k = \alpha_k$, $k \neq i, j$, and

$$G_{ij}(\theta)a = (\tilde{\alpha}_1, \ldots, \tilde{\alpha}_m)^T, \qquad \tilde{\alpha}_k = \begin{cases} c\alpha_i + s\alpha_j, & k = i; \\ -s\alpha_i + c\alpha_j, & k = j. \end{cases} \tag{1.10}$$

Thus a plane rotation may be multiplied into a vector at a cost of two additions and four multiplications. We can determine the rotation $G_{ij}(\theta)$ so that $\tilde{\alpha}_j$ becomes zero by taking

$$c = \alpha_i/\sigma, \quad s = \alpha_j/\sigma, \quad \sigma = (\alpha_i^2 + \alpha_j^2)^{1/2} \neq 0.$$

Premultiplication of a matrix $A \in R^{m \times n}$ with a Givens rotation G_{ij} will only affect the two *rows* i and j in A. The product $G_{ij}A$ requires $4n$ flops. It is essential to note that the matrix G_{ij} need never be explicitly formed.

Givens rotations can be used in several different ways to construct an orthogonal matrix Q such that $Q^Ta = \sigma e_1$. If G_{1k} is a Givens rotation which zeroes the k-th component in the vector a, then

$$G_{1m} \ldots G_{13}G_{12}a = \sigma e_1.$$

Note that G_{1k} will not destroy previously introduced zeros. Another possible sequence is $G_{k-1,k}$, $k = m, m-1, \ldots, 2$, where $G_{k-1,k}$ is chosen to zero the k-th component. This greater flexibility of Givens rotations compared to reflectors is particularly valuable for sparse problems.

Gram-Schmidt orthogonalization can also be used to compute the factorization (1.6), and produces both factors R and Q_1 explicitly. Methods based on the modified Gram-Schmidt are backward stable and produce equally accurate solutions to least squares problems if correctly used. It is important to note, however, that the equations (1.7) and (1.8) *have to be modified and should not be used* with the modified Gram-Schmidt method, see [12].

1.4. BANDED PROBLEMS

In many applications the nonzero elements of the matrix $A \in \mathbf{R}^{m \times n}$ are contained in a band. To describe such a structure we define f_i and l_i to be the column subscripts of the first and last nonzero in the i-th row of A, i.e.,

$$f_i = \min\{j \mid a_{ij} \neq 0\}, \quad l_i = \max\{j \mid a_{ij} \neq 0\}. \tag{1.11}$$

A matrix A is said to have row bandwidth w, where

$$w = \max_{1 \leq i \leq m} w_i, \quad w_i = (l_i(A) - f_i(A) + 1).$$

For this structure to have practical significance we need to have $w \ll n$.

If A has row bandwidth w then it follows from

$$|j - k| \geq w \Rightarrow (A^T A)_{jk} = \sum_{i=1}^{m} a_{ij} a_{ik} = 0$$

that $a_{ij} a_{ik} \neq 0 \Rightarrow |j - k| < w$. Hence the matrix of normal equations $A^T A$ has upper and lower bandwidth $r \leq w - 1$. As is well known, the Cholesky factor R has the same band structure as the upper triangle of $A^T A$. Thus R will again have upper bandwidth $w - 1$, and we can use a band Cholesky algorithm to solve the normal equations.

We now consider orthogonalization methods for the case when A has row bandwidth w. We assume in the following that the rows of A have been sorted so that the column indices $f_i, i = 1, 2, \ldots, m$ of the first nonzero element in each row form a nondecreasing sequence,

$$i \leq k \Rightarrow f_i \leq f_k.$$

Such a band matrix is said to be in **standard form**.

We first describe a scheme using Givens rotations to compute the QR factorization. Since R equals the unique Cholesky factor of $A^T A$ we know that R will have upper bandwidth $w - 1$. We initialize the upper triangular band matrix R to zero, and then update R adding row i of A, $i = 1, \ldots, m$ as follows:

for $j = f_i(A), \ldots, l_i(A)$
    if $a_{ij} \neq 0$ then
        construct Givens rotation from (r_{jj}, a_{ij});
        apply Givens rotation to annihilate a_{ij}

For band matrix A in standard form the updating of R by Givens method when a new row is added is basically identical to updating a *full triangular matrix* formed by rows and columns $f_i(A)$ to $l_i(A)$ of R by by the full row formed by elements $f_i(A)$ to $l_i(A)$ in the ith row. For a detailed discussion see Cox [17]. Very large problems can be handled since at each stage only a small part of the matrix need to be held in primary storage. Note that by initializing R to zero the description above is valid also for the processing of the first rows of A. For example, the first row $a_{1.}$ can just be inserted into R by a row permutation, which is a special case of a Givens rotation. Similarly, the number of rotations needed to process row i is at most equal to $\min(i - 1, w)$.

The Givens rotations can also be applied to one or several right hand sides b to produce

$$c = Q^T b = \begin{pmatrix} c_1 \\ c_2 \end{pmatrix}, \quad c_1 \in \mathbf{R}^n.$$

The least squares solution is then obtained from $Rx = c_1$ by back-substitution. The vector c_2 is normally not stored but used to accumulate the residual sum of squares $\|r\|_2^2 = \|c_2\|_2^2$.

It is clear from the above that the processing of row $a_{i.}$ uses at most $2w^2$ flops if 4-multiply Givens rotations are used. Thus the complete orthogonalization requires about $2mw^2$ flops, and can be performed in $\frac{1}{2}w(w + 3)$ locations of primary storage. If the rows of A are processed in random order, then we can only bound the operation count by $2mnw$ flops, which is a factor of n/w worse (see Cox [17]). Hence it almost invariably pays to sort the rows so that the matrix is in standard band form.

For the case when $m \gg n$ a more efficient algorithm based on Householder transformations can be used. Such a scheme for banded systems was first developed by Reid [51]. Lawson and Hanson [37, Ch.11] give a similar method and also provide Fortran subroutines implementing their algorithm. Here in the kth step all rows a_i, for which $f_i(A) = k$, are simultaneously merged into an upper triangle R. The reduction using this algorithm takes about $w(w + 1)(m + 3n/2)$ flops, which is approximately half as much as for the Givens method.

In the Householder method we first block the rows so that the kth block A_k consists of all rows for which $f_i(A) = k$, $k = 1, \ldots, p \leq n$. The algorithm proceeds in steps $k = 1, \ldots, p$. After the first $k - 1$ steps we have reduced the first $k - 1$ blocks by a sequence of Householder transformations to an upper trapezoidal matrix R_{k-1}. In step k we treat the kth block and compute

$$Q_k^T \begin{pmatrix} R_{k-1} \\ A_k \end{pmatrix} = \begin{pmatrix} R_k \\ 0 \end{pmatrix}.$$

where Q_k is a product of Householder transformations and R_k again upper trapezoidal. Note that because of the structure of the block A_k this (and later) steps will not involve the first $k-1$ rows and columns of R_{k-1}. It is essential that the Householder transformations are *subdivided as outlined above*, otherwise intermediate fill-in will occur and the operation count will increase greatly. The reader is encouraged to work through the example below, where the matrix A is a banded matrix with $w = 4$, with $m = 13$, $n = 8$. We show the matrix after the first $k = 3$ blocks have been reduced by Householder transformations $P_1, \ldots, P_9$. Elements which have been zeroed by P_j are denoted by j and fill-in elements by $+$. In step $k = 4$ only the indicated part of the matrix is involved.

$$\begin{pmatrix}
\times & \times & \times & \times & & & & \\
1 & \times & \times & \times & + & & & \\
1 & 2 & \times & \times & + & + & & \\
 & 3 & 4 & \times & \times & + & & \\
 & 3 & 4 & 5 & \times & + & & \\
 & & 6 & 7 & 8 & \times & & \\
 & & 6 & 7 & 8 & 9 & & \\
 & & 6 & 7 & 8 & 9 & & \\
 & & & \times & \times & \times & \times & \\
 & & & \times & \times & \times & \times & \\
 & & & & \times & \times & \times & \times \\
 & & & & \times & \times & \times & \times \\
 & & & & \times & \times & \times & \times
\end{pmatrix}$$

Some problems, for example, periodic spline approximation, lead to matrices which have an augmented band structure, $A = (\, A_1 \quad A_2 \,)$, where A_1 is a band matrix and A_2 a generally full matrix with a small number of columns. The band matrix algorithms are easily extended to matrices of this structure.

2. Generalized Linear Least Squares Problems

2.1. THE KKT SYSTEM

When interior point methods are applied to optimization problems the Karush-Kuhn-Tucker optimality condition gives rise to a linear system of the form

$$\begin{pmatrix} W & A \\ A^T & 0 \end{pmatrix} \begin{pmatrix} y \\ x \end{pmatrix} = \begin{pmatrix} b \\ c \end{pmatrix}. \tag{2.1}$$

where $A \in \mathbf{R}^{m \times n}$, and $W \in \mathbf{R}^{m \times m}$ is symmetric positive semidefinite. (In a more general case rank $(A, W) = m$, but both matrices A and W may be rank deficient.) Note that linear constraints can be treated simply by putting $e_i^T W = 0$, if the ith equation is to be treated like a constraint. The efficient solution of a sequence of systems of this form, where W and A are large and sparse, is of crucial importance in interior point methods.

The system (2.1) is also related to the general Gauss-Markoff linear model

$$Ax + \epsilon = b, \qquad V(\epsilon) = \sigma^2 W, \tag{2.2}$$

where ϵ is a random vector with zero mean and covariance matrix $\sigma^2 W$.

If W is positive definite (2.1) gives the condition for the solution of both problems:

1. *Generalized linear least squares problem* (GLLS)

$$\min_x (Ax - b)^T W^{-1} (Ax - b) + 2c^T x, \tag{2.3}$$

and $y = W^{-1}(b - Ax)$.

2. *Equality constrained quadratic optimization* (ECQO)

$$\min_y -2b^T y + y^T W y, \quad A^T y = c. \tag{2.4}$$

The solution x satisfies the generalized normal equations

$$A^T W^{-1} A x = A^T W^{-1} b. \tag{2.5}$$

A factorization of the form

$$W = BB^T, \qquad B \in \mathbf{R}^{m \times p}, \qquad p \le m. \tag{2.6}$$

with B nonsingular, can always be computed by Cholesky factorization. Then (2.5) is equivalent to

$$\min_x \|B^{-1}(Ax - b)\|_2, \tag{2.7}$$

In case B is given rather than W we can loose important information in forming W. If B is ill-conditioned the computed W can become singular even when B has full numerical rank. Therefore it is preferable to work with B instead of W.

The problem (2.7) can be written as a standard linear least squares problem by forming $(\bar{A}, \bar{b}) = B^{-1}(A, b)$, but when B is ill-conditioned this is not a stable computational approach. Stable and efficient methods can be developed instead on the observation that (2.7) is equivalent to the problem

$$\min_{v,x} \|v\|_2 \quad \text{subject to} \quad Bv + Ax = b. \tag{2.8}$$

For small dense matrices A and B this problem can be solved elegantly by computing the **generalized singular value decomposition** (GSVD) of A and B, see [46]. However, this method is not feasible for large and sparse problems. Paige [44], [45] has developed a method for solving (2.8) using successive QR decompositions, which we now describe.

For simplicity we assume that rank $(A) = n$, and begin with computing the QR decomposition of A,

$$Q^T A = \begin{pmatrix} R \\ 0 \end{pmatrix}, \quad Q = (Q_1, Q_2). \tag{2.9}$$

The orthogonal transformation Q^T is then applied also to b and B

$$Q^T b = \begin{pmatrix} c_1 \\ c_2 \end{pmatrix} \begin{matrix} \} \\ \} \end{matrix} \begin{matrix} n \\ m-n \end{matrix}, \quad Q^T B = \begin{pmatrix} C_1 \\ C_2 \end{pmatrix} \begin{matrix} \} \\ \} \end{matrix} \begin{matrix} n \\ m-n \end{matrix},$$

which allows the constraints in (2.8) to be written in partitioned form

$$\begin{pmatrix} C_1 \\ C_2 \end{pmatrix} v + \begin{pmatrix} R \\ 0 \end{pmatrix} x = \begin{pmatrix} c_1 \\ c_2 \end{pmatrix}. \tag{2.10}$$

For any vector $v \in R^m$ we can always determine x so that the first block of these equations are satisfied. We now determine an orthogonal matrix $P \in R^{m \times m}$ such that

$$P^T C_2^T = \begin{pmatrix} 0 \\ S^T \end{pmatrix} \begin{matrix} \} \\ \} \end{matrix} \begin{matrix} n \\ m-n \end{matrix}, \tag{2.11}$$

where the matrix S is upper triangular. By the nonsingularity of B it follows that C_2 will have linearly independent rows, and hence the matrix S will be nonsingular. The second set of constraints in (2.10) now becomes

$$S u_2 = c_2, \quad \text{where} \quad P^T v = u = \begin{pmatrix} u_1 \\ u_2 \end{pmatrix} \begin{matrix} \} \\ \} \end{matrix} \begin{matrix} n \\ m-n \end{matrix}. \tag{2.12}$$

Since P is orthogonal we have $\|v\|_2 = \|u\|_2$ and so the minimum in (2.8) is found by taking

$$u_1 = 0, \quad u_2 = S^{-1} c_2, \quad v = P_2 u_2,$$

where $P = (P_1, P_2)$. Finally x is obtained by solving the triangular system $Rx = c_1 - C_1 v$ in (2.10). Paige [44] obtains a perturbation analysis for the problem (2.7) by using the formulation (2.8), and gives a rounding error analysis to show that the above algorithm is numerically stable. The algorithm can be generalized in a straightforward way to rank deficient A and B. For details see Paige [44].

The algorithm as described here requires a total of about $2m^3/3 + m^2 n$ flops. If $m \gg n$ the work in the QR factorization of C_2 dominates the work. In interior point methods one usually solves a sequence of problems of the form (2.8), where A is constant but $B = B_k$, $k = 1, \ldots, p$. The QR decomposition (2.9) of A *can then be computed once and for all.* In case $m = n$ this reduces the work for solving a new problem from $5n^3/3$ to n^3.

The algorithm above does not take advantage of any special structure the matrix B may have. If B has been obtained from the Cholesky factorization $W = B^T B$ it is of upper triangular form. In this case, and also when G is diagonal it is advantageous to carry out the two QR decompositions in (2.9) and (2.11) together maintaining the lower triangular form throughout. Paige [45] has given such a variation of the algorithm using a "zero chasing technique". Using "fast" Givens rotations this reduces the total work when B is upper triangular to about $m^2n + 2mn^2 - 4n^3/3$ flops.

2.2. THE AUGMENTED SYSTEM METHOD

We now consider methods which rely on factorizing the symmetric indefinite system (2.1), which we write $Mz = d$, directly by Gaussian elimination. To enable a stable symmetric factorization of the form $M = LDL^T$ also 2×2 block pivots must be used. The constraint of symmetry allows row and column permutations, which use any element $d_1 = m_{rr}$, or any 2×2 submatrix

$$\begin{pmatrix} m_{rr} & m_{rs} \\ m_{sr} & m_{ss} \end{pmatrix}$$

as pivot. The latter choice is equivalent to a *double step* of Gaussian elimination, pivoting first on m_{rs} and then on m_{sr}. A factorization is then obtained in which D is block diagonal with, in general, a mixture of 1×1 and 2×2 blocks. L is unit lower triangular with $l_{k+1,k} = 0$ whenever $M^{(k)}$ is reduced by a 2×2 pivot.

To control element growth an efficient pivotal strategy is needed that does not require too much search. **Bunch-Kaufman** [15] have given a scheme where at most 2 columns need to be searched in each step, and at most n^2 comparisons are needed in all. With this strategy the element growth is bounded by

$$g_n \leq (1 + 1/\rho)^{n-1} < (2.57)^{n-1}.$$

No example is known where significant element growth occurs at every step.

Unfortunately the Bunch-Kaufman method is not a backward stable method for the system (2.1) since the perturbations introduced by round-off do not respect the structure of the I and 0 block in M. To make this clear we consider the special case $W = I$ in (2.1) and introduce the scaled vector $\alpha^{-1}y$. Then we get the scaled augmented system

$$M_\alpha = \begin{pmatrix} \alpha I & A \\ A^T & 0 \end{pmatrix}, \quad z = \begin{pmatrix} \alpha^{-1}y \\ x \end{pmatrix} \quad d = \begin{pmatrix} b \\ \alpha^{-1}c \end{pmatrix}. \tag{2.13}$$

Using any of the pivoting strategies referred to above the choice of pivots will depend on the value of α. Note also that for *a fixed pivot sequence* the numerical solution is *independent* of the scaling parameter α, and hence α only affects the accuracy through the choice of pivots. For sufficiently large values of α the Bunch-Kaufman strategy will choose the first m pivots from the diagonal block αI. Put $\alpha = 1$ and assume these pivots are chosen. Then the reduced block upper triangular system becomes

$$\begin{pmatrix} I & A \\ 0 & -A^T A \end{pmatrix} \begin{pmatrix} y \\ x \end{pmatrix} = \begin{pmatrix} b \\ c - A^T b \end{pmatrix}. \tag{2.14}$$

Hence, for this choice of pivots (and $c = 0$) we just recover the normal equations, $A^T A x = A^T b$. As discussed above this is not a backward stable method. Using smaller values of α will introduce 2×2 pivots, initially of the form

$$\begin{pmatrix} \alpha & a_{1r} \\ a_{1r} & 0 \end{pmatrix},$$

called *tile pivots* in [33], which improve the stability.

The above discussion raises the question of the *optimal choice* of α. In [11] a componentwise error analysis is carried out, to obtain separate error bounds for the computed solution vectors x and y. These upper bounds are functions of α, and the optimal α minimizing both these bounds is shown to be

$$\alpha = \alpha_{opt} = \left(\sigma_n \|y\|_2 / \|x\|_2 \right)^{1/2}. \tag{2.15}$$

(If $\alpha_{opt} \notin [\sigma_1(A), \sigma_n(A)]$ we take α as the closest endpoint of this interval.) For this choice of α the error bounds are the same as would be obtained for a backward stable method. Unfortunately, no efficient method to estimate α_{opt} a priori exists, and this is a serious drawback of the augmented system method.

We now consider the case when $W = \text{diag}(w_1, \ldots, w_m)$, and put

$$D = W^{-1/2} = \text{diag}(d_1, \ldots, d_m).$$

This corresponds to a **weighted** least squares problem, see also Section 2.3. By putting $\tilde{A} = DA$ this could be reduced to the standard case (2.13). However, if D is ill-conditioned $\kappa(DA)$ will be large, and this approach cannot be recommended in general. We recommend instead that A is *row equilibrated* so that

$$\max_{1 \le j \le n} |a_{ij}| = 1, \qquad i = 1, \ldots, m.$$

The necessary row scaling of A can be included in W, and then a scaling is incorporated by substituting αW for W (2.1). The analysis above can be carried through and error bounds depending on α derived. However, no longer can we give a closed expression for the optimum α, see Björck [10].

A drawback with the augmented system method is that it works with a system of order $m + n$, which may be much larger than n. Therefore, the main use of this method seems to be for sparse problems, where the sparsity of the blocks can be taken into account. See Arioli, Duff and de Rijk [1] for computational experience with this method. In case the LDL^T factorization is computed with a non-optimal α it is recommended that iterative refinement is used to improve the computed solution, see Sec. 2.4.

2.3. WEIGHTED LEAST SQUARES PROBLEMS

In this section we consider methods for the important special case when $W = \text{diag}(w_i) > 0$, the **weighted** linear least squares problem

$$\min_x \|D(Ax - b)\|_2, \qquad D = W^{-1/2} = \text{diag}(d_1, \dots, d_m). \qquad (2.16)$$

Here D corresponds to weighting of the rows of A and b. Obviously the weight matrix in (2.16) depends on the initial row scaling. If A is assumed to be row equilibrated, i.e., to have equal row norms, D becomes uniquely determined.

In many cases it is possible to solve (2.16) as a standard linear least squares problem for DA and Db. However, frequently the weights $d_i = w_i^{-1/2}$, $i = 1, 2, \dots, m$, are finite but vary widely in size. We assume in the following that the rows in A are ordered so that $d_1 \geq d_2 \geq \dots \geq d_m$. In the Gauss-Markoff model (2.2) $\gamma = d_1/d_m \gg 1$ corresponds to the case when some components of the error vector have much smaller variance than the rest. We call such weighted problems **stiff**. In the limit when some d_i tend to infinity, the ith equations will become exactly satisfied. Stiff problems of this form typically arise in interior point methods for linear programming.

For stiff problems $\kappa(DA)$ will be large, and an upper bound is given by

$$\kappa(DA) \leq \kappa(D)\kappa(A) = \gamma\kappa(A).$$

It is important to note that this need not imply that the problem of determining x from given data D, A, b is ill conditioned, if we consider small *relative* perturbations in the data. The perturbations in DA and Db will then have a very special form, and the usual norm-wise condition number $\kappa(DA) = \|DA\|_2\|(DA)^{-1}\|_2$ is not relevant. (The truth of this comment is obvious if one considers the case when $d_1 \to \infty$.) Special care is, however, needed in solving stiff weighted linear least squares problems. In particular the method of normal equations turns out to be sensitive to $\kappa(DA)$, and is not in general well suited for solving stiff problems. To illustrate this we consider the simple special case when only the first p equations are weighted with a factor γ. The problem is then of the form

$$\min_x \left\| \begin{pmatrix} \gamma A_1 \\ A_2 \end{pmatrix} x - \begin{pmatrix} \gamma b_1 \\ b_2 \end{pmatrix} \right\|_2^2, \qquad (2.17)$$

where $A_1 \in \mathbf{R}^{p\times n}$ and $A_2 \in \mathbf{R}^{(m-p)\times n}$, and the normal equations $(DA)^T(DA)x = (DA)^T Db$ are

$$(\gamma^2 A_1^T A_1 + A_2^T A_2)x = (\gamma^2 A_1^T b_1 + A_2^T b_2).$$

For γ sufficiently large and $A_1^T A_1$ dense then $(DA)^T(DA)$ will be completely dominated by the first term and the data contained in A_2 lost. However, if $p < n$ the solution depends critically on the unweighted data. We conclude that the method of normal equations may not be well behaved when $\gamma \gg 1$. There is an important exception to this observation. Suppose that after a permutation of columns $A_1 = (C, 0)$, $C \in \mathbf{R}^{p \times p}$ and nonsingular. Then p components of the solution x will be determined by the weighted equations, and there are $n - p$ normal equations not affected by the weighting to determine the remaining components.

We now consider the use of methods based on the QR decomposition of A for solving weighted problems. We first remark that it is essential that some form of *column pivoting* is employed. Further, the Householder QR algorithm can give poor accuracy for stiff problems unless also *row interchanges are used*. Consider the following example by Powell and Reid [50], where A and the reduced matrix $A^{(2)}$ obtained after the first Householder transformations (using exact arithmetic) are

$$A = \begin{pmatrix} 0 & 2 & 1 \\ \gamma & \gamma & 0 \\ \gamma & 0 & \gamma \\ 0 & 1 & 1 \end{pmatrix}, \qquad A_{22}^{(2)} = \begin{pmatrix} \frac{1}{2}\gamma - \sqrt{2} & -\frac{1}{2}\gamma - 1/\sqrt{2} \\ -\frac{1}{2}\gamma - \sqrt{2} & \frac{1}{2}\gamma - 1/\sqrt{2} \\ 1 & 1 \end{pmatrix}.$$

Clearly if $\gamma > u^{-1}$, where u denotes the machine precision, the terms $-\sqrt{2}$ and $-1/\sqrt{2}$ in the first and second rows are lost. This is equivalent to the *loss of all information present in the first row of A*. This loss is disastrous because the number of rows containing large elements is less than the number of components in x, so there is a substantial dependence of the solution x on the first row of A. Compared to the method of normal equations, which fails already when $\gamma > u^{-1/2}$, this is an improvement, but the method is still sensitive to $\kappa(DA)$. Powell and Reid [50] extend the Householder algorithm to include row interchanges as follows. In each step a pivot column is selected in the reduced matrix, and then the element of largest absolute value in this pivot column is permuted to the top. In most cases, instead of performing row pivoting, it suffices to sort the rows after decreasing row norm before the factorization.

An approach related to that of Powell and Reid is taken by Gullikson and Wedin [35]. They use scaled Householder transformations P of the form

$$P = I - 2Wvv^T/(v^TWv), \qquad P^2 = I,$$

i.e., P is a reflector. Note that $W^{-1/2}PW^{1/2}$ is an orthogonal reflector. A sequence of such reflectors, $Q^T = P_n \cdots P_2 P_1$, is used to transform $A\Pi$ (Π a permutation matrix) to upper triangular form. The resulting factorization is equivalent to the ordinary QR factorization

$$W^{-1/2}A\Pi = (W^{-1/2}QW^{1/2}) \begin{pmatrix} W^{-1/2}R \\ 0 \end{pmatrix}.$$

When $W > 0$ this method is equivalent to the algorithm of Powell and Reid. However, this approach generalizes more simply to the case when W has the form $W = \mathrm{diag}(0, W_2)$, which corresponds to a constrained least squares problem.

Another stable method for stiff problems is the Peters-Wilkinson method, see [48] and [13]. In the first step of this method the matrix A is reduced by Gaussian elimination to upper triangular form using row and column pivoting. This gives a factorization of the form

$$\Pi_1 A \Pi_2 = LU, \quad L \in \mathbf{R}^{m \times n}, \quad U \in \mathbf{R}^{n \times n}, \tag{2.18}$$

where L is unit lower trapezoidal, and U is upper triangular and nonsingular. In the second step we solve

$$\min_y \|Ly - \Pi_1 b\|_2, \quad U\Pi_2 x = y. \tag{2.19}$$

Any ill-conditioning from large weights is usually reflected in U, and *the least squares problem in L is well conditioned*. In the solution of the triangular system $U\Pi_2 x = y$ the weights are harmless and can be removed by row scaling.

We finally mention a method for solving the stiff least squares problem (2.17) by **updating** the solution to the unweighted problem. Let $x(g)$ be the solution to the problem with weight $\gamma > 1$, and let $r(\gamma) = b - Ax(\gamma)$ be the residual vector corresponding to the solution to the weighted problem. We now relate $x(\gamma)$ and $r_1(\gamma)$ to the corresponding quantities x and r_1 for the unweighted problem ($\gamma = 1$). If we put $\delta x = x(\gamma) - x$, and note that $A^T r = A_1^T r_1 + A_2^T r_2 = 0$, then the normal equations for (2.17) can be written

$$\left(A^T A + \hat{\gamma}^2 A_1^T A_1\right) \delta x = \hat{\gamma}^2 A_1^T r_1,$$

where $\hat{\gamma}^2 = \gamma^2 - 1$. It follows that

$$\delta x = \hat{\gamma}^2 (A^T A)^{-1} A_1^T r_1(\gamma), \quad r_1(\gamma) = r_1 - A_1 \delta x, \tag{2.20}$$

and solving for δx

$$\delta x = \hat{\gamma}^2 \left(A^T A + \hat{\gamma}^2 A_1^T A_1\right)^{-1} A_1^T r_1.$$

Using the Woodbury formula we obtain after some calculation

$$r_1(\gamma) = \left(\hat{\gamma}^{-2} I_p + A_1 (A^T A)^{-1} A_1^T\right)^{-1} r_1. \tag{2.21}$$

The formulas (2.20) and (2.21) can be used to compute the solution $x(\gamma)$. If $p \ll m$, then the extra work in the updating step is small. If $A^T A = R^T R$ (R from Cholesky or QR factorization) it follows that

$$A_1 (A^T A)^{-1} A_1^T = S_1 S_1^T, \quad R^T S_1^T = A_1^T,$$

where $S_1 \in \mathbf{R}^{p \times n}$ can be computed by forward substitution. Next we compute $r_1(\gamma)$ and the correction $x(\gamma) - x$ from

$$\left(\hat{\gamma}^{-2} I_p + S_1 S_1^T\right) \tilde{r}_1 = r_1, \quad R(x(\gamma) - x) = \hat{\gamma}^2 S_1^T r_1(\gamma). \tag{2.22}$$

Thus we only need to solve a (small) $p \times p$ system, form $S_1^T r_1(\gamma)$ and perform a back-substitution. If the matrix A is not too ill-conditioned the method of normal equations might be used to compute the solution to the unweighted problem and determine R. On the other hand, if the matrix A_1 is ill-conditioned, it is better to solve the $p \times p$ system in (2.22) by QR factorization, noting that

$$(\hat{\gamma}^{-2} I_p + S_1 S_1^T) = R_S^T R_S, \qquad \begin{pmatrix} S_1^T \\ \hat{\gamma}^{-1} I_p \end{pmatrix} = Q_S R_S.$$

Note that if $\operatorname{rank}(A_1) = p$, then for $\gamma \gg 1$ the residual $r_1(\gamma)$ will be proportional to γ^{-2}. If we let $\hat{\gamma} \to \infty$ we obtain a solution of a constrained problem which satisfies $A_1 x = b_1$ exactly. In this case the above algorithm simplifies, see also Section 3.3.

2.4. ITERATIVE REFINEMENT

A simple way to improve a computed solution $\bar{x}$ to a linear least squares problem is by iterative refinement. In this procedure the residual $\bar{r} = b - A\bar{x}$ of the current approximation is first computed. A correction δx is then obtained as the solution of a least squares problem with b replaced by $\bar{r}$. This may be repeated several times: Put $s := 1$, $x_s := \bar{x}$, and

$$
\begin{aligned}
&\text{for } s := 1, 2, \ldots \\
&\quad r_s := b - A x_s; \qquad \text{(in precision } u_2) \\
&\qquad \text{solve } \min_{\delta x_s} \| A \delta x_s - r_s \|_2; \quad \text{(in precision } u_1) \\
&\quad x_{s+1} := x_s + \delta x_s; \\
&\text{end}
\end{aligned}
$$

Each refinement step is cheap to perform since the same factorization used for computing the initial solution $\bar{x}$, can also be used to solve the least squares subproblem. We distinguish between *fixed precision* iterative refinement, $u_2 = u_1$, and *mixed precision* iterative refinement, $u_2 \ll u_1$.

This scheme has been analyzed in [34], where it was remarked that in mixed precision this scheme works well only for small residual problems. In general both r and x should be refined simultaneously using residuals of the augmented system of $m + n$ equations

$$\begin{pmatrix} I & A \\ A^T & 0 \end{pmatrix} \begin{pmatrix} \delta r_s \\ \delta x_s \end{pmatrix} = \begin{pmatrix} f_s \\ g_s \end{pmatrix}, \qquad \begin{pmatrix} f_s \\ g_s \end{pmatrix} = \mathrm{fl} \begin{pmatrix} b - r_s - A x_s \\ -A^T r_s \end{pmatrix}. \tag{2.23}$$

The solution of the system in (2.23) can be computed from a Householder QR factorization of A by

$$R^T h_s = g_s, \quad Q^T f_s = \begin{pmatrix} c_s \\ d_s \end{pmatrix}, \quad \delta r_s = Q \begin{pmatrix} h_s \\ d_s \end{pmatrix}, \quad R \delta x_s = c_s - h_s. \tag{2.24}$$

This scheme has been analyzed in [5] and [6]. It can be shown that the initial rate of reduction of the error in the solution is linear with rate

$$\rho = cu\kappa'(A), \qquad \kappa'(A) = \min_{D>0} \kappa(AD), \qquad (2.25)$$

where $c = c(m, n)$ is a generic error constant, and the minimum is taken over all diagonal matrices $D > 0$. This rate of convergence is achieved without actually carrying out the scaling of A by the optimal D, even for large residual problems. (Note that for large residual problems the condition number of the least squares problem includes a term proportional to $\kappa^2(A)$)

We now consider *fixed precision* iterative refinement. This procedure is usually of interest only if a method which is *not backward stable* is used. It can be very efficient for improving a solution obtained from the normal equations. If the corrections are also computed from the normal equations we obtain the algorithm:

Least Squares Iterative Refinement:
Set $x_0 = 0$, $r_0 = 0$. For $s = 0, 1, 2, \ldots$ until convergence do

$$r_s := b - Ax_s, \qquad R^T R \delta x_s = A^T r_s,$$
$$x_{s+1} := x_s + \delta x_s.$$

Here R is computed by Cholesky factorization of the matrix of normal equations $A^T A$. This algorithm only requires one matrix-vector multiplication each with A and A^T and the solution of two triangular systems. The first step, i.e. for $i = 0$, is identical to the normal equations. In this case one does not obtain the favorable rate of convergence in (2.25), but only

$$\bar{\rho} = cu\kappa'(A)^2 = \kappa'(A)\rho.$$

However, if several steps of refinement are carried out, this will for a large class of problems also lead to good accuracy.

Another application of fixed precision iterative refinement is to sparse problems, where R has been computed by a sparse QR factorization but the matrix Q has not been saved. Then the *semi-normal equation*

$$R^T R x = A^T b \qquad (2.26)$$

is often used to solve a least squares problem. It was shown in [8] that the solution computed by (2.26) in general is no more accurate than a solution obtained by the normal equations using Cholesky factorization. In this case we will achieve the rate of convergence factor ρ in (2.25), and one refinement step is usually sufficient to get the same error level as for a backward stable method. We refer to the algorithm with *one step* of refinement as the *corrected semi-normal equations* (CSNE). A detailed error analysis of this algorithm in [8] shows that, neglecting terms of higher order in $u\kappa$, the error in x_2 from CSNE is

$$\|x - \bar{x}_2\|_2 \leq \sigma u\kappa \left(c_2\|x\|_2 + n^{1/2}m\frac{\|b\|_2}{\|A\|_2}\right)$$
$$+ mn^{1/2}u\kappa\left(\|x\|_2 + \kappa\frac{\|r\|_2}{\|A\|_2}\right),$$

where

$$\sigma = c_3 u\kappa^2, \qquad c_3 \leq 2n^{1/2}(c_1 + 2n + m/2)$$

A comparison with the bounds for a backward stable method shows that semi-normal equation with one refinement step is *acceptable error stable* if $\sigma = O(1)$. (By acceptable error stable we mean that the error bound is no worse than the error bound for a backward stable method.)

We now make a comparison between fixed precision iterative refinement with R from Cholesky and QR factorization respectively. Denote by x_s the computed solution after i refinement steps. With R from QR the error $\|x - x_s\|$ initially behaves as

$$\|x - x_s\| \sim cu\kappa\kappa'(cu\kappa')^s.$$

Assuming that $c \approx 1$, $\kappa' = \kappa$, acceptable error stable level is achieved in p steps if $\kappa(A) < u^{-p/(p+1)}$. For example, with $u = 10^{-16}$, the maximum condition number for which acceptable error stable results are obtained after p refinements is

$$\kappa_{max}(p) = 10^8, 10^{10.7}, 10^{16}, \qquad p = 1, 2, \infty.$$

This can be compared with the result when R from Cholesky is used, (see also Foster [26]). Then

$$\|x - x_s\| \sim cu\kappa\kappa'(cu(\kappa')^2)^s,$$

and under the same simplifying assumptions acceptable error stable level is obtained in p steps if $\kappa(A) \leq u^{-p/(2p+1)}$, and

$$\kappa_{max}(p) = 10^{5.3}, 10^{6.4}, 10^8, \qquad p = 1, 2, \infty.$$

Hence for moderately ill-conditioned problems the normal equations combined with iterative refinement can give very good accuracy, but for more ill-conditioned problems QR and the semi-normal equations are much superior.

For sparse least squares a method based on the factorization of the augmented system has been considered by Arioli, Duff, and de Rijk [1]. As described in section 2.3, 1×1 and 2×2 diagonal pivots are used. Here the optimal scaling can not be obtained a priori. The corresponding loss of accuracy can often be compensated for by the use of fixed precision iterative refinement.

3. Linear Equality Constraints

3.1. PROBLEM LSE

In this section we consider least squares problems with linear equality constraints. One source of such problems is from least squares problems with *inequality* constraints. If solved by an active set method these give rise to a sequence of problems with equality constraints.

Problem LSE: Least Squares with Equality Constraints.

Given matrices $A \in \mathbf{R}^{m \times n}$ and $B \in \mathbf{R}^{p \times n}$ find a vector $x \in \mathbf{R}^n$ which solves

$$\min_x \|Ax - b\|_2 \quad \text{subject to} \quad Bx = d. \tag{3.1}$$

The problem (3.1) obviously has a solution if and only if the linear system $Bx = d$ is consistent. This is always the case if rank $(B) = p$. A solution to (3.1) is unique if and only if the null spaces of A and B intersect only trivially

$$\mathcal{N}(A) \cap \mathcal{N}(B) = \{0\}. \tag{3.2}$$

We note that (3.2) is equivalent to the rank condition rank $(A^T \ B^T) = n$. If (3.2) is not satisfied then we can seek a solution to problem LSE of minimum norm $\|x\|_2$.

We first note that if rank $(B) = p$ the constrained solution can be computed by an updating technique. Assume that we have computed the solution x_u to

$$\min_x \left\| \begin{pmatrix} A \\ B \end{pmatrix} x - \begin{pmatrix} b \\ d \end{pmatrix} \right\|_2.$$

Let R be the R-factor of the matrix $\begin{pmatrix} A \\ B \end{pmatrix}$ and solve the equation $R^T C^T = B^T$ for the matrix $C \in \mathbf{R}^{p \times n}$. Then the constrained solution is given by

$$x = x_u + R^{-1}w, \qquad w = C(CC^T)^{-1}(d - Bx_u).$$

Note that w is the solution to the minimum norm problem min $\|w\|$, subject to $Cw = d - Bx_u$. If $p \ll n$ the work in the updating step is small. This and similar updating techniques are particularly useful in *sparse problems*, see Sec. 6.3.

A robust algorithm for problem LSE should check for possible inconsistency of the constraint equations $Bx = d$. If it is not known a priori that the constraints are consistent then (3.1) may be reformulated as a **sequential** least squares problem

$$\min_{x \in S} \|Ax - b\|_2, \qquad S = \{x \mid \|Bx - d\|_2 = \min\} \tag{3.3}$$

This problem always has a unique solution of minimum norm. Many of the methods described in the following for solving problem LSE can be adopted to solve (3.3) with little modification.

The most natural way to solve problem LSE is to reduce the problem to an equivalent unconstrained least squares problem of dimension $n-p$. There are basically two different ways to perform this reduction: direct elimination and the null space method. We describe both these methods below.

3.2. METHOD OF DIRECT ELIMINATION

In the method of direct elimination we start by reducing the matrix B to upper trapezoidal form. It is essential that column pivoting is used in this step. In order to be able to solve also the more general problem (3.3) we will compute a QR decomposition of B. If $r = \text{rank}(B) \leq p$ there is an orthogonal matrix $Q_B \in \mathbf{R}^{p \times p}$ and a permutation matrix Π_B such that

$$Q_B^T B \Pi_B = \begin{pmatrix} R_{11} & R_{12} \\ 0 & 0 \end{pmatrix} \begin{matrix} \}r \\ \}p-r \end{matrix}, \tag{3.4}$$

where R_{11} is upper triangular and nonsingular. If we apply Q_B^T also to the vector d the constraints become

$$(R_{11}, R_{12})\bar{x} = \bar{d}_1, \qquad \bar{d} = Q_B^T d = \begin{pmatrix} \bar{d}_1 \\ \bar{d}_2 \end{pmatrix}, \tag{3.5}$$

where $\bar{x} = \Pi_B^T x$ and $\bar{d}_2 = 0$ if and only if the constraints are consistent.

We apply the permutation Π_B also to the columns of A and partition the resulting matrix conformingly with (3.4)

$$Ax - b = \bar{A}\bar{x} - b = (\bar{A}_1, \bar{A}_2) \begin{pmatrix} \bar{x}_1 \\ \bar{x}_2 \end{pmatrix} - b, \tag{3.6}$$

where $\bar{A} = A\Pi_B$. We now eliminate the variables $\bar{x}_1$ from (3.6) using (3.5). Substituting $\bar{x}_1 = R_{11}^{-1}(\bar{d}_1 - R_{12}\bar{x}_2)$ we get $Ax - b = \hat{A}_2 x_2 - \hat{b}$, where

$$\hat{A}_2 = \bar{A}_2 - \bar{A}_1 R_{11}^{-1} R_{12}, \qquad \hat{b} = b - \bar{A}_1 R_{11}^{-1} \bar{d}_1. \tag{3.7}$$

Hence, the reduced unconstrained least squares problem

$$\min_{\bar{x}_2} \|\hat{A}_2 \bar{x}_2 - \hat{b}\|_2, \qquad \hat{A}_2 \in \mathbf{R}^{m \times (n-r)} \tag{3.8}$$

is equivalent to the original problem LSE.

The solution to the unconstrained problem (3.8) can be obtained from the QR decomposition of $\hat{A}_2$. It is easy to show that if the condition (3.2) holds then $\text{rank}(\hat{A}_2) = n - r$ and (3.8) has a unique solution. Then we can compute the QR decomposition

$$Q_A^T \hat{A}_2 = \begin{pmatrix} R_{22} \\ 0 \end{pmatrix}, \qquad Q_A^T b = \begin{pmatrix} c_1 \\ c_2 \end{pmatrix},$$

where $R_{22} \in \mathbf{R}^{(n-r)\times(n-r)}$ is upper triangular and nonsingular. Then $x = \Pi_B \bar{x}$ solves problem LSE, where $\bar{x}$ is obtained from the triangular system

$$\begin{pmatrix} R_{11} & R_{12} \\ 0 & R_{22} \end{pmatrix} \bar{x} = \begin{pmatrix} \bar{d}_1 \\ c_1 \end{pmatrix}. \tag{3.9}$$

The coding of the algorithm outlined above can be kept remarkably compact, as is illustrated by the program given in Björck and Golub [14]. Note that the reduction in (3.7) can be interpreted as performing r steps of Gaussian elimination on the system

$$\begin{pmatrix} R_{11} & R_{12} \\ \bar{A}_1 & \bar{A}_2 \end{pmatrix} \begin{pmatrix} \bar{x}_1 \\ \bar{x}_2 \end{pmatrix} = \begin{pmatrix} \bar{d}_1 \\ b \end{pmatrix}.$$

Note that the set of vectors $x = \Pi_B \bar{x}$, where $\bar{x}$ satisfies (3.5) is exactly the set of vectors which minimize $\|Bx - d\|_2$. Thus, the algorithm outlined above actually solves the more general problem (3.3). If condition (3.2) is not satisfied, then the reduced problem (3.8) does not have a unique solution. Then column permutations are needed also in the QR decomposition of $\hat{A}_2$. In this case we can compute either a basic solution or a minimum norm solution to (3.3).

3.3 THE NULL SPACE METHOD

Assume that rank $(B) = p$, and first compute the QR decomposition

$$Q_B^T B^T = \begin{pmatrix} R_B \\ 0 \end{pmatrix}, \tag{3.10}$$

where $R_B \in \mathbf{R}^{p\times p}$ is upper triangular and nonsingular. Let

$$Q_B = (Q_1, Q_2), \qquad Q_1 \in \mathbf{R}^{n\times p}, \qquad Q_2 \in \mathbf{R}^{n\times(n-p)}.$$

Then Q_2 gives an orthogonal basis for the nullspace of B. Any vector $x \in \mathbf{R}^n$, which satisfies $Bx = d$ can then be represented as

$$x = x_1 + Q_2 y_2, \qquad x_1 = B^\dagger d = Q_1 R_B^{-T} d. \tag{3.11}$$

Hence $Ax - b = Ax_1 + AQ_2 y_2 - b$, $y_2 \in \mathbf{R}^{n-p}$, and it remains to solve the reduced system

$$\min_{y_2} \|(AQ_2)y_2 - (b - Ax_1)\|_2. \tag{3.12}$$

Let y_2 be the minimum length solution to (3.12), $y_2 = (AQ_2)^\dagger (b - Ax_1)$ and let x be defined by (3.11). Then since $x_1 \perp Q_2 y_2$ it follows that

$$\|x\|_2^2 = \|x_1\|_2^2 + \|Q_2 y_2\|_2^2 = \|x_1\|_2^2 + \|y_2\|_2^2$$

and x is the minimum norm solution to problem LSE.

Now assume that the condition (3.2) is satisfied. Then it follows that rank $(AQ_2) = (n - p)$, and we can compute the QR decomposition

$$Q_A^T(AQ_2) = \begin{pmatrix} R_A \\ 0 \end{pmatrix},$$

where R_A is upper triangular and nonsingular. The unique solution to (3.12) can then be obtained from

$$R_A y_2 = c_1, \qquad c = \begin{pmatrix} c_1 \\ c_2 \end{pmatrix} = Q_A^T(b - Ax_1). \tag{3.13}$$

The unique solution to problem LSE then is $x = x_1 + Q_2 y_2$.

A perturbation theory by Leringe and Wedin [38] shows that the problem LSE is well conditioned if $\kappa(B)$ and $\kappa(AQ_2)$ are small. It is important to note that these two condition numbers can be small even when $\kappa(A)$ is large. Any method which starts with minimizing $\|Ax - b\|_2$ will give bad results in such a case.

The method of direct elimination and the null space method both have good numerical stability. In a numerical comparison by Leringe and Wedin [38] they gave almost identical results. The operation count for the method of direct elimination is slightly lower because Gaussian elimination is used to derive the reduced unconstrained problem.

3.4. METHOD OF WEIGHTING AND DEFERRED CORRECTION

The method of weighting for solving problem LSE is based on the following simple observation. Assume that in a least squares problem we want some equations to be exactly satisfied. We can achieve that by giving these equations a large weight γ and solve the resulting unconstrained least squares problem. Hence to solve (3.1) we would compute the solution $x(\gamma)$ to the problem

$$\min_x \left\| \begin{pmatrix} \gamma B \\ A \end{pmatrix} x - \begin{pmatrix} \gamma d \\ b \end{pmatrix} \right\|_2^2. \tag{3.14}$$

Note that if (3.2) holds then (3.14) is a full rank least squares problem.

It follows from (2.20) that if rank $(B) = p$, then the residual $d - Bx(\gamma)$ is proportional to γ^{-2} for large values of γ, and hence $\lim_{\gamma \to \infty} x(\gamma) = x_{LSE}$. A more general analysis can be given in terms of the GSVD, see [52]. A detailed analysis of the method of weighting has also been given by Lawson and Hanson [37, Section 22].

The method of weighting is attractive for its simplicity. It allows the use of a subroutine or program for unconstrained least squares problem to be used to solve problem LSE. However, for large values of γ care must be exercised in the way (3.14) is solved because then the matrix in (3.14) is poorly conditioned. In particular, the method of normal equations will fail for large γ. Accurate solutions to (3.14) for large values of γ can be computed from a QR decomposition of the matrix

$$\begin{pmatrix} \gamma B \\ A \end{pmatrix}$$

provided that **both row and column permutations are used**, see Powell and Reid [50]. Note that in general it is not sufficient to initially order the constraints first, but row pivoting must be used, see [52].

For sparse problems the row and column pivoting needed may lead to unacceptable fill-in. It is then possible to us a method of deferred correction where a smaller value of γ is used in the factorization, and an extrapolation is performed, see [3] and [4].

4. Quadratic Constraints and Regularization

4.1. PROBLEM LSQI

In this section we consider methods for solving least squares problems with a quadratic inequality constraint,

$$\min_x \|Ax - b\|_2 \quad \text{subject to} \quad \|Bx - d\|_2 \leq \gamma, \tag{4.1}$$

where $A \in \mathbf{R}^{m \times n}, B \in \mathbf{R}^{p \times n}, \gamma > 0$. Problems of this form arise, e.g, in the regularization of ill-conditioned least squares problems, and in trust region methods for nonlinear least squares problems. Clearly (4.1) has a solution if and only if

$$\min_x \|Bx - d\|_2 \leq \gamma, \tag{4.2}$$

and in the following we assume that this condition is satisfied.

We define a B-generalized solution $x_{A,B}$ to the problem $\min_x \|Ax - b\|_2$ to be a solution to the problem

$$\min_{x \in S} \|Bx - d\|_2, \qquad S = \{x \in \mathbf{R}^n | \ \|Ax - b\|_2 = \min\}. \tag{4.3}$$

Notice that for $B = I$ and $d = 0$ we have $x_{A,I} = A^\dagger b$. Then the constraint in (4.1) is binding only if

$$\|Bx_{A,B} - d\|_2 > \gamma. \tag{4.4}$$

If we assume that problem LSQI has a solution, then using this observation it follows that either $x_{A,B}$ is a solution, or (4.4) holds and the solution occurs on the boundary of the constraint region. In the latter case the solution $x = x(\lambda)$ satisfies the generalized normal equations

$$(A^T A + \lambda B^T B)x(\lambda) = A^T b + \lambda B^T d, \tag{4.5}$$

where λ is determined by the **secular equation**

$$\|Bx(\lambda) - d\|_2 = \gamma. \tag{4.6}$$

As we shall see only positive values of λ are of interest. Note that (4.5) are the normal equations for the least squares problem

$$\min_x \left\| \begin{pmatrix} A \\ \mu B \end{pmatrix} x - \begin{pmatrix} b \\ \mu d \end{pmatrix} \right\|_2, \qquad \mu = \lambda^{1/2}. \tag{4.7}$$

In the following we assume that (4.4) holds so that the constraint is binding. Then there is a unique solution to problem LSQI if and only if the null spaces of A and B intersect only trivially, i.e., (3.2) is satisfied.

A numerical method for solving problem LSQI can be based on applying a method for solving the nonlinear secular equation

$$\phi(\lambda) = \gamma, \quad \text{where} \quad \phi(\lambda) = \|Bx(\lambda) - d\|_2,$$

and where $x(\lambda)$ is computed from (4.7). However, this means that for every function value $\phi(\lambda)$ we have to compute a new QR decomposition of (4.7). Methods which avoid this have been given by Eldén [25] and will be described later in this section.

4.2. PROBLEMS OF STANDARD FORM

A particularly simple but important case is when $B = I_n$, and $d = 0$. We will call this the **standard form** of LSQI. Let the SVD of A be

$$A = U \begin{pmatrix} D_A \\ 0 \end{pmatrix} V^T,$$

where U and V are orthogonal, and the singular values $\alpha_i = \sigma_i(A)$ are ordered so that $\alpha_1 \geq \alpha_2 \geq \ldots \geq \alpha_n \geq 0$. For the problem in standard form the rank condition (3.2) is trivially satisfied. The condition (4.4) simplifies to

$$\|A^\dagger b\|_2^2 = \sum_{i=1}^n (\tilde{b}_i/\alpha_i)^2 > \gamma^2, \qquad \tilde{b} = U^T b.$$

We assume that this condition is satisfied and determine λ by solving the secular equation

$$\phi^2(\lambda) = \sum_{i=1}^n y_i^2(\lambda) = \gamma^2, \qquad y_i(\lambda) = \alpha_i \tilde{b}_i/(\alpha_i^2 + \lambda). \tag{4.8}$$

It can be shown that $\phi(0) > \gamma$. Since $\phi(\lambda)$ is monotone decreasing for $\lambda > 0$ there exists a unique positive root to the secular equation, which is the desired root, see [27]. We finally obtain the solution from

$$x = Vy = \sum_{i=1}^n y_i(\lambda^*)v_i, \qquad V = (v_1, \ldots, v_n) \tag{4.9}$$

where λ^* is the unique positive solution to (4.8).

The SVD gives the most stable computational algorithm for the numerical solution of problem LSQI in standard form. However, more efficient algorithms which are almost as satisfactory can be devised which make use of simpler matrix decompositions, see [25]. In order to solve the secular equation $\|x(\lambda)\|_2 = \gamma$, we have to solve the least squares problem

$$\min_x \left\| \begin{pmatrix} A \\ \mu I_n \end{pmatrix} x - \begin{pmatrix} b \\ 0 \end{pmatrix} \right\|_2 \tag{4.10}$$

for a sequence of values of $\mu = \lambda^{1/2}$. To do this we first transform A to the bidiagonal form using Householder transformations,

$$A = Q \begin{pmatrix} B \\ 0 \end{pmatrix} P^T, \qquad \tilde{b} = Q^T b = \begin{pmatrix} \tilde{b}_1 \\ \tilde{b}_2 \end{pmatrix} \begin{matrix} \}n \\ \}m-n \end{matrix}, \tag{4.11}$$

where B is upper bidiagonal. This decomposition can be computed in only $mn^2 + n^3$ flops . If we put $x = Py$, (4.10) is transformed into

$$\min_x \left\| \begin{pmatrix} B \\ \mu I_n \end{pmatrix} y - \begin{pmatrix} \tilde{b}_1 \\ 0 \end{pmatrix} \right\|_2. \tag{4.12}$$

Since P is orthogonal the secular equation becomes $\|y(\lambda)\|_2 = \gamma$. We can now determine a sequence of Givens transformations $G_k = R_{k,n+k}$, $k = 1, \ldots, n$ and $J_k = R_{n+k,n+k+1}$, $k = 1, \ldots, n-1$ such that

$$G_n J_{n-1} \ldots G_2 J_1 G_1 \begin{pmatrix} B & \tilde{b}_1 \\ \mu I_n & 0 \end{pmatrix} = \begin{pmatrix} B_\mu & z_1 \\ 0 & z_2 \end{pmatrix}, \tag{4.13}$$

so that B_μ is again upper bidiagonal. Then $y(\mu)$ is determined by solving the bidiagonal system of equations $B_\mu y(\mu) = z_1$. The whole transformation in (4.13), including solving for $y(\lambda)$, takes only about $O(n)$ flops. Eldén [25] gives a detailed operation count, and also shows how to compute derivatives of $\phi(\lambda)$.

4.3. PROBLEMS OF GENERAL FORM

We consider here the more general form (4.1) of problem LSQI with $d = 0$. This problem can be analyzed using the generalized singular value decomposition (GSVD), see [9, Sec. 27]. We consider here some more efficient methods.

Let $B = L \in \mathbf{R}^{(n-t) \times n}$ be a banded matrix. We then have to solve

$$\min_x \left\| \begin{pmatrix} A \\ \mu L \end{pmatrix} x - \begin{pmatrix} b \\ 0 \end{pmatrix} \right\|_2 \tag{4.14}$$

for a sequence of values of μ. We can no longer start by transforming A to bidiagonal form since that would destroy the sparsity of L. (Note that the transformation P has to be applied to L from the right.) We compute instead the QR decomposition of A

$$Q_1^T A = \begin{pmatrix} R_1 \\ 0 \end{pmatrix}, \qquad Q_1^T b = \begin{pmatrix} c_1 \\ c_2 \end{pmatrix}.$$

Now (4.14) is equivalent to

$$\min_x \left\| \begin{pmatrix} R_1 \\ \mu L \end{pmatrix} x - \begin{pmatrix} c_1 \\ 0 \end{pmatrix} \right\|_2. \tag{4.15}$$

Some problems give rise to a matrix A which also has band structure. Then the matrix R_1 will be an upper triangular band matrix with the same band width w_1 as A, see Theorem 15.1 and the complete matrix in (4.15) will be of banded form. Problem (4.15) can be efficiently solved by first reordering the rows of the band matrix $\begin{pmatrix} R_1 \\ \mu L \end{pmatrix}$ so that it is in standard form. The QR factorization of the matrix can then be efficiently computed, see Sec. 1.4.

We now describe an algorithm for the case when R_1 does not have band structure. The idea is to transform (4.14) to a regularization problem of standard form. Note that if L was nonsingular we could achieve this by the change of variables $y = Lx$. However, normally $L \in \mathbf{R}^{(n-t) \times n}$ and is of rank $n - t < n$. The transformation to standard form can then be achieved using the pseudoinverse of L by a technique due to Eldén [25]. We compute the QR decomposition of L^T

$$L^T = (V_1, V_2) \begin{pmatrix} R_2 \\ 0 \end{pmatrix},$$

where V_2 spans the nullspace of L. We now set $y = Lx$. Then

$$x = L^\dagger y + V_2 w, \qquad L^\dagger = V_1 R_2^{-T}, \tag{4.16}$$

where $L^\dagger$ is the pseudoinverse of L, and $Ax - b = AL^\dagger y - b + AV_2 w$. If we form AV_2 and compute its QR decomposition

$$AV_2 = (Q_1, Q_2) \begin{pmatrix} U \\ 0 \end{pmatrix}, \qquad U \in \mathbf{R}^{t \times t}.$$

then

$$Q^T(Ax - b) = \begin{pmatrix} Q_1^T(AL^\dagger y - b) + Uw \\ Q_2^T(AL^\dagger y - b) \end{pmatrix} = \begin{pmatrix} r_1 \\ r_2 \end{pmatrix}.$$

If A and L have no nullspace in common then AV_2 has rank t and U is nonsingular. Then we can always determine w so that $r_1 = 0$ and (4.14) is equivalent to

$$\min_y \left\| \begin{pmatrix} \tilde{A} \\ \mu I \end{pmatrix} y - \begin{pmatrix} \tilde{b} \\ 0 \end{pmatrix} \right\|_2, \qquad \tilde{A} = Q_2^T A L^\dagger, \quad \tilde{b} = Q_2^T b, \tag{4.17}$$

which is of standard form. We then retrieve x from (4.16).

We remark that if m is substantially larger than n it is better to apply the above reduction to (4.15) instead of (4.14). Since the reduction involves the pseudoinverse of L it is numerically less stable than GSVD or the direct solution of (4.14). However, in practice it seems to give very similar results, see Varah [53, p. 104].

5. Sparse Least Squares Problems

5.1. INTRODUCTION

In this section we review methods for solving the linear least squares problem

$$\min_x \| Ax - b \|_2$$

which are effective when the matrix $A \in \mathbf{R}^{m \times n}$ is sparse, i.e., when the matrix A has relatively few nonzero elements. This is often the case in problems which arise as subproblems in optimization. Often sparse matrices occur with a very irregular pattern of nonzero elements.

We assume initially that $A \in \mathbf{R}^{m \times n}$ with $\mathrm{rank}\,(A) = n \leq m$. However, problems where $\mathrm{rank}\,(A) = m < n$ or $\mathrm{rank}\,(A) < \min(m, n)$ occur in practice. Other important variations include weighted problems, and/or linearly constrained problems. In some problems we have a sparse right hand side b. It may also be the case that only part of the components of the solution x are needed.

In order to solve large sparse matrix problems efficiently it is important that we only store and operate on the nonzero elements of the matrices. We must also try to minimize **fill-in** as the computation proceeds, which is the term used to denote the creation of new nonzeros. One of the main objectives of a sparse matrix data structure is to economize on storage while at the same time facilitating subsequent operations on the matrix. For a description of some suitable storage schemes for sparse vectors and matrices see [9].

Sparse least squares problems may be solved either by direct or iterative methods. Preconditioned iterative methods can often be considered as hybrids between these two classes of solution. In the method of normal equations, the system

$$A^T A x = A^T b, \tag{5.1}$$

is first formed. Then the (sparse) Cholesky factorization $A^T A = R^T R$ is computed, and finally two triangular systems $R^T z = A^T c$, $Rx = z$ are solved. The potential numerical instability of the method of normal equations is due to possible loss of information when forming $A^T A$ and $A^T b$, and to the fact that the condition number of $A^T A$ is the square of that of A. Note, however, that a great improvement in accuracy can be obtained by the use of iterative refinement, see the discussion in Sec. 2.4.

In interior linear programming methods a sequence of weighted sparse least squares problems with normal matrix $A^T D_k^2 A$, $k = 1, 2, \ldots$, are solved. This approach has been surprisingly successful in spite of potential numerical difficulties, see [54, Sec. 7]. One contributing factor to this observation could be that because in each step a **correction** to the solution of the previous problem is computed a similar effect to iterative refinement is achieved.

Orthogonalization methods avoid both these sources of trouble by working directly with A. An orthogonal matrix $Q \in \mathbf{R}^{m \times m}$ is used to reduce $A \in \mathbf{R}^{m \times n}$ of rank n and b to the form

$$Q^T A = \begin{pmatrix} R \\ 0 \end{pmatrix}, \quad Q^T b = \begin{pmatrix} c_1 \\ c_2 \end{pmatrix}, K \tag{5.2}$$

where $R \in \mathbf{R}^{n \times n}$ is upper triangular and $c_1 \in \mathbf{R}^n$. The solution is then obtained by solving the triangular system $Rx = c_1$ and the residual norm equals $\|c_2\|_2$. Note that mathematically R equals the Cholesky factor of $A^T A$, but numerically they differ.

A third approach uses the augmented system

$$\begin{pmatrix} I & A \\ A^T & 0 \end{pmatrix} \begin{pmatrix} r \\ x \end{pmatrix} = \begin{pmatrix} b \\ 0 \end{pmatrix}, \tag{5.3}$$

and computes a factorization of the symmetric *indefinite* matrix. For a discussion of this approach we refere to [21]).

5.2. PREDICTING THE STRUCTURE OF THE CHOLESKY FACTOR

The structure of a symmetric matrix B can conveniently be represented by the **undirected graph** of B, which we denote $G(B)$. The graph $G(B) = (X, E)$ consists of a finite set of nodes X together with a set E of edges, which are unordered pairs of nodes $\{u, v\}$. The nodes correspond to rows and columns of the matrix B and the edges correspond to nonzero elements in $B = B^T$. In an **ordered graph** the nodes are labeled $x_1, x_2, \ldots, x_n$, where x_j corresponds to the jth row and column of B and $\{x_i, x_j\} \in E$ if and only if $b_{ij} = b_{ji} \neq 0$, $i \neq j$. Two nodes x and y are adjacent in G if $\{x, y\} \in E$. The number of nodes adjacent to a node x is called the **degree** of x.

For any permutation matrix $P \in \mathbf{R}^{n \times n}$ the unlabeled graphs $G(B)$ and $G(P^T B P)$ are the same but the labelings are different. Hence, *the unlabeled graph represents the structure of B without any particular ordering*. Finding a good permutation for B is equivalent to finding a good labeling for its graph.

We first discuss the fill-in when the matrix $A^T A$ is formed. Partitioning A by rows we have (cf. (6.5))

$$A^T A = \sum_{i=1}^{m} a_i a_i^T \tag{5.4}$$

where a_i^T now denotes the i-th row of A. This expresses $A^T A$ as the sum of m matrices of rank one. We now make the important assumption that *no numerical cancellation occurs*, i.e., whenever two nonzero quantities are added or subtracted, the result is nonzero. The no-cancellation assumption makes it possible to determine the nonzero structure of $A^T A$ from that of A without any *numerical* computations. If the assumption is not satisfied this may considerably overestimate the number of nonzeros in $A^T A$, e.g., if A is orthogonal then $A^T A = I$ is sparse even when A is dense.

From the no cancellation assumption it follows that the graph $G(A^T A)$ is the direct sum of all the graphs $G(a_i a_i^T)$, $i = 1, 2, \ldots, m$, i.e., forming the union of all nodes and edges not counting multiple edges. Hence $G(A^T A)$ can be easily constructed directly from the structure of the matrix A. Note that the nonzeros in any row a_i^T will generate a subgraph where all pairs of nodes are connected. Such a subgraph is called a **clique** and corresponds to a full submatrix in $A^T A$. Because of the alternative characterization

$$(A^T A)_{jk} \neq 0 \iff a_{ij} \neq 0 \text{ and } a_{ik} \neq 0$$

for at least one row $i = 1, 2, \ldots, m$, the graph $G(A^T A)$ is also known as the **column intersection** graph of A.

From the no cancellation assumption it follows that if A contains at least one full row then $A^T A$ will be full even if the rest of the matrix is sparse. Sparse problems with only a few dense rows can be treated by updating the solution to the corresponding problem where the dense rows have been deleted, see end of this section.

Next we want to predict the structure of R using only the *structure* of $A^T A$. This can be done by performing the Cholesky factorization *symbolically* on the graph $G(A^T A)$ as first suggested by Parter [47]. We form an elimination graph sequence starting with $G_1 = G(A^T A)$. At step i, $i = 1, 2, \ldots, n-1$, we eliminate x_i. The graph G_{i+1} is formed from G_i by marking node i and connecting all of its neighboring nodes with edges. The final graph G_n is called the **filled graph** and gives the structure of the symmetric matrix $R^T + R$.

Instead of first constructing the explicit structure of $B = A^T A$ and then applying the symbolic Cholesky algorithm to B it is possible to perform the symbolic factorization of $A^T A$ directly from the structure of A. Hence, the step of determining the structure of $A^T A$ is redundant, unless the normal equations are going to be formed also numerically. Gilbert, Moler and Schreiber [32] describe an algorithm which works directly from $G(A)$ without forming $G(A^T A)$.

5.3. PERMUTATION TO BLOCK UPPER TRIANGULAR FORM

The matrix R in the QR factorization equals the Cholesky factor of $A^T A$, which is unique, apart from possible sign differences in some rows. Hence we may still use the symbolic steps 2 and 3 in Algorithm to determine a good column permutation P_c and set up a storage structure for the R factor associated with AP_c. However, this method may be too generous in allocating space for nonzeros in R because we begin not with the structure of $A^T A$, but with the structure of A. The elements in $A^T A$ are not independent and *structural cancellation* may occur. (Note that this is different from numerical cancellation, which occurs only for certain values of the nonzero elements in A.)

Coleman, Edenbrandt and Gilbert [16] exhibited a class of matrices for which symbolic factorization of $A^T A$ correctly predicts the structure of R, excluding numerical cancellations. From the above it follows that for this class also the Givens and Householder rules will give the correct result.

THEOREM

Let $A \in \mathbf{R}^{m \times n}$, $m \geq n$ have the **strong Hall property**, i.e., for every subset of k columns, $0 < k < m$, the corresponding submatrix has nonzeros in at least $k + 1$ rows. (Thus, when $m > n$, every subset of $k \leq n$ has the required property, and when $m = n$, every subset of $k < n$ columns has the property.) Then the structure of $A^T A$ will correctly predict that of R, excluding numerical cancellations.

Obviously the matrix A in Fig. 5.1. does not have the strong Hall property since the first column does only have one nonzero element. However, the matrix A' obtained by deleting the first column has this property.

$$
\begin{array}{ccccc}
\times & \times & \times & \times & \times \\
 & \times \\
 & & \times \\
 & & & \times \\
 & & & & \times
\end{array}
\qquad
\begin{array}{cccc}
\times & \times & \times & \times \\
\times \\
 & \times \\
 & & \times \\
 & & & \times
\end{array}
$$

$$A \qquad\qquad\qquad A'$$

Figure 5.1: Strong Hall property

An arbitrary rectangular matrix $A \in \mathbf{R}^{m \times n}$, $m \geq n$, can by row and column permutations be brought into a form called the generalized Dulmage-Mendelson decomposition. This is a block upper triangular form where the first diagonal block A_{11} may have more columns than rows, the last diagonal block $A_{k,k}$ may have more rows than columns,

$$
PAQ = \begin{pmatrix}
A_{11} & U_{12} & \cdots & U_{1,k-1} & U_{1,k} \\
 & A_{22} & \cdots & U_{2,k-1} & U_{2,k} \\
 & & \ddots & \vdots & \vdots \\
 & & & A_{k-1,k-1} & U_{k-1,k} \\
 & & & & A_{k,k}
\end{pmatrix}.
\tag{5.5}
$$

All the other diagonal blocks $A_{i,i}$, $i = 2, \ldots, k - 1$ are square and have nonzero diagonal elements. All the diagonal blocks are irreducible; this implies that A_{11}^T and A_{ii}, $i = 2, \ldots, k$ all have the strong Hall property, see [16]. (The matrix A in Fig. 5.1, can trivially be written in this form.) Algorithms which finds the partitioning in $O(n\tau)$ time, where $\tau = nnz(A)$ are given in [19], and [49].

We define the **structural rank** of A as the maximal rank over all possible assignments of numerical values to its nonzero elements. (Note that the mathematical rank is always less than or equal to its structural rank.) If A has structural rank equal to n, then A_{11} must be square, so all the blocks A_{ii}, $i = 1, 2, \ldots, k-1$ are square and have a nonzero diagonal.

The reordering to Dulmage-Mendelson form can save work and intermediate storage in solving least squares problems. If the original least squares problem is assumed to have numerical rank equal to n, it can after reordering be solved by a form of block back-substitution. First compute the solution of

$$\min_{\tilde{x}_k} \| A_{k,k} \tilde{x}_k - \tilde{b}_k \|_2, \tag{5.6}$$

where $\tilde{x} = Q^T x$ and $\tilde{b} = Pb$ have been partitioned conformally with PAQ in (18.2). Then $\tilde{x}_k, \ldots \tilde{x}_1$ is determined by

$$A_{i,i} \tilde{x}_i = \tilde{b}_i - \sum_{j=i+1}^{k} U_{ij} \tilde{x}_j, \quad i = k, \ldots, 2, 1. \tag{5.7}$$

Finally we have $x = Q\tilde{x}$. We can solve the subproblems in (5.6) and (5.7) by computing the factorizations $A_{i,i} = Q_i R_i$. Note that since $A_{11}, \ldots, A_{kk}$ have the strong Hall property the structures of the matrices R_i are correctly predicted by the structures of $A_{ii}^T A_{ii}$, $i = 1, \ldots, k$.

If the matrix A has structural rank less than n, then the block A_{11} may have more columns than rows. In this case we can still obtain the form (5.5) with a square block A_{11} by permuting the extra columns in the first block to the end, which will introduce zero columns in $A_{k,k}$ The solution is then not unique, but a unique solution of minimum length can be found as outlined in Section 7.

5.4. COLUMN REORDERINGS

Let P_r and P_c be permutation matrices of order m and n respectively and consider the matrix $P_r A P_c$. Then

$$(P_r A P_c)^T (P_r A P_c) = P_c^T A^T P_r^T P_r A P_c = P_c^T A^T A P_c.$$

Thus, the ordering of the rows of A has no effect on the matrix $A^T A$. Reordering the columns of A corresponds to a symmetric reordering of the rows and columns of $A^T A$. This will not affect the number of nonzeros in $A^T A$, only their positions, but may greatly affect the number of nonzeros in the Cholesky factor R. Before carrying out the Cholesky factorization numerically it is therefore important to find a permutation matrix P_c such that $P_c^T A^T A P_c$ has a sparse Cholesky factor R. It is important to emphasize that the reason why the ordering algorithm can be done symbolically, only working on the structure of $A^T A$, is that pivoting is not required for numerical stability in the Cholesky algorithm.

To find a permutation that minimizes the number of nonzeros in the Cholesky factor R is known to be an NP-complete problem. Hence, we have to use heuristic algorithms which in general only will give an approximate solution. The most important of these orderings are the **minimum degree ordering** and various **nested dissection** orderings. The minimum degree algorithm is based on a *local minimization* of fill.

The Minimum Degree Algorithm:

Let $G_1 = G(A^T A)$ and denote by G_i, $i = 2, \ldots, n$ the elimination graph after $i - 1$ nodes have been eliminated. In the ith step a node in G_i of *minimum degree* is determined and eliminated to get G_{i+1} from G_i.

This is the equivalent to choosing the ith pivot column in the Cholesky factorization as one with the minimum number of nonzero elements in the unreduced part of $A^T A$.

We now outline the structure of an algorithm based on the normal equations for solving sparse linear least squares problems:

Ordering Algorithm for Sparse Factorization:

1. Determine symbolically the structure of $A^T A$.

2. Determine a column permutation P_c such that $P_c^T A^T A P_c$ has a sparse Cholesky factor.

3. Perform the Cholesky factorization of $P_c^T A^T A P_c$ symbolically to generate a storage structure for R.

For details of the implementation of the minimum degree algorithm we refer to George and Liu [31, Ch. 5] and [20]. Remarkably fast symbolic implementations of the minimum degree algorithm exist, which use refinements of the elimination graph model of the Cholesky factorization described above. George and Liu [30] survey the extensive development of efficient versions of the minimum degree algorithm. Note that because of tie-breaking the minimum degree ordering is not unique. The way tie-breaking is done may have an important influence on the goodness of the ordering. Examples are known where minimum-degree will give very bad orderings if the tie-breaking is systematically done in the wrong way.

It should be noted that in interior linear programming methods a *sequence* of weighted sparse least squares problems with normal matrix $A^T D_k^2 A$ with D_k diagonal, are solved. These matrices have the same structure for all scaling matrices D_k, and hence the ordering for sparsity can be done once and for all. In this case it is therefore motivated to use a more efficient and time consuming ordering algorithm. In particular the minimum fill ordering algorithm, see [20], has been reported to work well.

6. Sparse Cholesky and QR-Factorization

6.1. CHOLESKY FACTORIZATION

Sparse Cholesky factorization has been widely studied and carefully implemented in several sparse matrix packages. An excellent introduction to theory and methods for sparse Cholesky factorization is given in the monograph in [31]. Excellent software packages for the method of normal equations for sparse least squares are MA27 (Duff and Reid [22]) YSMP (Eisenstat et al. [24]) and SPARSPAK (George and Liu [31]). In the next two subsections we study the more recent algorithms for sparse QR factorization.

6.2. SEQUENTIAL QR FACTORIZATION

For dense problems probably the most effective serial method to compute the QR-factorization is by using Householder reflections. In this method at the k-th step all the subdiagonal elements in the k-th column are annihilated. Then each column in the remaining unreduced part of the matrix which has a nonzero inner product with the column being reduced will take on the sparsity pattern of their union. In this way, even though the final R may be sparse, a lot of intermediate fill-in will take place with consequent cost in operations and storage.

As shown by George and Heath [29], by using a row oriented method employing Givens rotations it is possible to avoid the problem with intermediate fill in the orthogonalization method. We now describe their **sequential row orthogonalization algorithm** for the QR factorization of a general sparse matrix:

The rows a_i^T of A are processed sequentially, $i = 1, 2, \ldots, n$. We denote by $R_{i-1} \in \mathbf{R}^{n \times n}$ the upper triangular matrix obtained after processing rows $a_1^T, \ldots, a_{i-1}^T$. Initially we assume that R_0 has all elements equal to zero. The i-th row $a_i^T = (a_{i1}, a_{i2}, \ldots, a_{in})$ is processed as follows: We scan the nonzeros in a_i^T from left to right. For each $a_{ij} \neq 0$ a Givens rotation involving row j in R_{i-1} is used to annihilate a_{ij}. This may create new nonzeros both in R_{i-1} and in the row a_i^T. We continue until the whole row a_i^T has been annihilated. Note that if $r_{jj} = 0$, this means that this row in R_{i-1} has not yet been touched by any rotation and the entire j-th row must be zero. When this occurs the remaining part of row i will simply be inserted as the j-th row in R. We illustrate this process in Fig. 6.1, where circled elements are involved in the elimination of a_i^T. Nonzero elements created in R_{i-1} and a_i^T during the elimination are denoted by $\oplus$.

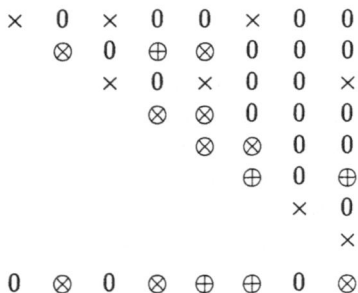

Figure 6.1: Processing of a row in the George-Heath algorithm

For simplicity we have not included the right hand side in Figure 6.1, but it should be processed simultaneously with the rows of A and carried along in parallel. In the implementation of George and Heath [29] the Givens rotations are not stored but discarded after use. Hence, only enough storage to hold the final R and a few extra vectors for the current row and right hand side(s) is needed in main memory.

Although the final R is independent of the ordering of the rows in A it is known that the number of operations needed in the sequential orthogonalization method depends on the row ordering. Unless the rows of A have widely differing norms, the row ordering does not affect numerical stability and can be chosen based on sparsity consideration only. The following is a heuristic algorithm for determining an a priori row ordering:

Denote the column index for the last nonzero element in row a_i^T by $l_i(A)$. Sort the rows so that the indices $l_i(A)$, $i = 1, \ldots, m$ form a monotone increasing sequence, i.e.,

$$l_i(A) \leq l_k, \quad \text{if} \quad i < k.$$

(This rule does not in general determine a unique ordering, and a method to resolve ties is needed.)

Ordering the rows after increasing $l_i(A)$ has been found to work well, see George and Heath [29]. With this row ordering we note that when row a_i^T is being processed only the columns $f_i(A)$ to $l_i(A)$ of R_{i-1} will be involved, since all the previous rows only have nonzeros in columns up to at most $l_i(A)$. Hence R_{i-1} will have zeros in column $l_{i+1}(A), \ldots, n$ and no fill will be generated in row a_i^T in these columns.

In some contexts, e.g., for banded systems, sorting rows after increasing values of $f_i(A)$, the column index of the first nonzero in row a_i^T, may be appropriate. For this ordering it follows that the rows $1, \ldots, f_i(A) - 1$ in R_{i-1} will not be affected when the remaining rows are processed. These rows therefore are the final first $f_i(A) - 1$ rows in R and may e.g. be transferred to auxiliary storage.

The Givens rotations are also applied to any right hand side(s), but normally then discarded to save storage. This creates a problem if we later wish to solve additional problems having the same matrix A but different right hand sides b. In this case the rotations could be saved on an external file for later use.

6.3. MULTIFRONTAL QR FACTORIZATION

A significant advance in direct methods for sparse matrix factorization is the **multifrontal method** by Duff and Reid [23]. This method reorganizes the factorization of a sparse matrix into a sequence of partial factorizations of small dense matrices. For the QR factorization such a scheme was first developed by Liu [40]. Liu called his method a general row merging scheme, but remarked that it essentially is equivalent to a multifrontal method. Lewis, Pierce and Wah [39] have implemented a multifrontal method for sparse QR factorization. Another implementation based on the Harwell MA27 code has been developed by Matstoms [42]. These codes can give a significant reduction in factorization time at a modest increase in working storage. An implementation in MATLAB is given in [43], where also a comparison with other methods is carried out.

There are several advantages with the multifrontal approach. The solution of the dense subproblems can more efficiently be handled by vector machines. Also, it leads to independent subproblems which can be solved in parallel. The good data locality of the multifrontal method gives fewer page faults on paging systems, and out-of-core versions can be developed.

We first describe the idea on a small example with $A \in \mathbf{R}^{12\times9}$ adopted from Liu [40]

$$A = \begin{pmatrix}
\times & & & & & \times & & \times & \times \\
\times & & & & & \times & & \times & \times \\
\times & & & & & \times & & \times & \times \\
& \times & & & & \times & & & \times & \times \\
& \times & & & & \times & & & \times & \times \\
& \times & & & & \times & & & \times & \times \\
& & \times & & & & \times & \times & \times \\
& & \times & & & & \times & \times & \times \\
& & \times & & & & \times & \times & \times \\
& & & \times & & \times & & \times & \times \\
& & & \times & & \times & & \times & \times \\
& & & \times & & \times & & \times & \times
\end{pmatrix}. \tag{6.1}$$

The graph $G(A)$ is a 3×3 mesh, and the ordering a nested dissection ordering, see Figure .

We can here perform a QR factorization of rows 1-3. Since there are nonzeros only in columns $\{1, 5, 7, 8\}$ this operation can be carried out as a QR factorization of a small dense matrix of size 3×4 by leaving out the zero columns. In this case the reduction to upper trapezoidal form does not produce any fill-in. The first row

$$\begin{array}{ccc} 1 & 5 & 2 \\ 7 & 8 & 9 \\ 3 & 6 & 4 \end{array}$$

Figure 6.1: The graph $G(A^T A)$ and a nested dissection ordering

equals the first of the final R of the complete matrix and can be stored away. The remaining two rows will be processed later.

The other three blocks rows 4-6, 7-9 and 10-12 can be reduced in the same way. Moreover, these tasks are independent and can be done in parallel. After this first stage these four blocks of rows have been transformed into four (partially filled) upper trapezoidal matrices, $R_1 - R_4$ of the form:

$$\begin{pmatrix}
\times & & & & \times & & \times & \times & \\
& & & & \times & & \times & \times & \\
& & & & & & \times & \times & \\
& \times & & & \times & & & \times & \times \\
& & & & \times & & & \times & \times \\
& & & & & & & \times & \times \\
& & \times & & & \times & \times & \times & \\
& & & & & \times & \times & \times & \\
& & & & & & \times & \times & \\
& & & \times & & \times & & \times & \times \\
& & & & & \times & & \times & \times \\
& & & & & & & \times & \times
\end{pmatrix}$$

Here the first row in each of $R_1 - R_4$ are final rows in R. These are removed to give $F_1 - F_4$. In the second stage we can simultaneously merge F_1, F_2 and F_3, F_4 into two upper trapezoidal matrices. Merging F_1 and F_2 we need to consider only columns $\{5, 7, 8, 9\}$. We first reorder the rows after the index of the first nonzero element, and then perform a QR factorization:

$$Q^T \begin{pmatrix} \times & \times & \times & \\ \times & & \times & \times \\ & \times & \times & \\ & & \times & \times \end{pmatrix} = \begin{pmatrix} \times & \times & \times & \times \\ & \times & \times & \times \\ & & \times & \times \\ & & & \times \end{pmatrix} = R_{1,2}.$$

The merging of F_3 and F_4 is performed similarly. Again, the first row in each reduced matrix is a final row in R, and is removed. In the final stage we merge two upper trapezoidal (in this example triangular) matrices to produce the final factor R. Note that this scheme can be considered as a generalization of the band matrix scheme described earlier.

The general row merge scheme described here can also be viewed as a special type of variable row pivot method as studied by Gentleman [28], Duff [18] and Zlatev [55]. However, as observed in [40] variable row pivoting has not been so popular because of the difficulty of generating good orderings for the rotations and because these schemes are complicated to implement. Also the dynamic storage structure needed tends to reduce the efficiency.

In general, the multifrontal scheme is best described by the **elimination tree** $T(A^T A)$. This can be constructed from the filled graph $G_F(A^T A)$, or equivalently from the structure of the Cholesky factor R. For the matrix A in the example above the graph $G_F(A^T A)$ equals $G(A^T A)$ with an added edge between nodes 7 and 9. This graph captures the column dependencies in the Cholesky factor, i.e., if we use directed edges in the filled graph (from lower to higher numbered nodes) then a directed edge from node j to node i indicates that column i depends on column j. To exhibit the column dependency relation we can simplify this directed graph by a *transitive reduction*. If there is a directed path from j to i of greater length than one, the edge from j to i is redundant and is removed. The removal of all such redundant edges generates precisely the elimination tree. This tree provides in compact form all information about the column dependencies. For the graph $G_F(A^T A)$ the result of the transitive reduction and the elimination tree is given in Figure 6.2.

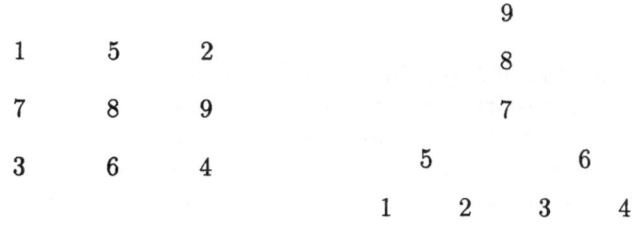

Figure 6.2: The transitive reduction and elimination tree $T(A^T A)$.

For a given rooted tree, we define a **topological ordering** on the tree to be an ordering which numbers children nodes before their parent node. It is known that all topological orderings are equivalent in the sense that they correspond to the same filled graph $G_F(A^T A)$. A **postordering** of a tree is a topological ordering such that the nodes in every subtree are numbered consecutively. The **assembly tree** defined by Duff and Reid [23] is a postordered elimination tree. The ordering of the nodes in the tree in Figure is an example of a postordering. Duff and Reid made the important observation that if a postordering is used in the multifrontal method, then the data management is simplified in that the sequence of update matrices can be managed in a last-in-first-out basis.

The basic computational subproblem in the multifrontal method is the merging of some of the original rows of A together with one or more upper trapezoidal matrices. The organization is based on the elimination tree and we visit the nodes in turn given by the ordering. Each node x_j in the tree is associated with the frontal matrix F_j, which consists of the set of rows A_j in A with the first nonzero in location j, together with one trapezoidal matrix contributed by each child node of x_j.

These are then transformed (merged) into an upper triangular matrix. The first row in this matrix is the jth row of the upper triangular factor R. The remaining rows form an **update matrix** U_j, which is stored until needed. Hence a formal description of the method is as follows.

Multifrontal sparse QR algorithm:

1. for $j := 1$ to n do

2. Form the matrix F_j by combining the set of rows A_j, and the update matrix U_s for each child x_s of the node x_j in the elimination tree $T(A^T A)$

3. Transform by an orthogonal transformation F_j into upper trapezoidal form $\bar{U}_j$. Remove the first row in $\bar{U}_j$, which is the jth row in the final matrix R. Store the rest of the matrix as an update matrix U_j.

4. end

It is easily verified that if Algorithm is applied to the matrix in the example above, this will with a trivial modification give the elimination order outlined before.

A problem with the multifrontal method as described above is that the frontal matrices tend to be too small to make it possible to efficiently utilize vector processors in the solution of the subproblems. Also non-numerical operations will make up a large part of the total factorization time. A first useful modification of the multifrontal method is therefore the idea of amalgamating several nodes into one supernode. Instead of eliminating one column in each node, the factorization of the frontal matrices now involves the elimination of several columns. In general nodes can be grouped together to form a supernode if they correspond to a block of contiguous columns in the Cholesky factor , where the diagonal block is full triangular and these rows all have identical off-block-diagonal column structures.

Because of the computational advantages of having large supernodes, it is possible to relax this condition and amalgamate also nodes which satisfy this condition if some local zeros are treated as nonzeros. However, a practical restriction is that if too many nodes are amalgamated then the frontal matrices become sparse. (In the extreme case that all nodes are amalgamated into one supernode, the frontal matrix becomes equal to the original sparse matrix.) For a discussion of these and other modifications of the multifrontal method see Liu [41]. Ashcraft [2] gives a very detailed description of the implementation of a supernodal multifrontal Cholesky code.

6.4. UPDATING SPARSE LEAST SQUARES SOLUTIONS

We remarked earlier that a single dense row in A will lead to a full matrix $A^T A$ and therefore, invoking the no cancellation assumption, a full Cholesky factor R. Problems where the matrix A is sparse except for a few dense rows can be treated by first solving the problem with the dense rows deleted. The effect of the dense rows into the solution is then incorporated by updating this solution. We stress that we only update the solution, **not** the Cholesky factor.

Consider the problem

$$\min_x \left\| \begin{pmatrix} A_s \\ A_d \end{pmatrix} x - \begin{pmatrix} b_s \\ b_d \end{pmatrix} \right\|_2, \tag{6.2}$$

where $A_s \in \mathbf{R}^{m_1 \times n}$ is sparse and $A_d \in \mathbf{R}^{m_2 \times n}$, $m_2 \ll n$, contains the dense rows. We assume for simplicity that $\operatorname{rank}(A_s) = n$. Denote by x_s the solution to the sparse problem $\min_x \|A_s x - b_s\|_2$, and let the corresponding Cholesky factor be R_s. The residual vectors in (6.2) corresponding to x_s are

$$r_s(x_s) = b_s - A_s x_s, \quad r_d(x_s) = b_d - A_d x_s.$$

We now wish to compute the solution $x = x_s + z$ to the full problem (6.2), and hence to choose z to solve

$$\min_z \left\{ \|r_s(x)\|_2^2 + \|r_d(x)\|_2^2 \right\},$$

where

$$r_s(x) = r_s(x_s) - A_s z, \quad r_d(x) = r_d(x_s) - A_d z.$$

Since $A_s^T r_s(x_s) = 0$, this is equivalent to the problem

$$\min_z \left\{ \|A_s z\|_2^2 + \|A_d z - r_d(x_s)\|_2^2 \right\}.$$

Letting $u = R_s z$, we have that since $\|A_s z\|_2 = \|u\|_2$ this reduces to

$$\min_u \left\{ \|u\|_2^2 + \|B_d u - r_d(x_s)\|_2^2 \right\}, \quad B_d = A_d R_s^{-1}. \tag{6.3}$$

Introducing

$$v = r_d(x_s) - B_d u, \quad C = (\, B_d \;\; I_{m_2} \,), \quad w = \begin{pmatrix} u \\ v \end{pmatrix},$$

we have $Cw = B_d u + v$, and thus (6.3) is equivalent to the minimum norm problem

$$\min \|w\|_2 \quad \text{subject to} \quad Cw = r_d(x_s). \tag{6.4}$$

Since C has full row rank we can use the normal equations to compute the solution

$$u = B_d^T (B_d B_d^T + I)^{-1} r_d(x_s). \tag{6.5}$$

Alternatively we can obtain u from the QR decomposition

$$C^T = \begin{pmatrix} B_d^T \\ I_{m_2} \end{pmatrix} = Q_C \begin{pmatrix} R_C \\ 0 \end{pmatrix}, \qquad \begin{pmatrix} u \\ v \end{pmatrix} = Q_C \begin{pmatrix} R_C^{-T} r_d \\ 0 \end{pmatrix}.$$

Finally solving for z by back-substitution in the sparse R factor, $R_s z = u$, we obtain $x = x_s + z$. Note the similarity with the updating for weighted problems in Sec. 2.3.

The updating scheme can be generalized to the case where the sparse subproblem has rank less than n, see [36]. A general scheme for updating equality constrained linear least squares solutions, where also the constraints are split in a sparse and a dense set is developed in [7]. It is important to point out that these updating algorithms cannot be expected to be stable in all cases. Stability will be a problem whenever the sparse subproblem is more ill-conditioned than the full problem.

References

[1] M. Arioli, J. Demmel, and I. S. Duff. Solving sparse linear systems with sparse backward error. *SIAM J. Matrix Anal. Appl.*, 10:165–190, 1989.

[2] C. Ashcraft. A vector implementation of the multifrontal method for large sparse positive definite systems. Technical Report ETA-TR-51, Engineering Technology Division, Boeing Computer Services, 1987.

[3] J. L. Barlow. Error analysis and implementation aspects of deferred correction for equality constrained least squares problems. *SIAM J. Numer. Anal.*, 25:1340–1358, 1988.

[4] J. L. Barlow and S. L. Handy. The direct solution of weighted and equality constrained least squares problems. *SIAM J. Sci. Statist. Comput.*, 9:704–716, 1988.

[5] Å. Björck. Iterative refinement of linear least squares solutions I. *BIT*, 7:257–278, 1967.

[6] Å. Björck. Comment on the iterative refinement of least squares solutions. *J. Amer. Statist. Assoc.*, 73:161–166, 1978.

[7] Å. Björck. A general updating algorithm for constrained linear least squares problems. *SIAM J. Sci. Statist. Comput.*, 5:394–402., 1984.

[8] Å. Björck. Stability analysis of the method of semi-normal equations for least squares problems. *Linear Algebra Appl.*, 88/89:31–48, 1987.

[9] Å. Björck. Least squares methods. In P. G. Ciarlet and J. L. Lions, editors, *Handbook of Numerical Analysis*, volume I: Finite Difference Methods-Solution of Equations in R^n, pages 466–647. Elsevier/North Holland, Amsterdam, 1990.

[10] Å. Björck. Algorithms for linear least squares problems. In E. Spedicato, editor, *Computer Algorithms for Solving Linear Algebraic Equations; The State of the Art.*, NATO ASI Series F: Computer and Systems Sciences, Vol. 77, pages 57–92, Berlin, 1991. Springer-Verlag.

[11] Å. Björck. Pivoting and stability in the augmented system method: In D. F. Griffiths and G. A. Watson, editors, *Numerical Analysis 1991: Proceedings of the 14th Dundee Conference, June 1991*, Pitman Research Notes in Mathematics 260, pages 1–16. Longman Scientific and Technical, 1992.

[12] Å. Björck. Numerics of Gram-Schmidt orthogonalization. *Linear Algebra and Appl.*, 197:to appear, 1994.

[13] Å. Björck and I. S. Duff. A direct method for the solution of sparse linear least squares problems. *Linear Algebra and Appl.*, 34:43–67, 1980.

[14] Å. Björck and G. H. Golub. Iterative refinement of linear least squares solution by Householder transformation. *BIT*, 7:322–337, 1967.

[15] J. R. Bunch and L. Kaufman. Some stable methods for calculating inertia and solving symmetric linear systems. *Mathematics of Computation*, 31:162–179, 1977.

[16] T. F. Coleman, A. Edenbrandt, and J. R. Gilbert. Predicting fill for sparse orthogonal factorization. *J. ACM*, 33:517–532, 1986.

[17] M. G. Cox. The least squares solution of overdetermined linear equations having band or augmented band structure. *IMA J. Numer. Anal.*, 1:3–22, 1981.

[18] I. S. Duff. Pivot selection and row orderings in Givens reduction on sparse matrices. *Computing*, 13:239–248, 1974.

[19] I. S. Duff. On algorithms for obtaining a maximum transversal. *ACM Trans. Math. Software*, 7:315–330, 1981.

[20] I. S. Duff, A. M. Erisman, and J. K. Reid. *Direct Methods for Sparse Matrices.* Oxford University Press, London, 1986.

[21] I. S. Duff, N. I. M. Gould, J. K. Reid, J. A. Scott, and K. Turner. The factorization of sparse symmetric indefinite matrices. *IMA J. Numer. Anal.*, 11:181–204, 1991.

[22] I. S. Duff and J. K. Reid. MA27—a set of Fortran subroutines for solving sparse symmetric sets of linear equations. Technical Report R.10533, AERE, Harwell, England, 1982.

[23] I. S. Duff and J. K. Reid. The multifrontal solution of indefinite sparse symmetric linear systems. *ACM Trans. Math. Software*, 9:302–325, 1983.

[24] S. C. Eisenstat, M. H. Schultz, and A. H. Sherman. Algorithms and data structures for sparse symmetric gaussian elimination. *SIAM J. Sci. Statist. Comput.*, 2:225–237, 1981.

[25] L. Eldén. Algorithms for the regularization of ill-conditioned least squares problems. *BIT*, 17:134–145, 1977.

[26] L. V. Foster. Modifications of the normal equations method that are numerically stable. In G. H. Golub and P. Van Dooren, editors, *Numerical Linear Algebra, Digital Signal Processing and Parallel Algorithms*, NATO ASI Series, pages 501–512, Berlin, 1991. Springer-Verlag.

[27] W. Gander. Least squares with a quadratic constraint. *Numer. Math.*, 36:291–307, 1981.

[28] W. M. Gentleman. Least squares computations by Givens transformations without square roots. *J. Inst. Maths. Applics.*, 12:329–336, 1973.

[29] J. A. George and M. T. Heath. Solution of sparse linear least squares problems using Givens rotations. *Linear Algebra and Its Application*, 34:69–83, 1980.

[30] J. A. George and J. H. W. Liu. The evolution of the minimum degree ordering algorithm. *SIAM Review*, 31:1–19, 1989.

[31] J. A. George and J. W. H. Liu. *Computer Solution of Large Sparse Positive Definite Systems*. Prentice-Hall, Englewood Cliffs, N.J., 1981.

[32] J. R. Gilbert, C. Moler, and R. Screiber. Sparse matrices in MATLAB: Design and implementation. *SIAM J. Matrix. Anal. Appl.*, 13:333–356, 1992.

[33] P. E. Gill, W. Murray, M. A. Saunders, and M. H. Wright. A Schur-complement method for sparse quadratic programming. In M.G. Cox and S. Hammarling, editors, *Reliable Numerical Computation*, pages 113–138, Oxford, 1990. Clarendon Press.

[34] G. H. Golub and J. H. Wilkinson. Note on the iterative refinement of least squares solution. *Numer. Math.*, 9:139–148, 1966.

[35] M. Gulliksson and P.-Å. Wedin. Modifying the QR decomposition to constrained and weighted linear least squares. *SIAM J. Matrix. Anal. Appl.*, 13:4:1298–1313, 1992.

[36] M. T. Heath. Some extensions of an algorithm for sparse linear least squares problems. *SIAM J. Sci. Statist. Comput.*, 3:223–237, 1982.

[37] C. L. Lawson and R. J. Hanson. *Solving Least Squares Problems*. Prentice Hall, Englewood Cliffs, New Jersey, 1974.

[38] Ö. Leringe and P.-Å. Wedin. A comparison between different methods to compute a vector x which minimizes $\|Ax - b\|_2$ when $Gx = h$. Technical Report, Department of Computer Science, Lund University, 1970.

[39] J. G. Lewis, D. J. Pierce, and D. C. Wah. A multifrontal Householder QR factorization. Technical Report ECA-TR-127, Engineering and Scientific Services Division, Boeing Computer Services, 1989.

[40] J. H. W. Liu. On general row merging schemes for sparse Givens transformations. *SIAM J. Sci. Statist. Comput.*, 7:1190–1211, 1986.

[41] J. H. W. Liu. The role of elimination trees in sparse factorization. *SIAM J. Matrix Anal. Appl.*, 11:134–172, 1990.

[42] P. Matstoms. *The Multifrontal Solution of Sparse Linear Least Squares Problems*. Lic. Thesis, Linköping University, 1991.

[43] P. Matstoms. Sparse QR factorization in MATLAB. *ACM Trans. Math. Software*, to appear, 1994.

[44] C. C. Paige. Computer solution and perturbation analysis of generalized linear least squares problems. *Math. Comp.*, 33:171–184, 1979.

[45] C. C. Paige. Fast numerically stable computations for generalized least squares problems. *SIAM J. Numer. Anal.*, 16:165–171, 1979.

[46] C. C. Paige. The general linear model and the generalized singular value decomposition. *Linear Algebra Appl.*, 70:269–284, 1985.

[47] S. V. Parter. The use of linear graphs in Gauss elimination. *SIAM Review*, 3:119–130, 1961.

[48] G. Peters and J. H. Wilkinson. The least squares problem and pseudo-inverses. *The Computer Journal*, 13:309–316, 1970.

[49] A. Pothen and C. J. Fan. Computing the block triangular form of a sparse matrix. *ACM Trans. Math. Software*, 16:303–324, 1990.

[50] M. J. D. Powell and J. K. Reid. On applying Householder's method to linear least squares problems. In A. J. M. Morell, editor, *Proceedings of the IFIP Congress 68*, pages 122–126, Amsterdam, 1965. North Holland.

[51] J. K. Reid. A note on the least squares solution of a band system of linear equations by Householder reductions. *Comput J.*, 10:188–189, 1967.

[52] C. F. Van Loan. Computing the CS and the generalized singular value decomposition. *Numer. Math.*, 46:479–492, 1985.

[53] J. M. Varah. A practical examination of some numerical methods for linear discrete ill-posed problems. *SIAM Review*, 21:100–111, 1979.

[54] M. H. Wright. Interior methods for constrained optimization. *Acta Numerica*, 1:341–407, 1992.

[55] Z. Zlatev. Comparison of two pivotal strategies in sparse plane rotations. *Comput. Math. Appl.*, 8:119–135, 1982.

Algorithms for Solving Nonlinear Systems of Equations[(*)]

José Mario Martínez

Department of Applied Mathematics
IMECC-UNICAMP, CP 6065
University of Campinas
13081 Campinas SP
Brazil

ABSTRACT: In this paper we survey numerical methods for solving nonlinear systems of equations $F(x) = 0$, where $F : \mathbb{R}^n \to \mathbb{R}^n$. We are especially interested in large problems. We describe modern implementations of the main local algorithms, as well as their globally convergent counterparts.

1. INTRODUCTION

Nonlinear systems of equations appear in many real - life problems. Moré [1989] has reported a collection of practical examples which include: Aircraft Stability problems, Inverse Elastic Rod problems, Equations of Radiative Transfer, Elliptic Boundary Value problems, etc.. We have also worked with Power Flow problems, Distribution of Water on a Pipeline, Discretization of Evolution problems using Implicit Schemes, Chemical Plant Equilibrium problems, and others. The scope of applications becomes even greater if we include the family of Nonlinear Programming problems, since the first-order optimality conditions of these problems are nonlinear systems.

Given $F : \mathbb{R}^n \to \mathbb{R}^n, F = (f_1, \ldots, f_n)^T$, our aim is to find solutions of

$$F(x) = 0. \tag{1.1}$$

We assume that F is well defined and has continuous partial derivatives on an open set of $\mathbb{R}^n$. We denote $J(x)$ the matrix of partial derivatives of F (Jacobian matrix). So,

---

[0(*)] Work supported by FAPESP (PT NQ 90-3724-6) FAEP-UNICAMP, FINEP and CNPq.

E. Spedicato (ed.), Algorithms for Continuous Optimization, 81–108.

described by

$$J(x^k)s^k = -F(x^k), \qquad (2.2)$$
$$x^{k+1} = x^k + s^k. \qquad (2.3)$$

At each iteration of Newton's method, we must compute the Jacobian $J(x^k)$ and solve the linear system (2.2). Using modern techniques of automatic differentiation (see Rall [1984, 1987], Griewank [1992], and references therein) we can compute $F(x)$ and $J(x)$ in a reliable and economical way. If, instead of the true Jacobian in (2.2), we use an approximation by differences of $J(x^k)$, which is generally expensive, we obtain the *Finite-Difference Newton's Method*, whose convergence properties are very similar to those of Newton's method.

Now, (2.2) is a linear system of equations. If n is small, this system can be solved using the LU factorization with partial pivoting or the QR factorization. See Golub and Van Loan [1989]. Using these linear solvers, the cost of solving (2.2) is $O(n^3)$ floating point operations. If n is large this cost becomes prohibitive. However, in many situations, where the matrix $J(x^k)$ is sparse, we can solve (2.2) using LU factorizations. In fact, many times the structure of the matrix is such that the factors L and U of its factorization are also sparse, and can be computed using a moderate amount of operations. Computer algorithms for sparse LU factorizations are surveyed in Duff, Erisman and Reid [1989]. In Gomes-Ruggiero, Martínez and Moretti [1992] we describe the first version of the NIGHTINGALE package for solving sparse nonlinear systems. In NIGHTINGALE, we use the sparse linear solver of George and Ng [1987]. The George-Ng method performs the LU factorization with partial pivoting of a sparse matrix A using a *static data structure* defined before beginning numerical computations. In Newton's method we solve a sequence of linear systems with the same structure, so, the *symbolic phase* that defines the data structure is executed only once.

The system (2.2) has a unique solution if and only if $J(x^k)$ is nonsingular. If the Jacobian is singular, the iteration must be modified. Moreover, if $J(x^k)$ is nearly singular, it is also convenient to modify the iteration in order to prevent numerical instability. Many modifications are possible to keep this phenomenon controlled. In the NIGHTINGALE package, when a very small pivot, relative to the size of the matrix, occurs, it is replaced by a nonzero scalar whose modulus is sufficiently large. Moreover, nearly singular or ill-conditioned matrices usually cause very large increments s^k. So, $||s^k||$ must also be controlled. Computer algorithms usually normalize the stepsize by

$$s^k \leftarrow \min\{1, \frac{\Delta}{||s^k||}\}s^k.$$

where Δ is a parameter given by the user.

The main convergence result relative to Newton's method is given by the following

$$J(x) \equiv F'(x) \equiv \begin{bmatrix} f_1'(x) \\ \vdots \\ f_n'(x) \end{bmatrix} \equiv \begin{bmatrix} \nabla f_1(x)^T \\ \vdots \\ \nabla f_n(x)^T \end{bmatrix} \equiv \begin{bmatrix} \frac{\partial f_1}{\partial x_1}(x) & \cdots & \frac{\partial f_1}{\partial x_n}(x) \\ \vdots & & \vdots \\ \frac{\partial f_n}{\partial x_1}(x) & \cdots & \frac{\partial f_n}{\partial x_n}(x) \end{bmatrix}.$$

We are mostly interested in problems where n is large and $J(x)$ is *structurally sparse*. This means that most entries of $J(x)$ are zero *for all x in the domain of F*. Sparsity is a particular case of the more general notion of *structure*. Jacobian matrices can be symmetric, antisymmetric, positive definite, combination of other matrices with some particular structure, etc.. Many times we can take advantage of particular structures of $J(x)$ in order to obtain efficient algorithms for solving (1.1).

Most popular methods for solving nonlinear systems are *local*. A local method is an iterative scheme that converges if the initial approximation is close enough to a particular solution. Frequently, we are also able to prove *rate of convergence* results for these methods, which tell something about the asymptotic velocity of convergence of the process. Fortunately, in many practical cases the domain of convergence of local methods is large, so that these methods are useful. However, when the initial estimate of the solution is very poor, local methods must be modified in order to improve their *global convergence properties*.

We say that a method for solving (1.1) is *globally convergent* if at least one limit point of the sequence generated by the method is a solution or, at least, a *stationary point*, where $\nabla \|F(x)\|^2 = 0$. Many times *all* the limit points are solutions or stationary points and, frequently, the whole sequence converges to a solution. In general, global methods are modifications of local methods that preserve the local convergence properties of the original algorithm.

This paper is organized as follows. In Section 2, 3, 4 and 5 we survey local methods, namely: Newton's method, Quasi-Newton methods, Inexact-Newton methods and Decomposition methods. In Sections 6 and 7 we survey the main procedures currently used for globalization: Optimization and Homotopies. Conclusions are given in Section 8.

2. NEWTON'S METHOD

Newton's method is the most widely used algorithm for solving nonlinear systems of equations. Given an initial estimation x^0 of the solution of (1.1), this method considers, at each iteration, the approximation

$$F(x) \approx L_k(x) \equiv F(x^k) + J(x^k)(x - x^k) \tag{2.1}$$

and computes x^{k+1} as a solution of the linear system $L_k(x) = 0$. This solution exists and is unique if $J(x^k)$ is nonsingular. Therefore, an iteration of Newton's method is

theorem.

Theorem 2.1. Let us assume that $F : \Omega \subset I\!\!R^n \to I\!\!R^n$, Ω an open and convex set, $F \in C^1(\Omega)$, $F(x^*) = 0$, $J(x^*)$ nonsingular, and that there exist $L, p > 0$ such that for all $x \in \Omega$

$$\|J(x) - J(x^*)\| \leq L\|x - x^*\|^p. \tag{2.4}$$

Then there exists $\varepsilon > 0$ such that if $\|x^0 - x^*\| \leq \varepsilon$, the sequence $\{x^k\}$ generated by (2.2) – (2.3) is well defined, converges to x^*, and satisfies

$$\|x^{k+1} - x^*\| \leq c\|x^k - x^*\|^{p+1}. \tag{2.5}$$

Proof. See Ortega and Rheinboldt [1970], Dennis and Schnabel [1983], etc. □

Property (2.5) (quadratic convergence if $p = 1$) depends on the Hölder condition (2.4). If (2.4) is not assumed, only superlinear convergence of $\{x^k\}$ can be proved $\left(\lim_{k \to \infty} \dfrac{\|x^{k+1} - x^*\|}{\|x^k - x^*\|} = 0 \right)$. Clearly, (2.5) implies superlinear convergence but the converse is not true.

3. QUASI-NEWTON METHODS

In this survey, we call Quasi-Newton methods those methods for solving (1.1) whose general form is

$$x^{k+1} = x^k - B_k^{-1} F(x^k). \tag{3.1}$$

Newton's method, studied in Section 2, belongs to this family. Most Quasi-Newton methods use less expensive iterations than Newton, but their convergence properties are not very different. In general, Quasi-Newton methods avoid either the necessity of computing derivatives, or the necessity of solving a full linear system per iteration or both tasks.

The most simple Quasi-Newton method is the *Stationary Newton Method*, where $B_k = J(x^0)$ for all $k \in I\!\!N$. In this method, derivatives are computed at the initial point and we only need the *LU* factorization of $J(x^0)$. A variation of this method is the *Stationary Newton Method with restarts*, where $B_k = J(x^k)$ if k is a multiple of a fixed integer m and $B_k = B_{k-1}$ otherwise. The number of iterations used by this method tends to increase with m, but the average computer time per iteration decreases. In some situations we can determine an optimal choice for m (Shamanskii [1967]).

An obvious drawback of the stationary Newton methods is that, except when $k \equiv 0 \pmod{m}$, B_k does not incorporate information about x^k and $F(x^k)$. Therefore, the adequacy of the model $L_k(x) \equiv F(x^k) + B_k(x - x^k)$ to the real function $F(x)$ can decrease rapidly as k grows. Observe that, due to (3.1), in Quasi-Newton methods x^{k+1} is defined as the solution of $L_k(x) = 0$, which exists and is unique if B_k is nonsingular. One way to incorporate new information about F on the linear model is to impose the interpolatory conditions

$$L_{k+1}(x^k) = F(x^k), \tag{3.2}$$
$$L_{k+1}(x^{k+1}) = F(x^{k+1}). \tag{3.3}$$

Defining

$$y^k = F(x^{k+1}) - F(x^k) \tag{3.4}$$

and substracting (3.2) from (3.3) we obtain the *Secant Equation*

$$B_{k+1}s^k = y^k. \tag{3.5}$$

Reciprocally, if B_{k+1} satisfies (3.5), L_{k+1} interpolates F at x^k and x^{k+1}. We give the name *Secant Methods* to the family of Quasi-Newton methods based on (3.1) and (3.5).

If $n \geq 2$, there exist infinite many possible choices of B_{k+1} satisfying (3.5). If, in addition to (3.5), we impose that

$$B_{k+1}s^{k-j} = y^{k-j}, j = 1, \ldots, n-1 \tag{3.6}$$

we obtain the Sequential Secant Method (Wolfe [1959], Barnes [1965], Gragg and Stewart [1976], Martínez [1979a], etc.). If the set of increments $\{s^k, s^{k-1}, \ldots, s^{k-n+1}\}$ is linearly independent there exists only one matrix B_{k+1} that satisfies (3.6). In this case,

$$B_{k+1} = (y^k, y^{k-1}, \ldots, y^{k-n+1})(s^k, s^{k-1}, \ldots, s^{k-n+1})^{-1} \tag{3.7}$$

and

$$B_{k+1}^{-1} = (s^k, s^{k-1}, \ldots, s^{k-n+1})(y^k, y^{k-1}, \ldots, y^{k-n+1})^{-1}. \tag{3.8}$$

B_{k+1}^{-1} can be obtained from B_k^{-1} using $O(n^2)$ floating point operations. However, in order to ensure numerical stability, the definition of the increments s^j that appear in (3.7) and (3.8) must sometimes be modified. When these modifications are not necessary, the Sequential Secant Method has the following interpolatory property:

$$L_{k+1}(x^{k+j}) = F(x^{k+j}), j = 0, -1, \ldots, -n. \tag{3.9}$$

The Sequential Secant Method is useful in many situations where n is small. When n is not very small, it is not worthwhile to waste time trying to preserve the interpolatory condition (3.9) for $j \approx -n$. It is more profitable to maintain the Secant

Equation (3.5), using the degrees of freedom inherent to this equation to guarantee numerical stability. Broyden's "good" method (Broyden [1965]) and the Column Updating Method (COLUM) (Martínez [1984]) are two examples of this idea. In both methods

$$B_{k+1} = B_k + \frac{(y^k - B_k s^k)(z^k)^T}{(z^k)^T s^k} \tag{3.10}$$

where

$$z^k = s^k \tag{3.11}$$

for Broyden's method and

$$z^k = e^{jk}, \tag{3.12}$$
$$|(e^{jk})^T s^k| = \|s^k\|_\infty \tag{3.13}$$

for COLUM. $\{e^1, \ldots, e^n\}$ is the canonical basis of $I\!R^n$.

Applying the Sherman-Morrison formula to (3.10) (Golub and Van Loan [1989, p. 51]) we obtain

$$B_{k+1}^{-1} = B_k^{-1} + \frac{(s^k - B_k^{-1} y^k)(z^k)^T}{(z^k)^T B_k^{-1} y^k} B_k^{-1}. \tag{3.14}$$

Formula (3.14) shows that B_{k+1}^{-1} can be obtained from B_k^{-1} using $O(n^2)$ floating point operations in the dense case. Moreover,

$$B_{k+1}^{-1} = (I + u^k(z^k)^T)B_k^{-1}, \tag{3.15}$$

where $u^k = (s^k - B_k^{-1} y^k)/(z^k)^T B_k^{-1} y^k$, so

$$B_k^{-1} = (I + u^{k-1}(z^{k-1})^T) \ldots (I + u^0(z^0)^T)B_0^{-1}, \tag{3.16}$$

for $k = 1, 2, 3 \ldots$

Formula (3.16) is used when n is large. In this case, the vectors $u^0, z^0, \ldots, u^{k-1}, z^{k-1}$ are stored and the product $B_k^{-1} F(x^k)$ is computed using (3.16). In this way, the computer time of iteration k is $O(kn)$ plus the computer time of computing $B_0^{-1} F(x^k)$. If k is large the process must be periodically restarted taking $B_k \approx J(x^k)$.

An intermediate method between Broyden's method and the Sequential Secant Method is *Broyden's Method with Projected Updates*, which was introduced by Gay and Schnabel [1978]. See also Martínez [1979b] and Lopes and Martínez [1980]. This method is probably more efficient than Broyden for small problems, but we have not heard about large-scale implementations.

Broyden's method is a particular case of the family of Least Change Secant Update (LCSU) methods (Dennis and Schnabel [1979, 1983], Dennis and Walker [1981],

Martínez [1990b, 1992a]), which include many algorithms that are useful for particular structures (Hart-Soul algorithms for boundary value problems (Hart and Soul [1973], Kelley and Sachs [1987]), Partitioned Quasi-Newton methods for separable problems (Griewank and Toint [1982a, 1982b, 1982c, 1984], Toint [1986]), Methods of Direct Updating of Factorizations (Dennis and Marwil [1982], Johnson and Austria [1983], Chadee [1985], Martínez [1990a]), BFGS and DFP algorithms for unconstrained minimization (see Dennis and Schnabel [1983]), etc.

Let us survey the main convergence results related to Quasi-Newton algorithms. We are going to assume, as in Theorem 2.1, that $F : \Omega \subset I\!\!R^n \to I\!\!R^n$, Ω open and convex, $F \in C^1(\Omega)$, $F(x^*) = 0$, $J(x^*)$ nonsingular and that the Hölder condition (2.4) is satisfied. The first result is the "Theorem of two neighborhoods".

Theorem 3.1. Given $r \in (0,1)$, there exists $\varepsilon, \delta > 0$ such that if $||x^0 - x^*|| \le \varepsilon$ and $||B_k - J(x^*)|| \le \delta$ for all $k \in I\!\!N$ then the sequence $\{x^k\}$ generated by (3.1) is well defined, converges to x^*, and satisfies

$$||x^{k+1} - x^*|| \le r||x^k - x^*|| \tag{3.17}$$

for all $k \in I\!\!N$.

Proof. See, for instance, Dennis and Walker [1981]. □.

Using Theorem 3.1 we can prove that the Stationary Newton Method and its variations with restarts have local convergence at a linear rate. The main tool for proving superlinear convergence of Quasi-Newton methods, is the following theorem, due to Dennis and Moré.

Theorem 3.2. Assume that the sequence $\{x^k\}$ generated by (3.1) is well defined and convergent to x^*. Then, the two following properties are equivalent.

(a) $$\lim_{k\to\infty} \frac{||[B_k - J(x^*)](x^{k+1} - x^k)||}{||x^{k+1} - x^k||} = 0. \tag{3.18}$$

(b) $$\lim_{k\to\infty} \frac{||x^{k+1} - x^*||}{||x^k - x^*||} = 0. \tag{3.19}$$

Proof. See Dennis and Moré [1974]. □

(3.18) is called the Dennis-Moré condition. Using (3.18), we can prove that the Stationary Newton Method with periodic restarts (for which $\lim_{k\to\infty} B_k = J(x^*)$) is superlinearly convergent. The Dennis-Moré condition says that the effect of $B_k - J(x^*)$ on the normalized increment tends to be null when $k \to \infty$. This condition is weaker than saying that B_k tends to $J(x^*)$. The Dennis-Moré condition is closely related to the Secant Equation. In fact, from (2.4) we can deduce that, for all $x \in \Omega$,

$$\|F(z) - F(x) - J(x^*)(z - x)\| \leq L\|z - x\| \max\{\|x - x^*\|^p, \|z - x^*\|^p\} \quad (3.20)$$

(Broyden, Dennis and Moré [1973]). So, writing $x = x^k$, $z = x^{k+1}$, and assuming that the sequence converges,

$$\lim_{k \to \infty} \frac{\|y^k - J(x^*)s^k\|}{\|s^k\|} = 0. \quad (3.21)$$

Therefore, if the Secant Equation (3.5) holds,

$$\lim_{k \to \infty} \frac{\|[B_{k+1} - J(x^*)]s^k\|}{\|s^k\|} = 0. \quad (3.22)$$

This allows us to prove the following result.

Lemma 3.3. If the sequence generated by a Secant Method converges to x^* and, in addition,

$$\lim_{k \to \infty} \|B_{k+1} - B_k\| = 0. \quad (3.23)$$

then the convergence is superlinear.

Proof. Use (3.18) and (3.22). □

Theorem 3.1 does not guarantee local convergence of all Secant Methods. In fact, the hypothesis of this theorem is that *all the B_k's* belong to a neighborhood of $J(x^*)$ of radius δ. Observe that, even if the first B_0 belongs to this neighborhood it could be possible that $\|B_k - J(x^*)\| \gg \|B_0 - J(x^*)\|$, destroying convergence. Fortunately, for LCSU methods (including Broyden) we are able to prove that exists $\delta' > 0$ such that $\|B_k - J(x^*)\| \leq \delta$ for all $k \in \mathbb{N}$, if $\|B_0 - J(x^*)\| \leq \delta'$. This is a *Bounded Deterioration Property*. Moreover, the Successive Projection scheme that characterizes LCSU methods guarantees also (3.23). Summing up, the following result holds for Broyden, and other LCSU methods.

Theorem 3.4. There exists $\varepsilon, \delta > 0$ such that, if $\|x^0 - x^*\| \leq \varepsilon$ and $\|B_0 - J(x^*)\| \leq \delta$, the sequence generated by Broyden's method is well defined, converges to x^* and satisfies (3.19).

Proof. See Broyden, Dennis and Moré [1973]. For an extension to "all" LCSU methods see Martínez [1990b, 1992a]. □

For Broyden's method, we also have the following result, which states that the convergence is 2n-quadratic.

Theorem 3.5. Under the hypotheses of Theorem 3.4, if $p \geq 1$, there exists $c > 0$ such that the sequence generated by Broyden's method satisfies

$$||x^{k+2n} - x^*|| \leq c||x^k - x^*||^2$$

for all $k \in I\!\!N$.

Proof. See Gay [1979]. □

Local superlinear convergence results also hold for suitable implementations of the Sequential Secant Method and its variations (Gay-Schnabel and Lopes-Martínez). Under slightly stronger assumptions, we can prove stronger convergence results for these methods. For the Sequential Secant Method and the Lopes-Martínez Method the R-order of convergence (Ortega and Rheinboldt [1970]) is the positive root of $t^{n+1} - t^n - 1 = 0$. The Gay-Schnabel method has R-order equal to the positive root of $t^{2n} - t^{2n-1} - 1 = 0$ (Martínez [1979b]). Both numbers tend to 1 when $n \to \infty$.

After nearly 10 years of practical experience, we verified that COLUM has practically the same behavior as Broyden, in terms of reliability and number of iterations. Many times, COLUM uses less computer time than Broyden because the iteration is slightly less expensive. However, COLUM is not an LCSU method, and local and superlinear convergence cannot be proved using the classical Bounded Deterioration techniques. The following convergence results have been proved for COLUM.

Theorem 3.6. Assume that the sequence $\{x^k\}$ is generated by COLUM, except that when $k \equiv 0 \pmod{m}$, $B_k = J(x^k)$. Then, there exists $\varepsilon > 0$ such that, if $||x^0 - x^*|| \leq \varepsilon$, the sequence converges superlinearly to x^*.

Proof. See Martínez [1984]. For a similar result, concerning an Inverse Column Updating Method, see Martínez and Zambaldi [1992]. □

Theorem 3.7. Assume that $n = 2$. Let $r \in (0,1)$. Then, there exists $\varepsilon, \delta > 0$ such that, if $||x^0 - x^*|| \leq \varepsilon$ and $||B_0 - J(x^*)|| \leq \delta$, the sequence $\{x^k\}$ generated by COLUM is well defined, converges to x^*, and satisfies (3.17).

Proof. See Martínez [1992c]. □

Theorem 3.8. Assume that the sequence $\{x^k\}$ generated by COLUM is well defined, converges to x^* and satisfies (3.17). Then

$$\lim_{k \to \infty} \frac{||x^{k+2n} - x^*||}{||x^k - x^*||} = 0 \tag{3.24}$$

and

$$\lim_{k \to \infty} ||x^k - x^*||^{1/k} = 0. \tag{3.25}$$

Proof. See Martínez [1992c]. Property (3.25) is called *R-superlinear convergence.* □

A large gap between practice and theory remains to be filled in relation to COLUM. Some of the relevant questions are:

Does COLUM with restarts (as in the hypothesis of Theorem 3.6), satisfy

$$||x^{k+m} - x^*|| \le c||x^k - x^*||^{p+q} \tag{3.26}$$

for some $c > 0$, q strictly greater than m? The motivation of this question is that the Stationary Newton method with restarts every m iterations satisfies (3.26) with $q = m$.

Does Theorem 3.7 hold for $n > 2$?

Is COLUM superlinearly convergent in the sense of (3.19)?

4. INEXACT-NEWTON METHODS

Many times, large nonlinear systems can be solved using Newton's method, employing a sparse LU factorization for solving (2.2). Most frequently, even if Newton's method is applicable, more efficient algorithms are obtained using COLUM or Broyden, with Newton restarts. In the NIGHTINGALE package, an automatic restart procedure has been incorporated, by means of which a Newton iteration is performed only when it is expected that its efficiency should be greater than the efficiency of previous Quasi-Newton iterations.

Sometimes, the structure of the Jacobian matrix is unsuitable for LU factorizations. That is, a lot of fill-in appears due to that structure and so the iteration becomes very expensive. In many of these cases a strategy of "false Jacobians" works well. By this we mean that we use a Quasi-Newton iteration with restarts, where, at the restarted iterations, B_k is not $J(x^k)$ but a "simplified Jacobian" $\tilde{J}(x^k)$ such that its LU factorization can be performed without problems.

Unhappily, in many cases, $||\tilde{J}(x^k) - J(x^k)||$ is excessively large, and the Quasi-Newton method looses its local convergence properties. In these cases, it is strongly recommendable to use Inexact-Newton methods.

The idea is the following. Since we cannot solve (2.2) using a direct (LU) method, we use an Iterative Linear Method for solving (2.2). Usually, iterative linear methods based on Krylov subspaces are preferred (Golub and Van Loan [1989], Hestenes and Stiefel [1952], Saad and Schulz [1986], etc). Iterative Linear Methods are interesting for solving large-scale systems of equations because of their low memory requirements.

When we solve (2.2) using an iterative linear method, we need a stopping criterion for deciding when to finish the calculation. A very reasonable stopping criterion is

$$||J(x^k)s^k + F(x^k)|| \leq \theta_k||F(x^k)||, \tag{4.1}$$

where $\theta_k \in (0,1)$. The condition $\theta_k < 1$ is necessary because, otherwise, the null increment $s^k \equiv 0$ could be accepted as an approximate solution of (2.2). On the other hand, if $\theta_k \approx 0$, the number of iterations needed by the Iterative Linear Method to obtain (4.1) could be excessively large. Therefore, in practice, an intermediate value $\theta_k \approx 0.1$ is recommended.

Dembo, Eisenstat and Steihaug introduced the criterion (4.1) and proved the main local convergence properties of the algorithms based on this criterion.

Theorem 4.1 Assume that $F(x^*) = 0, J(x^*)$ is nonsingular and continuous at x^*, and $\theta_k \leq \theta_{max} < \theta < 1$. Then there exists $\varepsilon > 0$ such that, if $||x^0 - x^*|| \leq \varepsilon$, the sequence $\{x^k\}$ obtained using (4.1) and (2.3) converges to x^* and satisfies

$$||x^{k+1} - x^k||_* \leq \theta||x^k - x^*||_* \tag{4.2}$$

for all $k \geq 0$, where $||y||_* = ||J(x^*)y||$. If $\lim_{k \to \infty} \theta_k = 0$ the convergence is superlinear.

Proof. See Dembo, Eisenstat and Steihaug [1982]. □

Krylov subspace methods for solving systems like (2.2) are usually implemented using some preconditioning scheme. See Axelsson [1985]. By this we mean that the original system is replaced by an equivalent one, which is easier to solve by the Iterative Linear Solver. In the case of (2.2), we wish to replace the linear system by

$$B_k^{-1}J(x^k)s^k = -B_k^{-1}F(x^k) \tag{4.3}$$

where B_k^{-1} (or, at least, the product $B_k^{-1}z$) must be easy to compute and $B_k \approx J(x^k)$. For general linear systems, many useful preconditioners B_k have been introduced. Most of them are based on Incomplete LU Factorizations, or on Stationary Linear iterations. A very cheap and popular procedure is to use the diagonal of the original matrix as preconditioner. Many other preconditioners for specific problems can be found in the papers published in Spedicato [1991]. A common feature to different preconditioning schemes applied to a linear system $Az = b$ is that the first iteration of the preconditioned Iterative Linear Solver is $z^1 = \lambda B^{-1}b$, where B is the preconditioner. So, in the case of the system (2.2), the first increment tried should be of the form $-\lambda B_k^{-1}F(x^k)$. This increment will be accepted if it satisfies (4.1). However, (2.2) is not an isolated linear system of equations. In fact, probably $J(x^k) \approx J(x^{k-1})$ specially when $k \to \infty$. Therefore, we are motivated to use information about $B_k, F(x^k), F(x^{k+1}), x^{k+1}, x^k$ when we choose the preconditioner B_{k+1}. This idea leads to impose a Secant Condition to the preconditioner. So, we would like to introduce an algorithm based on (4.1), where the sequence of preconditioners B_k are chosen in order to satisfy

$$B_{k+1}s^k = y^k \tag{4.4}$$

for all $k \in I\!N$.

We saw in Section 2 that there exist infinite many possible choices of B_{k+1} satisfying (4.4). Nazareth and Nocedal [1978] and Nash [1985] suggested to use the classical BFGS formula in order to precondition (2.2), when we deal with minimization problems. Our preference is to define

$$B_{k+1} = C_{k+1} + D_{k+1} \tag{4.5}$$

where C_{k+1} is a classical preconditioner and D_k is chosen to satisfy (4.4).

The main appeal of Secant Preconditioners is that it has been shown (Martínez [1992b]) that using them it is possible to obtain stronger convergence results than the one mentioned in Theorem 4.1. In fact, the main drawback of this result is the necessity of $\theta_k \to 0$ for obtaining superlinear convergence. The following Preconditioned Inexact Newton Method was introduced by Martínez [1992b] with the aim of obtaining superlinear convergence without imposing a precision tending to infinity in the iterative resolution of (2.2).

Algorithm 4.2. Let $\theta_k \in (0, \theta)$ for all $k \in I\!N, \theta \in (0, 1)$ and $\lim_{k \to \infty} \theta_k = 0$. Assume that $x^0 \in I\!R^n$ is an initial approximation to the solution of (1.1) and $B_0 \in I\!R^{n \times n}$ is an initial nonsingular preconditioner. Given $x^k \in I\!R^n$ and B_k nonsingular, the steps for obtaining x^{k+1}, B_{k+1} are the following.

Step 1. Compute

$$s_Q^k = -B_k^{-1}F(x^k). \tag{4.6}$$

Step 2. If

$$||J(x^k)s_Q^k + F(x^k)|| \le \theta||F(x^k)|| \tag{4.7}$$

define

$$s^k = s_Q^k. \tag{4.8}$$

Else, find an increment s^k such that (4.1) holds, using some iterative method.

Step 3. Define $x^{k+1} = x^k + s^k$. ⊙

The following theorem states the main convergence result relative to Algorithm 4.2.

Theorem 4.3. Assume that $F : \Omega \subset \mathbb{R}^n \to \mathbb{R}^n$, Ω an open and convex set, $F \in C^1(\Omega)$, $J(x^*)$ nonsingular, $F(x^*) = 0$ and (2.4) holds for some $L \geq 0, p \geq 1$. Suppose that $\|B_k\|$ and $\|B_k^{-1}\|$ are bounded and that the Dennis-Moré condition (3.18) is satisfied. Then, there exists $\varepsilon > 0$ such that, if $\|x^0 - x^*\| \leq \varepsilon$, the sequence $\{x^k\}$ generated by Algorithm 4.2 converges superlinearly to x^*. Moreover, there exists $k_0 \in \mathbb{N}$ such that $s^k = s_Q^k$ for all $k \geq k_0$.

Proof. See Martínez [1992b]. $\qquad\qquad\qquad\qquad\qquad\qquad\qquad\qquad\qquad\qquad\qquad\qquad\square$

Theorem 4.3 states that, if we use preconditioners satisfying the Dennis-Moré condition, superlinear convergence is obtained without $\lim\limits_{k\to\infty} \theta_k = 0$. In fact, the first iteration s_Q^k of the preconditioned iterative linear method will satisfy (4.6), and so it will be accepted as the new increment s^k, preserving superlinearity. Least-Change Secant-Update formulae can be used for obtaining preconditioners satisfying the hypotheses of Theorem 4.3.

Recently, Abaffy [1992] considered the possibility of using, for solving (2.2), iterative linear algorithms that work componentwise, that is, without evaluating the whole residual at each step. For this type of algorithms, he introduced a new stopping criterion, different from (4.1), which also ensures superlinear convergence.

5. DECOMPOSITION METHODS

The methods studied in the previous sections evaluate *all* the components of the function F at the same points. This is not always the best possible strategy. In many practical problems, given a guess of the solution, the evaluation of a few components of F is enough to suggest a new useful estimate. Methods that evaluate different components at different points are called *Decomposition Methods*.

The (Block) SOR-Newton method proceeds as follows. Assume that the components of F are divided into blocks $F_1, \ldots, F_m$ and that we decompose $x = \begin{pmatrix} x_1 \\ \vdots \\ x_m \end{pmatrix}$, where $x_i \in \mathbb{R}^{n_i}$, $F_i(x) \in \mathbb{R}^{n_i}$, $i = 1, \ldots, m$. We denote $\dfrac{\partial F_i}{\partial x_j}$ the matrix of partial derivatives of F_i with respect to the variables x_j. The SOR-Newton method is defined by

$$x_i^{k+1} = x_i^k - w \left(\frac{\partial F_i}{\partial x_i}\right)^{-1} (x_1^{k+1}, \ldots, x_{i-1}^{k+1}, x_i^k, \ldots, x_m^k) F_i(x_1^{k+1}, \ldots, x_{i-1}^{k+1}, x_i^k, \ldots, x_m^k)$$

$$(5.1)$$

for $i = 1, \ldots, m$, where $w \in (0, 2)$ is a *relaxation parameter*. A generalization of (5.1) is

$$x_i^{k+1} = x_i^k - w B_{i,k}^{-1} F_i(x_1^{k+1}, \ldots, x_{i-1}^{k+1}, x_i^k, \ldots, x_m^k). \qquad (5.2)$$

Methods of type (5.2) are called SOR-Quasi-Newton methods (Martínez [1992d, 1992e]). If $B_{i,k+1}$ satisfies

$$B_{i,k+1}(x_i^{k+1} - x_i^k) = F_i(x_1^{k+1}, \ldots, x_i^{k+1}, \ldots, x_m^k) - F_i(x_1^{k+1}, \ldots, x_i^k, \ldots, x_m^k) \qquad (5.3)$$

we say that (5.2) defines a SOR-Secant method (Martínez [1992e]). The local convergence analysis of (5.1), (5.2) and (5.3) has been made in Martínez [1992e]. Essentially, the condition for the local linear convergence of (5.1) - (5.3) is the same convergence condition of the linear SOR method for a linear system with matrix $J(x^*)$. If n_i is small for all $i = 1, \ldots, n$, SOR methods have low storage requirements and so, they are useful for large-scale problems.

The *asynchronous* generalization of (5.2) can be very useful in parallel computer environments. Assume that we work with m parallel processors, where processor i is specialized in the computation of new guesses of the component x_i. That is, processor i only computes $x_i - w B_i^{-1} F_i(x)$ for given B_i and x. Different processors use different times for their respective computations and, as soon as a processor produces a result, this result is communicated to the other processors. In this way, all the processors are working all the time and full advantage is taken from parallelism. See Bertsekas and Tsitsiklis [1989]. Let us now formalize these ideas. The index $k \in I\!N$ represents the times at which at least one component of x is changed. We define $T_i \subset I\!N$ by

$$T_i = \text{Set of times at which } x_i \text{ is updated.}$$

Then, the Asynchronous SOR method is defined by

$$x_i^{k+1} = x_i^k - w B_{i,k}^{-1} F_i(x_1^{\nu(i,1,k)}, \ldots, x_i^k, \ldots, x_m^{\nu(i,m,k)}) \qquad (5.4)$$

if $k + 1 \in T_i$, and

$$x_i^{k+1} = x_i^k \qquad (5.5)$$

if $k + 1 \notin T_i$, where $0 \le \nu(i, j, k) \le k$ for all $k \in I\!N, i, j = 1, \ldots, m$. (5.2) is a particular case of (5.4) - (5.5) defining $T_i = \{i + jm, j \in I\!N\}, \nu(i, j, k) = k$ for all $i, j, \ldots, n, i \ne j, k \in I\!N$. The Jacobi-Quasi-Newton method corresponds to $T_i = \{1, 2, 3, \ldots\}, \nu(i, j, k) = k$. Secant Asynchronous methods based on (5.4) can also be considered.

The SOR and Jacobi methods for linear systems are strongly related with the projection methods of Kaczmarz [1937] and Cimmino [1938] respectively. The advantage of Kaczmarz and Cimmino is that convergence is guaranteed for every linear system, while SOR and Jacobi require a condition on the spectral radius of the transformation. However, in general, when SOR and Jacobi converge, they are far more efficient than Kaczmarz and Cimmino. Nonlinear generalizations of Kaczmarz and Cimmino may be found in Tompkins [1955], McCormick [1977], Meyn [1983] , Martínez [1986a, 1986b, 1986c] and Diniz - Ehrhardt and Martínez [1992].

Different decomposition methods are motivated by direct methods for solving linear systems, for example, the family of Nonlinear ABS algorithms developed in Abaffy and Spedicato [1989], Abaffy, Broyden and Spedicato [1984], Abaffy, Galantai and Spedicato [1987], Spedicato, Chen and Deng [1992], etc.. These methods generalize classical algorithms due to Brown [1969], Brent [1973], Gay [1975] and Martínez [1979c, 1980]). The idea is the following. Divide the components of F into m groups $F_1, \ldots, F_m$. Assume that x^k has been computed. We generate $x^{k,1}, \ldots, x^{k,m}$ by:

$$x^{k,0} = x^k, \tag{5.6}$$

$$J_i(x^{k,i})(x^{k,i+1} - x^{k,i}) = -F_i(x^{k,i}), \tag{5.7}$$

$$J_j(x^{k,j})(x^{k,i+1} - x^{k,i}) = 0, j = 1, \ldots, i-1, \tag{5.8}$$

where $J_j(x) = F'_j(x), i = 0, 1, \ldots, m-1$.

$$x^{k+1} = x^{k,m}. \tag{5.9}$$

Clearly, the scheme (5.6) - (5.9) solves a linear system of equations in one cycle. However, ther are infinite many ways to choose the intermediate points $x^{k,1}, \ldots, x^{k,m-1}$. Different choices of these points originate different methods of the ABS class. The first motivation given in the papers of Brown and Brent for methods of type (5.6) - (5.9) was that, using suitable factorizations, the derivatives can be approximated by differences in a more economic way than in the Finite Difference Newton method. Methods of this class have, in general, the same local convergence properties of Newton, though the proofs are technically complicated.

Many other decomposition algorithms have been introduced with the aim of taking advantage of particular structures. For systems that are reducible to block lower triangular form see Eriksson [1976] and Dennis, Martínez and Zhang [1992]. For Block Tridiagonal Systems, see Hoyer, Schmidt and Shabani [1989]. Much theory about decomposition methods has been produced by the German school (Schmidt [1987], Burmeister and Schmidt [1988], Hoyer [1987], Hoyer and Schmidt [1984], Schmidt, Hoyer and Haufe [1985], etc.)

6. GLOBALIZATION BY OPTIMIZATION

In the previous sections we studied local methods, that is, algorithms that converge, usually with a high rate of convergence, if the initial point is close enough to the solution. Luckily, in many cases the domain of convergence of local algorithms is large enough to guarantee practical efficiency. However, locally convergent methods may not converge if the starting point is very poor, or if the system is highly nonlinear. By this reason, local methods are usually modified in order to improve their global convergence properties. The most usual way to do this is to transform (1.1) into an Optimization Problem, with the objective function $f(x) = \frac{1}{2}||F(x)||^2$. Then, (1.1) becomes the problem of finding a global minimizer of f.

However, the decision of merely using a method to minimize f, in order to solve (1.1), is not satisfactory. In fact, sometimes efficient local methods converge rapidly to a solution but the generated sequence $\{x^k\}$ does not exhibit monotonic behavior in $f(x^k)$. In these cases, the pure local method is much more efficient than the f-minimization method. Often, the minimization method converges to a local (non-global) minimizer of f, while the local method converges to a solution of (1.1). By these reasons, it is necessary to give a chance to the local method before calling the minimization algorithm. Different solutions have been proposed to this problem (Grippo, Lampariello e Lucidi [1986]). Here we describe a strategy that combines local algorithms and minimization methods introduced in the NIGHTINGALE package. We define "ordinary iterations" and "special iterations". By an ordinary iteration we understand an iteration produced by any of the methods described in sections 2 to 4 of this paper. Decomposition methods can also be considered, with some modifications. A special iteration is an iteration produced by a minimization algorithm applied to f. We define, for all $k \in I\!N$,

$$a^k = \text{Argmin } \{f(x^0), \ldots, f(x^k)\}. \tag{6.1}$$

Ordinary and special iterations are combined by the following strategy.

Algorithm 6.1. Initialize $k \leftarrow 0$, FLAG $\leftarrow 1$. Let $q \geq 0$ be an integer, $\gamma \in (0,1)$.

Step 1. If FLAG $= 1$, obtain x^{k+1} using an ordinary iteration. Else, obtain x^{k+1} using a special iteration.

Step 2. If

$$f(a^{k+1}) \leq \gamma f(a^{k-q}) \tag{6.2}$$

set FLAG $\leftarrow 1, k \leftarrow k + 1$ and go to Step 1. Else, re-define $x^{k+1} \leftarrow a^{k+1}$. Set FLAG $\leftarrow -1, k \leftarrow k + 1$ and go to Step 1. $\qquad\qquad\odot$

If the test (6.2) is satisfied infinite many times, then there exists a subsequence of $\{x^k\}$ such that $\lim_{k\to\infty} ||F(x^k)|| = 0$. So, if the sequence is bounded, we will be able to find a solution of (1.1) up to any prescribed accuracy. Conversely, if (6.2) does not hold for all $k \geq k_0$, then all the iterations starting from the k_0-th will be special, and the convergence properties of the sequence will be those of the minimization algorithm.

In principle, we can use as minimization algorithm to define special iterations any method for minimization described in the literature (Dennis and Schnabel [1983], Fletcher [1987], etc.). For large-scale problems, the NIGHTINGALE package uses a strategy based on Trust Regions derived from the Inexact Newton approach (Friedlander, Gomes-Ruggiero, Martínez and Santos [1993]). This strategy is described by the following algorithm.

Algorithm 6.2. Assume that $\Delta_{\min} > 0, \alpha \in (0,1)$ are given independently of the iteration k. Define $\psi_k(x) = ||F(x^k) + J(x^k)(x - x^k)||^2$, $\Delta \geq \Delta_{\min}$.

Step 1. Compute an approximate minimizer $\bar{x}$ of $\psi_k(x)$ on the box $||x - x^k||_\infty \leq \Delta$ such that $\psi_k(x) \leq \psi_k(x_Q^k)$, x_Q^k is the projection of $x^k - 2J(x^k)^T F(x^k)/M_k$ on the box and $M_k \geq 2||J(x_k)||_1||J(x_k)||_\infty$.

Step 2. If

$$||F(\bar{x})||^2 \leq ||F(x^k)||^2 + \alpha(\psi_k - \psi_k(\bar{x})) \tag{6.3}$$

define $x^{k+1} = \bar{x}$. Else, choose $\Delta_{\text{new}} \in [0.1||\bar{x} - x^k||, 0.9\Delta]$, replace Δ by Δ_{new} and go to Step 1. $\odot$

Most of the work of Algorithm 6.2 is concentrated on the approximate solution of

$$\left. \begin{array}{l} \text{Minimize } \psi_k(x) \\ s.t. \quad ||x - x^k||_\infty \leq \Delta \end{array} \right\} \tag{6.4}$$

(6.4) is the problem of minimizing a convex quadratic with box constraints. For this problem, algorithms based on combinations of Krylov Subspace methods with Gradient Projection strategies are currently preferred. In NIGHTINGALE, the approximate solution of (6.4) is defined as a point that satisfies $\psi_k(x) \leq \psi_k(x_Q^k)$ and where, in addition, the norm of projected gradient of $\psi_k(x)$ is less than $0.1||J(x^k)^T F(x^k)||$. We also choose: $\Delta_{\min} = 0.001\times$ (typical $||x||$), initial choice of $\Delta \equiv \Delta_0 = ||x^0||, \Delta_{\text{new}} = 0.5||\bar{x} - x^k||$, further choice is $\Delta = 4 \times \Delta$.

The convergence properties of Algorithm 6.2 were given in Friedlander, Gomes - Ruggiero, Martínez and Santos [1993]. Every limit point x^* of a sequence $\{x^k\}$ generated by this algorithm satisfies $J(x^*)^T F(x^*) = 0$. Therefore, x^* is a solution of (1.1) if $J(x^*)$ is nonsingular. Unhappily, if $J(x^*)$ is singular, it is possible that $F(x^*) \neq 0$. This is the main weakness of algorithms based on optimization for the globalization of (1.1).

An advantage of special iterations based on box trust regions is that they can be easily adapted to situations where we have natural bounds for the solution of (1.1). Other recent methods based on the inexact Newton approach with global convergent properties were given by Deuflhard [1991] and Eisenstat and Walker [1993].

7. GLOBALIZATION BY HOMOTOPIES

In Section 6 we saw that local methods for solving $F(x) = 0$ can be "globalized" through their transformation into minimization problems. In this section we study another popular technique to solve (1.1) when the initial approximation is poor. This technique is based on homotopies. A homotopy associated to this problem is a function $H(x,t) : I\!R^n \times I\!R \to I\!R$ such that

$$\left. \begin{array}{rcl} H(x,1) & = & F(x) \\ H(x^0,0) & = & 0 \end{array} \right\} \tag{7.1}$$

If H satisfies (7.1) we can expect that $\Gamma \equiv \{(x,t) \in I\!R^n \times I\!R \mid H(x,t) = 0, 0 \leq t \leq 1\}$ will be an arc connecting the initial approximation x^0 with a solution x^*. Homotopic techniques consist in tracing Γ from $t = 0$ to $t = 1$ in a reliable and efficient way. The first to propose the homotopic idea were Lahaye [1935] and Davidenko [1955]. The homotopic principle is very popular. In fact, this is one of the techniques mentioned in the article *Numerical Analysis* of *Encyclopaedia Britannica*.

Sometimes, the homotopy has an interest by itself. In other cases we are interested only in the solution of $H(x,1) = 0$. In this section we deal with the latter situation. This has a practical consequence. If we are interested only in the solution of $H(x,1) = 0$, it makes sense to interrupt the tracing of Γ when t is close to 1, trying to apply a local method from the current point. In general, this will be more efficient than to insist in carefully tracing Γ.

Unhappily, Γ could not be an arc of finite length. It is possible that, if we trace Γ starting from $(x^0,0)$, we never arrive to $(x^*,1)$. However, some classical results of Differential Geometry help us to identify situations in which tracing Γ leads to the solution of (1.1). If $H'(x,t) \equiv (H'_x(x,t), H'_t(x,t))$ has full rank for all $(x,t) \in H^{-1}(\{0\})$, then $H^{-1}(\{0\})$ is a discrete union of curves homomorphic to $I\!R$ or to the one-dimensional sphere S^1. See Milnor [1969]. In this case, each component of $H^{-1}(\{0\})$ can be traced efficiently using numerical methods. The condition that $H'(x,t)$ has full rank results, in many practical cases, from the application of a theorem of Chow, Mallet-Paret and Yorke [1979]. See also Watson [1979]. However, this does not guarantee that the component of $H^{-1}(\{0\})$ that passes through $(x^0,0)$ should arrive to $(x^*,1)$.

Now, if the unique solution of $H(x,0) = 0$ is x^0 and $H'_x(x^0,0)$ is nonsingular, it can be proved that the component of $H^{-1}(\{0\})$ that passes through $(x^0,0)$ is homomorphic to $I\!R$. In fact, even weaker hypotheses are sufficient (Ortega and Rheinboldt [1970, Ch. 6]). These hypotheses are easy to verify and guarantee that, if we trace Γ starting from $(x^0,0)$, it is not possible to return to the x^0. However this does not guarantee yet that $(x^*,1)$ will be reached.

To be sure that this homotopy is useful we must verify, in addition to the previous assumptions, that Γ is bounded. In this case, Γ will be necessarily a segment of arc joining $(x^0,0)$ with $(x^*,1)$. When we want to make sure that a given homopopy will solve a problem, the difficult part is precisely to prove boundedness of Γ.

Sometimes, "natural" homotopies are used. This mean, homotopies where the parameter t that is suggested by the problem under consideration. However, there exist useful "artificial" homotopies. Let us mention here two of them. The homotopy of Reduction of Residual is

$$H(x,t) = F(x) + (t-1)F(x^0).$$

The "regularizing homotopy", used in the well known package HOMPACK (Watson, Billups and Morgan [1987]) is

$$H(x,t) = tF(x) + (1-t)(x - x^0).$$

After the choice of H, we must use a numerical method to trace Γ. First, Γ must be parametrized. Frequently, we describe Γ as a function of the parameter t. However, if, for some t_0, we have that $H'_x(x, t_0)$ is singular, x cannot be expressed as a function of t in a neighborhood of t_0, and, instead of increasing t, we must decrease this parameter, in order to advance in Γ. By this reason, it is usual to trace Γ using s, the arclength, as parameter.

If this is the case, the procedure generally recommended to trace Γ is of Predictor-Corrector type. Given a set of points $(x(s_1), t(s_1)), \ldots, (x(s_m), t(s_m))$ computed consecutively on Γ, and an increment $\Delta > 0$, we calculate a polynomial interpolating those points and, using this polynomial, we compute a predictor point $(\overline{x}(s_m + \Delta), \overline{t}(s_m + \Delta))$. This new point does not necessarily belong to Γ. So, using this point as initial approximation, we compute a point on Γ using a local method for the nonlinear system $H(x,t) = 0$. If, in this system, we consider that t is also a variable, we have n equations with $n+1$ unknowns. Special local algorithms for underdetermined nonlinear systems were developed by Walker and Watson [1989], Martínez [1991], etc. This is the Corrector Phase. When we arrive to a point (x, t) with t close to 1, we try the direct application of a local method for $F(x) = 0$.

An interesting discussion on homotopy methods is given in a recent survey by Forster [1993].

8. CONCLUSIONS

Many practical applications give rise to large scale nonlinear systems of equations. From the local point of view, the most interesting techniques for this case are variations of the Inexact Newton method. To develop preconditioners for Krylov Subspace solvers, taking into account the structure of these methods and problems, is a challenging problem. Since solving a linear system is a particular case of minimizing a quadratic on a box, it turns out that to solve efficiently this last problem is crucial. Moreover the problem appears again in the development of methods for globalization using optimization, and in the Corrector Phase of Homotopy methods.

For very small problems Newton's method continues to be the best choice, and for medium to large problems a combination Newton-Quasi-Newton seems to be better. Many decomposition methods are interesting when they are induced by characteristics of the problem structure , or when decomposition is suggested by the computer architecture.

In this paper, we surveyed methods based on first-order approximations of the nonlinear system. Methods based on approximations of higher order have also been developed (Schnabel and Frank [1984]) and are useful in the presence of singularities of the Jacobians. Large - scale implementations of these methods seem to be hard.

References

Abaffy, J. [1992]: Superlinear convergence theorems for Newton-type methods for nonlinear systems of equations, *JOTA* , 73, pp. 269 - 277.

Abaffy, J.; Broyden, C.G.; Spedicato, E. [1984]: A class of direct methods for linear equations, *Numerische Mathematik* 45, pp. 361-376.

Abaffy, J.; Galantai, A. ; Spedicato, E. [1987]: The local convergence of ABS method for nonlinear algebraic system, *Numerische Mathematik* 51, pp. 429 - 439.

Abaffy, J.; Spedicato, E. [1989]: *ABS Projection Algorithms, Mathematical Techniques for Linear and Nonlinear Equations*, Ellis Horwood, Chichester.

Axelsson, O. [1985]: A survey of preconditioned iterative methods for linear systems of equations, *BIT* 25, pp. 166 - 187.

Barnes, J.G.P. [1965]: An algorithm for solving nonlinear equations based on the secant method, *Computer Journal* 8, pp. 66 - 72.

Bertsekas, D.P.; Tsitsiklis, C. [1989]: *Parallel and Distributed Computation. Numerical Methods*, The Johns Hopkins University Press, New York.

Brent, R.P. [1973]: Some efficient algorithms for solving systems of nonlinear equations, *SIAM Journal on Numerical Analysis* 10, pp. 327 - 344.

Brown, K.M. [1969]: A quadratically convergent Newton - like method based upon Gaussian elimination, *SIAM Journal on Numerical Analysis* 6, pp. 560 - 569.

Broyden, C.G. [1965]: A class of methods for solving nonlinear simultaneous equations, *Mathematics of Computation* 19, pp. 577-593.

Broyden, C.G.; Dennis Jr., J.E.; Moré, J.J. [1973]: On the local and superlinear convergence of quasi-Newton methods, *Journal of the Institute of Mathematics and its Applications* 12, pp. 223-245.

Burmeister, W.; Schmidt, J.W. [1978]: On the k-order of coupled sequences arising in single - step type methods, *Numerische Mathematik* 33, pp. 653 - 661.

Chadee, F.F. [1985]: Sparse quasi-Newton methods and the continuation problem, T.R. S.O.L. 85-8, Department of Operations Research, Stanford University.

Chow, S.N.; Mallet-Paret, J.; Yorke, J.A. [1978]: Finding zeros of maps: Homotopy methods that are constructive with probability one, Mathematics of Computation 32, pp. 887-899.

Cimmino, G. [1938]: Calcolo approsimato per la soluzione dei sistemi di equazioni lineari, *La Ricerca Scientifica Ser II, Ano IV 1* pp. 326-333.

Coleman,T. F.; Garbow, B. S.; Moré,J. J. [1984]: Software for estimating sparse Jacobian matrices, *ACM Trans. Math. Software* 11, pp. 363-378.

Coleman,T. F.; Moré,J. J. [1983]: Estimation of sparse Jacobian matrices and graph coloring problems, *SIAM Journal on Numerical Analysis* 20, pp. 187-209.

Curtis,A. R.; Powell, M. J. D.; Reid, J. K. [1974], On the estimation of sparse Jacobian matrices, *Journal of the Institute of Mathematics and its Applications* 13, pp. 117-120.

Davidenko, D.F. [1953]: On the approximate solution of nonlinear equations, *Ukrain. Mat. Z.* 5, pp. 196 -206.

Dembo, R.S.; Eisenstat, S.C.; Steihaug, T. [1982]: Inexact Newton methods, *SIAM Journal on Numerical Analysis* 19, pp. 400–408.

Dennis Jr., J.E.; Martínez, J.M.; Zhang, X. [1992]: Triangular Decomposition Methods for Reducible Nonlinear Systems of Equations , Technical Report, Department of Mathematical Sciences, Rice University.

Dennis Jr., J.E.; Marwil, E.S. [1982]: Direct secant updates of matrix factorizations, *Mathematics of Computation* 38, pp. 459–476.

Dennis Jr., J.E.; Moré, J.J. [1974]: A characterization of superlinear convergence and its application to quasi - Newton methods, *Mathematics of Computation* 28, pp. 549 -560.

Dennis Jr., J.E.; Moré, J.J. [1977]: Quasi-Newton methods, motivation and theory, *SIAM Review* 19, pp. 46-89.

Dennis Jr.,J.E.; Schnabel,R.B. [1979]: Least change secant updates for quasi-Newton methods, *SIAM Review* 21, pp. 443-459.

Dennis Jr.,J.E.; Schnabel,R.B. [1983]:*Numerical Methods for Unconstrained Optimization and Nonlinear Equations*, Prentice-Hall, Englewood Cliffs.

Dennis Jr., J.E. ; Walker, H.F. [1981]: Convergence theorems for least-change secant update methods, *SIAM Journal on Numerical Analysis* 18, pp. 949-987.

Deuflhard, P. [1991]: Global inexact Newton methods for very large scale nonlinear problems *Impact of Computing in Science and Engineering* 3, pp. 366–393.

Diniz-Ehrhardt, M.A.; Martínez, J.M. [1992]: A parallel projection method for overdetermined nonlinear systems of equations, to appear in *Numerical Algorithms*.

Duff, I.S. [1977]: MA28 – a set of Fortran subroutrines for sparse unsymmetric linear equations. AERE R8730, HMSO, London.

Duff, I.S.; Erisman, A.M.; Reid, J.K. [1989]:*Direct Methods for Sparse Matrices*, Oxford Scientific Publications.

Eisenstat, S.C.; Walker, H.F. [1993]: Globally convergent inexact Newton methods, to appear in *SIAM Journal on Optimization*.

Eriksson, J. [1976]: A note on the decomposition of systems of sparse nonlinear equations, *BIT* 16, pp. 462 - 465.

Fletcher, R. [1987]: *Practical Methods of Optimization* (2nd edition), John Wiley and Sons, New York.

Forster, W. [1993]: Homotopy methods, to appear in *Handbook of Global Optimization*, Kluwer.

Friedlander, A.; Gomes-Ruggiero, M.A.; Martínez, J.M.; Santos, S.A. [1993]: A globally convergent method for solving nonlinear systems using a trust - region strategy, Technical Report, Department of Applied Mathematics, University of Campinas.

Gay, D.M. [1975]: Brown's method and some generalizations with applications to minimization problems, Ph D Thesis, Computer Science Department, Cornell University, Ithaca , New York.

Gay, D.M. [1979]: Some convergence properties of Broyden's method, *SIAM Journal on Numerical Analysis* 16, pp. 623 - 630.

Gay, D.M.; Schnabel, R.B. [1978]: Solving systems of nonlinear equations by Broyden's method with projected updates, in *Nonlinear Programming 3*, edited by O. Mangasarian, R. Meyer and S. Robinson, Academic Press, New York, pp. 245-281.

George, A.; Ng, E. [1987]: Symbolic factorization for sparse Gaussian elimination with partial pivoting, *SIAM Journal on Scientific and Statistical Computing* 8, pp. 877-898.

Golub, G.H.; Van Loan, Ch.F. [1989]: *Matrix Computations*, The Johns Hopkins University Press, Baltimore and London.

Gomes–Ruggiero, M.A.; Martínez, J.M. [1992]: The Column–Updating Method for solving nonlinear equations in Hilbert space, *Mathematical Modelling and Numerical Analysis* 26, pp 309-330.

Gomes–Ruggiero, M.A.; Martínez, J.M.; Moretti, A.C. [1992]: Comparing algorithms for solving sparse nonlinear systems of equations, *SIAM Journal on Scientific and Statistical Computing* 13, pp. 459 - 483.

Gragg, W.B.; Stewart, G.W. [1976]: A stable variant of the secant method for solving nonlinear equations, *SIAM Journal on Numerical Analysis* 13, pp. 127 - 140.

Griewank, A. [1986]: The solution of boundary value problems by Broyden based secant methods, *CTAC – 85. Proceedings of the Computational Techniques and Applications Conference*, J. Noye and R. May (eds.), North Holland, Amsterdam.

Griewank, A. [1992]: Achieving Logarithmic Growth of Temporal and Spacial Complexity in Reverse Automatic Differentiation, *Optimization Methods and Software* 1, pp. 35 - 54.

Griewank, A.; Toint, Ph.L. [1982a]: On the unconstrained optimization of partially separable functions, in *Nonlinear Optimization 1981*, edited by M.J.D. Powell, Academic Press, New York.

Griewank, A.; Toint, Ph.L. [1982b]: Partitioned variable metric for large structured optimization problems, *Numerische Mathematik* 39, pp. 119 - 137.

Griewank, A.; Toint, Ph.L. [1982c]: Local convergence analysis for partitioned quasi-Newton updates, *Numerische Mathematik* 39, pp. 429-448.

Griewank, A.; Toint, Ph.L. [1984]: Numerical experiments with partially separable optimization problems, in *Numerical Analysis Proceedings Dundee 1983*, edited by D.F. Griffiths, Lecture Notes in Mathematics vol. 1066, Springer - Verlag, Berlin, pp. 203-220.

Grippo, L.; Lampariello, F.; Lucidi, S. [1986]: A nonmonotone line search technique for Newton's method, *SIAM Journal on Numerical Analysis* 23, pp. 707 - 716.

Hart, W.E.; Soul, S.O.W. [1973]: Quasi-Newton methods for discretized nonlinear boundary value problems, *J. Inst. Math. Applics.* 11, pp. 351 - 359.

Hestenes, M.R.; Stiefel, E. [1952]: Methods of conjugate gradients for solving linear systems, *Journal of Research of the National Bureau of Standards* B49, pp. 409 - 436.

Hoyer, W. [1987]: Quadratically convergent decomposition algorithms for nonlinear systems with special structure, Inform. Tech. Univ. Dresden 07 , pp. 23-87.

Hoyer, W.; Schmidt, J.W. [1984]: Newton-type decomposition methods for equations arising in network analysis, *Z. Angew-Math. Mech.* 64, pp. 397 - 405.

Hoyer, W.; Schmidt, J.W.; Shabani, N. [1989]: Superlinearly convergent decomposition methods for block - tridiagonal nonlinear systems of equations, *Numerical Functional Analysis and Optimization* 10 (9 & 10), pp. 961 - 975.

Iri, M. [1984]: Simultaneous computations of functions, partial derivatives and estimates of rounding errors. Complexity and Practicality , *Japan Journal of Applied Mathematics* 1, pp. 223 - 252.

Johnson, G.W.; Austria, N.H. [1983]: A quasi-Newton method employing direct secant updates of matrix factorizations, *SIAM Journal on Numerical Analysis* 20, pp. 315-325.

Kaczmarz, S. [1937]: Angenaherte Auflösung von Systemen linearer Gleichungen, *Bull. Acad. Polon. Sci. Lett. A35*, pp. 355-357.

Kelley, C.T.; Sachs, E.W. [1987]: A quasi-Newton method for elliptic boundary value problems, *SIAM Journal on Numerical Analysis* 24, pp. 516 - 531.

Lahaye, E. [1934]: Une méthode de résolution d'une catégorie d'equations transcendantes, Comptes Rendus Acad. Sci. Paris 198, pp. 1840-1842.

Lopes, T.L.; Martínez, J.M. [1980]: Combination of the Sequential Secant Method and Broyden's method with projected updates, *Computing* 25, pp. 379-386.

Martínez, J.M. [1979a]: Three new algorithms based on the sequential secant method, *BIT* 19, pp. 236-243.

Martínez, J.M. [1979b]: On the order of convergence of Broyden - Gay - Schnabel's method, *Commentationes Mathematicae Universitatis Carolinae* 19, pp. 107-118.

Martínez, J.M. [1979c]: Generalization of the methods of Brent and Brown for solving nonlinear simultaneous equations, *SIAM Journal on Numerical Analysis* 16, pp. 434 - 448.

Martínez, J.M. [1980]: Solving nonlinear simultaneous equations with a generalization of Brent's method, *BIT* 20, pp. 501 - 510.

Martínez, J.M. [1983]: A quasi–Newton method with a new updating for the LDU factorization of the approximate Jacobian, *Matemática Aplicada e Computacional* 2, pp. 131–142.

Martínez, J.M. [1984]: A quasi–Newton method with modification of one column per iteration, *Computing* 33, pp. 353–362.

Martínez, J.M. [1986a]: The method of Successive Orthogonal Projections for solving nonlinear simultaneous equations, *Calcolo* 23, pp. 93 - 105.

Martínez, J.M. [1986b]: Solving systems of nonlinear simultaneous equations by means of an accelerated Successive Orthogonal Projections Method, *Computational and Applied Mathematics* 165, pp. 169 - 179.

Martínez, J.M. [1986c]: Solution of nonlinear systems of equations by an optimal projection method, *Computing* 37, pp. 59 - 70.

Martínez, J.M. [1987]: Quasi–Newton Methods with Factorization Scaling for Solving Sparse Nonlinear Systems of Equations, *Computing* 38, pp. 133–141.

Martínez, J.M. [1990a]: A family of quasi-Newton methods for nonlinear equations with direct secant updates of matrix factorizations, *SIAM Journal on Numerical Analysis* 27, pp. 1034-1049.

Martínez, J.M. [1990b]: Local convergence theory of inexact Newton methods based on structured least change updates, *Mathematics of Computation* 55, pp. 143-168.

Martínez, J.M. [1991]: Quasi-Newton Methods for Solving Underdetermined Nonlinear Simultaneous Equations, *Journal of Computational and Applied Mathematics* 34, pp. 171–190.

Martínez, J.M. [1992a]: On the relation between two local convergence theories of least change secant update methods, *Mathematics of Computation* 59, pp. 457–481.

Martínez, J.M. [1992b]: A Theory of Secant Preconditioners, to appear in *Mathematics of Computation*.

Martínez, J.M. [1992c]: On the Convergence of the Column-Updating Methods, Technical Report, Department of Applied Mathematics, University of Campinas.

Martínez, J.M. [1992d]: Fixed-Point Quasi-Newton Methods, *SIAM Journal on Numerical Analysis* 29, pp. 1413–1434.

Martínez, J.M. [1992e]: SOR - Secant Methods, to appear in *SIAM Journal on Numerical Analysis*.

Martínez, J.M.; Zambaldi, M.C. [1992]: An inverse Column-Updating Method for solving Large-Scale Nonlinear Systems of Equations, to appear in *Optimization Methods and Software*.

Matthies, H.; Strang, G. [1979]: The solution of nonlinear finite element equations, *International Journal of Numerical Methods in Engineering* 14, pp. 1613 - 1626.

Mc Cormick, S.F. [1977]: The method of Kaczmarz and row-orthogonalization for solving linear equations and least-squares problems in Hilbert space, *Indiana University Mathematical Journal* 26, pp. 1137 - 1150.

Meyn, K-H [1983]: Solution of underdetermined nonlinear equations by stationary iteration, *Numerische Mathematik* 42, pp. 161-172.

Milnor, J.W. [1969]: *Topology from the differential viewpoint*, The University Press of Virginia, Charlottesville, Virginia.

Moré, J.J. [1989]: A collection of nonlinear model problems, Preprint MCS - P60 - 0289, Mathematics and Computer Science Division, Argonne National Laboratory, Argonne, Illinois.

Nazareth, L.; Nocedal, J. [1978]: A study of conjugate gradient methods, Report SOL 78-29, Department of Operations Research, Stanford University.

Nash, S.G. [1985]: Preconditioning of Truncated Newton methods, *SIAM Journal on Scientific and Statistical Computing* 6, pp. 599 -616.

Ortega, J.M.; Rheinboldt, W.G. [1970]: *Iterative Solution of Nonlinear Equations in Several Variables*, Academic Press, NY.

Ostrowski, A.M. [1973]: *Solution of Equations in Euclidean and Banach Spaces*, Academic Press, New York.

Rall, L.B. [1984]: Differentiation in PASCAL - SC: Type Gradient, *ACM Transactions on Mathematical Software* 10, pp. 161-184.

Rall, L.B. [1987]: Optimal Implementation of Differentiation Arithmetic, in *Computer Arithmetic, Scientific Computation and Programming Languages*, U. Külisch (ed.), Teubner, Stuttgart.

Rheinboldt, W.C. [1986]: *Numerical Analysis of Parametrized Nonlinear Equations*, J. Wiley, Interscience, New York.

Saad, Y.; Schultz, M.H. [1986]: GMRES: A generalized minimal residual algorithm for solving nonsymmetric linear systems, *SIAM Journal on Numerical Analysis* 7, pp. 856-869.

Schmidt, J.W. [1987]: A class of superlinear decomposition methods in nonlinear equations, *Numerical Functional Analysis and Optimization* 9, pp. 629 - 645.

Schmidt J.W.; Hoyer, W.; Haufe, Ch. [1985]: Consistent approximations in Newton - type decomposition methods, *Numerische Mathematik* 47, pp. 413 - 425.

Schnabel, R.B.; Frank, P.D. [1984]: Tensor methods for nonlinear equations, *SIAM Journal on Numerical Analysis* 21, pp. 815 - 843.

Shamanskii, V.E. [1967]: A modification of Newton's method, *Ukrain Mat. Z.* 19, pp. 133-138.

Schwandt, H. [1984]: An interval arithmetic approach for the construction of an almost globally convergent method for the solution of the nonlinear Poisson equation on the unit square, *SIAM Journal on Scientific and Statistical Computing* 5, pp. 427 - 452.

Schwetlick, H. [1978]: *Numerische Losung Nichtlinearer Gleichungen*, Deutscher Verlag der Wissenschaften, Berlin.

Spedicato, E. [1991] (editor): *Computer Algorithms for Solving Linear Algebraic Equations. The State of Art*, NATO ASI Series, Series F: Computer and Systems Sciences, Vol. 77, Springer Verlag, Berlin.

Spedicato, E.; Chen, Z.; Deng, N. [1992]: A class of difference ABS - type algorithms for a nonlinear system of equations, Technical Report, Department of Mathematics, University of Bergamo.

Tewarson, R.P. [1988]: A new Quasi-Newton algorithm, *Applied Mathematics Letters* 1, pp. 101 - 104.

Tewarson, R.P.; Zhang, Y. [1987]: Sparse Quasi-Newton LDU update, *International Journal on Numerical Methods in Engineering* 24, pp. 1093-1100.

Tompkins, C. [1955]: Projection methods in calculation, *Proceedings Second Symposium on Linear Programming*, Washington D.C., pp. 425 - 448.

Walker, H.F.; Watson, L.T. [1989]: Least - Change Update Methods for underdetermined systems, Research Report, Department of Mathematics, Utah State University.

Watson, L.T. [1979]: An algorithm that is globally convergent with probability one for a class of nonlinear two–point boundary value problems, *SIAM Journal on Numerical Analysis* 16, pp. 394–401.

Watson, L.T. [1980]: Solving finite difference approximations to nonlinear two-point boundary value problems by a homotopy method, *SIAM Journal on Scientific and Statistical Computing* 1, pp. 467-480.

Toint, Ph.L. [1986]: Numerical solution of large sets of algebraic nonlinear equations, *Mathematics of Computation* 16, pp. 175 - 189.

Watson, L.T. [1983]: Engineering applications of the Chow–Yorke algorithm, in *Homotopy Methods and Global Convergence* (B.C. Eaves and M.J. Todd eds.), Plenum, New York.

Watson, L.T.; Billups, S.C.; Morgan, A.P. [1987]: Algorithm 652: HOMPACK: A suite of codes for globally convergent homotopy algorithms, *ACM Trans. Math. Software* 13, pp. 281–310.

Watson, L.T.; Scott, M.R. [1987]: Solving spline-collocation approximations to nonlinear two-point boundary value problems by a homotopy method, *Applied Mathematics and Computation* 24, pp. 333-357.

Watson, L.T.; Wang, C.Y. [1981]: A homotopy method applied to elastica problems, *International Journal on Solid Structures* 17, pp. 29-37.

Wolfe, P. [1959]: The secant method for solving nonlinear equations ,*Communications ACM* 12, pp. 12 - 13.

Zambaldi, M.C. [1990]: *Estruturas estáticas e dinâmicas para resolver sistemas não lineares esparsos*, Tese de Mestrado, Departamento de Matemática Aplicada, Universidade Estadual de Campinas, Campinas, Brazil.

Zlatev, Z.; Wasniewski, J.; Schaumburg, K. [1981]: *Y12M. Solution of large and sparse systems of linear algebraic equations*, Lecture Notes in Computer Science 121, Springer-Verlag, New York, Berlin, Heidelberg and Tokyo.

AN OVERVIEW OF UNCONSTRAINED OPTIMIZATION

R. FLETCHER
Dept. of Mathematics and Computer Science,
University of Dundee,
Dundee DD1 4HN,
Scotland, U.K.

ABSTRACT. Developments in the theory and practice of unconstrained optimization are described. Both line search and trust region prototypes are explained and various algorithms based on the use of a quadratic model are outlined. The properties and implementation of the BFGS method are described in some detail and the current preference for this approach is discussed. Various conjugate gradient methods for large unstructured systems are given, but it is argued that limited memory methods are considerably more effective without a great increase in storage or time. Further developments in this area are described. For structured problems the possibilities for updates that retain sparsity are described, including a recent proposal which maintains positive definite matrices and reduces to the BFGS update in the dense case. The alternative use of structure in partially separable optimization is also discussed.

1. Background and Introduction

The development of modern numerical methods for unconstrained optimization has taken place over the last thirty or forty years and considerable progress has been made. We now have good methods for many classes of problem and many insights into why these methods are successful. There is a wealth of literature available. Nevertheless there are a number of open questions of interest, and new ideas continue to enrich the field.

This paper reviews the progress made in *practical* methods of unconstrained optimization, and the relevant theory. Relevant considerations include the following.

- Ease of implementation, reliability, insensitivity to roundoff errors.
- Practical experience on actual problems and test problems.
  Relative performance measures. Ease of use.
- Global convergence to a solution from a remote starting point.
- Local convergence. Rate of convergence.
- Idealized behaviour. Worst case performance.
- Effects of size and/or structure.

The presentation reflects the author's view of the field, but many other interesting contributions are omitted due to limitations of space.

The problem of interest can be stated as that of finding a local solution x^* to the problem

$$\text{minimize} \quad f(x) \qquad x \in \mathbb{R}^n. \tag{1.1}$$

Usually x^* exists and is locally unique, although there are some pathological exceptions. There are two particular caveats that must be made. One is that these

E. Spedicato (ed.), Algorithms for Continuous Optimization, 109–143.
© 1994 *Kluwer Academic Publishers.*

methods do not guarantee to find a global solution to (1.1), but one which is optimal only in some neighbourhood of x^*. Global optimization is an order of magnitude more difficult and is reviewed in the paper of Evtushenko in these Proceedings. Another caveat is that the objective function $f(x)$ must be sufficiently smooth in some sense. Nonsmooth functions, such as arise when max or modulus operations are used to define $f(x)$, require a different approach more related to constrained optimization. The minimization of nonsmooth functions is reviewed in the paper of Zowe in these Proceedings. In what follows it is assumed that the *gradient vector* denoted by $g(x) = \nabla f(x)$ and also the *Hessian matrix* denoted by $G(x) = \nabla^2 f(x) = [\partial^2 f/\partial x_i \partial x_j]$ are continuously differentiable functions of x.

A necessary condition for x^* to solve (1.1) is that x^* is a *stationary point*, that is $g^* = 0$. (The notation $g^* = g(x^*)$ etc. is used.) This is equivalent to having zero slope $s^T g* = 0$ in any direction s from x^*. Another necessary condition is that the matrix G^* is positive semi-definite, which is equivalent to nonnegative curvature $s^T G^* s \geq 0$ in any direction. It is then convenient to refer to x^* as a *semi-definite point*. A sufficient condition for x^* to solve (1.1) is that both $g^* = 0$ and G^* is positive definite. These conditions play a fundamental part in the development of methods for unconstrained optimization. The proof of these results and more details pertaining to the subject can be found for example in Fletcher [21].

Methods for unconstrained optimization differ according to how much information the user feels able to provide. The most desirable situation from the point of view of providing useful information is that the user provides subroutines from which $f(x)$, $g(x)$ and $G(x)$ can be evaluated for any x (a *second derivative method*). However programming the Hessian can be quite demanding for the user and often the lesser requirement of providing subroutines for $f(x)$ and $g(x)$ is made (a *gradient method*). Most of this paper relates to gradient methods. Even less demanding is that the user only provides a subroutine for $f(x)$ (a *no-derivative method*). Such methods have proved considerably less effective and robust than gradient methods and I shall not discuss them here. Probably the best advice is to use high precision arithmetic and estimate first derivatives by finite differences, in conjunction with a gradient method. Recent methods for optimization may also ask for information about the structure of $f(x)$ in some way, and this can lead to significant improvements in efficiency. Also recent developments in automatic differentiation (see Dixon's paper in these Proceedings) make second derivative methods much more attractive.

Methods for unconstrained optimization are generally iterative methods in which the user typically provides an initial estimate $x^{(1)}$ of x^*, and possibly some additional information. A sequence of iterates $\{x^{(k)}\}$ is then generated according to some algorithm. Usually the algorithm is such that the sequence of function values $\{f^{(k)}\}$ is monotonically decreasing ($f^{(k)}$ denotes $f(x^{(k)})$ etc.). A practical algorithm must be well-defined and converge from any $x^{(1)}$, which is the issue of *global convergence*. This issue is usually addressed by aiming to show theoretically that accumulation points of the sequence $\{x^{(k)}\}$ satisfy the necessary conditions for a local solution, and in particular that they are stationary points. Another important theoretical issue is that of *local convergence* and involves a study of the rate at which the sequence converges. In practice the sequence is terminated after a finite number of iterations, if for example the gradient vector is sufficiently small, or negligible progress is being made. In addition to theoretical studies it is important that methods are shown to work well in practice. To this end, libraries of suitable *test problems* (e.g. Moré, Garbow and Hillstrom [43]) have been compiled, which can be used for algorithm development.

The usual practice is to design a method on the basis of some *model* of the objective function. Most fruitful has been the use of a *quadratic model* in which $f(x)$ is approximated on the kth iteration by a Taylor series approximation about $x^{(k)}$, that is

$$f(x^{(k)} + \delta) \approx q^{(k)}(\delta) = f^{(k)} + \delta^T g^{(k)} + \tfrac{1}{2}\delta^T B^{(k)}\delta \qquad (1.2)$$

where δ represents a displacement to $x^{(k)}$, and $B^{(k)}$ is either the Hessian $G^{(k)}$ or some approximation to it. A fundamental algorithm known as *Newton's method* calculates the displacement $\delta^{(k)}$ on iteration k as the minimizer of $q^{(k)}(\delta)$. Newton's method is a second derivative method so that $B^{(k)} = G^{(k)}$ is used. A unique minimizer of $q^{(k)}(\delta)$ exists if and only if $G^{(k)}$ is positive definite and Newton's method is only well-defined in this case. Then $\delta^{(k)}$ is obtained by finding the stationary point of $q^{(k)}(\delta)$, which requires the solution of the linear system

$$G^{(k)}\delta^{(k)} = -g^{(k)}. \qquad (1.3)$$

This is most readily done by finding the LL^T (Choleski) or LDL^T factors of $G^{(k)}$, which also enables the positive definite requirement to be checked. The next iterate is then defined by

$$x^{(k+1)} = x^{(k)} + \delta^{(k)}. \qquad (1.4)$$

Usually $G^{(k)}$ is positive definite when $x^{(k)}$ is in some neighbourhood of x^*. It can then be proved that the sequence $\{x^{(k)}\}$ converges, and the order of convergence is second order. These local convergence properties represent the ideal local behaviour which other algorithms aim to emulate as far as possible. In fact *superlinear convergence* of any algorithm is obtained if and only if the step $\delta^{(k)}$ is asymptotically equal to the step given by solving (1.3). This fundamental result due to Dennis and Moré [17] emphasizes the importance of the Newton step for local convergence.

When $x^{(k)}$ is remote from x^* then Newton's method may not converge, and may not be defined (when $G^{(k)}$ is not positive definite). Globally convergent methods are often designed on the basis of some *prototype algorithm* which specifies an overall approach within which some variation might be possible. The two main prototypes are *line search methods* and *trust region methods*. In a line search method the kth iteration involves the following steps

- determine a direction of search $s^{(k)}$
- find $\alpha^{(k)}$ to minimize $f(x^{(k)} + \alpha s^{(k)})$ with respect to α
- set $x^{(k+1)} = x^{(k)} + \alpha^{(k)} s^{(k)}$.

Different methods correspond to different ways of choosing $s^{(k)}$. This choice is of significant importance in obtaining an efficient method and we return to it below. It is necessary (when $g^{(k)} \neq 0$) that $s^{(k)}$ is a *descent direction* so that a reduction in $f(x)$ can be guaranteed for sufficiently small $\alpha > 0$. This is equivalent to the condition

$$s^{(k)^T} g^{(k)} < 0, \tag{1.5}$$

showing that the slope of $f(x)$ along the line $x = x^{(k)} + \alpha s^{(k)}$ is negative. Finding $\alpha^{(k)}$ is the *line search subproblem* which is carried out by repeatedly sampling $f(x)$ and $g(x)$ (if available) at different points along the line $x = x^{(k)} + \alpha s^{(k)}$. In the form stated above, this step is idealized (an *exact line search*) insofar as the exact minimizer cannot usually be calculated in a finite number of arithmetic operations. It is also not efficient to find $\alpha^{(k)}$ to high accuracy when $x^{(k)}$ is remote from x^*. In practice an *inexact line search* is used which enables a significant reduction in $f(x)$ to be made at the cost of only a few evaluations of $f(x)$ and $g(x)$. Algorithms for the line search subproblem are quite sophisticated and are described in more detail for example in [21].

An ideal choice for $s^{(k)}$ when $G^{(k)}$ is positive definite is the Newton direction obtained by solving (1.3). Likewise when $B^{(k)}$ approximates $G^{(k)}$ then $s^{(k)}$ can be obtained by solving

$$B^{(k)} s^{(k)} = -g^{(k)} \tag{1.6}$$

by analogy with (1.3). Such methods are referred to as *Newton-like methods*. It is important that $B^{(k)}$ is positive definite which is a sufficient condition for the descent property (1.5) (because $s^{(k)^T} g^{(k)} = -s^{(k)^T} B^{(k)} s^{(k)} < 0$). Otherwise it is possible that $s^{(k)^T} g^{(k)} = 0$ when $g^{(k)} \neq 0$, and the method fails to make further progress. A simple choice is $B^{(k)} = I$ (the unit matrix) giving $s^{(k)} = -g^{(k)}$. This is known as the *steepest descent method* on account of the steepest slope property of the gradient vector. Despite this apparently attractive property, the worst case behaviour of the method is to exhibit very slow linear convergence and it is this that is usually observed in practice. When $G^{(k)}$ is available but is indefinite, a good strategy is to choose $B^{(k)} = G^{(k)} + D^{(k)}$ where $D^{(k)}$ is a sufficiently large diagonal matrix. A popular method is that of Schnabel and Eskow [63], and the method of (1.10) below can also be interpreted in this way. In the case of a gradient method, an attractive way of specifying $B^{(k)}$ is that used in the class of *quasi-Newton methods*. Another type of line search method that uses quadratic information in a different way is the *conjugate gradient method*. Both these developments are described in detail below.

A different prototype is the *trust region method*, which may be regarded as addressing the difficulties posed by the lack of a minimizer for the quadratic model

(1.2) when $B^{(k)}$ is indefinite. On iteration k a *trust region* of radius $\rho^{(k)}$ about $x^{(k)}$ is defined by the set

$$\Omega^{(k)} = \{x : \|x - x^{(k)}\| \leq \rho^{(k)}\}. \tag{1.7}$$

and $x^{(k+1)}$ is constrained to lie within this set. Thus $x^{(k+1)} = x^{(k)} + \delta^{(k)}$ is determined by choosing $\delta^{(k)}$ as the solution of the *trust region subproblem*

$$\text{minimize} \quad q^{(k)}(\delta) \qquad \text{subject to} \quad \|\delta\| \leq \rho^{(k)}, \tag{1.8}$$

and it is unnecessary to require that $B^{(k)}$ is positive definite. The aim is to choose the radius so that $q^{(k)}(\delta)$ agrees with $f(x^{(k)} + \delta)$ to a certain extent for $x^{(k)} + \delta$ in $\Omega^{(k)}$, hence the term trust region. The algorithm adjusts $\rho^{(k)}$ adaptively to obtain a certain level of agreement between the actual and predicted reduction in $f(x)$ that is obtained in the algorithm, whilst keeping $\rho^{(k)}$ as large as possible. If the Newton-like step $\delta^{(k)}$ defined by solving (1.3) or (1.6) is feasible in (1.8) then it is taken by the trust region algorithm. Otherwise, and when the 2-norm is used in (1.7), then (1.8) reduces to the solution of the nonlinear equation (in λ)

$$\|\delta(\lambda)\|_2^2 = \rho^{(k)2} \tag{1.9}$$

in which $\delta(\lambda)$ is defined by solving the linear system

$$(B^{(k)} + \lambda I)\delta = -g^{(k)}. \tag{1.10}$$

In general (1.9) cannot be solved in a finite number of operations, but an acceptable approximation can usually be obtained in two or three inner iterations, each of which requires the solution of (1.10) (see Moré [42]). If the ∞-norm is used in (1.7) then (1.8) becomes a quadratic programming problem with simple bounds. Although this can be solved finitely, it is usually more expensive in practice, and if $B^{(k)}$ is not positive definite then the existence of local but not global solutions to (1.10) may cause potential difficulty.

An attractive feature of trust region methods is the strong global convergence properties that can be proved. It is only necessary to assume that $\|B^{(k)}\|$ is $O(k)$, in order to show that the sequence $\{x^{(k)}\}$ has an accumulation point which is stationary, and this result allows $B^{(k)}$ to be indefinite. If $B^{(k)} = G^{(k)}$ then it can also be proved that the accumulation point is semi-definite. For a line search descent method the global convergence results are less strong and some aspects of these are now described. First of all it is necessary to ensure that a sufficient decrease in f is made in the line search. A possible strategy is to accept a step $\alpha^{(k)}$ if it satisfies the *Wolfe conditions*

$$f(x^{(k)} + \alpha^{(k)}s^{(k)}) \leq f^{(k)} + \rho\alpha^{(k)}g^{(k)T}s^{(k)} \tag{1.11}$$

$$-g(x^{(k)} + \alpha^{(k)}s^{(k)})^T s^{(k)} \leq -\sigma g^{(k)T}s^{(k)} \tag{1.12}$$

where $0 \leq \rho < \sigma \leq 1$ are preset parameters. A modified form (the *Strong Wolfe conditions*) is obtained by taking the modulus of the left hand side of (1.12). For some algorithms this is preferable as it allows an exact line search to be approached by taking σ arbitrarily close to zero. It is possible to develop line search algorithms which find such an $\alpha^{(k)}$ in a finite number of operations (see e.g. [21]).

To prove global convergence for a line search algorithm also requires an assumption that the angle $\theta^{(k)}$ between the vectors $s^{(k)}$ and $-g^{(k)}$ does not approach $\frac{\pi}{2}$ too rapidly. It follows from (1.11), (1.12) and Lipchitz continuity of $g(x)$ that

$$\sum_{k=1}^{\infty} \cos^2 \theta^{(k)} \|g^{(k)}\|^2 < \infty. \tag{1.13}$$

If the sequence $\{s^{(k)}\}$ is such that there is an $\varepsilon > 0$ for which

$$\cos \theta^{(k)} \geq \varepsilon \tag{1.14}$$

then it readily follows from (1.13) that $g^{(k)} \to 0$. A sufficient condition for a Newton-like method to satisfy (1.14) is that the condition number is uniformly bounded ($\|B^{(k)}\| \|B^{(k)-1}\| \leq \varepsilon^{-1}$). Unfortunately for most practical algorithms (1.14) cannot be asserted. However if the lesser aim of proving $g^{(k)} \to 0$ on a subsequence is considered, then it is only necessary to show that

$$\sum_{k=1}^{\infty} \cos^2 \theta^{(k)} = \infty. \tag{1.15}$$

It has been possible to show that some practical algorithms satisfy (1.15) but there are many other algorithms for which the question is open.

Line search algorithms are less appropriate when second derivatives are available in that for example they allow convergence to a saddle point which is not a semi-definite point. More significantly, they do not adequately use the knowledge that $G^{(k)}$ may have negative eigenvalues. Trust region algorithms provide one solution to this problem, but another advance has been the development of trajectory search algorithms which use directions of negative curvature. The idea is to determine a *descent pair* of directions $(s^{(k)}, d^{(k)})$ on each iteration, where loosely speaking, $s^{(k)}$ represents a descent direction calculated on the basis of positive curvature information in $G^{(k)}$, and $d^{(k)}$ is a *negative curvature descent direction* with $d^{(k)T} G^{(k)} d^{(k)} < 0$. If $G^{(k)}$ is positive semi-definite then $d^{(k)} = 0$. This descent pair is used in a search along the trajectory

$$x = x^{(k)} + \alpha d^{(k)} + \alpha^2 s^{(k)}. \tag{1.16}$$

These ideas are formalised by Moré and Sorensen [44] who show that if the sequences $\{s^{(k)}\}$ and $\{d^{(k)}\}$ are bounded and satisfy

$$g^{(k)^T} s^{(k)} \to 0 \quad \Rightarrow \quad g^{(k)} \to 0 \quad \text{and} \quad s^{(k)} \to 0 \tag{1.17}$$

and

$$d^{(k)^T} G^{(k)} d^{(k)} \to 0 \quad \Rightarrow \quad \min\{\lambda_{\min}(G^{(k)}), 0\} \to 0 \quad \text{and} \quad d^{(k)} \to 0 \tag{1.18}$$

then accumulation points of the sequence $\{x^{(k)}\}$ are both stationary and semi-definite. Moré and Sorensen give analogues of the Wolfe conditions for finding an acceptable step $\alpha^{(k)}$ in the trajectory search. Within the scope of (1.17) and (1.18) there are many ways of defining the descent pair $(s^{(k)}, d^{(k)})$. Moré and Sorensen define a method based on the Bunch-Parlett factors of $G^{(k)}$. A recent paper of Forsgren, Gill and Murray [27] gives a method based on partial Choleski factors of $G^{(k)}$, which reduces to the regular Choleski factors when $G^{(k)}$ is positive definite.

The rest of this section is relevant to gradient methods, and examines different ways in which quadratic information can be estimated when the Hessian $G^{(k)}$, and hence the quadratic model (1.2), is not directly available. One idea has been to derive methods with the property of *quadratic termination*, in which the method is able to locate the minimizer of a quadratic function in a known finite number of iterations, yet which can be applied iteratively to minimize non-quadratic functions. A particular way of obtaining quadratic termination is to invoke the concept of the *conjugacy* of a set of non-zero vectors $s^{(k)}$, $k = 1, 2, ..., n$ relative to a given positive definite matrix G. This is the property that

$$s^{(i)^T} G s^{(j)} = 0 \qquad \forall\, i \neq j. \tag{1.19}$$

A *conjugate direction method* is a line search method which generates such directions when applied to a quadratic function with Hessian G. It is readily shown in this case that a conjugate direction method terminates in at most n iterations if exact line searches are made (e.g. [21]). (Subsequent references in this paper to termination properties for quadratic functions will assume that the Hessian is positive definite.) Of particular importance are *conjugate gradient methods* in which directions are generated by

$$\begin{aligned} s^{(1)} &= -g^{(1)} \\ s^{(k)} &= -g^{(k)} + \beta^{(k)} s^{(k-1)} \qquad \forall\, k > 1. \end{aligned} \tag{1.20}$$

The merit of this approach lies in the fact that the method can be implemented with only a few vectors of storage. Various choices of the scalar $\beta^{(k)}$ exist which give different performance on non-quadratic functions, yet are equivalent for quadratic functions. Such methods, and developments thereof, are considered in some detail in Section 3.

Another general approach is to determine a matrix $B^{(k)}$ which approximates to $G^{(k)}$. This approximation can then be used in various ways, e.g. by using (1.2) in a trust region algorithm or (1.6) in a line search algorithm. Mostly we shall concentrate on the latter case. One possibility is to estimate $B^{(k)}$ by finite differences using repeated evaluation of the gradient vector. This is disadvantageous both on the grounds of cost and the possible lack of symmetry and positive definiteness in $B^{(k)}$. These disadvantages are removed in the very important class of quasi-Newton methods. In these methods, $B^{(k)}$ is initially set to an arbitrary symmetric positive definite matrix (often $B^{(1)} = I$), and $B^{(k)}$ is updated after each iteration to obtain a new and hopefully better approximation $B^{(k+1)}$. The information available to do this is the pair of difference vectors

$$\delta^{(k)} = x^{(k+1)} - x^{(k)} \tag{1.21}$$

and

$$\gamma^{(k)} = g^{(k+1)} - g^{(k)}. \tag{1.22}$$

These vectors are related by a Taylor series expansion

$$\gamma^{(k)} = \bar{G}^{(k)} \delta^{(k)} \tag{1.23}$$

where

$$\bar{G}^{(k)} = \int_0^1 G(x^{(k)} + \theta \delta^{(k)}) \, d\theta \tag{1.24}$$

is the average Hessian matrix along the step. By analogy with (1.23), $B^{(k+1)}$ is chosen to satisfy

$$\gamma^{(k)} = B^{(k+1)} \delta^{(k)} \tag{1.25}$$

known as the *secant equation* or *quasi-Newton condition*. Within this framework, many possible *updating formulae* are possible, and it is important to make a choice which is both readily computed and effective in practice. Over the last 20 years or so the *BFGS updating formula* has generally been regarded as an effective choice. We study this formula and the claims of alternative formulae in some detail in Section 2.

A special case of unconstrained optimization of considerable importance occurs when the objective function is a sum of m squared terms

$$f(x) = \sum_{i=1}^{m} [r_i(x)]^2. \tag{1.26}$$

This is the *nonlinear least squares problem* as it can be viewed as an attempt to solve the system of m equations in n unknowns

$$r_i(x) = 0 \qquad i = 1, 2, ..., m. \tag{1.27}$$

When $m > n$ the system is *over-determined* and an exact solution cannot usually be found, but an approximate solution obtained by solving (1.26) is often of interest.

The problem can be solved by any available gradient or second derivative method, and derivatives of $f(x)$ are given by

$$g(x) = 2Ar \qquad (1.28)$$

and

$$G(x) = 2AA^T + 2\sum_{i=1}^{m} r_i \nabla^2 r_i, \qquad (1.29)$$

where r is the column vector with elements $r_i(x)$, and $A = A(x)$ is the *Jacobian matrix* whose columns are the vectors $\nabla r_i(x)$ for $i = 1, 2, ..., m$. Using (1.29) directly in Newton's method is somewhat inconvenient because the Hessian matrices $\nabla^2 r_i$ for all the functions r_i must be evaluated. However it is often the case (on account of (1.27)) that the functions $r_i(x)$ are close to zero at the solution. This suggests that a good approximation to $G(x)$ might be obtained by neglecting the final term in (1.29) to give

$$B^{(k)} = 2A^{(k)}A^{(k)^T}, \qquad (1.30)$$

where $A^{(k)}$ denotes $A(x^{(k)})$. Use of (1.30) in a line search method is referred to as the *Gauss-Newton method*. The important feature is that using only the information (r and A) required to determine the gradient vector, it is often possible to obtain a good approximation to the Hessian matrix. Conversely a quasi-Newton method might take up to n iterations to build up a good approximation. Also (1.30) usually gives a positive definite matrix (unless $A^{(k)}$ is rank deficient).

Although the Gauss-Newton method often works well, there are some cases in which it works badly (e.g. problems with large residuals or in which $A^{(k)}$ becomes rank deficient). Various suggestions have been made for methods which improve the approximation given by (1.30), and some of these are reviewed in [21]. My current preference is for a *hybrid method* in which $B^{(k)}$ is either the Gauss-Newton matrix (1.30), or the result of applying the BFGS updating formula to $B^{(k-1)}$, depending on the outcome of a certain test.

There are many other ideas relating to unconstrained optimization that are interesting and might occur in future methods. These include developments of simplex methods (Torczon [74], Wood [75]), non-monotone line searches (Grippo, Lampariello and Lucidi [35]), and steepest descent steps without line search (Barzilai and Borwein [3]) to name but a few.

2. Quasi-Newton Methods

In this section we examine quasi-Newton methods as described above, and look in particular at the many possible updating formulae that have been suggested. Mostly the context is that of a line search method, but the ideas are also relevant to use in trust region methods. The current preference for the BFGS formula is critically debated.

The quasi-Newton updating formula prescribes $B^{(k+1)}$ by making a change to $B^{(k)}$ so that the secant equation (1.25) is satisfied. The aim is to provide an update which is readily computed, and effective in practice. It is often convenient that the update should preserve positive definiteness in $B^{(k)}$. A particularly simple update is obtained if the correction is taken to be a symmetric rank-one matrix. The secant equation then determines the update uniquely as

$$B^{(k+1)} = B^{(k)} + \frac{(\gamma^{(k)} - B^{(k)}\delta^{(k)})(\gamma^{(k)} - B^{(k)}\delta^{(k)})^T}{(\gamma^{(k)} - B^{(k)}\delta^{(k)})^T \delta^{(k)}}, \qquad (2.1)$$

and this is known as the *Symmetric Rank 1 (SR1)* formula. This update is suggested independently by various authors around 1967-8 (see [21] for references). The good and bad features of the update are discussed below, but it can be noted from (1.6) that the vector $\gamma^{(k)} - B^{(k)}\delta^{(k)}$ is a linear combination of $g^{(k)}$ and $g^{(k+1)}$. It turns out that the update preserves invariance under an affine transformation of the variables x, if the correction to $B^{(k)}$ is composed out of gradient vectors. The significance of this fact is discussed in more detail in [21], Chapter 3.3, but it is an indication that the resulting method is not unduly affected by, for example, changes in the relative scaling of the variables.

This consideration suggests that we might try an update in which the correction is a symmetric rank 2 matrix composed from $g^{(k+1)}$ and $g^{(k)}$, or equivalently (using (1.6)) from $\gamma^{(k)}$ and $B^{(k)}\delta^{(k)}$. This gives an update with three free parameters, and imposing the secant equation (1.25) fixes two of these. Thus a one-parameter family (the *Broyden family*) (Broyden [5]) is obtained

$$B^{(k+1)} = B^{(k)} - \frac{B^{(k)}\delta^{(k)}\delta^{(k)^T}B^{(k)}}{\delta^{(k)^T}B^{(k)}\delta^{(k)}} + \frac{\gamma^{(k)}\gamma^{(k)^T}}{\delta^{(k)^T}\gamma^{(k)}} + \theta^{(k)}w^{(k)}w^{(k)^T} \qquad (2.2)$$

where $\theta^{(k)}$ is the free parameter and

$$w^{(k)} = (\delta^{(k)^T}B^{(k)}\delta^{(k)})^{\frac{1}{2}} \left(\frac{\gamma^{(k)}}{\delta^{(k)^T}\gamma^{(k)}} - \frac{B^{(k)}\delta^{(k)}}{\delta^{(k)^T}B^{(k)}\delta^{(k)}} \right). \qquad (2.3)$$

The *DFP formula* given by $\theta^{(k)} = 1$ (Davidon [14], Fletcher and Powell [25]) was the first such formula to be discovered, and showed the effectiveness of this approach.

However the *BFGS formula*, (Broyden [5], Fletcher [20], Goldfarb [30], Shanno [64]), corresponding to $\theta^{(k)} = 0$ in the Broyden family, is now the usual formula of choice and we look at the reasons for this below. An alternative way of presenting the family is to define $H = B^{-1}$ throughout (i.e. $H^{(k)} = B^{(k)^{-1}}$ etc.) in which case (2.2) becomes

$$H^{(k+1)} = H^{(k)} - \frac{H^{(k)}\gamma^{(k)}\gamma^{(k)^T}H^{(k)}}{\gamma^{(k)^T}H^{(k)}\gamma^{(k)}} + \frac{\delta^{(k)}\delta^{(k)^T}}{\delta^{(k)^T}\gamma^{(k)}} + \phi^{(k)}v^{(k)}v^{(k)^T} \qquad (2.4)$$

where $\phi^{(k)}$ is now the free parameter and

$$v^{(k)} = (\gamma^{(k)^T}H^{(k)}\gamma^{(k)})^{\frac{1}{2}}\left(\frac{\delta^{(k)}}{\delta^{(k)^T}\gamma^{(k)}} - \frac{H^{(k)}\gamma^{(k)}}{\gamma^{(k)^T}H^{(k)}\gamma^{(k)}}\right). \qquad (2.5)$$

It is indicated in [21], p.76 that $\theta^{(k)}$ and $\phi^{(k)}$ are related by

$$\theta^{(k)} = (\phi^{(k)} - 1)/(\phi^{(k)} - 1 - \phi^{(k)}\mu^{(k)}) \qquad (2.6)$$

where $\mu^{(k)} = \gamma^{(k)^T}H^{(k)}\gamma^{(k)}\delta^{(k)^T}B^{(k)}\delta^{(k)}/(\delta^{(k)^T}\gamma^{(k)})^2$ ($\mu^{(k)} \geq 1$ by Cauchy's inequality). Thus in (2.5) $\phi^{(k)} = 1$ corresponds to the BFGS formula and $\phi^{(k)} = 0$ to the DFP formula. Note that the DFP update for H is obtained from the BFGS update for B by interchanging $H \leftrightarrow B$ and $\delta \leftrightarrow \gamma$. Formulae related in this way are said to be *dual*. The correction in (2.5) reduces to being rank 1 when $\phi^{(k)} = \delta^{(k)^T}\gamma^{(k)}/(\delta^{(k)^T}\gamma^{(k)} - \gamma^{(k)^T}H^{(k)}\gamma^{(k)})$ giving the SR1 formula, which is seen to be self-dual from (2.6). It can often be useful to write the BFGS update of $H^{(k)}$ ($\phi^{(k)} = 1$ in (2.4)) in the form

$$H^{(k+1)} = \left(I - \frac{\delta^{(k)}\gamma^{(k)^T}}{\delta^{(k)^T}\gamma^{(k)}}\right)H^{(k)}\left(I - \frac{\gamma^{(k)}\delta^{(k)^T}}{\delta^{(k)^T}\gamma^{(k)}}\right) + \frac{\delta^{(k)}\delta^{(k)^T}}{\delta^{(k)^T}\gamma^{(k)}} \qquad (2.7)$$

and the DFP update of $B^{(k)}$ is the dual of this.

The property of a formula that it maintains positive definite matrices is important, and a necessary condition from (1.25) is that $\delta^{(k)^T}\gamma^{(k)} > 0$. For a line search method, this inequality follows from (1.12). There is a critical value $\bar{\theta}^{(k)} = 1/(1 - \mu^{(k)}) < 0$ such that $B^{(k+1)}$ is singular, and $B^{(k)}$ is positive definite implies $B^{(k+1)}$ is positive definite if $\theta^{(k)} > \bar{\theta}^{(k)}$. Thus both the BFGS and DFP formulae retain positive definite matrices. The same is true for any formulae in the so-called *convex class* $\theta^{(k)} \in [0, 1]$. However the SR1 is never in the convex class and may give rise to an indefinite matrix. It is also possible for SR1 that $B^{(k)}$ might be singular ($\theta^{(k)} = 1/(1 - \mu^{(k)})$) or infinite ($\phi^{(k)} = 1/(1 - \mu^{(k)})$). These aspects of the SR1 formula have deterred many users, but there is currently some resurgence of interest. There are other self-dual formulae, some of which are in the convex class (e.g. $\phi^{(k)} = \delta^{(k)^T}\gamma^{(k)}/(\delta^{(k)^T}\gamma^{(k)} + \gamma^{(k)^T}H^{(k)}\gamma^{(k)})$ and $\phi^{(k)} = 1/(1 + \mu^{(k)\frac{1}{2}})$), the latter following by setting $\theta^{(k)} = \phi^{(k)}$ in (2.6)). Although self-duality seems a desirable symmetry property for positive definite updates, such formulae have attracted little interest and little is known about their properties.

We now turn to the properties of quasi-Newton methods that use updates from the Broyden family. It is assumed that the updates do not break down due to $B^{(k)}$ or $H^{(k)}$ becoming singular. If exact line searches are carried out then the sequence of iterates $\{x^{(k)}\}$ is independent of the choice of $\phi^{(k)}$. If $f(x)$ is quadratic, then the method terminates at the solution in at most n iterations, and $B^{(n)} = G$. Also the search directions are conjugate, and are equivalent to directions generated by the conjugate gradient method if $B^{(1)} = I$. A remarkable result due to Dixon [19] shows that non-dependence on $\phi^{(k)}$ (assuming an exact line search) carries over to non-quadratic functions. Thus Powell's result [55] that DFP converges globally if the line search is exact and $f(x)$ is strictly convex, implies that the same holds for any update from the Broyden family. Practical comparisons on test problems (e.g. [21]) support this theory, showing that if the parameter σ in the strong Wolfe conditions is small ($\sigma \leq 0.1$ say) then there is little to choose between the various update formulae.

It is usually the case however that quasi-Newton methods require fewer function and gradient evaluations overall, when a very weak line search criterion (say $\sigma = 0.9$) is used. In this case the DFP method often performs quite badly, which is not true of the BFGS method. Much effort has been expended in trying to explain this, and in promoting the claims of other updating formulae. Local convergence theory (see for example Dennis and Schnabel [18]) does not hint at any difference between DFP and BFGS. If $x^{(1)}$ is sufficiently close to a local minimizer x^*, and $B^{(1)}$ is sufficiently close to G^*, then the iterates of both the BFGS and DFP methods with steps $\alpha^{(k)} = 1$ converge to x^* superlinearly. In fact it has become clear that the inefficiency of the DFP method is caused by the inability to reduce large eigenvalues in $B^{(k)}$. The main contribution here is due to Powell [59] and is described in the next paragraph. However a similar conclusion is also drawn from global convergence results. Powell [56] proves global convergence of the BFGS method when $f(x)$ is convex and the Wolfe conditions are satisfied in the line search. Byrd, Nocedal and Yuan [9] extend this result for all $\theta^{(k)} \in [0,1)$ but *not* for $\theta^{(k)} = 1$ (the DFP method). The method of proof in the latter case fails because a term which reduces large eigenvalues of $B^{(k)}$ is no longer present. Convergence of the DFP method under these conditions is an open question of some interest *. Another open question which has defied researchers for many years is the global convergence of

---

* Some discussion at the conference suggested that Sargent has proved this using techniques developed by Griewank and Toint, but there does not seem to be a paper available.

the BFGS method for *non-convex* functions. Nonetheless in most implementations of a line search method, the BFGS formula has usually been the update formula of choice, due to its good theoretical properties in the convex case, and its good performance on test problems and in practice. Later in this section we discuss whether this preference is likely to continue in the future, and what possibilities there are for alternative formulae.

Powell [59] isolates a critical difference between the DFP and BFGS methods by considering the worst case behaviour on quadratic functions. Because of invariance under linear transformations, he assumes without loss of generality that $G = I$. He also assumes that steps $\alpha^{(k)} = 1$ are taken, and that the parameter σ in the Wolfe conditions is such that the step is accepted. Because $G = I$ it follows that $\gamma^{(k)} = \delta^{(k)}$ and (1.25) then shows that $B^{(k+1)}$ always has a unit eigenvalue. It is known (Fletcher [20]) that under these circumstances the eigenvalues of $B^{(k)}$ converge monotonically towards 1 for any formula in the convex class. Powell considers the simple case that $n = 2$ and assumes that $B^{(k)}$ has eigenvalues 1 and $\lambda^{(k)}$ for all k. He then analyses the convergence of the DFP and BFGS methods to see how fast $\lambda^{(k)}$ approaches 1. The initial approximation $x^{(1)}$ is chosen so that the worst case behaviour is attained for the formula being studied. Ultimately both formulae give rise to superlinear convergence, but there are significant differences in the number of iterations required to reach a neighbourhood in which superlinear convergence becomes apparent. When $\lambda^{(1)}$ is much smaller than 1 then both methods are very efficient and only a few iterations are required. Conversely when $\lambda^{(1)}$ is much larger than 1 then the DFP method may become very inefficient, and in the worst case Powell shows that

$$\lambda^{(k+1)} \approx \lambda^{(k)} - 1 \tag{2.8}$$

when $\lambda^{(k)}$ is large. Thus for example if $\lambda^{(1)} = 1000$ then about 1000 iterations are required to solve the problem. The behaviour of the resulting method is that a sequence of very short steps are taken and the iterates $x^{(k)}$ approach the solution in a very slowly converging spiral. Similar behaviour has been observed for the DFP method in non-quadratic problems and it is likely that a similar mechanism is at work. In contrast the BFGS method does not exhibit this behaviour and for the $n = 2$ quadratic it requires only about $2.39 \log_{10} \lambda^{(1)}$ iterations in the worst case.

It is nonetheless the case that the BFGS method is better at correcting small eigenvalues than large eigenvalues, which holds out the interesting possibility of using negative values of $\theta^{(k)}$ in a controlled way. Of course it would be necessary to have $\theta^{(k)} > \bar{\theta}^{(k)}$ to ensure that positive definite matrices are retained. Zhang and Tewarson [76] perform numerical tests with fixed values of $\theta^{(k)} < 0$ and report a moderate but consistent improvement over the BFGS method. Experience of Luksan [41] supports this observation, and I also obtained encouraging results with a limited test using the formula

$$\theta^{(k)} = \max(-1, \bar{\theta}^{(k)}/2). \tag{2.9}$$

Zhang and Tewarson are also able to modify Powell's global convergence theorem [56] to allow values of $\theta^{(k)}$ in the range $(1 - \nu)\bar{\theta}^{(k)} \leq \theta^{(k)} \leq 0$ where $\nu \in (0,1)$ is a fixed constant. Byrd, Liu and Nocedal [7] investigate the case $\theta^{(k)} < 0$ in some detail. They observe that if the iterates converge q-superlinearly and $\{H^{(k)}\}$ is bounded, then $\bar{\theta}^{(k)} \rightarrow -\infty$. This appears to give plenty of scope for choosing negative values of $\theta^{(k)}$. However they also show that a necessary condition for superlinear convergence is $\theta^{(k)}/\bar{\theta}^{(k)} \rightarrow 1$. These observations would admit a constant value $\theta^{(k)} = C$, but Byrd et al. also show that superlinear convergence may not occur if $C < -1$ (hence the choice in (2.9) above). Byrd and Nocedal do not suggest any obvious choice of the parameter $\theta^{(k)} < 0$, but give some limited numerical tests with one particular choice, compared against the BFGS update. A noticeable improvement over BFGS is obtained when the initial eigenvalues are large, with no overall loss of performance for small eigenvalues. Unfortunately the value of $\theta^{(k)}$ chosen requires the Hessian matrix $G^{(k)}$ to be available, so it is not a practical option for a gradient method, but it does give a further indication that negative values of $\theta^{(k)}$ are effective. Davidon's choice [15] is also tested (Davidon chooses $\theta^{(k)}$ to minimize the condition number of $H^{(k)}B^{(k+1)}$, which allows the SR1 formula to be selected in some cases where it maintains positive definiteness) but this does not show much improvement on BFGS.

Next the implementation of the BFGS method itself is considered. In early implementations $H^{(k)}$ rather than $B^{(k)}$ is stored, so that (1.6) can be written in the form

$$s^{(k)} = -H^{(k)}g^{(k)} \tag{2.10}$$

and solved in $n^2 + O(n)$ flops. Updating $H^{(k)}$ as in (2.7) takes a further $2n^2 + O(n)$ flops. Later Gill and Murray (e.g. [29]) pioneered the idea of updating Choleski factors

$$B^{(k)} = L^{(k)}L^{(k)T} \tag{2.11}$$

($L^{(k)}$ is lower triangular), which also allows (1.6) to be solved in $n^2 + O(n)$ flops. Although no practical advantage has been reported, even on ill-conditoned problems (Grandinetti [32]), (2.11) does provide an assurance that $B^{(k)}$ is positive definite in the presence of round-off errors. For the BFGS formula ($\theta^{(k)} = 0$ in (2.2)) the update of $L^{(k)}$ can be computed in an attractive way. The BFGS formula can be written as

$$B^{(k+1)} = L^{(k)}\left(I - \frac{L^{(k)T}\delta^{(k)}\delta^{(k)T}L^{(k)}}{\delta^{(k)T}L^{(k)}L^{(k)T}\delta^{(k)}}\right)L^{(k)T} + \frac{\gamma^{(k)}\gamma^{(k)T}}{\delta^{(k)T}\gamma^{(k)}}. \tag{2.12}$$

Note that the vector $L^{(k)T}\delta^{(k)} = -\alpha^{(k)}L^{(k)-1}g^{(k)}$ is a by-product of the computation of $s^{(k)}$, since we assume that $s^{(k)}$ is computed using (1.6) and (2.11). To update $L^{(k)}$, an orthogonal matrix P is determined such that

$$P^T L^{(k)T}\delta^{(k)} = \pm e_1\|L^{(k)T}\delta^{(k)}\| \tag{2.13}$$

(e_1 denotes the first coordinate vector). P^T is calculated by applying a backward sequence of plane rotation operations to $\delta^{(k)^T} L^{(k)}$ in positions $(n, n-1), (n-1, n-2), \ldots, (2, 1)$ in turn. This gives

$$B^{(k+1)} = L^{(k)} P (I - e_1 e_1^T) P^T L^{(k)^T} + \frac{\gamma^{(k)} \gamma^{(k)^T}}{\delta^{(k)^T} \gamma^{(k)}}. \qquad (2.14)$$

Calculating $L^{(k)} P$ by applying the same plane rotations to $L^{(k)}$ gives rise to a lower Hessenberg matrix. Then the $I - e_1 e_1^T$ operation deletes column 1 of $L^{(k)} P$ which can be overwritten by the vector $\gamma^{(k)} / \sqrt{\delta^{(k)^T} \gamma^{(k)}}$. Finally the resulting lower Hessenberg matrix can be restored to lower triangular form by another forward sequence of plane rotations, giving $L^{(k+1)}$. The cost of updating $L^{(k)}$ is $2n^2 + O(n)$ rots (a *rot* is defined as the computation involved in $z = c*x + s*y$ for scalar x and y). If necessary the rotations can be implemented using square-root free Givens rotations, so that the cost of updating $L^{(k)}$ reduces to close on $2n^2$ flops, giving a total of about $3n^2$ flops per iteration, which is the same as for the update of $H^{(k)}$. In fact this figure is not quite optimal and Goldfarb [31] shows how to update $L^{(k)}$ so that the total cost is about $2.5n^2$ flops.

A plausible idea to improve quasi-Newton methods is to attempt to correct $B^{(k)}$ by scaling, rather than through the use of low rank corrections. This for example might alleviate difficulties caused by large eigenvalues in $B^{(k)}$. An idea which combines both scaling and low rank corrections (Oren and Luenberger [51]) is simply to replace $B^{(k)}$ by the scaled matrix $\tau^{(k)} B^{(k)}$ in the update formula (2.2). It can be shown that quadratic termination and conjugate gradient properties with exact line searches remain valid. Various choices for $\tau^{(k)}$ have been proposed (see also Oren and Spedicato [52], Oren [50]), but the use of

$$\tau^{(k)} = \frac{\gamma^{(k)^T} \delta^{(k)}}{\delta^{(k)^T} B^{(k)} \delta^{(k)}} = \frac{\delta^{(k)^T} \bar{G}^{(k)} \delta^{(k)}}{\delta^{(k)^T} B^{(k)} \delta^{(k)}} \qquad (2.15)$$

is attractive in that the curvature of $\tau^{(k)} B^{(k)}$ along $\delta^{(k)}$ is equated to that of the average Hessian $\bar{G}^{(k)}$ in (1.24). Unfortunately practical experience (e.g. Shanno and Phua [65]) has often been disappointing. One explanation (Nocedal and Yuan [48]) is that the unit step $\alpha^{(k)} = 1$ no longer ensures that superlinear convergence takes place. Another pointer in the quadratic function/exact line search case is that the conditions (1.25) from previous iterations

$$B^{(k+1)} \delta^{(j)} = \gamma^{(j)} \qquad j = 1, 2, ..., k-1 \qquad (2.16)$$

are not preserved by the scaled updates (in contrast to say BFGS). Consequently the property $B^{(n+1)} = G$ no longer holds, which suggests that the method may be less effective at approximating the Hessian for a non-quadratic problem. Thus scaling is commonly used only to scale $B^{(1)}$ after the initial line search, which can be quite effective. However there is still much interest in scaling, see for example Luksan [41].

The above discussion suggests that it would be better if scaling could be done whilst preserving (2.16), as well as (1.25). An implementation of BFGS which allows this is given by Powell [60] who works with the factors

$$H^{(k)} = Z^{(k)} Z^{(k)T} \tag{2.17}$$

where $Z^{(k)}$ is nonsingular. The idea is readily explained as an instance of (2.12) to (2.14), but without the final sequence of forward plane rotations that keeps $L^{(k+1)}$ lower triangular. Thus $L^{(k)}$ is in general not triangular and we denote

$$Z^{(k)} = L^{(k)-T}. \tag{2.18}$$

The same rotations which define the matrix P are calculated, again using the fact that $L^{(k)T} \delta^{(k)}$ is proportional to $Z^{(k)T} g^{(k)}$ which is available from the computation of $s^{(k)}$. In the update of $L^{(k)}$ it is possible to write

$$L^{(k+1)} = \left[\frac{\gamma^{(k)}}{\sqrt{\delta^{(k)T} \gamma^{(k)}}} \,\middle|\, L^{(k)} P J \right] \tag{2.19}$$

where J is the matrix obtained by deleting column 1 of I, so that $L^{(k)} P J$ expresses the operation of deleting column 1 of $L^{(k)} P$. It is readily verified that the inverse transpose of (2.19) is given by

$$Z^{(k+1)} = \left[\frac{\delta^{(k)}}{\sqrt{\delta^{(k)T} \gamma^{(k)}}} \,\middle|\, \left(I - \frac{\delta^{(k)} \gamma^{(k)T}}{\delta^{(k)T} \gamma^{(k)}} \right) Z^{(k)} P J \right]. \tag{2.20}$$

Thus the corresponding update of $Z^{(k)}$ is obtained by applying the same rotations to $Z^{(k)}$, deleting column 1, multiplying by $I - \delta^{(k)} \gamma^{(k)T} / \delta^{(k)T} \gamma^{(k)}$ (a rank one change), and finally introducing $\delta^{(k)} / \sqrt{\delta^{(k)T} \gamma^{(k)}}$ as a new first column.

It is readily observed from (2.20) that at the start of iteration k, the contents of columns $k, k+1, ..., n$ of $Z^{(k)}$ are an orthogonal linear combination of the columns of $Z^{(1)}$, which is arbitrary, whilst 'new' information has only filtered into columns 1 to $k-1$ of $Z^{(k)}$. In fact it can be shown in the quadratic function/exact line search case that columns 1 to $k-1$ of $Z^{(k)}$ are the vectors $\delta^{(j)} / \sqrt{\delta^{(j)T} \gamma^{(j)}}$ for $j = k-1, ..., 2, 1$ (note the reverse order). This interpretation makes it clear that previous secant conditions (2.16) are preserved if scaling is applied only to columns $k, k+1, ..., n$ of $Z^{(k)}$ before applying the update (2.20). Siegel [67] describes a strategy of this type and reports results on a single test problem with increasingly ill-conditioned choices of $Z^{(1)}$, which show an improved performance over BFGS. Lalee and Nocedal [38] propose an alternative strategy, and give conditions on the scaling parameters for the algorithm to exhibit global and superlinear convergence. Such column scaling strategies seem to provide an appropriate way of implementing scaling in BFGS, albeit at the cost of doubling the storage requirement of the algorithm ($Z^{(k)}$ is square whilst $L^{(k)}$ in (2.11) is triangular). The operation count is also higher if the algorithm is implemented as indicated by (2.20). However the operation count can be improved using the idea of Siegel [65,67]. Essentially the idea is to use rotations so that $g^{(k+1)T} Z^{(k)} P$ is zero except in columns 1 and 2.

Finally in this section we examine the claims of other formulae. Recent interest has been shown in the SR1 formula (2.1) although researchers are ambivalent about its merits. It has already been observed that the update is not in the convex class, does not preserve positive definite matrices, and may break down altogether. On the other hand it has some redeeming features. It shares the invariance properties of the Broyden class, but does not need the condition $\delta^{(k)T} \gamma^{(k)} > 0$ to hold. This makes it suitable for situations such as trust region algorithms, partially separable optimization (see Section 4), and nonlinear programming where this condition may not be readily obtained. It also has a unique quadratic termination property, not dependent on exact line searches. If the difference vectors $\delta^{(k)}$ $k = 1, 2, ..., n$ are linearly independent then $B^{(n+1)} = G$ is obtained and the method terminates on the next Newton step. Conn, Gould and Toint [10] report good practical experience with a trust region algorithm. They also give a theorem [11] which shows that $B^{(k)} \rightarrow G^*$ if the denominator in (2.1) is not too close to zero, and the difference vectors are uniformly linearly independent. However Khalfan, Byrd and Schnabel [37] question whether the latter assumption is realistic, and provide alternative local convergence results.

Another formula that can be used when $\delta^{(k)T} \gamma^{(k)} \leq 0$ is the PSB formula (see (4.2) below), the motivation for which is discussed in more detail in Section 4. Like the SR1 formula, positive definite matrices are not preserved by the PSB formula. Also a termination result does not hold in the quadratic case, although it can be shown that the error in the Hessian approximation is monotonically decreasing. Moreover, and in contrast to SR1, the update does not fail in any way. Unfortunately practical experience has been disappointing and this is usually ascribed to a lack of invariance to linear transformations such as holds for the Broyden family. There is no doubt that researchers would welcome the development of a formula which works as well as BFGS when $\bar{G}^{(k)}$ in (1.24) is positive definite, but allows indefinite matrices to arise when $\bar{G}^{(k)}$ is indefinite, and is numerically stable like the PSB update.

3. Conjugate Gradients and Extensions

The class of conjugate gradient methods (1.20) is important in that it may provide the only practicable approach for unconstrained optimization problems in which n is very large and only a few vectors can be stored. In this section different versions of the method are described, along with recent developments in convergence theory. Various ideas for extending the method are described when rather more storage is available, but not enough to implement a quasi-Newton method. Another important application of conjugate gradients is in approximately solving the Newton equations ((1.3) or (1.8)) in a line search or trust region method, and this is also considered. A summary of current recommendations is given.

The conjugate gradient (CG) method is a line search method in which directions are given by

$$
\begin{aligned}
s^{(1)} &= -g^{(1)} \\
s^{(k)} &= -g^{(k)} + \beta^{(k)} s^{(k-1)} \qquad \forall \, k > 1.
\end{aligned}
\tag{3.1}
$$

This formula derives from the CG method for solving a linear system $Ax = b$ in which A is positive definite (Hestenes and Stiefel [36]). This is equivalent to minimizing the quadratic function $\frac{1}{2} x^T A x + b^T x + c$. A method for non-quadratic optimization is then obtained by using the same parameter choice for $\beta^{(k)}$, and choosing $\alpha^{(k)}$ by a line search. This development gives rise to the *Fletcher-Reeves method* [26] in which $\beta^{(k)}$ is defined by

$$
\beta_{FR}^{(k)} = \frac{g^{(k)^T} g^{(k)}}{g^{(k-1)^T} g^{(k-1)}}.
\tag{3.2}
$$

The correspondence with linear conjugate gradients ensures that the method has the property of quadratic termination if exact line searches are carried out. In this case it can also be proved (whilst $g^{(k)} \neq 0$) that

$$
\begin{aligned}
&\{g^{(k)}\} \text{ is an orthogonal set} \\
&\{s^{(k)}\} \text{ is a } G\text{--conjugate set.}
\end{aligned}
\tag{3.3}
$$

The use of CG methods for minimizing non-quadratic functions gives rise to a number of issues which are not easily resolved. One is the type of line search to use, and how accurate it should be. Experimental studies indicate that a fairly accurate line search works best (say $\sigma = 0.1$ in the the strong Wolfe conditions). Another question is whether there may be an advantage in periodically resetting $s^{(k)}$ to the steepest descent direction. The termination result perhaps suggests that this should be done every n iterations. However the methods are unlikely to be used for small n since the BFGS method is usually much more efficient and robust. For large n, resetting every n iterations is largely irrelevant since that number of iterations is unlikely to be exceeded.

The most interesting question is that of which of many alternative formulae for $\beta^{(k)}$ to use. It follows directly from (3.3) that, for quadratic functions, (3.2) is equivalent to

$$\beta_{PR}^{(k)} = \frac{g^{(k)^T}(g^{(k)} - g^{(k-1)})}{g^{(k-1)^T}g^{(k-1)}}, \qquad (3.4)$$

due to Polak and Ribière [53]. Various other equivalent formulae can also be generated from the properties of quadratic functions (see [21]). We only mention one of these

$$\beta^{(k)} = \frac{g^{(k)^T}(g^{(k)} - g^{(k-1)})}{s^{(k-1)^T}\gamma^{(k-1)}}, \qquad (3.5)$$

corresponding to a search direction (on iteration $k + 1$)

$$s^{(k+1)} = -\left(I - \frac{\delta^{(k)}\gamma^{(k)^T}}{\delta^{(k)^T}\gamma^{(k)}}\right)g^{(k+1)}. \qquad (3.6)$$

We draw some parallels using this formula later in the section.

In practice the Fletcher-Reeves method can be implemented with three n-vectors of storage, whilst the Polak-Ribière method requires four. However experimental studies favour the Polak-Ribière method. We now look at what is known theoretically about these two methods. An important result due to Al-Baali [1] is that if the Fletcher-Reeves method is implemented in conjunction with the strong Wolfe conditions, with $\sigma < \frac{1}{2}$, then the search directions satisfy a sufficient slope property of the form

$$-g^{(k)^T}s^{(k)} \geq c\|g^{(k)}\|^2 \qquad (3.7)$$

where $c > 0$ is a constant independent of k. This result is established by an elegant inductive argument. Al-Baali uses (3.7) to deduce (1.15) and hence proves global convergence of the Fletcher-Reeves algorithm. This result can be extended to any method for which

$$|\beta^{(k)}| \leq \beta_{FR}^{(k)} \qquad (3.8)$$

(Gilbert and Nocedal [28]) and this result is tight in a certain sense.

Practical experience with the Fletcher-Reeves method shows a propensity to take short steps remote from the solution, leading to slow convergence and even stalling (despite the global convergence proof). Powell [57] offers a possible explanation for this, and suggests why the Polak-Ribière method may not suffer from this disadvantage. In this case, if $g^{(k)} \approx g^{(k-1)}$, then $\beta_{PR}^{(k)} \approx 0$ and the Polak-Ribière method tends to reset $s^{(k)} \approx -g^{(k)}$ to the steepest descent direction. Powell also shows that the Fletcher-Reeves direction can be much worse than the steepest descent direction, thus indicating a preference for the Polak-Ribière method. This argument would lead us to expect a global convergence result for the Polak-Ribière method. Powell also shows that this can be proved if the assumption is made that the steps $\|d^{(k)}\|$ converge to zero. However this assumption cannot be removed (when an exact line search is used) as Powell [58] shows in a later paper. He constructs a remarkable counter-example with $n = 3$ in which the sequence $\{x^{(k)}\}$ is bounded and has six accumulation points, none of which is a stationary point.

As yet however I do not think we have a really convincing explanation of why the Polak-Ribière method is more successful than Fletcher-Reeves. Since we know that the steepest descent method is often very inefficient, it is not clear that the ability to reset $s^{(k)} = -g^{(k)}$ will necessarily result in an effective method. My feeling is that the Polak-Ribère formula succeeds because it is closely related to (3.5), and (3.5) is a formula which forces conjugacy to the previous direction for non-quadratic functions (in the sense that $s^{(k)^T} \bar{G}^{(k)} s^{(k+1)} = \gamma^{(k)^T} s^{(k+1)} / \alpha^{(k)} = 0$ from (3.6)). This local conjugacy property seems to be an important factor that has largely been ignored in discussing the effectiveness of CG methods.

The counter-example of Powell [58] requires that $\beta^{(k)} < 0$ occurs infinitely, leading Powell to suggest a modification

$$\beta^{(k)} = \max(\beta_{PR}^{(k)}, 0) \tag{3.9}$$

of the Polak-Ribière formula. Gilbert and Nocedal [28] show that this leads to a globally convergent method if (a) strong Wolfe conditions are used, and (b) the strong descent condition (3.7) holds. Gilbert and Nocedal show that, because $\beta^{(k)} \geq 0$, it is always possible to design a line search to satisfy both (a) and (b).

A feature of a number of papers has been to restart the CG sequence when a significant deviation from the quadratic model is detected. Powell [57] uses the test

$$|g^{(k)^T} g^{(k-1)}| \geq 0.2 \|g^{(k)}\|^2 \tag{3.10}$$

as an indication to restart. He also uses the three term recurrence of Beale [4] in order to avoid restarting with a steepest descent search. Shanno and Phua [65] and Buckley and LeNir [6] also suggest algorithms based on the use of restarts. Algorithms such as these have proved more effective than the simple Polak-Ribière method, albeit at the cost of requiring more vectors of storage. However there are other ways in which extra storage can usefully be used, which have given rise to even more effective algorithms. There is also a worry with restart algorithms that restarts may be triggered too often, therefore degrading the overall efficiency of the method.

We now consider how fast is the local convergence of CG methods. For exact line searches Crowder and Wolfe [13] show that the order of convergence is linear and cannot be Q-superlinear. This is not encouraging, and there are other results that also do not give cause for hope. Yet the Polak-Ribière method can be spectacularly successful for certain large problems. The following argument in the quadratic case provides some explanation. It is possible to deduce from (3.1) that

$$g^{(k+1)} = P_k(G)g^{(1)} \tag{3.11}$$

where P_k is a polynomial of degree k with $P_k(0) = 1$. If G has eigenvalues λ_j and orthonormal eigenvectors ξ_j, then $g^{(1)}$ can be expanded as

$$g^{(1)} = \sum_{j=1}^{n} \rho_j \xi_j \tag{3.12}$$

where $\rho_j = \xi_j^T g^{(1)}$. It follows from (3.11) that

$$g^{(k+1)} = \sum_{j=1}^{n} \rho_j P_k(\lambda_j) \xi_j . \tag{3.13}$$

A term in (3.13) can be small (i.e. close to zero) either if ρ_j is small or if $P_k(\lambda_j)$ is small. In regard to the latter case, if the number of distinct eigenvalues of G is k then it is a property of the polynomial P_k that $P_k(\lambda_j) = 0$ for $j = 1, 2, ..., n$. Thus $g^{(k+1)}$ can be close to zero if the problem has certain symmetries such that either many of the ρ_j are near-zero, or there are many multiple or near-multiple eigenvalues in G so that $P_k(\lambda_j) \approx 0$ for $k \ll n$. Under these circumstances the CG method will approximately solve the quadratic problem in at most $k \ll n$ iterations.

A method that is related to (3.1) is the *preconditioned conjugate gradient (PCG) method* defined by

$$\begin{aligned} s^{(1)} &= -Hg^{(1)} \\ s^{(k)} &= -Hg^{(k)} + \beta^{(k)} s^{(k-1)} \qquad \forall\, k > 1 \end{aligned} \tag{3.14}$$

where $\beta^{(k)}$ is given either by

$$\beta^{(k)} = \frac{g^{(k)^T} H g^{(k)}}{g^{(k-1)T} H g^{(k-1)}} \tag{3.15}$$

(c.f. (3.2)) or by a similar formula derived from (3.4) for example. Here H is a *fixed* symmetric positive definite matrix, and the PCG method is equivalent (see [21]) to the CG method applied in a transformed coordinate system ($H = I$ gives the CG method). For quadratic functions and exact line searches the BFGS method with $H^{(1)} = H$ generates the same sequences $\{x^{(k)}\}$ and $\{s^{(k)}\}$ as the PCG method. If $H = G^{-1}$ then the PCG method terminates in one iteration. Thus the rationale of the PCG method is to select a readily available matrix H which is a good approximation to G^{-1} in the sense that $H^{\frac{1}{2}} G H^{\frac{1}{2}}$ has many near-multiple eigenvalues. Then by virtue of the argument of the previous paragraph applied to the transformed problem, there is the possibility of obtaining rapid convergence.

We are now in a position to consider the case in which substantially more storage is available than the $4n$ locations needed for the Polak-Ribière method, but less than the $\frac{1}{2}n^2 + O(n)$ locations for BFGS. Methods applicable to this case are referred to as *limited memory methods*. Various such methods have been suggested but there are two early methods worthy of note. In the method of Nocedal [47], a parameter m ($m \ll n$) is selected depending on the storage available $((2m+4)n+O(m)$ locations are required). The most recent m difference pairs $\delta^{(i)}$, $\gamma^{(i)}$, $i = k-m, ..., k-2, k-1$ are stored, and a matrix $H^{(k)}$ is defined implicitly by applying the BFGS formula to a given diagonal matrix D for each difference pair in turn. Nocedal gives a recurrence relation derived from (2.7) that enables $s^{(k)}$ in (2.10) to be computed in $4nm + O(n)$ flops without calculating $H^{(k)}$. Usually D is chosen as a multiple of the unit matrix, using for example (2.15). After each iteration (once $k > m$) the oldest difference pair is deleted from storage and the new difference pair is introduced. In practice a suitable choice for m has been found to be $m = 5$ and it is interesting and somewhat surprising that larger values of m do not seem to give significant increases in performance, even on very large problems. Another interesting observation is that in the case $m = 1$, if the line search is exact, then it follows from (2.7) that the direction $s^{(k+1)}$ for Nocedal's method is the same as that for the CG method (3.6), so that these methods are equivalent.

Another limited memory method is that of Buckley and LeNir [6]. This is a two-phase method, starting with the *QN-phase* in which for $k = 1, 2, ..., m$, $H^{(k)}$ is built up as in Nocedal's method, and the available storage is filled. The *CG-phase* follows, in which for $k = m+1, m+2, ...$, $H^{(m)}$ is used as the fixed matrix H in the PCG method. The CG phase is terminated using the test (3.10). All the difference pairs except for $\delta^{(k-1)}$, $\gamma^{(k-1)}$ are then deleted and the QN-phase is re-entered. Both methods are effective, but Liu and Nocedal [40] show a preference for Nocedal's method, based on a wide selection of large dimension problems. This may be an indication that restarting the Buckley and LeNir method by deleting most of the current difference pairs is a source of inefficiency.

For this reason there is considerable interest in limited memory methods which specify $H^{(k)}$ on the basis of the most recent m difference pairs, which we may express as columns of the matrices

$$\Delta^{(k)} = \left[\delta^{(k-m)}, ..., \delta^{(k-2)}, \delta^{(k-1)} \right] \tag{3.16}$$

and

$$\Gamma^{(k)} = \left[\gamma^{(k-m)}, ..., \gamma^{(k-2)}, \gamma^{(k-1)} \right]. \tag{3.17}$$

Nocedal's recurrence may be regarded as providing a useful approximate solution to the problem of finding a symmetric positive definite matrix $H^{(k)}$ which satisfies the *multiple secant equation*

$$H^{(k)}\Gamma^{(k)} = \Delta^{(k)}. \tag{3.18}$$

This problem also arises when finding the compliance matrix of an elastic structure, given measurements of loading and displacement. The matrix $M^{(k)} = \Gamma^{(k)^T}\Delta^{(k)}$

plays an important role. First we consider the case that $M^{(k)}$ is symmetric and positive definite, which arises either if $m = 1$ or if the differences are sampled from a quadratic function and rank($\Delta^{(k)}) = m$. In this case exact solutions to (3.18) exist, and we can write down analogues of all the common update formulae in an obvious way. For example the *Multiple BFGS formula* (applied to a matrix D) can be defined by analogy with (2.7) as

$$H^{(k)} = \left(I - \Delta^{(k)}M^{(k)-1}\Gamma^{(k)T}\right) D \left(I - \Gamma^{(k)}M^{(k)-1}\Delta^{(k)T}\right) + \Delta^{(k)}M^{(k)-1}\Delta^{(k)T}.$$

(3.19)

In general this gives a different matrix $H^{(k)}$ to that computed by Nocedal's recurrence, although the amount of computation is the same. However it is interesting to observe that if the columns of $\Delta^{(k)}$ are G-conjugate, then the same matrix $H^{(k)}$ is obtained. Unfortunately, for nonquadratic functions with $m > 1$, $M^{(k)}$ is almost always nonsymmetric. Schnabel [62] considers some ways in which $\Delta^{(k)}$ and $\Gamma^{(k)}$ might be modified so that $M^{(k)}$ is symmetric and positive definite, but does not strongly support any particular scheme. Allwright [1] considers finding a symmetric positive semi-definite matrix $H^{(k)}$ that best solves (3.18). However this matrix might be singular so there is also doubt about its merit. Given that Nocedal's recursion implicitly defines a symmetric positive definite $H^{(k)}$ without any restriction other than $\delta^{(k)T}\gamma^{(k)} > 0$, it is not likely to be improved upon significantly.

Some recent work has looked at the possibility of limited memory algorithms which store only one $n \times m$ matrix and not two as in (3.16) and (3.17). Fletcher [22] describes a method in which $B^{(k)}$ is represented as

$$B^{(k)} = \left[S \big| S^\perp\right] \begin{bmatrix} S^T \\ S^{\perp T} \end{bmatrix}$$

(3.20)

(omitting superscript k), where $S \in \mathbb{R}^{n \times k}$ and columns of $S^\perp$ form an orthonormal basis for the null space of S. The BFGS method (with $B^{(1)} = I$) can then be expressed as follows. Initially $S^{(1)} = g^{(1)}/\|g^{(1)}\|$ and iteration k takes the form

(i) calculate search direction $s := (SS^T)^+ g$
(ii) line search to give $x^{(k+1)}$ and $g^{(k+1)}$
(iii) calculate $g^\perp := S^\perp S^{\perp T} g^{(k+1)}$
(iv) if $g^\perp \neq 0$ extend $[S|g^\perp/\|g^\perp\|] =: \hat{S}$ else $\hat{S} := S$
(v) update $\hat{S}$ to give $S^{(k+1)}$ as for the BFGS method.

(3.21)

The algorithm is implemented by storing QL factors of S so that s and $g^\perp$ are readily computed (using $(SS^T)^+ = QL^{-T}L^{-1}Q^T$ and $S^\perp S^{\perp T} = I - QQ^T$). Siegel [68] follows a similar approach and points out that algorithm (3.21) can be interpreted as one which updates a reduced Hessian $\hat{B} = Q^T B^{(k)} Q$. An attractive way to implement this is to update Choleski factors $\hat{B} = LL^T$ of the reduced Hessian matrix using reduced difference vectors, in much the same way as for (2.14).

This indicates how to update S in step (v) of (3.21). Siegel injects another good idea, which is that Q should not be stored directly: rather Δ (c.f. (3.16)) should be stored and Q determined implicitly from QR factors of Δ. This enables the dominant computational cost to be reduced to $2nk$ flops per iteration, where k refers to the number of columns in Δ. (Using k also to denote the iteration number assumes that the *else* option in step (iv) is not activated.)

Algorithm (3.21) is exactly equivalent to BFGS and can be continued until the available nm storage locations (storing either S or Q or Δ) are filled. At this stage it is necessary to deviate from the BFGS algorithm. Fletcher proceeds in the same fashion as Buckley and LeNir, using the matrix $H = B^{(m)-1}$ as a fixed preconditioner in PCG. Once test (3.10) fails then all the stored information is discarded and the algorithm is restarted. Siegel proceeds differently when the storage is filled and throws away only the oldest column of Δ. However it can be shown that both these algorithms retain the quadratic termination property. I have a student following up these ideas (Lefebure [39]) and hope to report in more detail at a later date.

Finally mention should be made of the *Truncated Newton* or *Truncated conjugate gradient* method. This type of method has an outer iteration in which the Newton equation (1.3) is solved approximately by applying an inner linear conjugate gradient iteration (Toint [73], Steihaug [70]). This cuts down the computational effort for solving (1.3), whilst rapid convergence is still possible if the accuracy requirement is made more severe as the outer iteration count increases (Dembo, Eisenstat and Steihaug [16]). Both trust region and line search approaches are possible. To make the methods efficient, a PCG inner iteration should be used, e.g. Nash [45], although the 'best' preconditioner is likely to be problem dependent. For the method to be applicable as a gradient method, vectors of the form $G^{(k)}p$ required in the inner PCG iteration must be estimated, and this may be done using the finite difference quotient

$$G^{(k)}p \approx \frac{g(x^{(k)} + hp) - g^{(k)}}{h} \tag{3.22}$$

where h is a small stepsize (O'Leary [49]). This does require an additional gradient evaluation on every inner PCG iteration. As more inner iterations are usually required as the outer iteration counter k increases, so the cost of each outer iteration also increases. A comparison of Truncated Newton and Nocedal's limited memory method is given by Nash and Nocedal [46]. Both methods are effective and neither is the outright winner. The tentative conclusion is that Truncated Newton is better on nearly quadratic problems, whilst Nocedal's method is better on highly nonlinear problems.

4. Structured Large Scale Optimization

The methods described in the previous section are most suitable for *unstructured* large scale optimization. Frequently however a large scale optimization problem has some structure that can be exploited, such as might be reflected in the Hessian matrix being sparse. In this case, and when second derivatives are available, there is little doubt that Newton's method (suitably modified in a line search or trust region method as described in Section 1) is a very attractive method which takes advantage of structure. Providing the organizational problems can be handled, this must be the method of choice. The increasing availability of automatic differentiation packages reinforces this conclusion. The only negative factor remains the need to account for an indefinite Hessian. However as we shall see below, the same is true for most other alternatives.

The rest of this section concerns the case that a gradient method is required. First of all it should be realized that use of a finite difference approximation to the Hessian becomes much more efficient in the sparse case. It is possible by a careful choice of directions $d^{(i)}$, $i = 1, 2, ..., p$ to use the vectors

$$y^{(i)} = (g(x^{(k)} + hd^{(i)}) - g^{(k)})/h \approx G^{(k)}d^{(i)} \tag{4.1}$$

to estimate the non-sparse elements of $G^{(k)}$. For example if $G^{(k)}$ is a band matrix with p bands on or below the diagonal ($p = 2$ for a tridiagonal matrix) then only the p directions

$$d^{(i)} = (e_i + e_{i+p} + e_{i+2p} + ...) \qquad\qquad i = 1, 2, ..., p \tag{4.2}$$

are required. (Here e_i denotes the unit vector that is column i of I.) Powell and Toint [61] give a algorithm that extends this idea to general sparsity patterns. Moreover advantage can be taken of known fixed but non-zero elements in $G^{(k)}$ to further reduce the amount of differencing required.

We now go on to consider the possibilities for quasi-Newton methods that respect sparsity. First we digress to point out that many updating formulae possess a *variational property* from which they can be derived. A significant result due to Goldfarb [30] (based on the original idea due to Greenstadt [33]) is that the correction $E = H^{(k+1)} - H^{(k)}$ in the BFGS formula satisfies a minimum property with respect to a weighted Frobenius norm of the form

$$\|E\|_W^2 = (\text{trace}(EWEW))^{1/2} \tag{4.3}$$

where $W > 0$ and $W\delta^{(k)} = \gamma^{(k)}$. This result can be interpreted as showing that $H^{(k)}$ is changed by the minimum amount (in the sense of (4.3)) required to satisfy (1.25) and symmetry. This ensures that previous information accumulated in $B^{(k)}$ is disturbed as little as possible. The well-known DFP formula can also be interpreted in a similar way. Another formula that satisfies a minimum correction property is the PSB formula (Powell [54])

$$B^{(k+1)} = B^{(k)} + \frac{\eta\delta^T + \delta\eta^T}{\delta^T\delta} - \frac{\eta^T\delta}{\delta^T\delta^2}\delta\delta^T \tag{4.4}$$

where $\eta = \gamma - B^{(k)}\delta$ and where δ and γ denote $\delta^{(k)}$ and $\gamma^{(k)}$ respectively. In this case it is the Frobenius norm ($W = I$ in (4.3)) of the correction to $B^{(k)}$ that is minimized (subject to (1.25) and symmetry). Unfortunately the PSB update does not in general preserve $B^{(k)} > 0$ and practical experience has been disappointing as discussed in Section 2.

Quasi-Newton methods become less attractive when n is very large because of the storage and computational requirements associated with large dense matrices. However if the Hessian is a sparse matrix it is attractive to look for updating formulae which preserve the same sparsity in $B^{(k)}$. Thus we express the sparsity conditions on B as

$$B_{ij} = 0 \quad \forall\, (i,j) \in \mathcal{S} \tag{4.5}$$

where $\mathcal{S}$ is a set of pairs of integers in the range $[1:n]$. Because of symmetry it is assumed that $(i,j) \in \mathcal{S}$ if and only if $(j,i) \in \mathcal{S}$. It is also assumed that $G(x)$ satisfies (4.5) for all $x \in \mathbb{R}^n$. The complementary set of index pairs not in $\mathcal{S}$ is denoted by $\mathcal{S}^\perp$. Because we are concerned with positive definite matrices it is assumed that

$$(i,i) \in \mathcal{S}^\perp \quad i = 1, 2, ..., n. \tag{4.6}$$

For such problems it is fruitful to determine a minimum correction update formula which is constrained by (4.5) in addition to the quasi-Newton condition and symmetry. In a seminal paper Toint [71] shows that it is reasonably straightforward to compute a minimum correction to $B^{(k)}$ in the Frobenius norm subject to these conditions. To present Toint's update we define the projection operator $\mathcal{G}(M) : \mathbb{R}^{n \times n} \to \mathbb{R}^{n \times n}$ by

$$\mathcal{G}(M)_{ij} = \begin{cases} 0 & (i,j) \in \mathcal{S} \\ M_{ij} & (i,j) \in \mathcal{S}^\perp \end{cases} \tag{4.7}$$

This has been colourfully dubbed the *gangster operator* since it shoots holes in M according to the sparsity pattern defined by (4.5). It is convenient to introduce the notation

$$\delta_{[i]} = \mathcal{G}(\delta e_i^T)e_i \tag{4.8}$$

in which the sparsity pattern of column i of the Hessian is imposed on δ. Toint shows that the resulting minimum correction satisfies

$$B^{(k+1)} = B^{(k)} + \mathcal{G}(\delta\lambda^T + \lambda\delta^T) \tag{4.9}$$

where $\lambda \in \mathbb{R}^n$ is obtained by solving the linear system

$$Q\lambda = r \tag{4.10}$$

in which $r = \gamma - B^{(k)}\delta$. It follows from (4.9) and (1.25) that column i of the matrix Q is defined by

$$Qe_i = \delta_i \delta_{[i]} + \delta_{[i]}^T \delta_{[i]} e_i. \tag{4.11}$$

It is easily shown that Q is symmetric positive semi-definite and that $\mathcal{G}(Q) = Q$ (i.e. Q satisfies the sparsity conditions (4.5)). In addition Q is positive definite if

$$\delta_{[i]} \neq 0 \quad i = 1, 2, ..., n \tag{4.12}$$

in which case sparse LDL^T factors of Q are calculated and (4.10) can readily be solved to obtain λ. If Q is singular then $\delta_{[i]} = 0$ for some i. However it then follows from (1.23) that $\gamma_i = 0$ and hence $r_i = 0$. Thus (4.10) is consistent and can be solved by deleting row and column i from Q (ignoring the effects of round-off error).

As with the dense PSB update, the condition $B^{(k)} > 0$ is not preserved by this update, and likewise practical performance has not been outstanding. In view of this it is natural to enquire what happens when the sparsity conditions (4.5) are included in the minimum correction property that defines the BFGS or DFP formula. Unfortunately as Toint [71] points out the use of a weighted Frobenius norm leads to formulae that are intractable in both cases (see also [72] for more details). However Toint [72] proves that if

$$\delta_i \neq 0 \quad i = 1, 2, ..., n, \tag{4.13}$$

if G is irreducible, and if $\delta^T \gamma > 0$ then there does exist a symmetric update which preserves positive definiteness. The condition $\delta^T \gamma > 0$ is clearly required, else (2.3) would imply $\delta^T B \delta \leq 0$, contradicting $B > 0$. The assumption of irreducibility (that is G cannot be reduced to block diagonal form by a symmetric permutation) is not a serious restriction because if G is reducible then (1.1) can be decomposed into two or more problems which can be solved separately. It is assumed throughout what follows that G is irreducible. On the other hand, (4.13) is critical and Sorensen [69] shows that a positive definite update may not exist if $\delta_i = 0$ for some i, and that serious growth in B can occur in a neighbourhood of this situation. We return to these points below.

More recently an alternative approach to an optimal BFGS-like formula that preserves sparsity and positive definiteness has been described by Fletcher [24]. This arises from an observation of Fletcher [23] that the BFGS and DFP formulae can be derived by a variational argument using the measure function

$$\psi(A) = \text{trace}(A) - \ln \det(A) \tag{4.14}$$

that is introduced by Byrd and Nocedal [8] in the convergence analysis of quasi-Newton methods. This function is strictly convex on the set of positive definite matrices and is minimized by $A = I$. The function becomes unbounded as A becomes singular or infinite and so acts as a barrier function that keeps A positive definite. A suitable variational property is to minimize $\psi(H^{(k)}B)$ since in the absence of any constraints the solution is just $B = H^{(k)-1}$. Introducing the constraints of symmetry and (1.25) leads to an update in which $H = B^{-1}$ stays close to $H^{(k)}$ in this sense. Fletcher shows that it is the BFGS formula that solves this variational problem.

A theorem extending this result to include the sparsity conditions (4.5) is set out in [24]. It is shown that the solution is characterized by the existence of $\lambda \in \mathbb{R}^n$ such that

$$\mathcal{G}(H) = \mathcal{G}(H^{(k)} + \lambda \delta^T + \delta \lambda^T). \tag{4.15}$$

In the dense case it is straightforward to find λ such that H satisfies $H\gamma = \delta$ (that is (1.25)). In the sparse case no direct solution for λ appears to be possible but Fletcher shows how an iterative solution can be determined. Given a vector λ, the sequence of operations

- Calculate $\mathcal{G}(H)$ from (4.15)
- Find $B > 0$ such that $\mathcal{G}(B^{-1}) = \mathcal{G}(H)$. $\hspace{1cm}$ (4.16)
- Calculate $r := B\delta - \gamma$.

defines r as a function of λ ($\mathbb{R}^n \to \mathbb{R}^n$) and the update is determined by finding λ such that

$$r(\lambda) = 0, \tag{4.17}$$

which is a nonlinear system of n equations in n variables. Moreover it can be shown that $r(\lambda)$ is the gradient of a concave programming problem derived by using the Wolfe dual and this enables the solution of $r(\lambda) = 0$ to be undertaken in a reliable way. The Jacobian $Q(\lambda)$ of this system is important and plays a similar role to the matrix Q that arises in the linear system (4.10) in Toint's sparse PSB update. If a certain assumption related to fill-in is valid, the matrix Q satisfies the same structural and definiteness conditions as for Toint's matrix. Thus the nonlinear system can be solved by a few iterations of analogous complexity to (4.10). Fletcher also suggests alternative possibilities for computing the update via what might be thought of as primal methods. The dual update has been implemented for tridiagonal systems and some numerical experiments show a significant reduction in the number of quasi-Newton iterations, as compared with limited memory, conjugate gradient, and dense BFGS codes. However the amount of computation required to compute the update is significantly greater than for the dense case.

A question of some interest is whether the information present in $\mathcal{G}(H)$ does indeed determine $B > 0$ (as required by (4.16)), and whether the outcome is unique. It is hopeful that $\mathcal{G}(H)$ contains the same number of non-zero elements as B. If the elements of $\mathcal{S}^\perp$ are chosen to include elements that fill-in when the LDL^T

factors of B are calculated, then it is also shown in [24] that B is well-determined by $\mathcal{G}(H)$ and an algorithm is given for computing the LDL^T factors of B. This algorithm is shown to be particularly efficient when the sparsity pattern of B is formed from dense overlapping blocks on the diagonal. One feature of these results is particularly surprising. If B is an irreducible sparse matrix then $H = B^{-1}$ is generally dense. It is therefore most unexpected to find that it is only the elements of $\mathcal{G}(H)$ in (4.15) that determine the sparsity pattern of B.

Assumption (4.13) is critical for the existence of a positive definite sparse update. This is made clear by Sorensen [69] who essentially cites the following example in which B is required to solve

$$\begin{bmatrix} a & b & \\ b & c & * \\ & * & * \end{bmatrix} \begin{pmatrix} -1 \\ \varepsilon \\ 1 \end{pmatrix} = \begin{pmatrix} 1 \\ 0 \\ 2 \end{pmatrix}. \tag{4.18}$$

When $\varepsilon \neq 0$ the first equation implies $b = (a+1)/\varepsilon$, and $a > 0$ is required for positive definiteness. Thus b grows without limit as ε goes to zero. Moreover the inequality $ac > b^2$ implies that ac increases like ε^{-2} so the rate of growth is quadratic. If $\varepsilon = 0$ then the only solution has $a = -1$ and a positive definite update does not exist. Yet $\delta^T \gamma = 1 > 0$.

It is important to realise that such examples can only occur when the average Hessian matrix $\bar{G}$ in (1.24) is indefinite. Even if $\bar{G}$ is indefinite then a satisfactory update may yet be obtained (this happens for example in the dense case). Thus it may be that these difficulties arise relatively infrequently and an ad-hoc solution may be adequate. However when difficulties do arise then their effects are severe due to the ε^{-2} growth. Generally speaking, we have a choice to relax either the positive definite condition or the secant condition if we wish to retain sparsity. In the context of Fletcher's sparse update, a simple solution is to terminate the dual iteration if some preset upper limit on the size of B and H is exceeded. In fact the high cost of Fletcher's update may make it necessary to take only a small number of dual iterations, so that the secant equation is never satisfied exactly. It is difficult to say whether this will adversely affect the convergence of the outer quasi-Newton iteration.

Other research into structured unconstrained optimization has avoided the requirement that $B^{(k)}$ should be positive definite. Most promising has been the approach of Griewank and Toint (e.g. [34]) based on the *partially separable* optimization problem

$$\text{minimize} \quad f(x) = \sum_{j=1}^{n_e} f_j(x_{[j]}) \tag{4.19}$$

in which there are n_e *nonlinear element functions* $f_j(x_{[j]})$, each of which depends on a usually small subset $x_{[j]}$ of the components of x. Then the Hessian of $f(x)$ can be decomposed into a sum of Hessians of the $f_j(x_{[j]})$, the nontrivial submatrices of which can be treated as dense matrices. Similar remarks apply for the gradient vector. Griewank and Toint suggest that these element Hessian submatrices are approximated by the use of dense updating techniques. There are however some difficulties that have to be overcome. The element Hessian submatrices may not be positive definite so it is not possible to rely on the analogue of the condition $\delta^T \gamma > 0$ holding for each submatrix. Thus the BFGS formula cannot be used, and Toint uses the symmetric rank 1 formula in the Harwell Subroutine Library code VE08. Consequently the overall Hessian approximation obtained by summing the submatrix approximations is also not in general positive definite. Also the possibility of zero in the denominator of the updates has to be allowed for.

A more flexible format has recently been used by Conn, Gould and Toint (e.g. [12]) in the LANCELOT code. Here the *group partially separable function*

$$f(x) = \sum_{j=1}^{n_g} g_i(\alpha_i(x)) \tag{4.20}$$

is defined in which there are n_g *group functions* $g_i \in C^2 : \mathbb{R} \to \mathbb{R}$. The argument of each group function is a weighted sum

$$\alpha_i(x) = \sum_{j \in \mathcal{J}_i} w_{ij} f_j(x_{[j]}) + a_i^T x - b_i \tag{4.21}$$

over a subset $\mathcal{J}_i$ of the nonlinear element functions, together with a linear term. It is envisaged that the vectors a_i in the linear term should usually be sparse.

It is difficult to predict whether such a format will become generally accepted. On the one hand it is very flexible and most large scale problems can be effectively expressed in this way. Numerical experience with the LANCELOT code on unconstrained problems is very satisfactory and often approaches the performance of Newton's method. This suggests that the difficulties of working with indefinite Hessians and updating using the SR1 formula have largely been overcome. On the other hand the organizational problems are more severe than those required merely to define the sparsity conditions in (4.6), and this may particularly deter the non-specialist user. It may be that there is still some merit in developing an effective positive definite sparse update, should this be possible.

5. References

[1] Al-Baali M. (1985) Descent property and global convergence of the Fletcher-Reeves method with inexact line search, *J. Inst. Maths. Applns.*, 5, 121-124.

[2] Allwright J.C. (1988) Positive semi-definite matrices: Characterization via conical hulls and least-squares solution of a matrix equation, *SIAM J. Control Optim.*, **26**, 537-556 (see also errata in *SIAM J. Control Optim.*, **28**, (1990), 250-251).

[3] Barzilai J. and Borwein J.M. (1988) Two-point step size gradient methods, *IMA J. Numer. Anal.*, **8**, 141-148.

[4] Beale E.M.L. (1972) A derivation of conjugate gradients, in *Numerical Methods for Nonlinear Optimization*, ed. F.A.Lootsma, Academic Press, London.

[5] Broyden C.G. (1967) Quasi-Newton methods and their application to function minimization, *Maths. of Computation*, **21**, 368-381.

[6] Buckley A. and LeNir A. (1983) QN-like variable storage conjugate gradients, *Math. Programming*, **27**, 155-175.

[7] Byrd R.H., Liu D.C. and Nocedal J. (1990) On the behavior of Broyden's class of quasi-Newton methods, Report NAM-01, Northwestern Univ., Dept. EECS.

[8] Byrd R.H. and Nocedal J. (1989) A tool for the analysis of quasi-Newton methods with application to unconstrained minimization, *SIAM J. Numer. Anal.*, **26**, 727-739.

[9] Byrd R.H., Nocedal J. and Yuan Y. (1987) Global convergence of a class of quasi-Newton methods on convex problems, *SIAM J. Numer. Anal.*, **24**, 1171-1190.

[10] Conn A.R., Gould N.I.M. and Toint Ph.L. (1988) Testing a class of methods for solving minimization problems with simple bounds on the variables, *Maths. of Computation*, **50**, 399-430.

[11] Conn A.R., Gould N.I.M. and Toint Ph.L. (1991) Convergence of quasi-Newton matrices generated by the symmetric rank one update, *Math. Programming*, **50**, 177-195.

[12] Conn A.R., Gould N.I.M. and Toint Ph.L. (1992) *LANCELOT: a Fortran package for large-scale nonlinear optimization (Release A)*, Springer Series in Computational Mathematics 17, Springer-Verlag, Heidelberg.

[13] Crowder H.P. and Wolfe P. (1972) Linear convergence of the conjugate gradient method, *IBM J. Res. and Dev.*, **16**, 431-433.

[14] Davidon W.C. (1959) Variable metric methods for minimization, Argonne Nat. Lab. Report ANL-5990 (rev.), and in *SIAM J. Optimization*, **1**, (1991), 1-17.

[15] Davidon W.C. (1975) Optimally conditioned optimization algorithms without line searches, *Math. Programming*, **9**, 1-30.

[16] Dembo R.S., Eisenstat S.C. and Steihaug T. (1982) Inexact Newton methods, *SIAM J. Numer. Anal.*, **19**, 400-408.

[17] Dennis J.E. and Moré J.J. (1974) A characterization of superlinear convergence and its application to quasi-Newton methods, *Maths. of Computation*, **28**, 549-560.

[18] Dennis J.E. and Schnabel R.B. (1983) *Numerical Methods for Unconstrained Optimization ana Nonlinear Equations*, Prentice-Hall Inc., Englewood Cliffs, NJ.

[19] Dixon L.C.W. (1972) Quasi-Newton algorithms generate identical points, *Math. Programming*, **2**, 383-387.

[20] Fletcher R. (1970) A new approach for variable metric algorithms, *Computer J.*, **13**, 317-322.

[21] Fletcher R. (1987) *Practical Methods of Optimization, (2nd. Edition)*, John Wiley, Chichester.

[22] Fletcher R. (1990) Low storage methods for unconstrained optimization, in *Computational Solution of Nonlinear Systems of Equations*, eds. E.L.Allgower and K.Georg, Lectures in Applied Mathematics Vol. 26, AMS Publications, Providence, RI.

[23] Fletcher R. (1991) A new result for quasi-Newton formulae, *SIAM J. Optimization*, **1**, 18-21.

[24] Fletcher R. (1992) An optimal positive definite update for sparse Hessian matrices, Report NA/145, Univ. of Dundee, to appear in *SIAM J. Optimization*.

[25] Fletcher R. and Powell M.J.D. (1963) A rapidly convergent descent method for minimization, *Computer J.*, **3**, 163-168.

[26] Fletcher R. and Reeves C.M. (1964) Function minimization by conjugate gradients, *Computer J.*, **7**, 149-154.

[27] Forsgren A., Gill P.E. and Murray W. (1993) Computing modified Newton directions using a partial Cholesky factorization, Royal Inst. of Tech., Sweden, Dept. of Math. Report TRITA/MAT-93-09.

[28] Gilbert J.C. and Nocedal J. (1990) Global convergence properties of conjugate gradient methods for optimization, Report 1268, INRIA.

[29] Gill P.E. and Murray W. (1972) Quasi-Newton methods for unconstrained optimization, *J. Inst. Maths. Applns.*, **9**, 91-108.

[30] Goldfarb D. (1970) A family of variable metric methods derived by variational means, *Maths. of Computation*, **24**, 23-26.

[31] Goldfarb D. (1976) Factorized variable metric methods for unconstrained optimization, *Maths. of Computation*, **30**, 796-811.

[32] Grandinetti (1979) Factorization versus nonfactorization in quasi-Newtonian algorithms for differentiable optimization, in *Methods for Operations Research*, Hain Verlay.

[33] Greenstadt J.L. (1970) Variations of variable metric methods, *Maths. of Computation*, **24**, 1-22.

[34] Griewank A. and Toint Ph.L. (1982) Partitioned variable metric updates for large structured optimization problems, *Numerische Math.*, **39**, 429-448.

[35] Grippo L., Lampariello F. and Lucidi S. (1990) A quasi-discrete Newton algorithm with a nonmonotone stabilization technique, *J. Optim. Theo. Applns.*, **64**, 485-500.

[36] Hestenes M.R. and Stiefel E. (1952) Methods of conjugate gradients for solving linear systems, *J. Res. NBS*, **49**, 409-436.

[37] Khalfan H., Byrd R.H. and Schnabel R.B. (1990) A theoretical and experimental study of the symmetric rank one update, Report CU-Cs-489-90, Univ. of Colorado at Boulder, Dept. CS.

[38] Lalee M. and Nocedal J. (1991) Automatic scaling strategies for quasi-Newton updates, Report NAM-04, Northwestern Univ., Dept. EECS.

[39] Lefebure B. (1993) Investigating new methods for large scale unconstrained optimization. M.Sc. Thesis, Univ. of Dundee, Dept. of Maths. and CS.

[40] Liu D.C. and Nocedal J. (1989) On the limited memory BFGS method for large scale optimization, *Math. Programming*, **45**, 503-528.

[41] Luksan L. (1993) Computational experience with known variable metric updates, Working paper.

[42] Moré J.J. (1983) Recent developments in algorithms and software for trust region methods, in *Mathematical Programming, The State of the Art*, eds. A.Bachem, M Grotschel and G.Korte, Springer-Verlag, Berlin.

[43] Moré J.J., Garbow B.S. and Hillstrom K.E. (1981) Testing unconstrained optimization software, *ACM Trans. Math. Softw.*, **7**, 17-41.

[44] Moré J.J. and Sorensen D.C. (1979) On the use of directions of negative curvature in a modified Newton method, *Math. Programming*, **16**, 1-20.

[45] Nash S.C. (1985) Preconditioning of truncated-Newton methods, *SIAM J. S.S.C.*, **6**, 599-618.

[46] Nash S.C. and Nocedal J. (1989) A numerical study of the limited memory BFGS method and the truncated-Newton method for large scale optimization, Report NAM-02, Northwestern Univ., Dept. EECS.

[47] Nocedal J. (1980) Updating quasi-Newton matrices with limited storage, *Maths. of Computation*, **35**, 773-782.

[48] Nocedal J. and Yuan Y. (1991) Analysis of a self-scaling quasi-Newton method, Report NAM-02, Northwestern Univ., Dept. EECS.

[49] O'Leary D.P. (1982) A discrete Newton algorithm for minimizing a function of many variables, *Math. Programming*, **23**, 20-33.

[50] Oren S.S. (1974) On the selection of parameters in self-scaling variable metric algorithms, *Math. Programming*, **7**, 351-367.

[51] Oren S.S. and Luenberger D.G (1974) Self-scaling variable metric (SSVM) algorithms: I Criteria and sufficient conditions for scaling a class of algorithms, *Management Sci.*, **20**, 845-862.

[52] Oren S.S. and Spedicato E. (1976) Optimal conditioning of self-scaling variable metric algorithms, *Math. Programming*, **10**, 70-90.

[53] Polak E. and Ribière G. (1969) Note sur la convergence de methodes de directions conjugées, *Rev. Française Informat Recherche Operationnelle, 3e Année*, **16**, 35-43.

[54] Powell M.J.D. (1970) A new algorithm for unconstrained optimization, in *Nonlinear Programming*, eds. J.B.Rosen, O.L.Mangasarian and K.Ritter, Academic Press, New York.

[55] Powell M.J.D. (1971) On the convergence of the variable metric algorithm, *J. Inst. Maths. Applns.*, **7**, 21-36.

[56] Powell M.J.D. (1976) Some global convergence properties of a variable metric algorithm for minimization without exact line searches, in *Nonlinear Programming, SIAM-AMS Proceedings Vol. IX*, eds. R.W.Cottle and C.E.Lemke, SIAM

Publications, Philadelphia.

[57] Powell M.J.D. (1977) Restart procedures for the conjugate gradient method, *Math. Programming*, **12**, 241-254.

[58] Powell M.J.D. (1984) Nonconvex minimization calculations and the conjugate gradient method, in *Numerical Analysis, Dundee 1983*, ed. D.F.Griffiths, Lecture Notes in Mathematics 1066, Springer Verlag, Berlin.

[59] Powell M.J.D. (1986) How bad are the BFGS and DFP methods when the objective function is quadratic?, *Math. Programming*, **34**, 34-47.

[60] Powell M.J.D. (1987) Updating conjugate directions by the BFGS formula, *Math. Programming*, **38**, 29-46.

[61] Powell M.J.D. and Toint Ph. L. (1979) On the estimation of sparse Hessian matrices, *SIAM J. Numer. Anal.*, **16**, 1060-1074.

[62] Schnabel R.B. (1983) Quasi-Newton methods using multiple secant equations, Report CU-CS-247-83, Univ. of Colorado at Boulder, Dept. CS.

[63] Schnabel R.B. and Eskow E. (1990) A new modified Cholesky factorization, *SIAM J. Sci. Stat. Comp.*, **11**, 1136-1158.

[64] Shanno D.F. (1970) Conditioning of quasi-Newton methods for function minimization, *Maths. of Computation*, **24**, 647-656.

[65] Shanno D.F. and Phua K.H. (1980) Remark on algorithm 500, minimization of unconstrained multivariate functions, *ACM Trans. Math. Softw.*, **6**, 618-622.

[66] Siegel D. (1991) Updating of conjugate direction matrices using members of Broyden's family, Rep. DAMTP 1991/NA4, Univ. of Cambridge, Dept. AMTP.

[67] Siegel D. (1991) Modifying the BFGS update by a new column scaling technique, Report DAMTP 1991/NA5, Univ. of Cambridge, Dept. AMTP.

[68] Siegel D. (1992) Implementing and modifying Broyden class updates for large scale optimization, Report DAMTP 1992/NA12, Univ. of Cambridge, Dept. AMTP.

[69] Sorensen D.C. (1982) Collinear scaling and sequential estimation in sparse optimization algorithms, *Math. Programming Study*, **18**, 135-159.

[70] Steihaug T. (1983) The conjugate gradient method and trust regions in large scale optimization, *SIAM J. Numer. Anal.*, **20**, 626-637.

[71] Toint Ph. L. (1977) On sparse and symmetric updating subject to a linear equation. *Maths. of Computation*, **31**, 954-961.

[72] Toint Ph. L. (1981) A note on sparsity exploiting quasi-Newton methods. *Math. Programming*, **21**, 172-181.

[73] Toint Ph. L. (1981) Towards an efficient sparsity exploiting Newton method for minimization, in *Sparse Matrices and Their Uses*, ed. I.S.Duff, Academic Press, London.

[74] Torczon V. (1991) On the convergence of the multidimensional search algorithm, *SIAM J. Optimization*, **1**, 123-145.

[75] Wood G.R. (1992) The bisection method in higher dimensions, *Math. Programming*, **55**, 319-338.

[76] Zhang Y. and Tewarson R.P. (1988) Quasi-Newton algorithms with updates from the pre-convex part of Broyden's family, *IMA J. Numer. Anal.*, **8**, 487-509.

Nonquadratic Model Methods in Unconstrained Optimization

N.Y. Deng   and   Z.F. Li

Beijing Agricultural Engineering University
P.O. Box 71
100083 Beijing
P.R.China

Abstract: Standard methods for unconstrained optimization base each iteration on a quadratic model of the objective function. Recently, methods using two generalizations of the standard models have been proposed. Nonlinear scaling invariance methods use a model that is a nonlinear regular scaling of a quadratic function, and conic methods use a model that is the ratio of a quadratic function divided by the square of a linear function. This paper will attempt to survey some interesting developments in these fields.

1. Introduction

Our problem is to minimize an objective function of n variables

$$\min f(x), \quad x \in R^n, \tag{1.1}$$

where $f \in C^2$, and its gradient $g(x) = \nabla f(x)$ is available. Let us review some standard methods first. Let the current point be x_k, and let the unconstrained optimization algorithms have an iteration of the form

$$x_{k+1} = x_k + \lambda_k d_k, \tag{1.2}$$

where d_k is a search direction, and λ_k is a step length obtained by means of a line search, for example an exact line search or an inexact line search with Wolfe conditions. Wolfe conditions can be described as follows:

$$f(x_k + \lambda_k d_k) \leq f(x_k) + \alpha \lambda_k g_k^T d_k, \tag{1.3}$$

and

$$g(x_k + \lambda_k d_k)^T d_k \geq \beta g_k^T d_k, \tag{1.4}$$

where $g_k = g(x_k) = \nabla f(x_k)$, $g(x_k + \lambda_k d_k) = \nabla f(x_k + \lambda_k d_k)$, α and β are constants that satisfy $0 < \alpha < \beta < 1$ and $\alpha < 1/2$.

145

E. Spedicato (ed.), Algorithms for Continuous Optimization, 145–168.
© 1994 *Kluwer Academic Publishers.*

Various choices of the direction d_k give rise to distinct algorithms. Many algorithms are based on the idea of minimizing an approximation to the objective function around the current point x_k. The approximate function is usually a quadratic model

$$m_k(x) = f(x_k) + g_k^T(x - x_k) + \frac{1}{2}(x - x_k)^T B_k(x - x_k), \tag{1.5}$$

where B_k is an approximation to $\nabla^2 f(x_k)$. When B_k is positive definite, $m_k(x)$ has a unique minimizer at $x_k + d_k^*$, where $d_k^* = -B_k^{-1}g_k$. Therefore, a broad class of methods defines the search direction by

$$d_k = -B_k^{-1}g_k. \tag{1.6}$$

The Newton direction is obtained when $B_k = \nabla^2 f(x_k)$. The variable metric method is also of the form (1.6), but in this case B_k is not only a function of x_k, but depends also on B_{k-1} and x_{k-1}. In fact, after a step from x_k to x_{k+1}, the approximation B_k to $\nabla^2 f(x_k)$ is updated into an approximation B_{k+1} to $\nabla^2 f(x_{k+1})$. To make the quadratic model

$$m_{k+1}(x) = f(x_{k+1}) + g_{k+1}^T(x - x_{k+1}) + \frac{1}{2}(x - x_{k+1})^T B_{k+1}(x - x_{k+1})$$

to be an approximation to $f(x)$, we require that

$$\nabla m_{k+1}(x_k) = \nabla f(x_k). \tag{1.7}$$

In addition, $m_{k+1}(x)$ satisfies

$$m_{k+1}(x_{k+1}) = f(x_{k+1}), \quad \nabla m_{k+1}(x_{k+1}) = \nabla f(x_{k+1}). \tag{1.8}$$

Thus, from the interpolation condition (1.7) and (1.8), B_{k+1} satisfies the quasi-Newton equation

$$B_{k+1}s_k = y_k, \tag{1.9}$$

where $s_k = x_{k+1} - x_k$, $y_k = g_{k+1} - g_k$. In addition, B_{k+1} is chosen to be symmetric since $\nabla^2 f(x)$ is symmetric. Still many symmetric B_{k+1} satisfying (1.9) exist; the most popular update is the Broyden's class given by

$$B_{k+1} = B_k - \frac{B_k s_k s_k^T B_k}{s_k^T B_k s_k} + \frac{y_k y_k^T}{s_k^T y_k} + \alpha_k v_k v_k^T, \tag{1.10}$$

where

$$v_k = (s_k^T B_k s_k)^{\frac{1}{2}} \left(\frac{y_k}{s_k^T y_k} - \frac{B_k s_k}{s_k^T B_k s_k} \right)$$

and α_k is a parameter. When α_k is restricted in some interval, e.g. $\alpha_k \geq 0$, the update (1.10) has a very useful property: B_{k+1} is positive definite provided B_k is positive definite and the inequality

$$s_k^T y_k > 0 \tag{1.11}$$

holds, which follows from the line search being sufficiently accurate.

If $\alpha_k = 0$ in (1.10), we obtain the famous BFGS (Broyden-Fletcher-Goldfarb-Shanno) update

$$B_{k+1} = B_k - \frac{B_k s_k s_k^T B_k}{s_k^T B_k s_k} + \frac{y_k y_k^T}{s_k^T y_k}, \tag{1.12}$$

and if $\alpha_k = 1$, then we obtain the DFP (Davidon-Fletcher-Powell) update.

The variable metric method is very efficient and robust. Usually, it has nice convergence properties — global convergence and superlinear convergence rate. Particularly, when the objective function is quadratic

$$q(x) = \delta + r^T x + \frac{1}{2} x G^T x, \tag{1.13}$$

where $\delta \in R^1, r \in R^n, G \in R^{n \times n}$ is positive definite, the variable metric method generates conjugate search directions $\{d_k\}$ and terminates at the minimizer x^* within at most n iterations provided the line searches are exact.

Another important class of methods are the conjugate gradient methods. Some of them can be obtained from the $BFGS$ update with $B_1 = I$. Assume that the objective function is of the form of (1.13) and the line searches are exact. Then it is not difficult to show that the direction d_k in the BFGS method has the following simple expression

$$d_k = -g_k + \theta_k d_{k-1}, \quad k = 1, 2, \cdots, \tag{1.14}$$

where

$$\theta_k = \begin{cases} 0, & \text{for} \quad k = 1; \\ \| g_k \|^2 / \| g_{k-1} \|^2, & \text{for} \quad k \geq 2, \end{cases} \tag{1.15}$$

or

$$\theta_k = \begin{cases} 0, & \text{for} \quad k = 1; \\ g_k^T (g_k - g_{k-1}) / \| g_{k-1} \|^2, & \text{for} \quad k \geq 2. \end{cases} \tag{1.16}$$

The famous Fletcher-Reeves method is based on (1.14) and (1.15) while the Polak-Ribiere method is based on (1.14) and (1.16). Both methods terminate at the minimizer x^* of (1.13) within at most n iterations.

Other choices of the parameter θ_k in (1.14) are also possible, and give rise to distinct conjugate gradient algorithms. Since the conjugate gradient method requires storage of only a few vectors, it is very useful for solving large scale problems, even if the rate of convergence is only linear.

In this paper, we discuss two classes of algorithms based upon nonquadratic models for solving the problem (1.1): nonlinear scaling invariance methods and conic methods. Both classes contain interesting innovations and seem to offer some advantages over the standard methods. The nonlinear scaling invariance methods base each step on a model of the form

$$f(x) = F(q(x)), \tag{1.17}$$

where F is a function of one variable and $q(x)$ is defined as (1.13). Conic model methods base each iteration on a model of the form

$$\hat{m}(x_k + d) = f(x_k) + \frac{g_k^T d}{1 + b_k^T d} + \frac{1}{2} \frac{d^T A_k d}{\left(1 + b_k^T d\right)^2}, \tag{1.18}$$

where $A_k \in R^{n \times n}$ is symmetric positive definite, and $b_k \in R^n$.

The remainder of the paper is organized as follows. In section 2, we discuss the variable metric method and the conjugate gradient method with nonlinear scaling invariance. Conic-variable metric methods and conic-conjugate gradient methods are discussed in Section 3 and in Section 4 respectively.

Schnabel (1989) and Schnabel et al. (1991) introduced a new type of methods for unconstrained optimization, called tensor methods. Since a thorough treatment is not included in this paper, we give a brief description here. The tensor methods are based upon a fourth order model of the objective function. This model consists of the quadratic portion of the Taylor series, plus low-rank third and fourth order terms that cause the model to interpolate the function values and the gradients computed during one or more previous iterations. They also show that the cost of forming, storing, and solving the tensor model is not significantly more than the cost for a standard method based upon a quadratic Taylor series model. It seems that their tensor method is very promising.

There has been other interesting work on nonquadratic models, e.g. the methods based upon homogeneous functions investigated by Jacobson et al.(1972), Charalambous (1973),Kowalik et al. (1976), the method based upon L-function of Davison et al. (1975), and the dominant degree method of Biggs (1971) etc. We shall not deal with them in this paper because of space limitation.

2. Nonlinear scaling invariance methods

The success of the standard methods mentioned above is attributed to the facts that the objective function can be approximated by a quadratic function, and for a quadratic function the standard methods terminate within finite iterations and behave well. To enhance the efficiency, it is useful to consider a larger class of functions than quadratics and establish efficient algorithms for these functions. Spedicato(1976) first considered this problem. He constructed a larger class of functions by expanding every element in the class of quadratic functions into a set of functions as follows. The following definition, for the sake of simplicity, is slightly different from the original one by Spedicato.

Definition 2.1 A function $f(x)$ is a nonlinear regular scaling of $q(x)$ if the following condition holds

$$f = F(q(x)), \quad \frac{dF}{dq} > 0 \quad \text{for } x \neq x^*, \tag{2.1}$$

where F is a differentiable function of one variable, $q(x)$ is defined by (1.13) and x^* is the minimizer of $f(x)$.

Note that the nonlinear scaling of the function $q(x)$ does not modify the contour lines, i.e. both the nonlinear scaling $f(x)$ of $q(x)$ and $q(x)$ itself have the same contour lines. One wants to construct algorithms so that the sequence $\{x_k\}$ generated for the nonlinear scaling $f(x)$ is identical to the one generated for $q(x)$. This leads to the following definition.

Definition 2.2 An algorithm has the invariancy to nonlinear regular scaling with respect to the function $q(x)$ defined by (1.13) if the sequence $\{x_k\}$ generated by the algorithm is the same for every function which is a nonlinear regular scaling of $q(x)$ when the starting point x_1 is the same.

To describe an invariant algorithm, we propose a restriction on the line search which is weaker than exact line search.

Definition 2.3 A line search is consistent if the terminal point x_{k+1} is the same for any function with the same contours when the starting point x_k and the search direction d_k are the same respectively.

There are two classes of algorithms with invariance to nonlinear scaling, one is a modification of the variable metric methods, the other is a modification of the conjugate gradient methods.

2.1 THE VARIABLE METRIC METHODS DERIVED FROM INVARIANCY

Spedicato (1976) showed that the variable metric method can be modified to satisfy the invariance to nonlinear scaling. He proved the following

Theorem 2.1 A sufficient condition for the variable metric method with Broyden update and consistent line searches to be invariant to nonlinear regular scaling with respect to the function $q(x)$ defined by (1.13) is that the vector y_k in (1.10) is modified by

$$\tilde{y}_k = \frac{\nabla f(x_{k+1})}{F'_{k+1}} - \frac{\nabla f(x_k)}{F'_k} = \frac{1}{F'_{k+1}} \left[g_{k+1} - \frac{F'_{k+1}}{F'_k} g_k \right], \qquad (2.2)$$

where

$$F'_k = \frac{dF}{dq}\Big|_{q=q(x_k)} .$$

It follows from Theorem 2.1 that, since the variable metric method with Broyden's update and exact line searches finds the minimizer of $q(x)$ within finite iterations, the above modified method does the same on every function which is a nonlinear regular scaling of $q(x)$.

Such a modification of Broyden update requires the determination of the derivatives $(dF/dq)_k$. Spedicato interpreted these derivatives as parameters

$$\phi_k = \frac{1}{F'_{k+1}}, \quad \sigma_k = \frac{F'_{k+1}}{F'_k}.$$

Therefore, the following extended Broyden update is obtained

$$B_{k+1} = B_k - \frac{B_k s_k s_k^T B_k}{s_k^T B_k s_k} + \frac{y_k y_k^T}{s_k^T y_k} + \alpha_k v_k v_k^T, \tag{2.3}$$

where α_k, ϕ_k and σ_k are parameters, and

$$s_k = x_{k+1} - x_k, \quad y_k = \phi_k(g_{k+1} - \sigma_k g_k),$$

$$v_k = (s_k^T B_k s_k)^{1/2} \left(\frac{y_k}{s_k^T y_k} - \frac{B_k s_k}{s_k^T B_k s_k} \right).$$

The corresponding algorithm is as follows.

Algorithm 2.1

1^0 Select positive series $\{\alpha_k\}, \{\phi_k\}$ and $\{\sigma_k\}$. Select a starting point x_1 and a positive definite matrix B_1. Set $k = 1$;

2^0 Compute $g_k = \nabla f(x_k)$. If $g_k = 0$, stop; otherwise go to step 3^0;

3^0 Set $d_k = -B_k^{-1} g_k$;

4^0 Compute the step length λ_k by line search (e.g. exact line search or inexact line search with Wolfe conditions (1.3) and (1.4));

5^0 Set $x_{k+1} = x_k + \lambda_k d_k$;

6^0 Compute $g_{k+1} = \nabla f(x_{k+1})$. If $g_{k+1} = 0$, stop; otherwise compute B_{k+1} by (2.3);

7^0 Increase k by one, and go to step 3^0.

To consider the properties of the above algorithm, we suppose that the following assumption on ϕ_k and σ_k holds.

Assumption A For all k, ϕ_k and σ_k are positive and satisfy

$$| 1 - \phi_k | \le \phi' \parallel s_k \parallel, \quad | 1 - \sigma_k | \le \sigma' \parallel s_k \parallel, \tag{2.4}$$

and

$$\sigma_k > \beta, \tag{2.5}$$

where both ϕ' and σ' are positive scalars, and β is defined as in (1.4).

Note that a result of Spedicato (1976) suggests that, in general $\sigma_k \ge 1$, so the assumption (2.5) is reasonable.

If λ_k satisfies (1.3) and (1.4), using (2.5), we have

$$s_k^T y_k = \phi_k(s_k^T g_{k+1} - \sigma_k s_k^T g_k)$$
$$\ge \phi_k(\beta s_k^T g_k - \sigma_k s_k^T g_k) = \phi_k(\sigma_k - \beta)(-s_k^T g_k) > 0$$

for $s_k^T g_k < 0$. Therefore, if B_1 is positive definite, the matrices $B_k(k = 1, 2, 3, \cdots)$ generated by Algorithm 2.1 are positive definite. This implies that Algorithm 2.1 is well-defined.

The convergence properties of Algorithm 2.1 was first discussed by Flachs (1982), and further studied by Deng and Li (1993a, 1993b). Flachs showed the global convergence of Algorithm 2.1 with exact line searches. Deng and Li (1993a,1993b) proved the following

Theorem 2.2 Let x_1 be a starting point for which f is twice continuously differentiable and uniformly convex on the level set $D = \{x : f(x) \leq f(x_1)\}$. Suppose in addition that Assumption A holds. Then for any positive definite B_1, the sequence $\{x_k\}$ generated by Algorithm 2.1 when $\alpha_k = 0$ with exact line searches or inexact line searches with Wolfe conditions (1.3) and (1.4), converges to the minimizer x^* of $f(x)$ superlinearly.

Broyden update (1.10) for B_k has the corresponding one for $H_k = B_k^{-1}$. By $y_k = \phi_k(g_{k+1} - \sigma_k g_k)$, Spedicato(1976) generalized the Broyden update for H_k to the following one, after a multiplication by ϕ_k which does not affect the search direction

$$H_{k+1} = \phi_k H_k + \frac{s_k^T \hat{y}_k + \phi_k(1 - \alpha_k)\hat{y}_k^T H_k \hat{y}_k}{(s_k^T \hat{y}_k)^2} s_k s_k^T$$
$$- \frac{\phi_k \alpha_k}{\hat{y}_k^T H_k \hat{y}_k} H_k \hat{y}_k \hat{y}_k^T H_k - \frac{\phi_k(1 - \alpha_k)}{s_k^T \hat{y}_k} (H_k \hat{y}_k s_k^T + s_k \hat{y}_k^T H_k), \tag{2.6}$$

where

$$\hat{y}_k = g_{k+1} - \sigma_k g_k. \tag{2.7}$$

From stability rules, he suggested the following strategy for selecting the parameters σ_k, ϕ_k and α_k in (2.6): σ_k is defined by

$$\sigma_k = \max \left\{ (\varepsilon_k \bar{\tau}_k / \bar{\rho}_k^2)^{1/2}, \ (\varepsilon_k - 2\lambda_k \bar{\rho}_k)/\varepsilon_k \right\}, \tag{2.8}$$

where λ_k is the step length and

$$\bar{\rho}_k = s_k^T(g_{k+1} - g_k), \quad \bar{\tau}_k = (g_{k+1} - g_k)^T H_k(g_{k+1} - g_k), \quad \varepsilon_k = \lambda_k^2 g_k^T H_k g_k$$

The parameters ϕ_k and α_k are defined by Oren and Spedicato (1976)

$$\phi_k = (\varepsilon_k/\tau_k)^{1/2}, \quad \alpha_k = 1 - 1/\left[1 + (\varepsilon_k \tau_k/\rho_k^2)^{1/2}\right], \tag{2.9}$$

where ε_k is the same as that in (2.8) and

$$\rho_k = s_k^T(g_{k+1} - \sigma_k g_k), \quad \tau_k = (g_{k+1} - \sigma_k g_k)^T H_k(g_{k+1} - \sigma_k g_k),$$

As an alternative choice to the parameter σ_k, Spedicato(1978) proved the following

Theorem 2.3 Consider a nonlinear regular scaling $f(x)$ defined by (2.1). Let x be a point such that $d^T g(x) \neq 0$, and $x' = x + \lambda d$, where λ is the exact line search step length. Take a point $\hat{x}$, arbitrary except for $g(\hat{x})^T d \neq 0$, on the line connecting x and x'. Define $\hat{\lambda}$ by the relation $\hat{x} = x + \hat{\lambda} d$. Then

$$F'|_{q=q(x')}/F'|_{q=q(x)} = \lim_{\hat{x} \to x'} \frac{g(\hat{x})^T d \cdot \lambda}{\hat{\lambda} \cdot g(x)^T d}. \tag{2.10}$$

This theorem provides a practical way of approximately determining σ_k. However, Spedicato suggested that this approach is useful only when the objective is actually related to a quadratic one by nonlinear scaling and fairly accurate line searches are available. For the general case, it may be better to select σ_k by (2.8) or by the methods described in the next paragraph.

2.2 THE CONJUGATE GRADIENT METHODS WITH INVARIANCY TO NONLINEAR SCALING

Boland et al. (1979) showed how to alter the standard gradient method to exhibit the invariance to nonlinear scaling. They proved the following

Theorem 2.4 A sufficient condition for the conjugate gradient method with consistent line searches to be invariant to nonlinear regular scaling with respect to the function $q(x)$ defined by (1.13) is that the scalar θ_k in (1.14) is defined by

$$\theta_k = \begin{cases} 0, & \text{for} \quad k = 1; \\ \mu_k \| g_k \|^2 / \| g_{k-1} \|^2, & \text{for} \quad k \geq 2, \end{cases} \tag{2.11}$$

or

$$\theta_k = \begin{cases} 0, & \text{for} \quad k = 1; \\ g_k^T(\mu_k g_k - g_{k-1}) / \| g_{k-1} \|^2, & \text{for} \quad k \geq 2, \end{cases} \tag{1.12}$$

where

$$\mu_k = \frac{F'_{k-1}}{F'_k}, \quad F'_k = \frac{dF}{dq} \Big|_{q=q(x_k)} . \tag{2.13}$$

In particular, since the conjugate gradient method with exact line searches finds the minimizer of $q(x)$ within finite iterations, the corresponding method with the above modification does the same on every function which is a nonlinear regular scaling of $q(x)$.

The essential piece of information in this process is the scalar μ_k. A calculation formula is given by Sloboda (1980) using determinants as follows

$$\mu_k = \tau \left| \begin{matrix} g_{k-1}^T g_{k-1} & g_t^T g_{k-1} \\ g_{k-1}^T g_t & g_{k-1}^T g_{k-1} \end{matrix} \right| / \left| \begin{matrix} g_k^T g_{k-1} & g_t^T g_{k-1} \\ g_k^T g_t & g_t^T g_t \end{matrix} \right|$$

where $x_k = \tau x_{k-1} + (1 - \tau)x_t$, $\tau \neq 0, 1$. Note that this scheme requires the gradients of the objective function at three points along a search line, i.e. x_t as well as x_{k-1} and x_k. Many unconstrained optimization codes are against to this. To make the modified conjugate gradient method more practical, it is better to compute μ_k by knowledge of the objective function at only two points. Unfortunately, for general nonlinear scaling, we seem unable to obtain such a formula. However, Boland et al. (1979) considered a special class of nonlinear regular scaling of the form

$$\begin{cases} f(x) = F(q(x)), \\ F(q) = \epsilon_0 + \epsilon_1 q + \epsilon_2 q^2 / 2, \ \frac{dF}{dq} > 0, \end{cases} \tag{2.14}$$

where ϵ_i is a scalar for $i = 1, 2, 3$ and $q(x)$ is defined by (1.13). They proved the following

Theorem 2.5 Consider the problem (1.1), where $f(x)$ has the form (2.14). Let x_k be the exact minimizer along the ray $\{x_{k-1} + \lambda d_{k-1}, \ \lambda \geq 0\}$. Then μ_k defined by (2.13) can be computed by

$$\mu_k = \frac{-g_{k-1}^T s_{k-1}}{g_{k-1}^T s_{k-1} + 4(f(x_{k-1}) - f(x_k))}, \qquad (2.15)$$

where $s_{k-1} = x_k - x_{k-1}$.

Thus we get the following modified conjugate gradient algorithm.
Algorithm 2.2
1^0 Select a starting point $x_1 \in R^n$. Set $k = 1$;
2^0 Compute $g_k = \nabla f(x_k)$. If $g_k = 0$, stop; otherwise go to step 3^0;
3^0 If $k = 1$, set $d_k = -g_k$; otherwise set

$$d_k = -g_k + \theta_k d_{k-1} \qquad (2.16)$$

where θ_k is defined by (2.11) or (2.12) while μ_k is defined by (2.15);
4^0 Find the smallest positive number λ_k such that the function of one variable

$$\phi_k(\lambda) = f(x_k + \lambda d_k), \quad \lambda > 0$$

has a local minimum at $\lambda = \lambda_k$;
5^0 Set $x_{k+1} = x_k + \lambda_k d_k$;
6^0 Increase k by one, and go to step 2^0.

If θ_k in (2.16) is defined by (2.11), we obtain the modified Fletcher-Reeves algorithm; if θ_k is defined by (2.12), we obtain the modified Polak-Ribiere algorithm. The global convergence and the convergence rate of these two algorithms with exact line searches are studied by Deng and Li (1993c). They proved the following

Theorem 2.6 Let x_1 be a starting point for which f is twice continuously differentiable and uniformly convex on the level set $D = \{x : f(x) \leq f(x_1)\}$. Then the sequence $\{x_k\}$ generated by Algorithm 2.2 with exact line searches converges to x^* at least linearly.

When the line searches are not exact, i.e. $g_k^T s_{k-1} \neq 0$, μ_k should satisfy a quadratic equation as shown by the following theorem, which was proved by Shirey (1982) and Deng and Chen (1987):

Theorem 2.7 Consider the problem (1.1), where $f(x)$ is a special nonlinear scaling of the form (2.14). Then μ_k defined by (2.13) satisfies the following quadratic equation with one-variable

$$(s_{k-1}^T g_k)\mu^2 + [s_{k-1}^T g_k + s_{k-1}^T g_{k-1} + 4(f(x_{k-1}) - f(x_k))]\mu + s_{k-1}^T g_{k-1} = 0. \qquad (2.17)$$

If $s_{k-1}^T g_k < 0$, μ_k is the larger root of (2.17), and if $s_{k-1}^T g_k > 0$, there are only the following two possibilities:

(i) If one root of (2.17) is larger than 1 and the other one is less than 1, then μ_k is the larger one.

(ii) If both of the two roots of (2.17) are larger than 1, then any one can be μ_k. Therefore, it is impossible to give a correct choice unless extra information, such as a function value at an extra point, is provided.

Based on the above theorem, the corresponding modified conjugate gradient algorithm can be established and analyzed.

3. The conic-variable metric methods

3.1 THE NORMAL CONIC FUNCTION

Davidon (1980) was the first to exploit the conic function model in unconstrained optimization. His paper provides the essential mathematical background for developing the conic algorithms. The conic function is defined by

$$c(s) = f + \frac{g^T s}{1 + b^T s} + \frac{\frac{1}{2} s^T A s}{(1 + b^T s)^2}, \qquad s \in X, \tag{3.1}$$

where $f \in R^1, g \in R^n, A \in R^{n \times n}, b \in R^n$ and $X = \{s : 1 + b^T s \neq 0\}$. The vector b is called the horizon vector and the hyperplane $H = \{s : 1 + b^T s = 0\}$ is called the singular hyperplane. If $b = 0$, $c(s)$ is reduced to a quadratic function. Therefore, conic functions form a larger class which includes quadratic functions as a subclass.

Consider the variable transformation

$$w = \frac{s}{1 + b^T s},$$

or equivalently

$$s = S(w) = \frac{w}{1 - b^T w}, \tag{3.2}$$

where S is a map: $W \to X$ and $W = \{w : 1 - b^T w \neq 0\}$. Under the above transformation, the conic function (3.1) is transformed into a quadratic function of the variable w

$$h(w) \equiv c(S(w)) = f + g^T w + \frac{1}{2} w^T A w. \tag{3.3}$$

The map S also translates the properties of the quadratic function h in the w-space into the properties of the conic function c in the s-space. Note that if A is positive definite, $w^* = -A^{-1} g$ is the unique minimizer of h. By (3.2), we know that the vector

$$d^* = \frac{-A^{-1} g}{1 + b^T A^{-1} g} \tag{3.4}$$

is the unique minimizer of the conic fuction (3.1) if A is positive definite and $1 + b^T A^{-1} g \neq 0$. A conic function satisfying these two conditions is called a normal conic function.

A typical geometric graph of a normal conic function in 2-dimensions is shown in Fig.1, where the contours of $c(s) = k = e^2/(b^T b)$ are depicted and

$$c(s) = \frac{s^T s}{(1 + b^T s)^2}, \quad b = (-1, 0)^T.$$

The line $1 + b^T s = 1 - s_1 = 0$ is the singular line; and the contour $c(s) = 1$ (corresponding to $e = 1$) is a parabola. These two lines divide the (s_1, s_2)-plane into 3 parts with different geometric behavior. The contours $c(s) = e^2$ in the left side are ellipses with increasing eccentricity when e increases from 0 to 1. In the other parts, the contours are hyperbolas; two branches of which lie in different parts.

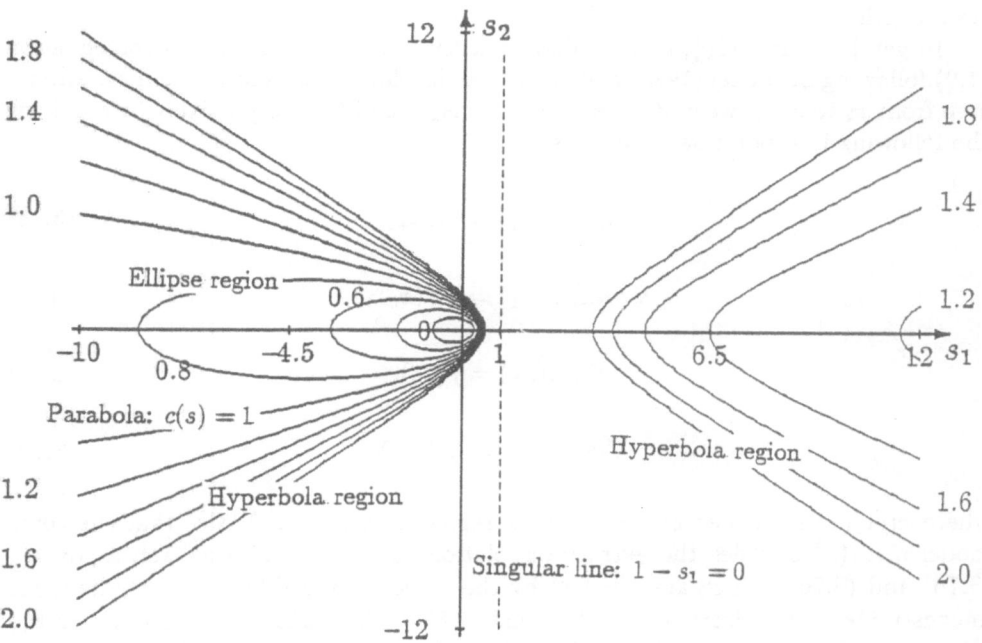

Fig.1. The contours of $c(s) = (s_1^2 + s_2^2)/(1 - s_1)^2 = k$

3.2 CONIC-VARIABLE METRIC ALGORITHMS

The variable metric methods for (1.1) via a conic model can be established in a way similar to the standard variable metric methods. Assume that the current point is x_k. Here a conic function, which corresponds to the quadratic (1.5), is used to approximate $f(x)$ around x_k

$$f(x) \approx \hat{m}_k(x) = f(x_k) + \frac{g_k^T(x - x_k)}{1 + b_k^T(x - x_k)} + \frac{1}{2}\frac{(x - x_k)^T A_k(x - x_k)}{(1 + b_k^T(x - x_k))^2}. \qquad (3.5)$$

It has been shown that if A_k is a positive definite and $1 + b_k^T A_k^{-1} g_k \neq 0$, the unique minimizer of the conic function occurs at $x_k + d_k^*$, where

$$d_k^* = \frac{-A_k^{-1} g_k}{1 + b_k^T A_k^{-1} g_k}. \qquad (3.6)$$

The point $x_k + d_k^*$ is a candidate for x_{k+1}. But, to guarantee convergence, we execute a line search along $d_k = -A_k^{-1} g_k$, and take $x_k + \lambda_k d_k$ as x_{k+1}, where $\lambda_k > 0$ is the step length.

To get $\{A_k\}$ and $\{b_k\}$, now we derive an analogue of the quasi-Newton equation (1.9) following Schnabel (1983) first, and then develop updates for A_k and b_k. After a step from x_k to x_{k+1}, we update A_k and b_k to A_{k+1} and b_{k+1} respectively by requiring the following interpolation conditions

$$\hat{m}_{k+1}(x_{k+1}) = f(x_{k+1}), \qquad (3.7a)$$

$$\nabla \hat{m}_{k+1}(x_{k+1}) = \nabla f(x_{k+1}), \qquad (3.7b)$$

$$\hat{m}_{k+1}(x_k) = f(x_k), \qquad (3.7c)$$

$$\nabla \hat{m}_{k+1}(x_k) = \nabla f(x_k), \qquad (3.7d)$$

where $\hat{m}_{k+1}(x)$ is defined by (3.5) with k replaced by $k + 1$. Notice that the conic model $\hat{m}_{k+1}(x)$ satisfies the extra interpolation condition (3.7c) as well as (3.7a), (3.7b) and (3.7d), which are satisfied by the standard variable metric methods for unconstrained minimization (see (1.7) and (1.8)). This causes the direction to the minimizer of the conic model to depend on function values as well as gradients, whereas in the standard variable metric methods the search direction is determined only by gradients. Comparing with the standard quadratic model, more flexibility and more accurate approximation to the objective function may be expected for the conic model.

Let
$$p_k = x_{k+1} - x_k. \tag{3.8}$$

The interpolation condition (3.7c) requires

$$f(x_k) = f(x_{k+1}) - \frac{p_k^T g_{k+1}}{1 - p_k^T b_{k+1}} + \frac{p_k^T A_{k+1} p_k}{(1 - p_k^T b_{k+1})^2}, \tag{3.9}$$

while the condition (3.7d) is satisfied if

$$g_k = \frac{1}{1 - p_k^T b_{k+1}}[I + \frac{b_{k+1} p_k^T}{1 - p_k^T b_{k+1}}][g_{k+1} - \frac{A_{k+1} p_k}{1 - p_k^T b_{k+1}}]. \tag{3.10}$$

Let

$$\beta_k = 1 - b_{k+1}^T p_k. \tag{3.11}$$

Taking the inner product of (3.10) with p_k and then using (3.9) to eliminate the term $p_k^T A_{k+1} p_k$ yields

$$\beta_k[\beta_k^2 p_k^T g_k + 2\beta_k[f(x_k) - f(x_{k+1})] + p_k^T g_{k+1}] = 0. \tag{3.12}$$

The equation (3.12) has nonzero real roots if and only if the inequality

$$\rho_k^2 \equiv (f(x_k) - f(x_{k+1}))^2 - (p_k^T g_{k+1})(p_k^T g_k) \geq 0 \tag{3.13}$$

holds.

It is reasonable to require the matrices $\{A_k\}_1^\infty$ to be positive definite. This ensures $g_k^T p_k < 0$ and therefore, the condition $\rho_k^2 > 0$ as well as $f(x_{k+1}) < f(x_k)$ can be satisfied by a step length selection. If these conditions are satisfied and $\rho_k \equiv \sqrt{\rho_k^2}$, then we obtain

$$p_k^T A_{k+1} p_k = \pm 2\rho_k \beta_k^2 \tag{3.14}$$

from (3.9) by taking $\beta_k = \beta_\pm = [f(x_k) - f(x_{k+1}) \pm \rho_k]/(-g_k^T p_k)$ as the roots of (3.12). Clearly, we should choose the plus sign in (3.14) if A_{k+1} is positive definite. Therefore, we have

$$\beta_k = \frac{f(x_k) - f(x_{k+1}) + \rho_k}{-g_k^T p_k}. \tag{3.15}$$

Eliminating $p_k^T A_{k+1} p_k$ in (3.10) by (3.9), we get

$$A_{k+1} p_k = \beta_k g_{k+1} - \beta_k^2[I - b_{k+1} p_k^T]g_k, \tag{3.16}$$

which is the analogue of the quasi-Newton equation (1.9) in the standard variable metric methods and can be called conic-quasi-Newton equation. In a word, to make the interpolation conditions (3.7a-d) to be valid for the model (3.5) with k replaced by $k + 1$, it is sufficient that A_{k+1} and b_{k+1} satisfy (3.16) and (3.11) respectively.

Note that

$$b_{k+1} = \frac{(1 - \beta_k)a_k}{a_k^T p_k}, \tag{3.17}$$

for any $a_k \in R^n$ with $a_k^T p_k \neq 0$, will satisfy (3.11). Considering the special choice $a_k = g_k$ as in Ariyawansa (1990), we have

$$b_{k+1} = \frac{(1 - \beta_k)}{g_k^T p_k} g_k. \tag{3.18}$$

Taking (3.18) into (3.16), we get

$$A_{k+1} p_k = q_k, \tag{3.19}$$

where

$$q_k = \beta_k g_{k+1} - \beta_k^3 g_k. \tag{3.20}$$

Observe that the equation (3.19) corresponds to (1.9) in the standard variable metric methods. So, update formulae of A_k can be derived similarly to the standard updates. For example, we have the conic-BFGS update as follows

$$A_{k+1} = A_k - \frac{A_k p_k p_k^T A_k}{p_k^T A_k p_k} + \frac{q_k q_k^T}{q_k^T p_k}, \tag{3.21}$$

which corresponds to the BFGS update (1.11) in the standard variable metric methods.

It is obvious that the positive definiteness of A_{k+1} in (3.12) is related with the quantity $p_k^T q_k$. Since β_k is computed by (3.15), we have

$$p_k^T q_k = \beta_k p_k^T (g_{k+1} - \beta_k^2 g_k) = \beta_k [p_k^T g_{k+1} - \frac{(f(x_k) - f(x_{k+1}) + \rho_k)^2}{(p_k^T g_k)^2} p_k^T g_k]. \tag{3.22}$$

$$= 2\rho_k \beta_k^2 > 0$$

So, the choice of β_k ensures that (3.22) holds for all $k \geq 1$, and then all the matries $A_k, k = 1, 2, 3, \cdots$ generated by (3.21) are positive definite if A_1 is positive definite. Therefore the following conic-BFGS algorithm is well defined.

Algorithm 3.1

1^0 Select $b_1 \in R^n$. Select a starting point x_1 and a positive definite matrix A_1. Set $k = 1$;

2^0 Compute $g_k = \nabla f(x_k)$. If $g_k = 0$, stop; otherwise, go to step 3^0;

3^0 Set $d_k = -A_k^{-1} g_k$, $\delta_k = b_k^T d_k$;

4^0 Compute the step length $\lambda_k > 0$ such that $x_{k+1} = x_k + \lambda_k d_k$ satisfies that $\rho_k^2 = (f(x_k) - f(x_{k+1}))^2 - \lambda_k^2 (g_{k+1}^T d_k)(g_k^T d_k) > 0$ and $f(x_{k+1}) < f(x_k)$. If $1 - \delta_k > 0$, begin by the trial step length $1/(1 - \delta_k)$;

5^0 Set $x_{k+1} = x_k + \lambda_k d_k$;

6^0 Compute $g_{k+1} = \nabla f(x_{k+1})$. If $g_{k+1} = 0$, stop; otherwise, go to step 7^0;

7^0 Set $p_k = x_{k+1} - x_k$, $\rho_k = +\sqrt{\rho_k^2}$, $\beta_k = (f(x_k) - f(x_{k+1}) + \rho_k)/(-g_k^T p_k)$ and set $b_{k+1} = (1 - \beta_k)g_k/(p_k^T g_k)$;

8^0 Set $q_k = \beta_k g_{k+1} - \beta_k^3 g_k$ and

$$A_{k+1} = A_k - \frac{A_k p_k p_k^T A_k}{p_k^T A_k p_k} + \frac{q_k q_k^T}{q_k^T p_k};$$

9^0 Increase k by one, and go to step 3^0.

Remark: Due to the form of the conic-quasi-Newton equation (3.16), together with (3.17), (3.15) and (3.13), the above approach provides a rather large class of conic-variable metric methods. On the one hand, the vector a_k (or b_{k+1}) in this class can be considered as a parameter vector. On the other hand, many update formulae can be used since A_{k+1} only needs to satisfy (3.16). Therefore, we now have two degrees of freedom.

When the vector a_k is specified as the gradient $g_k = \nabla f(x_k)$, the conic-quasi-Newton equation (3.16) is reduced to (3.19). Together with (3.20), (3.15) and (3.13), there are still many conic-variable metric methods. The above conic-BFGS method (Algorithm 3.1) is derived from the BFGS update. Similarly, as extension of Broyden(1967), Oren(1972) and Spedicato(1976) classes, we can get the conic-Broyden class, conic-Oren self-scaling class and conic-Spedicato class with three parameters etc.

It is easy to see that there is a close relationship between the conic-variable metric methods and the variable metric methods with nonlinear scaling invariance discussed in Section 2.1, e.g. the conic-Broyden class is a special case of the Algorithm 2.1 when $\phi_k = \beta_k$, $\sigma_k = \beta_k^2$.

3.3 OTHER ALGORITHMS RELATED TO ALGORITHM 3.1

Sorensen (1980) indicated how the collinear scaling algorithms may be derived as a natural generalization of the standard variable metric methods. Consider problem (1.1). At the current point x_k, a local collinear scaling $S_k : w \to X$ of the form

$$x = x_k + \frac{J_k w}{1 + h_k^T w}, \quad w \in W_k, \tag{3.23}$$

is used to scale X, where $J_k \in R^{n \times n}$ is nonsingular, $W_k = \{w : 1 + h_k^T w \neq 0\}$ and X is the domain of $f(x)$. Then a local quadratic function ψ_k is used to approximate $\Phi_k(w) = f(S_k(w))$

$$f(x_k + J_k w/(1 + h_k^T w)) = \Phi_k(w) \approx \psi_k(w) = f(x_k) + [J_k^T g_k]^T w + \frac{1}{2} w^T B_k w, \tag{3.24}$$

where $B_k \in R^{n \times n}$, which together with the map S_k (3.23) yields the local conic approximation to $f(x)$

$$f(x) \approx \Psi_k(x) = f(x_k) + \frac{g_k^T(x - x_k)}{1 - h_k^T J_k^{-1}(x - x_k)} + \frac{1}{2} \frac{(x - x_k)^T J_k^{-1} B_k J_k^{-1}(x - x_k)}{[1 - h_k^T J_k^{-1}(x - x_k)]^2}. \quad (3.25)$$

Sorensen (1980) chooses J_k, h_k, B_k by means of appropriate updating formulae so that the local conic approximation (3.25) interpolates the function values and gradients of f at the two most recent iterates and possibly at several additional past iterates. His analysis very critically depends on the specific formula used for updating B_k, J_k, h_k. A typical and also the only pratical one of Sorensen's algorithms is the collinear scaling BFGS algorithm (his Algorithm 6.1). It should be pointed out that Sorensen's technique can not obtain a similar collinear scaling DFP formula. This is because it seems impossible to obtain the analogue of Lemma 4.1 in Sorensen(1980) to provide a closed-form of update formula for $C_k = J_k A_k^{-1} J_k^T$

Xu et al.(1991) extended Sorensen's collinear scaling BFGS algorithm to a class of collinear scaling Broyden algorithms, including collinear scaling DFP algorithm. They used the same local collinear scaling and local quadratic approximation as the ones in (3.23) and (3.24) respectively. But they established a closed form of update formula for $C_k = J_k A_k^{-1} J_k^T$ and therefore overcame the difficulty in Sorensen's technique.

Mainly for tutorial introduction, Sorensen (1982) provided another derivation of collinear scaling algorithms in which underlying local conic approximations interpolate the function values and the gradients only at the two most recent iterates. In this derivation he used the local collinear scaling S_k of (3.23) at x_k to scale X, and the local quadratic approximation ψ_k (3.4) with $B_k = I$ for all k to approximate $\Phi_k(w) = f(S_k(w))$. Moreover, in the process of forcing the local conic approximation at x_k to interpolate function values and gradients at x_k and x_{k-1}, he did not use the consistency condition (the equation (2.7) in Sorensen(1980)) but rather used the condition $x_{k-1} = S_k(-\bar{v}_{k-1})$ for appropriately chosen $\bar{v}_{k-1}$ (the equation (6.4) in Sorensen(1982)).

Ariyawansa (1990) also derived collinear scaling algorithms whose underlying local conic approximation at x_k were forced to interpolate the function values and gradients of f at x_{k-1} and x_k only. In particular, he used the local collinear scaling (3.23) with $J_k = I$ for all k and replaced the consistency condition, $x_{k-1} = S_k(-v_{k-1})$, where v_{k-1} is such that $x_k = S_{k-1}(v_{k-1})$, with the condition $x_{k-1} = S_k(-\tilde{v}_{k-1})$, where $\tilde{v}_{k-1}$ is chosen appropriatly. Letting $p_k = x_{k+1} - x_k$, he got the following equations which B_{k+1} and h_{k+1} should satisfy

$$B_{k+1}\tilde{v}_k = r_k, \quad (3.26)$$

$$h_{k+1}^T \tilde{v}_k = 1 - \gamma_k, \quad (3.27)$$

where

$$r_k = g_{k+1} - \frac{1}{\gamma_k}[I + h_{k+1}s_k^T]g_k, \quad (3.28)$$

$$\tilde{v}_k = \gamma_k p_k, \tag{3.29}$$

and γ_k is a root of the quadratic equation

$$(g_{k+1}^T p_k)\gamma_k^2 + 2[f(x_k) - f(x_{k+1})]\gamma_k + g_k^T p_k = 0, \tag{3.30}$$

i.e.

$$\gamma_k = \frac{-g_k^T p_k}{f(x_k) - f(x_{k+1}) + \rho_k}, \tag{3.31}$$

where

$$\rho_k = \sqrt{(f(x_k) - f(x_{k+1}))^2 - (g_k^T p_k)(g_{k+1}^T p_k)}. \tag{3.32}$$

This is the basis of Ariyawansa's Algorithmic Schema 2.1. Ariyawansa (1990) has shown that if the parameter h_{k+1} in his Algorithmic Schema 2.1 is chosen so that $h_{k+1} = (1 - \gamma_k)g_k/(\gamma_k s_k^T g_k)$ for all k, then this special case and the class of algorithms implicit in Sorensen(1982), pp.154-156, are equivalent. Indeed the class of algorithms of Sorensen may be treated as a factored version of the subclass of Ariyawansa's Algorithmic Schema 2.1. Moreover a member of Ariyawansa's algorithms related to BFGS method is equivalent to Sorensen's Algorithm 6.1 in Sorensen(1980).

Now we point out that our algorithms described in the Remark of Section 3.2 essentially include all of the methods mentioned above although our derivation is natural and simple. In fact, for the same points x_k and x_{k+1}, β_k defined by (3.15) with (3.13) and γ_k defined by (3.31) with (3.32) have the following relationship: $\beta_k = 1/\gamma_k$. Moreover if $h_{k+1} = -b_{k+1}$, it is easy to verify that the equation (3.16) is equivalent to the equation (3.26). So, our algorithm class implicit in the Remark of Section 3.2 is the same as the Algorithmic Schema 2.1 in Ariyawansa(1990). Particularly, we have the following conclusion: The algorithm class with the limitation $a_k = g_k$ implicit in the Remark of Section 3.2, Sorensen's algorithm class implicit in Sorensen(1982), pp. 154-156, and the Algorithmic Schema 2.1 in Ariyawansa(1990) are equivalent if the horizon vector used in Ariyawansa's algorithms takes the form similar to (3.18).

3.4 THE CONVERGENCE OF THE CONIC-VARIABLE METRIC METHODS

For simplicity, we shall mainly discuss the convergence properties of the conic-BFGS algorithm. The local convergence of conic-variable metric algorithms (or collinear scaling algorithms) was first discussed by Sorensen (1980). He showed that his Algorithm 6.1 with direct iterations in Sorensen(1980) is locally and superlinearly convergent. Ariyawansa et al. (1992) extended this result to the algorithms related to Broyden's class. Following the discussion in Section 3.3, we immediately get the local convergence of Algorithm 3.1 above with direct iterations, where 'direct iterations' means that the step length $\lambda_k = 1/(1 + b_k^T A_k^{-1} g_k)$ is chosen for all k. More precisely, we have the following

Theorem 3.1 Assume that $f(x)$ is twice continuously differentiable and uniformly convex on the neighborhood of x^*, where x^* is a strong local minimizer of $f(x)$.

Suppose that the sequence $\{x_k\}$ is generated by Algorithm 3.1 with direct iterations from initial quantities x_1, b_1 and symmetric positive definite matrix A_1. Then there is a positive scalar δ such that if

$$\| x_1 - x^* \| < \delta, \quad \| A_1 - \nabla^2 f(x^*) \| < \delta \quad \text{and} \quad | 1 - \lambda_1 | < \delta,$$

the sequence $\{x_k\}$ is well-defined and converges to x^* superlinearly.

The global convergence of the conic-variable metric algorithms, (or the collinear scaling algorithms) was discussed by Deng and Li (1993a). They proved the following

Theorem 3.2 Let x_1 be a starting point for which f is twice continuously differentiable and uniformly convex on the level set $D = \{x : f(x) \leq f(x_1)\}$. Then for any positive definite matrix A_1, the sequence $\{x_k\}$ generated by Algorithm 3.1 with exact line searches is well defined and converges to the minimizer x^* of $f(x)$ superlinearly.

However, Theorem 3.2 is useful only in theory. Practically, it is best to be quite tolerant in the criterion determining when to finish the line searches. Deng and Li (1993b) investigated the conic-variable metric method with Wolfe conditions (1.3) and (1.4). Note that the new iterate x_{k+1} obtained by the Wolfe line search might destroy the validity of the inequality (3.13). This induces the algorithm not to be well-defined. They overcame this difficulty by presenting two remedy approaches: (i) discard the desire that $f(x)$ is approximated by a conic function around x_{k+1}, and simply make a quadratic approximation to $f(x)$; (ii) insist on making a conic approximation to $f(x)$ around x_{k+1}, but require the new conic approximation only to satisfy the interpolation (3.7a), (3,7b) and (3.7d). More precisely, they suggested to replace Step 4^0 and Step 7^0 in Algorithm 3.1 by the following Step 4' and Step 7' respectively:

4' Compute the step length $\lambda_k > 0$ such that both (1.3) and (1.4) are satisfied. If $1 - \delta_k > 0$, begin by the trial step length $1/(1 - \delta_k)$;

7' Set $p_k = x_{k+1} - x_k$. If $\rho_k^2 = (f(x_k) - f(x_{k+1}))^2 - (g_{k+1}^T p_k)(g_k^T p_k) > 0$, set $\rho_k = \sqrt{\rho_k^2}$, $\beta_k = (f(x_k) - f(x_{k+1}) + \rho_k)/(-g_k^T p_k)$; otherwise set $\beta_k = 1$ or $\beta_k = \sqrt{(g_{k+1}^T p_k)/(g_k^T p_k)} + \epsilon$, where ϵ is a small positive scalar. Set $b_{k+1} = (1-\beta_k)g_k/(p_k^T g_k)$.

Deng and Li (1993b) proved that the algorithm with the above modification is well-defined, and is also globally and superlinearly convergent under suitable conditions.

4. The conic-conjugate gradient methods

Conjugate gradient methods based on conic model were first investigated by Davidon (1982) and then by Gourgeon et al. (1985). They are generalizations of the conjugate gradient methods for minimizing a convex quadratic and find the minimizer of a normal conic function in at most n iterations. Liu et al. (1989) developed the method for general unconstrained optimization. Now we introduce the conic-conjugate gradient algorithm described in Liu et al. (1989) briefly. Given a reference point $x_1 \in R^n$, we write any $x \in R^n$ as $x = x_1 + s$. With respect to this reference point we define the normal conic function

$$f(x) = f(x_1 + s) = f_1 + \frac{g_1^T s}{1 + b^T s} + \frac{1}{2} \frac{s^T A s}{(1 + b^T s)^2}, \tag{4.1}$$

where $g_1, b \in R^n, A$ is an $n \times n$ positive definite matrix with $1 + b^T A^{-1} g \neq 0$. Our purpose is to establish a conjugate gradient-type algorithm for the objective function (4.1). Consider the transformation introduced in Section 3.1

$$w = \frac{s}{1 + b^T s}, \tag{4.2}$$

or equivalently

$$s = \frac{w}{1 - b^T w}. \tag{4.3}$$

Then the conic function (4.1) is transformed into a quadratic function in w

$$h(w) = f_1 + g_1^T w + \frac{1}{2} w^T A w. \tag{4.4}$$

To simplify the formulae below we define

$$\gamma(x) = 1 + b^T s = \frac{1}{1 - b^T w}, \tag{4.5}$$

so that $w = s/\gamma$.

The transformations (4.2) and (4.3) provide an approach for minimizing the normal conic function. In fact, the standard conjugate gradient methods can be applied to the strictly convex quadratic function h (4.4) to generate the iterates in w-space. Then, to find the minimizer of $f(x)$, we only need to transform these iterates, given so far in terms of the variable w, back into the original variable s by the transformation (4.3).

Let $w_1 = 0$, $z_0 = 0$. While minimizing h, the displacement at the k-th step is

$$w_{k+1} - w_k = \nu_k z_k \equiv u_k, \quad k = 1, 2, \ldots, \tag{4.6}$$

where ν_k is the step length and z_k is the search direction. According to (1.14) – (1.16), z_k is defined by

$$z_k = \begin{cases} -\nabla h(w_k), & \text{for } k = 1; \\ -\nabla h(w_k) + \theta_k z_{k-1}, & \text{for } k \geq 2, \end{cases} \tag{4.7}$$

where

$$\theta_k = \frac{\| \nabla h(w_k) \|^2}{\| \nabla h(w_{k-1}) \|^2}, \quad \text{or} \quad \theta_k = \frac{\nabla h(w_k)^T (\nabla h(w_k) - \nabla h(w_{k-1}))}{\| \nabla h(w_{k-1}) \|^2}. \tag{4.8}$$

Now we need to express some quantities in the w-space by the ones in the x-space. It follows from (4.1) and the chain rule that

$$\nabla h(w) = \gamma(x)(I + bs^T)g(x), \tag{4.9}$$

where $g(x) = \nabla f(x)$. From (4.2) and (4.5) we get

$$
\begin{aligned}
u_k &= \frac{s_{k+1}}{\gamma_{k+1}} - \frac{s_k}{\gamma_k} \\
&= \frac{s_{k+1} - s_k}{\gamma_{k+1}} - \frac{s_k}{\gamma_k} + \frac{s_k}{\gamma_{k+1}} \\
&= \frac{s_{k+1} - s_k}{\gamma_{k+1}} - s_k\Big(\frac{1 + b^T s_{k+1} - 1 - b^T s_k}{\gamma_k \gamma_{k+1}}\Big) \\
&= \frac{1}{\gamma_{k+1}}\Big(I - \frac{s_k b^T}{\gamma_k}\Big)(s_{k+1} - s_k),
\end{aligned}
\tag{4.10}
$$

where $\gamma_k = \gamma(x_k)$, $\gamma_{k+1} = \gamma(x_{k+1})$. Now express γ_{k+1} in terms of quantities evaluated at x_k. From (4.5) and (4.6), we have

$$
\begin{aligned}
\gamma_{k+1} &= \frac{1}{1 - b^T w_{k+1}} = \frac{1}{1 - b^T w_k - b^T u_k} \\
&= \frac{1}{\frac{1}{\gamma_k} - b^T u_k} = \frac{\gamma_k}{1 - \gamma_k b^T u_k}.
\end{aligned}
\tag{4.11}
$$

Noting that the inverse of $(I - s_k b^T/\gamma_k)$ is $(I + s_k b^T)$, from (4.10) and (4.11) we obtain

$$s_{k+1} = s_k + \frac{\gamma_k}{1 - \gamma_k b^T u_k}(I + s_k b^T)u_k. \tag{4.12}$$

This formula defines the iterates in terms of the original variable s, or equivalently, in terms of x.

According to (4.5), (4.9), (4.7), (4.6) and (4.12), we can now describe an algorithm for minimizing the normal conic function $f(x)$ defined by (4.1) assuming that the horizon vector b is known.

Algorithm 4.1
1^0 Select a starting point x_1. Set $s_1 = 0$ and $k = 1$;
2^0 Compute

$$
\begin{aligned}
\gamma_k &= (1 + b^T s_k), \\
\nabla h(w_k) &= \gamma_k(I + bs_k^T)g_k,
\end{aligned}
$$

where $g_k = \nabla f(x_k)$;

3^0 If $k = 1$, set $z_k = -\nabla h(w_k)$; otherwise set

$$z_k = -\nabla h(w_k) + \theta_k z_{k-1},$$

$$\theta_k = \frac{\| \nabla h(w_k) \|^2}{\| \nabla h(w_{k-1}) \|^2}, \quad \text{or} \quad \theta_k = \frac{\nabla h(w_k)^T(\nabla h(w_k) - \nabla h(w_{k-1}))}{\| \nabla h(w_{k-1}) \|^2};$$

4^0 Compute the search direction

$$d_k = \frac{\gamma_k}{1 - \gamma_k b^T z_k}(I + s_k b^T)z_k$$

and the step length λ_k by line search for $f(x_1 + s_k + \lambda d_k)$;
5^0 Set

$$s_{k+1} = s_k + \lambda_k d_k, \quad x_{k+1} = x_1 + s_{k+1}.$$

6^0 If the solution has been obtained, stop; otherwise increase k by one, and go to step 2^0.

It is easy to see from (4.12) that $\lambda_k = 1$ corresponds to $\nu_k = 1$, where ν_k is defined in (4.6). Due to the quadratic termination of the standard CG method, and the fact that minimizing h and f is equivalent, it follows that Algorithm 4.1 minimizes $f(x)$ defined by (4.1) within at most n steps, if the line searches are exact. To make Algorithm 4.1 practical for minimizing general functions, we need to eliminate all quantities dependent on the horizon vector b. The following two theorems enable us to compute b using the function values and gradients of (4.1) at three collinear points.

Theorem 4.1 Assume that $\gamma_i = \gamma(x_i)$, $i = k, k+1$, is defined by (4.5). Then the ratio γ_{k+1}/γ_k can be computed by the function values and gradients at x_k and x_{k+1}

$$\frac{\gamma_{k+1}}{\gamma_k} = \frac{-g_k^T(x_{k+1} - x_k)}{f(x_k) - f(x_{k+1}) + \rho_k} \tag{4.13}$$

where $\gamma_1 = 1$ and $\rho_k = \sqrt{(f(x_k) - f(x_{k+1}))^2 - g_{k+1}^T(x_{k+1} - x_k)g_k^T(x_{k+1} - x_k)}$.

Theorem 4.2 The horizon vector b can be computed by

$$b = -\frac{(\lambda_t - \lambda_k)\gamma_k^2 g_k + \lambda_k \gamma_t^2 g_t - \lambda_t \gamma_{k+1}^2 g_{k+1}}{(\lambda_t - \lambda_k)\gamma_k^2 s_k^T g_k + \lambda_k \gamma_t^2 s_t^T g_t - \lambda_t \gamma_{k+1}^2 s_{k+1}^T g_{k+1}} \tag{4.14}$$

where $s_i = x_i - x_1$, $\gamma_i = \gamma(x_i)$ is defined by (4.5) and can be obtained by (4.13), $i = k, t, k+1$; $x_{k+1} = x_k + \lambda_k d_k$, $x_t = x_k + \lambda_t d_k$, $\lambda_t \neq \lambda_k$, $\lambda_t, \lambda_k \neq 0$.

Due to formulae (4.13) and (4.14), Algorithm 4.1 can be generalized to minimize the general objective functions. See detail in Liu et al. (1989).

Lukšan (1986a) presented another algorithm for minimizing the normal conic function

$$c(s) = f + \frac{g^T s}{1 + b^T s} + \frac{\frac{1}{2}s^T A s}{(1 + b^T s)^2} \tag{4.15}$$

where A is positive definite with $1 + b^T A^{-1} g \neq 0$. The basic idea of his algorithm is rather simple. Note that $c(s)$ restricted on the superplane $1 + b^T s = \mu_1$ is a strictly convex quadratic, where $\mu_1 \neq 0$. So, the linearly constrained optimization problem

$$P(\mu_1): \min_{1 - a^T s = \mu_1} c(s) \tag{4.15}$$

is easily solved. It has been shown that if $s_i * *$ is a minimizer of $P(\mu_i)$, where $\mu_i \neq 0$ $(i = 1, 2)$ and $\mu_1 \neq \mu_2$, then the minimizer of $c(s)$ lies on the line connecting s_1^* and s_2^*. Therefore, we can find the minimizer of (4.1) by a line search at last.

Further generalization has also been made to the extended conic model $f(s) = \chi(q(s), l(s))$ in Luksăn (1986b) and $f(s) = \chi(q(s), l_1(s)), \cdots, l_m(s))$ in Zhu et al. (1992) respectively, where $q(x)$ is a strictly convex function, l, l_i $(i = 1, \ldots, m)$: $R^n \to R$ are linear functions and $\chi : R^2 \to R$ is a twice continuously differentiable function.

References

[1] Ariyawansa, K.A. (1990) 'Deriving collinear scaling algorithms as estension of quasi-Newton methods and the local convergence of DFP and $BFGS$-related collinear scaling algorithms,' Math. Prog. 49, 23-48.

[2] Ariyawansa, K.A. and Lau, D.T.M. (1992) 'Local and Q-superlinear convergence of a class of collinear scaling algorithms that extends quasi-Newton methods with Broyden's bounded-Φ class of updates,' Optimization 23, 323-339.

[3] Biggs, M.C. (1971) 'Minimization algorithms making use of non-quadratic properties of the objective function', J. Inst. Maths. Applics. 8, 315-327.

[4] Boland, W.R., Kamgnia, E.R. and Kowalik, J.S. (1979) 'A conjugate-gradient optimization method invariant to nonlinear scaling,' J. Optim. Theory Appl. 27, 221-230.

[5] Broyden, C.G. (1967) 'Quasi-Newton methods and their application to function minimization' Maths. Comput., 21, 368-381.

[6] Charalambous, C. (1973) 'Unconstrained optimization based on homogeneous models', Math. Prog. 5, 189-198

[7] Davidon, W.C. (1980) 'Conic approximations and collinear scalings for optimizers,' SIAM J. Numer. Anal. 17, 268-281.

[8] Davidon, W.C. (1982) 'Conjugate directions for conic functions', in M.J.D. Powell (ed.), Nonlinear Optimization 1981, Academic Press, New York, 23-28.

[9] Davison, E.J. and Wong, P. (1975) 'A robust conjugate-gradient algorithm which minimizes L-Functions', Automatica 11, 297-308.

[10] Deng, N.Y. and Chen, Z. (1987) 'On the algorithms invariant to nonlinear scaling with inexact line searches,' Chinese Annals of Mathematics 8b, 65-69.

[11] Deng, N.Y. and Li, Z.F. (1993a) 'On the global convergence of conic-variable metric methods for unconstrained minimization,' to appear.

[12] Deng, N.Y. and Li, Z.F. (1993b) 'Some global convergence of conic-variable metric methods with inexact line search,' to appear.

[13] Deng, N.Y. and Li, Z.F. (1993c) 'On the convergence of the extended conjugate gradient method with invariance to nonlinear scaling,' to appear.

[14] Flachs, J. (1982) ' On the convergence, invariance and related aspects of a modification of Huang's algorithm', J. Optim. Theory Appl.37, 315-341.

[15] Gourgeon, H. and Nocedal, J. (1985) 'A conic algorithm for optimization,' SIAM J. Sci. Stat. Comput. 6, 253-267.

[16] Jacobson, D.H. and Oksman, W. (1972) 'An algorithm that minimizes homogeneous functions of N variables in $N+2$ iterations and rapidly minimizes general functions', J. Math. Anal. and Appl. 38, 535-552.

[17] Kowalik, J. S. and Ramakrishnan, K. G.(1976) ' A numerically stable optimization method based on a homogeneous function', Math. Prog. 11, 50-66.

[18] Liu, D.C. and Nocedal, J. (1989) 'Algorithms with conic termination for nonlinear optimization,' SIAM J. Sci. Stat. Comput. 10, 1-17.

[19] Luksăn, L. (1986a) 'Conjugate gradient algorithms for conic functions,' Aplikace Matematiky 31, 427-440.

[20] Luksăn, L. (1986b) 'Conjugate direction algorithms for extended conic functions,' Kybernetika 22, 31-46.

[21] Oren S.S., (1972) 'Self-scaling variable metric algorithm for unconstrained minimization', Ph. D. Dissertation, Stanford University.

[22] Oren, S.S. and Spedicato, E. (1976) 'Optimal conditioning of self-scaling variable metric algorithms', Math. Prog., 10, 70-90.

[23] Schnabel, R.B. (1983) 'Conic methods for unconstrained minimization and tensor methods for nonlinear equations,' in A. Bachem, M. Grotschel and B. Korte (eds.), Math. Prog., The State of the Art, Berlin, Springer-Verlag, 417-438.

[24] Schnabel, R.B. (1989) 'Sequential and parallel methods for unconstrained optimization,' in M. Iri and K. Tanabe (eds.), Mathematical Programming, KTK Scientific Publishers, Tokyo, 227-261.

[25] Schnabel, R.B. and Chao, T. (1991) 'Tensor methods for unconstrained optimization using second derivatives', SIAM J. Optimization 1, 293-315.

[26] Shirey, J.E. (1982) 'Minimization of extended quadratic functions,' Numer. Math. 39, 157-161.

[27] Sloboda, F. (1980) 'A generalized conjugate gradient algorithm for minimization,' Numer. Math. 35, 223-230.

[28] Sorensen, D.C. (1980) 'The Q-superlinear convergence of a collinear scaling algorithm for unconstrained optimization,' SIAM J. Numer. Anal. 17, 84-114.

[29] Sorensen, D.C. (1982) 'Collinear scaling and sequential estimation in sparse optimization algorithms,' Math. Prog. Study 18, 135-159.

[30] Spedicato, E. (1976) 'A variable metric method for function minimization derived from invariancy to nonlinear scaling', J. Optim. Theory Appl. 20, 315-329.

[31] Spedicato, E., (1978) 'A note on the determination of the scaling parameters in a class of quasi-Newton methods for unconstrained minimization', J. Inst. Maths. Applics., 21, 285-291.

[32] Xu, H.F. and Sheng, S.B. (1991) 'Broyden family of collinear scaling algorithm for unconstrained optimization', Numer. Math., A Journal of Chinese Universities, 13, 318-330.

[33] Zhu, M.F. and Xue, Y. (1992) 'An algorithm for unconstrained optimization with non-quadratic model', in Y. Wang et al. (eds.) OR and Decision Making, Chengdu University of Science and Technology Press, 409-416.

ALGORITHMS FOR GENERAL CONSTRAINED NONLINEAR OPTIMIZATION

M.C. BARTHOLOMEW-BIGGS
Numerical Optimisation Centre
Mathematics Division
University of Hertfordshire
Hatfield, England

ABSTRACT   This paper gives an overview of some widely used methods for constrained optimization.   In particular it deals with the special cases of quadratic programming and linear constraints before describing penalty function and sequential QP approaches to the general nonlinearly constrained problem.   The paper also highlights some current areas of research and mentions some algorithms which are still under development.

1. Introduction

The general constrained optimization problem is

$$\text{Minimize} \quad F(x)$$
$$\text{s.t.} \quad c_i(x) = 0 \quad i = 1,\dots, m_e \qquad c_i(x) \geq 0 \quad i = m_e+1,\dots, m \tag{1}$$

where x is the vector $(x_1,\dots,x_n)$ of independent variables and the objective function, F and constraints, $c_1,\dots,c_m$, are assumed to be twice continuously differentiable.   The equality and inequality constraints together define a *feasible region*.   A local solution, x^*, of (1) is characterized by several *optimality conditions* as follows.

First of all x^* must be feasible, i.e.

$$c_i(x^*) = 0 \quad i = 1,\dots, m_e \text{ and } c_i(x^*) \geq 0 \quad i = m_e+1,\dots, m \tag{2}$$

In general, the inequality constraints will not all be satisfied as equalities at x^* - i.e. a solution will lie on some constraint boundaries and inside others.   The term *binding constraints* refers to the equalities and those inequalities such that $c_i(x^*) = 0$.

We must also have $F(x^*) \leq F(x)$ for all feasible points x near x^*.   This implies that the gradient $\nabla F(x^*)$ must lie in the space spanned by the binding constraint normals, $\nabla c_i(x^*)$ and hence there exist scalars $\lambda_1^*,\dots,\lambda_m^*$ such that

$$\nabla F(x^*) - \sum_{i=1}^{m} \lambda_i^* \nabla c_i(x^*) = 0 \tag{3}$$

The λ_i^* are known as *Lagrange multipliers* and the function

$$L^*(x) = F(x) - \sum_{i=1}^{m} \lambda_i^* c_i(x) \tag{4}$$

is called the *Lagrangian*.   Clearly, (3) is equivalent to the stationarity condition

$$\nabla L^*(x^*) = 0 \tag{5}$$

169

E. Spedicato (ed.), Algorithms for Continuous Optimization, 169–207.
© 1994 *Kluwer Academic Publishers.*

The Lagrange multipliers associated with inequality constraints in (1) must also satisfy

$$\lambda_i^* \geq 0 \quad \text{and} \quad \lambda_i^* c_i(x^*) = 0 \qquad i = m_e+1, ..., m \tag{6}.$$

The first condition in (6) implies that F cannot be reduced by a move off an inequality constraint boundary into the interior of the feasible region while the second (*complementarity*) condition implies $\lambda_i^* = 0$ for all non-binding constraints.

The stationarity condition (3) can be expressed in terms of the *null space* of A^*, the Jacobian matrix whose rows are the normals $(\nabla c_i(x^*))^t$ of the binding constraints. If Z is an orthonormal matrix such that $A^* Z = 0$ then the columns of Z span the subspace tangential to the binding constraints. The vector $Z^t \nabla F(x^*)$ is called the *reduced gradient* and the condition

$$Z^t \nabla F(x^*) = 0 \tag{7}$$

ensures that F is stationary for all feasible moves away from x^*.

The matrix Z also appears in a second order condition which distinguishes a constrained local minimum from a maximum or saddle point. The matrix $Z^t \nabla^2 L^*(x^*) Z$ is called the *reduced Hessian* of the Lagrangian in the subspace tangential to the binding constraints and it must be positive semi-definite if x^* is a solution of (1).

Numerical methods for solving constrained minimization problems are iterative and typically generate a sequence of estimates (x, λ) which are intended to converge to (x^*, λ^*). The optimality conditions (2) and (3) provide convenient measures which can be used in practice to terminate the iterations. For instance, let $T(x,\lambda)$ be defined by

$$T(x,\lambda) = \|\nabla L\| + \|v\| \tag{8},$$

where L is an approximate Lagrangian and v is a vector of constraint violations so that

$$L = F(x) - \sum_{i=1}^{m} \lambda_i c_i(x)$$

and

$$\begin{aligned} v_i \quad &= c_i(x) \quad i = 1,..., m_e \\ &= min(0, c_i(x)) \quad i = m_e+1,..., m. \end{aligned}$$

If we assume that the second order optimality condition holds then we can say that an iterative scheme for solving (1) has converged when $T(x,\lambda)$ is sufficiently small.

In the following sections we shall describe the main features of some well-known approaches to the constrained optimization problem. More information about these techniques can be found in texts by Gill et al (1982) and Fletcher (1987).

2. Quadratic Programming

When F is quadratic and all the c_i are linear then (1) is a quadratic programming (QP) problem. We shall begin by considering some QP algorithms because they embody ideas relevant to more general problems and also because (1) can be solved via a sequence of QPs.

2.1 EQUALITY CONSTRAINED QP (EQP)

In this section we consider the problem

$$\text{Minimize} \quad (x^t G x)/2 + g^t x \quad \text{s.t.} \quad a_i^t x + b_i = 0 \quad i=1,\dots, t \tag{9}.$$

We let A be the t-by-n matrix whose rows are a_i^t and b the t-vector whose elements are b_i. We shall assume that the rows of A are linearly independent. We also assume, for the moment, that G is positive definite, which implies (9) has a unique solution that can be found in a number of ways, all based on the optimality conditions.

EQP Algorithm 1

Determine x^*, λ^* by solving the linear system

$$G x^* - A^t \lambda^* = -g \tag{10}$$

$$-A x^* = b \tag{11}$$

Solving this $(n+t)$-by-$(n+t)$ system for x^* and λ^* simultaneously may be suitable when the matrices G and A are sparse; but for dense problems it is usually more efficient to consider expressions which give x^* and λ^* separately. The following technique is obtained by multiplying (10) by AG^{-1} and then making use of (11) to eliminate x^*.

EQP Algorithm 2

Determine x^* and λ^* from

$$\lambda^* = (AG^{-1}A^t)^{-1}(AG^{-1}g - b) \tag{12}$$

$$x^* = G^{-1}(A^t\lambda^* - g) \tag{13}$$

Note that (13) corresponds to the use of Newton's method to solve $\nabla L^* = 0$. Combining (12) and (13) gives an expression for x^* alone, i.e.

$$x^* = G^{-1}(A^t(AG^{-1}A^t)^{-1}AG^{-1}g - g) - G^{-1}A^t(AG^{-1}A^t)^{-1}b \qquad (14).$$

If h denotes the first term on the right of (14) and v the second then it is easy to verify that $Ah = 0$ and $Av = b$. It is the component v which ensures that x^* is feasible since h is in the null space of A. The component h causes the stationarity condition to be satisfied at x^* and it is sometimes called the *projected Newton step*.

A third EQP method obtains x^* and λ^* in terms of orthonormal basis matrices Y and Z which span the range and the null space of A. One way of calculating Y and Z is via *QR* factorization of A which produces an n-by-n orthogonal matrix Q such that $AQ = R = (L : 0)$ where L is t-by-t lower triangular. If Y consists of the first t columns of Q and Z denotes the remaining $n-t$ columns we have

$$AZ = 0, \quad Y^tZ = 0 \quad \text{and} \quad AY = L \qquad (15).$$

EQP Algorithm 3 determines x^* from its components in Y- and Z-space, i.e.

$$x^* = Yy^* + Zz^*. \qquad (16).$$

EQP Algorithm 3

Compute Y and Z to satisfy (15).

Determine y^* by solving the lower triangular system $\quad AYy^* + b = 0 \qquad (17)$

Determine z^* by solving $\quad Z^tGZz^* = -Z^tg - Z^tGYy^* \qquad (18)$

Calculate x^* from (16) and find λ^* by solving the upper triangular system

$$Y^tA^t\lambda^* = Y^tg + Y^tGx^* \qquad (19)$$

Equations (18), (19) are derived by premultiplying (10) by Z^t, Y^t respectively and using (16). Hence EQP Algorithm 3 computes x^* in terms of a *vertical* step

$$v^* = Yy^* = -Y(AY)^{-1}b,$$

in the space of the constraint normals and a *horizontal* step in the tangent space, i.e.

$$h^* = Zz^* = -(Z^tGZ)^{-1}Z^t(g + Gv^*).$$

Another similar approach involves basis matrices which are *conjugate* with respect to G. Matrices U and W can be constructed so that

$$AU = 0, \quad U^tGW = 0 \quad \text{and} \quad AW = T,$$

where T is t-by-t lower triangular. This is done using a QR factorization of AJ^{-t}, where J is lower triangular and $JJ^t = G$. After finding Q' such that $AJ^{-t}Q' = (T : 0)$ we can set

$$U = J^{-t}Z' \quad \text{and} \quad W = J^{-t}Y',$$

where Y' and Z' denote the first t and the last n-t columns of Q'. It should be noted that the columns of U span the null space of A.

EQP Algorithm 4

Compute conjugate basis matrices U and W.

Obtain w^* from the lower triangular system $AWw^* + b = 0$

Obtain u^* by solving $U^tGUu^* = -U^tg$ and set $x^* = Uu^* + Ww^*$

Obtain λ^* from the upper triangular system $W^tA^t\lambda^* = W^tg + W^tGx^*$

In this algorithm the components u^* and w^* - unlike y^* and z^* - are calculated independently of each other. A useful consequence of this will appear in a later section. EQP Algorithm 4, however, requires G to be positive definite because it depends on the existence of the factorization $G = JJ^t$, whereas we only require positive definiteness of the reduced Hessian Z^tGZ to ensure that x^* and λ^* exist and can be found via EQP algorithms 1 - 3. If Z^tGZ is not positive definite then (9) has no solution. This point is relevant to the solution of inequality constrained QPs.

2.2 INEQUALITY CONSTRAINED QP (IQP)

In this section we consider the problem

$$\text{Minimize} \quad (x^tGx)/2 + f^tx \quad \text{s.t.} \quad a_i^tx + q_i \geq 0 \quad i = 1,..., m \qquad (20).$$

If G is positive definite then (20) has a unique solution which can be found by an iterative algorithm which solves a sequence of EQPs. Each iteration starts from an estimate of the solution, x, and involves the calculation of a *search direction, p^** and a *step size, s,* so that $x + sp^*$ is a better approximation to x^*.

IQP Algorithm 1 (Positive definite IQP)

Step 0: Choose an initial point x and set $\lambda_1 = ... = \lambda_m = 0$.

Step 1: Stop if x satisfies optimality conditions $Ax + q \geq 0$; $Gx + f - A^t\lambda^* = 0$; $\lambda^* \geq 0$

Step 2: Identify *active* constraints, i.e. those for which

$$a_i^t x + q_i < 0 \quad \text{or} \quad (a_i^t x + q_i = 0 \text{ and } \lambda_i^* \geq 0)$$

Renumber constraints so active set is $i = 1,..., t$

Set $g = Gx + f$ and $b_i = a_i^t x + q_i$ for $i = 1,..., t$

Step 3: Find p^* and μ^* to solve the EQP

$$\text{Minimize } (p^t Gp)/2 + g^t p \quad \text{s.t. } a_i^t p + b_i = 0 \quad i = 1,..., t \tag{21}$$

Set $\lambda_i^* = \mu_i^*$ $(i = 1,..., t)$ and $\lambda_i^* = 0$ $(i = t+1,..., m)$

Step 4: Set $s = 1$

For $i = t+1,..., m$ (i.e. for all *inactive* constraints)

if $a_i^t p^* < 0$ then set $s = min(s, - (a_i^t x + q_i)/a_i^t p^*)$

Replace x by $x + sp^*$ and return to Step 1.

If the active set found in Step 2 is the binding set for (20) then $x + p^*$ solves the IQP. Otherwise the active set will change either because the step sp^* encounters a new constraint or because the signs of the Lagrange multipliers, λ^*, indicate that some constraints may be dropped. Particular implementations of IQP algorithm 1 may differ in the method of solving (21) and also in rules about how many constraints may be dropped on a single iteration.

A more complicated process is needed when G is indefinite because, even when (20) has a unique solution, it may happen that (21) cannot be solved for some choice of active set in Step 2. One way to deal with this possibility is for Step 3 to use a variant of EQP Algorithm 3 which includes a check on the positive definiteness of $Z^t GZ$.

IQP Algorithm 2 (Indefinite IQP)

Identical with IQP Algorithm 1 except for

Step 3: Find Y and Z satisfying (15) and obtain y^* from (17).

If $Z^t GZ$ is not positive definite then

Obtain ξ by solving $\quad Z^t \hat{G} Z \xi = -Z^t g - Z^t GY y^* \qquad\qquad$ (22)

where $\hat{G}$ is a modification to G so that $Z^t \hat{G} Z$ is positive definite.

Set $z^* = \sigma\xi \qquad$ where $\sigma = -\xi^t Z^t (g + GY y^*)/\xi^t Z^t GZ\xi \quad$ if $\xi^t Z^t GZ\xi > 0$

$$\sigma = 1 \qquad \text{otherwise}$$

If $Z^t GZ$ is positive definite then obtain z^* from (18).

Calculate $p^* = Y y^* + Z z^*$. Find μ^* by solving $Y^t A^t \mu^* = Y^t g + Y^t G p^*$

Set $\lambda_i^* = \mu_i^* \ (i = 1,..., t)$ and $\lambda_i^* = 0 \ (i = t+1,..., m)$

This procedure finds the solution to EQP (21) if it exists; but in any case it provides p^* as a useful search direction so that a better estimate of the solution of (20) can be found by moving away from x along p^*.

Indefiniteness of $Z^t GZ$ is detected when negative diagonal elements arise during the Cholesky factorization involved in the solution of (18). Such negative elements can be corrected as they occur (see Gill et al, 1982) and this strategy implicitly creates a modified reduced Hessian $Z^t \hat{G} Z$ whose positive definiteness ensures that the horizontal step $Z\xi$ is a descent direction for the objective function. The step length σ is chosen to ensure that the distortion of second derivative information does not cause the new point to overshoot a one-dimensional minimum along the direction $Z\xi$.

There are other techniques for solving indefinite QPs which do not modify the Hessian but seek p^* as a direction of negative curvature when $Z^t GZ$ is found to be non-positive definite. As with IQP Algorithm 2, however, their aim is to produce a robust iterative scheme which gives good search directions to enable the correct active set to be identified.

IQP algorithms can of course deal with mixed equality/inequality QPs.

3. Projected and Reduced Gradient methods

In this section we shall chiefly be concerned with solving (1) when the constraints are linear but the objective function is non-quadratic. A common strategy in constrained minimization, as in unconstrained, is to use an iterative scheme based on local quadratic approximations to F. Near to a solution estimate, x, we suppose that

$$F(x + d) \cong Q(d) = F + d^t \nabla F(x) + (d^t H d)/2 \qquad (23)$$

where H is the Hessian matrix, $\nabla^2 F(x)$, or else some approximation to it. It is probably true to say that the majority of current algorithms use *quasi-Newton* techniques to construct an approximation to the Hessian which is updated on every iteration. One of the most successful methods of this kind uses the BFGS formula

$$H = H^- + H^- \gamma \gamma^t H^- / \gamma^t H \gamma + \delta \delta^t / \delta^t \gamma \qquad (24).$$

In (24) H^- denotes the Hessian approximation at a previous point x^-, δ denotes the step $x - x^-$ and γ is the corresponding change in gradient $\nabla F(x) - \nabla F(x^-)$. This formula ensures that H satisfies the quasi-Newton condition $H\delta = \gamma$.

Advantages of obtaining H by updating rather than setting $H = \nabla^2 F$ are (i) users do not have to obtain and code second derivative expressions; and (ii) it is easy to ensure that H is positive definite. Remark (i) may become less important as methods for *automatic differentiation* (Dixon, 1993) become more widely used; but remark (ii) remains quite persuasive. We have already seen that extra steps must be introduced into an IQP algorithm to deal with indefiniteness in the Hessian and we would like to be able to avoid such complications. The BFGS formula (24) gives a positive definite H provided H^- is positive definite and $\delta^t \gamma > 0$. Modified forms of (24) have also been proposed which yield positive definite H even when $\delta^t \gamma \leq 0$. One of these (Powell, 1978) involves replacing γ by $\bar{\gamma}$, where $\bar{\gamma} = \gamma + \theta H^- \delta$ and θ is chosen so that $\delta^t \bar{\gamma} > 0$.

In the unconstrained case it is easy to justify use of an update which makes H positive definite. For constrained problems, however, $\nabla^2 F(x^*)$ is usually indefinite and so a positive definite updating scheme may not be consistent with making H agree with the exact Hessian. Fortunately this need not inhibit convergence of quasi-Newton methods for linearly constrained minimization since it turns out that second order information is only significant in the null space of the binding constraint normals; and in this subspace the optimality conditions require F to have positive curvature.

Descriptions of algorithms are simplified if a positive definite Hessian estimate is assumed to be available and so we shall usually discuss methods in quasi-Newton form. In practice, however, the benefits of using exact second derivatives sometimes outweigh the drawbacks of having to deal with indefinite matrices.

If a search direction d is obtained using (23), the non-quadraticity of F implies that a *line search* should be used to calculate a step length α so that a new point $x + \alpha d$ is an improvement on x. This would be the case even if H were the exact Hessian at x.

A *perfect* line search finds α to minimize $F(x + \alpha d)$ (i.e. to satisfy the condition $\nabla F(x + \alpha d)^t d = 0$). A *weak* search merely seeks a value of α such that $F(x + \alpha d) - F(x)$ is negative and bounded away from zero. More details are given by Fletcher (1987).

We now give an outline of the projected and reduced gradient methods for linearly constrained minimization. Both are *feasible point* techniques - i.e. they generate points which satisfy the constraints on every iteration - and therefore they require a feasible starting approximation x.

Projected Gradient Algorithm (linear constraints)

Step 0: Choose an initial *feasible* point x and set $\lambda_1 = ... = \lambda_m = 0$.
Choose H as a positive definite estimate of $\nabla^2 F(x)$.

Step 1: Let $g = \nabla F(x)$ and N = the Jacobian matrix whose rows are $\nabla c_i(x)^t$.
Stop if x, λ satisfy optimality conditions $g - N^t \lambda = 0$; $\lambda_i \geq 0$ for $i = m_e + 1, ... m$

Step 2: Identify *active* inequalities i.e. those for which $c_i(x) = 0$ and $\lambda_i \geq 0$

Renumber constraints so active inequalities are $i = m_e + 1, ..., t$
Let A be the matrix whose rows are $\nabla c_i(x)^t$ for $i = 1, ..., t$

Step 3: Find d and μ from
$$d = H^{-1}(A^t(AH^{-1}A^t)^{-1}AH^{-1}g - g) \quad \text{and} \quad \mu = (AH^{-1}A^t)^{-1}AH^{-1}g$$

Set $\lambda_i = \mu_i$ ($i = 1, ..., t$) and $\lambda_i = 0$ ($i = t+1, ..., m$)

Step 4: Use a line search to find α so that $F(x + \alpha d) < F(x)$. Set $s = \alpha$.

For $i = t+1, ..., m$ (i.e. for all *inactive* constraints)
if $\nabla c_i(x)^t d < 0$ then set $s = min(s, -c_i(x)/(\nabla c_i(x)^t d))$
Set $x^- = x$, and $x = x^- + sd$.
Set $H^- = H$, update H and return to Step 1.

Clearly this method has features in common with IQP algorithms. Step 3 gives d as a projected quasi-Newton direction (similar to h in EQP Algorithm 2) with $Ad = 0$ in order to keep feasibility with respect to the active linear constraints. The value of s in Step 4 ensures that F is reduced and that at most one new constraint is introduced into the new active set. Note that the form of the calculations in Step 3 suggests that it might be appropriate to use a quasi-Newton update to approximate the *inverse* Hessian of F.

The reduced gradient approach to linearly constrained optimization is similar except that the search direction calculation involves orthogonal basis matrices as in EQP Algorithm 3.

Reduced Gradient Algorithm (linear constraints)

Identical to the projected gradient algorithm except for

Step 3: Calculate Y and Z satisfying (15)

Determine z from $Z^t HZz = -Z^t g$ and set $d = Zz$

Find μ by solving $Y^t A^t \mu = Y^t g + Y^t Hd$

Set $\lambda_i = \mu_i$ $(i = 1,..., t)$ and $\lambda_i = 0$ $(i = t+1,..., m)$

This method proceeds in much the same way as the previous one: each iteration makes a horizontal move in the subspace satisfying the current active constraints, the step size being chosen so as to reduce the objective function while encountering at most one new inequality. (Note that no vertical move is needed since all iterations start with a feasible point.)

The search direction for the reduced gradient approach is obtained from a positive definite system of $(n-t)$ equations while the projected gradient approach involves a system of t equations. Hence the reduced gradient algorithm is likely to use less arithmetic in obtaining d when the number of active constraints is close to n. Conversely, the projected gradient approach may be more efficient when t is small. While these remarks are rather simplistic (they ignore the overhead costs of finding Y and Z and of determining Lagrange multipliers) they do indicate the sort of practical consideration which makes one method preferable to another for a particular problem.

One case where the calculation of Z is rather simple and where the reduced gradient approach is especially attractive is when (1) involves no equality constraints and all the inequalities are simple bounds

$$l_i \leq x_i \leq u_i \quad i = 1,...,n.$$

In this situation it is a simple matter to split the variables at the start of each iteration into those which are "fixed" - i.e. on their bounds - and those which are "free". The Z matrix whose columns span the space of the free variables can then be taken simply as a partition of the identity matrix.

Reduced gradient Algorithm (simple bounds)

Step 0: Choose an initial feasible point x

Choose H as a positive definite estimate of $\nabla^2 F(x)$.

Step 1: Let $g = \nabla F(x)$ and set Z to be the $n \times n$ identity matrix

Identify the *active* bounds i.e. those for which

$$(x_i = l_i \text{ and } g_i > 0) \text{ or } (x_i = u_i \text{ and } g_i < 0)$$

and for each one, delete the i-th column of Z.

Step 2: Stop if $Z^t g = 0$. Otherwise set $d = Zz$ where $Z^t H Z z = -Z^t g$

Step 3: Use a line search to find s so that $F(x + sd) < F(x)$

For each variable x_i which is not on one of its bounds

Set $\quad s = min(s, (u_i - x_i)/d_i) \quad$ if $d_i > 0$

$\qquad\quad = min(s, (l_i - x_i)/d_i) \quad$ if $d_i < 0$

Set $x^- = x$, and $x = x^- + sd$

Set $H^- = H$, update H and return to Step 1.

We now consider the extension of projected and reduced gradient methods to deal with nonlinear constraints; and an extremely important point concerns the matrix H appearing in (23). In order to include second derivatives of the constraint functions in Q it is usual to treat H as an approximation to the Hessian of the Lagrangian

$$\nabla^2 L^* = \nabla^2 F - \sum_{i=1}^{m} \lambda_i \nabla^2 c_i.$$

It is easy to verify that if x^* and λ^* solve the nonlinearly constrained problem

$$\text{Minimize} \quad F(x^*) + d^t\nabla F(x^*) + d^t\nabla^2 F(x^*)d/2$$

$$\text{s.t.} \quad d^t\nabla c_i(x^*) + d^t\nabla^2 c_i(x^*)d/2 \geq 0 \quad i = 1,\ldots, m$$

then they also satisfy optimality conditions for the linearly constrained problem

$$\text{Minimize} \quad F(x^*) + d^t\nabla F(x^*) + d^t\nabla^2 L^*(x^*)d/2$$

$$\text{s.t.} \quad d^t\nabla c_i(x^*) \geq 0 \quad i = 1,\ldots, m.$$

Therefore, it is still possible for an optimization algorithm to make use of linear approximations to nonlinear constraints provided the quadratic model of the objective function (23) is modified to include second derivatives of the c_i as well as F.

A quasi-Newton method which constructs H to approximate $\nabla^2 L^*$ is obtained by redefining γ in (24) as $\gamma = \nabla L(x) - \nabla L(x^-)$, where L is a local approximation to L^* based on Lagrange multiplier estimates, λ_i, determined at the current point x, so that

$$\nabla L(x) = \nabla F(x) - \sum_{i=1}^{m} \lambda_i \nabla c_i(x).$$

The chief difficulty encountered by projected and reduced gradient methods in nonlinearly constrained problems is that of maintaining feasibility because the condition $Ad = 0$ does not now ensure that $c_i(x + sd) = 0$ for each active constraint c_i. In order for each new iteration to begin from a point which does not violate any constraints it is necessary to include an extra *restoration step*.

If, for instance, a reduced gradient algorithm is applied to a problem with nonlinear constraints then the basic horizontal move must be followed by a vertical restoration step in the space spanned by the columns of Y. A first estimate of this step back onto the boundary of the feasible region can be obtained by defining

$$c_i^+ = c_i(x + sd) \quad i = 1,\ldots, t,$$

finding y to solve the lower triangular system

$$AYy = -c^+$$

and then setting $d^+ = Yy$. If the constraints in (1) are nearly linear then the point

$$x = x^- + sd + d^+$$

may be suitable for the start of a new iteration; but when the c_i are highly nonlinear the calculation of a suitable restoration step may itself be an iterative process.

Conjugate bases, as used in EQP Algorithm 4, may have some advantages over orthogonal ones in the context of restoration steps. If F is quadratic and if the line search in Step 4 yields a point $x + sd$ such that $Z^t g(x + sd) = 0$ then the effect of the extra orthogonal step d^+ is given by

$$Z^t g(x + sd + d^+) = Z^t g(x + sd) + Z^t \nabla^2 F Y y.$$

Since the second term on the right need not be zero it follows that stationarity in Z-space is lost after the restoration move $d^+ = Yy$. However it is easy to see that the property $Z^t g = 0$ can be preserved if we use a restoration step $Y'y$ obtained by solving $AY'y = -c^+$, where Y' is a matrix with columns spanning the space of the active constraint normals and satisfying the conjugacy condition $Z^t \nabla^2 F Y' = 0$.

4. Penalty and Barrier function methods

4.1 PENALTY FUNCTIONS

Difficulties associated with choosing an active set and maintaining feasibility with respect to nonlinear constraints can be avoided in the *penalty function* approach to the solution of (1).

The classical form of penalty function is

$$P(x, r) = F(x) + [\sum_{i=1}^{m_e} \{c_i(x)\}^2 + \sum_{i=m_e+1}^{m} \{min(0, c_i(x))\}^2]/r \qquad (25).$$

The positive scalar r is called the *penalty parameter* and it is easy to see that $P(x, r) = F(x)$ when x is a feasible point of (1) and also that, when x is infeasible, (25) includes an extra contribution or "penalty" proportional to the squared constraint violations. If v is the vector of constraint violations introduced at the end of section 1, i.e.

$$v_i \quad = c_i(x) \quad i = 1,..., m_e$$
$$= min(0, c_i(x)) \quad i = m_e+1,..., m.$$

then (25) can be written

$$P(x, r) = F + \sum_{i=1}^{m} v_i^2/r.$$

Now suppose that $\hat{x}(r)$ is the unconstrained minimum of $P(x, r)$ for some particular choice of penalty parameter. Then it can be proved, under quite mild conditions, that

$$\lim_{r \to 0+} \hat{x}(r) = x^*.$$

This important result means (1) can be solved via a sequence of unconstrained minimizations.

Classical Penalty function algorithm

Step 1: Choose a penalty parameter $r^{(0)}$, a constant $\beta < 1$ and a tolerance ε.

Step 2: Repeat for $k = 0,1,2,...$

   Minimize $P(x, r^{(k)})$ to obtain $x^{(k)} = \hat{x}(r^{(k)})$

   Set $r^{(k+1)} = \beta r^{(k)}$

   until $|v_i(x^{(k)})| \le \varepsilon$ for $i = 1,...,$

   Dropping explicit dependence on x, the gradient of the penalty function can be written

$$\nabla P = \nabla F + 2[\sum_{i=1}^{m} v_i \nabla c_i]/r \tag{26}$$

and it follows, by comparing $\nabla P = 0$ with the optimality condition $\nabla L^* = 0$, that the Lagrange multipliers λ^* satisfy

$$\lambda_i^* = \lim_{r \to 0} \{ -2v_i(\hat{x}(r))/r\}   i = 1,..., m.$$

and hence

$$\lambda_i^* = \lim_{r \to 0} \{-2c_i(\hat{x}(r))/r\}   i = 1,..., m_e$$

$$\lambda_i^* = \lim_{r \to 0} \{-min(0, 2c_i(\hat{x}(r))/r\}   i = m_e+1,..., m$$

$$\tag{27}.$$

Penalty function methods were very popular when first introduced in the 1960's because they enabled constrained minimization to be attempted using any unconstrained techniques that were to hand. Experience, however, revealed practical drawbacks. One minor difficulty concerns the initial choice of penalty parameter, since $P(x, r)$ may not have a minimum if r is not small enough. Much more serious is the fact that the unconstrained minimizations become numerically difficult as $r \to 0$ because $\nabla^2 P$ tends to be ill-conditioned.

To see how this ill-conditioning arises we consider the Hessian of $P(\hat{x}(r), r)$. We assume, for convenience, that the violated inequalities at $\hat{x}(r)$ are numbered from m_e+1 to t. If A is the matrix whose rows are $(\nabla c_i(\hat{x}(r)))^t$ $(i = 1,..., t)$ we have

$$\nabla^2 P(\hat{x}(r), r) = \nabla^2 F(\hat{x}(r)) + 2[\sum_{i=1}^{t} c_i(\hat{x}(r)) \, \nabla^2 c_i(\hat{x}(r))]/r + 2A^t A/r.$$

It follows from (27) and (5) that, as r tends to zero, the first two terms on the right tend to $\nabla^2 L^*$ while the last term may be unbounded. Near to x^*, therefore, $\nabla^2 P$ can have arbitrarily large curvature in the space of the constraint normals which makes it hard to compute good search directions by solving the Newton equations.

The difficulty of minimizing $P(x, r)$ when r is small can be partly alleviated by extrapolating a good estimate of $\hat{x}(r^{(k+1)})$ from the preceding two minima $\hat{x}(r^{(k)})$ and $\hat{x}(r^{(k-1)})$. However the ill-conditioning can be largely avoided by using a *shifted* penalty function involving extra parameters $\theta_1,...\theta_m$, i.e.

$$M(x, \theta, r) = F(x) + [\sum_{i=1}^{m_e} \{c_i(x) - \theta_i\}^2 + \sum_{i=m_e+1}^{m} \{min(0, c_i(x) - \theta_i)\}^2]/r \qquad (28).$$

If w is defined by

$$w_i = c_i(x) - \theta_i \qquad i = 1,..., m_e$$

$$= min(0, c_i(x) - \theta_i) \qquad i = m_e+1,..., m$$

then we can write

$$M = F + \sum_{i=1}^{m} w_i^2/r$$

and

$$\nabla M = \nabla F + 2[\sum_{i=1}^{m} w_i \nabla c_i]/r.$$

Hence $\nabla M(x^*) = 0$ if the parameters θ_i are chosen so that $w_i(x^*) = -r\lambda_i^*/2$. This choice of w_i implies that $\theta_i = r\lambda_i^*/2$ $(i=1,..., m)$ because $c_i(x^*) = 0$ for all binding constraints and $\lambda_i^* = 0$ for non-binding constraints.

It can be shown that there is a threshold ρ (depending on $\nabla^2 F$ and $\nabla^2 c_i$) such that $\nabla^2 M(x^*, r\lambda^*/2, r)$ is positive definite if $r < \rho$. Hence a shifted penalty function algorithm can avoid ill-conditioning because r does not need to approach zero once the θ_i have been adjusted to obtain a particular M whose unconstrained minimum is at x^*.

Shifted Penalty function algorithm

Step 1: Choose $r^{(0)}$ and $\theta^{(0)}$ and also a constant $\beta < 1$ and a tolerance ε

Step 2: Repeat for $k = 0,1,2,...$

$\qquad$ Find $x^{(k)}$ to minimize $M(x, \theta^{(k)}, r^{(k)})$

$\qquad$ Set $\quad \theta_i^{(k+1)} = \theta_i^{(k)} - c_i(x^{(k)})$ if $i \le m_e$ or $i > m_e$ and $c_i(x^{(k)}) < \theta_i^{(k)}$

$\qquad\qquad\qquad = 0$ otherwise.

$\qquad$ Set $\quad r^{(k+1)} = \beta r^{(k)}$

$\qquad$ until $|v_i(x^{(k)})| \le \varepsilon$ for $i = 1,..., m$

The update for $\theta_i^{(k+1)}$ is equivalent to setting $\theta_i^{(k+1)} = r\lambda_i^{(k)}/2$ where

$$\lambda^{(k)} = -2w(x^{(k)})/r$$

gives a local approximation to the Lagrange multipliers λ^*.

In the case when (1) contains only equality constraints we can write

$$M(x, \theta^{(k+1)}, r) = F(x) - \sum_{i=1}^{m_e} 2\theta_i^{(k+1)}c_i(x)/r + \sum_{i=1}^{m_e} \{c_i(x)\}^2/r + \sum_{i=1}^{m_e} \{\theta_i^{(k+1)}\}^2/r$$

$$= F(x) - \sum_{i=1}^{m_e} c_i(x)\lambda_i^{(k)} + \sum_{i=1}^{m_e} \{c_i(x)\}^2/r + \sum_{i=1}^{m_e} \{\theta_i^{(k+1)}\}^2/r \qquad (29).$$

Because (29) includes the Lagrange multiplier approximations, $\lambda_i^{(k)}$, M is often called the *Augmented Lagrangian*. As $\lambda^{(k)}$ tends to λ^*, the right hand side of (29) resembles the Lagrangian function L^* augmented by some extra terms. The last of these additional terms is irrelevant as regards the location of the minimum of M; but the term involving $\{c_i(x)\}^2$ has the effect of counteracting negative curvature in L^*. The optimality conditions require L^* to have positive curvature at x^* in the subspace tangential to the binding constraints and so the contribution from the third term on the right of (29) to the second derivatives in the space of the constraint normals can make $\nabla^2 M$ positive definite when r is sufficiently small.

4.2. BARRIER FUNCTIONS

Penalty function methods typically generate solution estimates which approach x^* from outside the feasible region. It is sometimes desirable - perhaps because the objective function is undefined when certain constraints are violated - to consider methods which tend to the solution through the interior of the feasible region. Barrier function methods have this property. It should be noted straightaway that they are only applicable to problems in which all the constraints are inequalities.

Two forms of barrier function are

$$B(x, r) = F(x) + r[\sum_{i=1}^{m} \{c_i(x)\}^{-1}] \qquad (30);$$

and
$$B(x, r) = F(x) - r[\sum_{i=1}^{m} log\{c_i(x)\}] \qquad (31).$$

In both (30) and (31) the term involving the constraints becomes very large near the boundary of the feasible region, thus creating a "barrier" to constraint violation. It is the *log-barrier* form (31) which is more widely used. If $\hat{x}(r)$ denotes the minimum of (31) for some particular r then, as with the penalty function, it can be proved that

$$\lim_{r \to 0+} \hat{x}(r) = x^* \quad \text{and} \quad \lim_{r \to 0+} \{r/c_i(\hat{x}(r))\} = \lambda_i^* \quad i=1,..., m.$$

The barrier function algorithm below resembles the penalty approach. Since it does not allow constraints to be violated, however, the iterations cannot be terminated in the same way as in the classical penalty technique. Instead we use the relationship between λ_i^*, r and $c_i(\hat{x}(r))$ which implies that, near the solution, $r^{(k)} \cong \lambda_i^* c_i(\hat{x}^{(k)})$. Optimality condition (6) states that $\lambda_i^* c_i(\hat{x}^{(k)})$ should approach zero which justifies a termination criterion based on $r^{(k)}$ being sufficiently small.

Classical Barrier function algorithm (inequality constraints only)

Step 1: Choose a penalty parameter $r^{(0)}$, a constant $\beta < 1$ and a tolerance ε.

Step 2: Repeat for $k = 0,1,2,...$

>   Find $x^{(k)}$ to minimize $B(x, r^{(k)})$

>   Set $r^{(k+1)} = \beta r^{(k)}$

>   until $r^{(k)} < \varepsilon$

The line search in the unconstrained minimization procedure used to find $x^{(k)}$ may need slight modification to prevent infeasible points being chosen.

Ill-conditioning in $\nabla^2 B(x, r)$ can cause practical difficulties for this algorithm similar to those noted in the penalty approach. A remedy is to use a shifted barrier function,

$$K(x, \theta, r) = F(x) - [\sum_{i=1}^{m} r_i log\{c_i(x) + \theta_i\}] \qquad (32).$$

The function K has a stationary point at x^* if

$$\theta_i = r_i/\lambda_i^* \quad \text{for each } i \text{ such that } c_i(x^*) = 0$$

and
$$\theta_i \geq 0, \quad r_i = 0 \quad \text{for each } i \text{ such that } c_i(x^*) > 0.$$

This stationary point is a minimum if r is less than some problem-dependent threshold (which means it need not tend to zero).

An algorithm using (32) is as follows.

Shifted Barrier function algorithm (inequality constraints only)

Step 1: Choose $r^{(0)}$ and $\theta^{(0)}$ and also a constant $\beta < 1$ and a tolerance ε

Step 2: Repeat for $k = 0,1,2,...$

Find $x^{(k)}$ to minimize $K(x, \theta^{(k)}, r^{(k)})$

Set $\lambda_i^{(k)} = r_i^{(k)}/(c_i(x^{(k)}) + \theta_i)$ and $\theta_i^{(k+1)} = \theta_i^{(k)} + c_i(x^{(k)})$ $(i = 1,..., m)$

Set $r^{(k+1)} = \beta r^{(k)}$

until $|c_i(x^{(k)})\lambda_i^{(k)}| < \varepsilon$ for $i = 1,..., m$

Here the update for $\theta_i^{(k+1)}$ is equivalent to setting $\theta_i^{(k+1)} = r_i^{(k)}/\lambda_i^{(k)}$.

Note that the shifted barrier function approach is not a feasible point method - which was one of the advantages claimed for techniques based on (30) and (31).

Mixed penalty/barrier function approaches can be used for problems with both equalities and inequalities.

4.3 EXACT PENALTY FUNCTIONS

Consider the function

$$E(x, r) = F(x) + \left\{ \sum_{i=1}^{m_e} |c_i(x)| + \sum_{i=m_e+1}^{m} |min(0, c_i(x))| \right\}/r \qquad (33).$$

This is called the l_1 penalty function and it has a minimum at x^* for all r sufficiently small. Hence a solution of (1) can be found via a single minimization of (33), and for this reason E is called an "exact" penalty function because it does not involve parameters requiring iterative adjustment. In making this remark, of course, we assume that r has been chosen suitably. In fact there is a threshold condition

$$r < 1/\|\lambda^*\|_\infty ;$$

but normally this cannot be used in practice because λ will not be known in advance.

The function E has the undesirable property of being *non-smooth* since its derivatives are discontinuous across any surface for which $c_i(x) = 0$. This fact may cause difficulties for many unconstrained minimization algorithms which assume continuity of first derivatives.

For equality constrained problems there is a penalty function which is both exact and smooth. This has the form

$$E'(x, r) = F - A^t(AA^t)^{-1}Ag + c^tc/r \qquad (34),$$

where c is the vector of constraints $c_i(x)$, A is the matrix whose rows are the constraint normals $\{\nabla c_i(x)\}^t$ and g is the gradient vector $\nabla F(x)$. The second term on the right of (34) can be viewed as a continuous approximation to the Lagrange multipliers. This follows because stationarity condition (5) at x^* can be written as

$$g - A^t\lambda^* = 0$$

and so λ^* can be obtained by solving

$$(AA^t)\lambda^* = -Ag.$$

Hence E' is a form of Augmented Lagrangian function in which λ varies continuously instead of being adjusted at periodic intervals.

The exact penalty function E' also has a disadvantage. Since the right hand side of (34) involves first derivatives of the function and constraints it follows that second derivatives of F and c_i have to be obtained if E' is to be minimized by a gradient method. Worse still, third derivatives will be needed if we wish to use a Newton algorithm! This probably explains why E' has not been more widely used: but it is possible that the availability of automatic differentiation software will change this state of affairs in the near future.

5. Sequential Quadratic Programming

5.1 QUADRATIC/LINEAR APPROXIMATIONS OF (1)

In essence this is a straightforward approach in which a QP subproblem is constructed at every iteration, using information at the current point, x^-. This subproblem consists of a quadratic approximation to F together with linearisations of the constraints; and from it we can obtain new solution and Lagrange multiplier estimates x, λ.

The process terminates when an optimality measure such as $T(x, \lambda)$, given by (8), becomes sufficiently small. For this simple approach to be successful, however, care must be taken over details such as the line search and the method of obtaining the second derivative matrix in the quadratic approximation to F.

SQP Algorithm 1

Step 1: Choose an initial point x and an initial Hessian matrix H.

Step 2: Obtain p and λ by solving the QP subproblem

$$\text{Minimize} \quad p^t H p/2 + p^t \nabla F(x)$$
$$\text{s.t.} \quad c_i(x) + \nabla c_i(x)^t p = 0 \quad i = 1,..., m_e$$
$$c_i(x) + \nabla c_i(x)^t p \geq 0 \quad i = m_e+1,..., m$$

Step 3: Set $x^- = x$ and $H^- = H$.

   Obtain a new point $x = x^- + sp$ via a line search.

Step 4: Stop if $T(x,\lambda)$ is sufficiently small.

   Otherwise obtain H by updating H^- and return to step 2.

The line search may be based on ensuring that $P(x) < P(x^-)$, where P denotes some penalty function; and various choices for P have been tried. Some authors recommend the l_1 exact penalty function while others use a version of the Augmented Lagrangian. The line search is important, because it forces a reduction in some composite *merit* function involving both F and the c_i to ensure that the new point x is measurably "better" than the old one x^-, thereby providing a basis for a global convergence proof.

   It is also important for the QP subproblems to involve H as an estimate of $\nabla^2 L^*$ rather than of $\nabla^2 F$ as might be supposed. The reasons for this are the same as those outlined at the end of section 3. The update in Step 4 of the algorithm is typically performed using (24) with modifications to keep H positive definite.

5.2 SQP METHODS BASED ON PENALTY FUNCTIONS

In SQP Algorithm 1 there is no necessary connection between the calculation of the search direction and the merit function used in the line search. In the methods which follow, however, the QP subproblem and the step length calculation are more closely related because both are derived from the same penalty function.

To simplify the explanation we suppose that (1) involves only equality constraints and that A is the matrix with rows $\nabla c_i(x)^t$, $i = 1,...,m_e$ while g denotes $\nabla F(x)$. For the classical penalty function (25) a Taylor expansion of ∇P about a typical point, x, gives

$$\nabla P(x+p, r) = g + 2A^t c/r + 2A^t Ap/r + Gp + ... \qquad (36)$$

where
$$G = \nabla^2 F + 2[\sum_{i=1}^{m} \nabla^2 c_i c_i]/r \qquad (37).$$

It follows that the Newton step towards the minimum of P is given by

$$(G + 2A^t A/r)p = -g - 2A^t c/r \qquad (38).$$

We know from section 4 that the matrix $G + 2A^t A/r$ is ill-conditioned when r is small. We therefore seek a way of finding p which does not use (38). Assuming G is nonsingular we can premultiply (36) by AG^{-1} and obtain, after some re-arrangement,

$$Ap = -ru/2 - c \qquad (39)$$

where u is given by
$$(rI/2 + AG^{-1}A^t)u = AG^{-1}g - c \qquad (40).$$

From (39) we obtain a first order estimate of the constraints at the minimum of $P(x, r)$

$$c(\hat{x}(r)) \cong c + Ap = -ru/2.$$

Recalling the relation (27) between c, r and λ^* we see that u can be regarded as a vector of Lagrange multiplier approximations. It also follows from (27) and (37) that, near to x^*, the matrix G tends to the Hessian of the Lagrangian function. If we now consider the QP

$$\text{Minimize } p^t Gp/2 + p^t g \quad \text{s.t. } Ap = -ru/2 - c \qquad (41)$$

it is easy to verify that its a solution is

$$p = G^{-1}(A^t u - g) \qquad (42)$$

where u, given by (40), is the vector of Lagrange multipliers. (Hence (41) has the unusual property that its formulation includes its own multipliers!).

Expressions (40) and (42) together are algebraically equivalent to (38) but provide a well-conditioned way of calculating the Newton step p.

The preceding ideas can be extended to the case of inequality constraints, as illustrated by the following algorithm.

SQP Algorithm 2

Step 1: Choose initial values for x, H and the penalty parameter r.

Choose a scaling factor $\beta < 1$. Set $\lambda = 0$, $T^- = T(x,\lambda)$, from (8).

Step 2: Obtain p and λ by solving the QP subproblem

$$\text{Minimize} \quad p^t H p/2 + p^t \nabla F(x)$$
$$\text{s.t.} \ c_i(x) + \nabla c_i(x)^t p = -r\lambda_i/2 \quad i = 1,..., m_e \tag{43}$$
$$c_i(x) + \nabla c_i(x)^t p \geq -r\lambda_i/2 \quad i = m_e+1,..., m$$

Step 3: Set $x^- = x$, $H^- = H$ and $r^- = r$.

Obtain a new point $x = x^- + sp$ via a line search to give $P(x, r) < P(x^-, r)$.

Step 4: If $T(x,\lambda) < T^-$ then set $r = \beta r^-$ and $T^- = T(x,\lambda)$

Step 5: Stop if $T(x, \lambda)$ is sufficiently small.

Otherwise obtain H by updating H^- and return to step 2.

Since the QP subproblem in this algorithm is derived from the penalty function $P(x, r)$ it is natural for a line search along p also to be based on $P(x, r)$. The inequality constrained QP (43) can be solved via repeated use of expressions (40) and (42) within an active set strategy like IQP Algorithm 1 in Section 2. (The form of (40) and (42) mean that in practice it is more efficient to work with a quasi-Newton estimate of the *inverse* matrix G^{-1}.)

SQP algorithm 2 can be viewed as a way of constructing a controlled approximation to the "trajectory" of true penalty function minima generated by the sequential unconstrained minimization techniques in section 4.1. The penalty parameter r is reduced at Step 4 whenever a better approximation to an optimal point is found, rather than after a complete minimization of $P(x, r)$; and in practice this gives a much more rapid approach to x^* than is obtained by conventional penalty function techniques.

A similar algorithm can be based on the shifted penalty function.

SQP Algorithm 3

Step 1: Choose initial values for x, H, r and a parameter vector μ.

Choose a scaling factor $\beta < 1$. Set $\lambda = 0$, $T^- = T(x,\lambda)$, from (8).

Step 2: Obtain p and λ by solving the QP subproblem

$$\text{Minimize} \quad p^tHp/2 + p^t\nabla F(x)$$
$$\text{s.t.} \quad c_i(x) + \nabla c_i(x)^tp = -r(\lambda_i - \mu_i)/2 \quad i = 1,..., m_e \tag{44}$$
$$c_i(x) + \nabla c_i(x)^tp \geq -r(\lambda_i - \mu_i)/2 \quad i = m_e+1,..., m$$

Step 3: Set $x^- = x$, $H^- = H$ and $r^- = r$. Set $\theta = r\mu/2$.

Obtain a new point $x = x^- + sp$ via a line search to give $M(x, \theta, r) < M(x^-, \theta, r)$.

Step 4: If $T(x,\lambda) < T^-$ then set $r = \beta r^-$, $\mu = \lambda$ and $T^- = T(x,\lambda)$

Step 5: Stop if $T(x, \lambda)$ is sufficiently small.

Otherwise obtain H by updating H^- and return to step 2.

The QP (44) comes from the shifted penalty function $M(x, \theta, r)$ in much the same way as (43) comes from $P(x, r)$ and is solved via expressions similar to (40) and (42). Since both μ and λ approximate the multipliers λ^* we can see that the QP subproblem (44) causes $x + p$ to approach the feasible region without requiring r to tend to zero.

Finally we give an SQP algorithm derived from the l_1 exact penalty function. The QP subproblem in this approach involves a quadratic approximation to (33)

$$QE(p) = p^tHp/2 + p^t\nabla F(x) + \{ \sum_{i=1}^{m_e} |c_i(x)+p^t\nabla c_i(x)| + \sum_{i=m_e+1}^{m} | min(0, c_i(x)+p^t\nabla c_i(x))|\}/r$$

The algorithm is based on the use of a *trust region* rather than a line search to ensure global convergence. Trust region methods have the property that a maximum allowable step size (the trust region radius) is fixed at the start of an iteration and a search direction is then computed to provide the "best" move subject to this limit. This contrasts with the strategy used in previous algorithms where it is the search direction which is determined first, followed by an appropriate step size. The term "trust region" expresses the fact that its radius reflects the extent to which we trust the local approximations to the function and constraints about the current point, x.

SQP Algorithm 4

Step 1: Choose initial values for x, H and r. Choose a trust region radius R.

Choose parameters ρ_1, ρ_2 $(0 < \rho_1 < \rho_2 < 1)$, α (< 1) and β (> 1).

Step 2: Obtain p and λ by solving the QP subproblem

$$\text{Minimize } QE(p) \quad \text{s.t.} \quad \|p\|_\infty \le R \qquad (45).$$

Step 3: Set $x^- = x$, $H^- = H$, $R^- = R$ and $\Gamma = (E(x+p)-E(x))/(QE(p)-QE(0))$

If $\Gamma < \rho_1$ set $R = \alpha R^-$ and return to step 2

If $\Gamma \ge \rho_1$ set $x = x^- + p$

If $\Gamma > \rho_2$ set $R = \beta R^-$

Step 4: Stop if $T(x, \lambda)$ is sufficiently small.

Otherwise obtain H by updating H^- and return to step 2.

Step 3 of this algorithm adjusts the size of the trust region on the basis of a comparison between the actual and the quadratically predicted change in E. If the ratio Γ is small or negative then p does not reduce E sufficiently and so it is rejected. The calculation is then repeated with a smaller trust region. If $\rho_1 < \Gamma < \rho_2$ then p is acceptable and the same trust region is retained for the next iteration. Finally if $\Gamma > \rho_2$ then agreement between E and the quadratic model is good enough to permit a larger step on the next iteration.

Penalty function based SQP methods all have one significant advantage over SQP Algorithm 1 - namely that QP subproblems (43), (44) and (45) are guaranteed to have a solution. The same cannot be said for the subproblem in SQP Algorithm 1 because, for instance, it can happen that linearisations of nonlinear constraints are inconsistent even when the original constraints give a well-defined feasible region.

6. Computational considerations

SQP techniques are, at present, probably the most widely used of the methods we have described for solving small to medium sized nonlinear programming problems. This may be because they handle nonlinear constraints without the need for restoration steps of the kind described in section 3 and also because the QP subproblem is regarded as being easier than the unconstrained minimizations used by penalty and barrier methods.

This is not to say that the methods of sections 3 and 4 are invariably inferior to SQP techniques. Indeed many different SQP implementations exist, and some of them borrow ideas from reduced gradient and penalty function methods. If we consider behaviour near a solution then it can be shown that, under certain conditions (see below), both the reduced gradient algorithm and SQP algorithm 1 are capable of one-step superlinear convergence on linearly constrained problems. When the constraints are nonlinear, one-step superlinear convergence is still possible for SQP methods; but a reduced gradient technique with the orthogonal restoration step described at the end of section 3 may only permit two-step superlinear convergence. For penalty and barrier methods, convergence of the iterates $x^{(k)}$ is usually linear, although the shifted versions can give superlinear convergence if the $\theta^{(k)}$ are chosen by a second-order method similar to that used to obtain Lagrange multipliers in a QP subproblem. Similarly the penalty-based SQP algorithm 2 also has ultimately linear convergence while SQP algorithm 3 permits superlinearly convergent.

We shall now look more closely at some algorithmic details. In particular we consider the two conditions which must be fulfilled if superlinear convergence of methods from section 3 and section 5 is to be achieved in practice. Firstly there must be agreement between the matrix H and the true Lagrangian Hessian $\nabla^2 L^*$ in the subspace tangential to the binding constraints. This subspace is spanned by the columns of the matrix Z in (7); and if δ denotes $x - x^*$ then $\|Z^t(H - \nabla^2 L^*)ZZ^t\delta\|/\|\delta\|$ must tend to zero as δ tends to zero. (Note that the use of exact second derivatives in H can give quadratic convergence.) Secondly, in the neighbourhood of the solution, the line search strategy must be able to accept the *unit step* $x = x^- + p$ on every iteration. These conditions are addressed in the next two sections.

6.1 LINE SEARCH STRATEGIES NEAR THE SOLUTION

The nature of the penalty functions used in line searches for SQP can prevent a unit step being taken because constraint nonlinearities may be magnified when the penalty parameter is small. If P is a line search merit function - e.g. the Augmented Lagrangian (Gill et al, 1986) or the l_1 exact penalty function (Han, 1977, Powell, 1978) - then we may find $P(x^- + p) > P(x^-)$ due to small increases in the violated constraints not predicted by the quadratic/linear model in the QP subproblem. This phenomenon is called the *Maratos effect* and attempts to overcome it include the use of second order correction steps (Fletcher, 1987) and curvilinear searches (Mayne & Polak, 1982).

The "watchdog technique" has been proposed by Chamberlain et al (1982) which on some iterations will accept a point which increases the merit function provided a sufficient decrease is obtained on a subsequent step. Powell & Yuan (1984) have also considered a line search based on the differentiable exact penalty function (34) in which careful attention is paid to control of the penalty parameter.

6.2 MATRIX UPDATING

We have already noted that it is acceptable to force the approximate Hessian to be positive definite by making arbitrary modifications in the space of the active constraint normals since ultimately we only require H to provide a good estimate of the reduced Hessian $Z^t \nabla^2 L^* Z$. Hence there is interest in methods which only seek to approximate this t-by-t matrix rather than the full n-by-n Hessian of the Lagrangian.

Coleman & Conn (1984) discuss a suitable updating strategy in connection with a method which takes horizontal and vertical steps, i.e.

$$x = x^- + h + v.$$

If H^- is the current estimate of $Z^t \nabla^2 L^* Z$ then H may be obtained from update (24) with

$$\delta = Z^t h \text{ and } \gamma = Z^t(g^+ - A^{+t}\lambda - g^-)$$

where g^+, A^+ are the gradient and Jacobian at the "halfway point" $x^+ = x^- + h$.

With such a strategy it is suitable to force H to be positive definite at points away from the solution because, as in the unconstrained case, it now approximates a matrix which must be positive definite at the solution. It should be noted, however, that the Coleman-Conn update just described costs an extra gradient and Jacobian evaluation at every iteration when compared with the methods outlined in previous sections. Moreover, it must be remembered that the situation is not quite as simple as may appear at first, because for inequality constrained problems the active constraint set may change from iteration to iteration along with the dimensionality of Z and H.

Methods which are based on the reduced Hessian and could therefore employ updated estimates of it have been mentioned in section 3. We can also use the reduced Hessian in the context of SQP algorithms if the QP subproblem is solved using EQP Algorithm 3 or 4.

Interest also continues in updates where $H \cong \nabla^2 L^*$ with a view to maintaining the convenience of positive definiteness in H while distorting second derivative information as little as possible in the subspace tangential to the constraints (e.g. Nguyen 1989).

6.3 EQP vs IQP SUBPROBLEMS

The SQP algorithms in section 5 all perform a full solution of an inequality constrained QP to determine a search direction. In other words, the QP subproblem includes linearizations of all the constraints in (1) and its solution is an iterative process which implicitly identifies which of them are to be regarded as active at the current point. Some SQP algorithms have been proposed which use a separate active set strategy at the start of each iteration (as in the methods of section 3) so that the QP subproblem includes fewer than m constraints *which can be expressed as equalities*. This can reduce the amount of work involved in determining a search direction to something of the order of a solution of one set of linear equations.

One of the original penalty function based QP methods (Biggs, 1972) made use of an EQP subproblem; and a discussion of the merits of EQP and IQP formulations has been given by Gill et al (1982) and Bartholomew-Biggs (1982). A drawback of the EQP approach is that the preliminary active set strategy is often based on *ad hoc* rules which may give misleading choices of constraints in the QP with consequently poor search directions. The IQP formulation has been more widely used in the last decade, but there has been a recent revival of interest in the potential economy of the EQP approach (e.g. Kleinmichel et al, 1992, Facchinei & Lucidi, 1993).

6.4 LARGE SCALE PROBLEMS

It is interesting to consider which of the techniques we have described are most suitable when problem (1) involves many variables and constraints. A full discussion of this important topic is given by Conn (1993) and we shall make only a few brief observations. We assume that the main cost in the optimization calculations is in the formation and solution of systems of equations for computing search directions and/or Lagrange multipliers (although this might not be the case when the evaluation of function and constraints is very expensive).

We have already seen for the methods in sections 2 and 3, for instance, that for n variables and t constraints a search direction calculation may involve the solution systems of $n+t$ or t or $n-t$ equations. The same is true of SQP methods, depending on how the QP subproblem is handled. Penalty and barrier function methods usually deal with a set of n equations. In choosing a method when n and t are both large we need to consider not only the dimensionality of these systems but also how far they permit the exploitation of sparsity.

The most well-established techniques for large scale nonlinear programming are of the reduced gradient type (e.g. Murtagh & Saunders 1982, Smith & Lasdon 1992). These are successful when $n-t$ is small compared with both n and t so that the solution of a system like (18) is relatively inexpensive, even though Z^tGZ is a dense matrix.

The LANCELOT package (Conn et al, 1992) for large-scale optimization is a shifted penalty function technique. It performs a sequence of approximate minimizations of (28), imposing additional simple bound constraints to represent a trust region, as in SQP algorithm 4. Much of the efficiency of this method for large scale problems comes from techniques used to obtain second derivative information by exploiting structure and *separability* in the function and constraints.

The extension of SQP methods to large problems has been considered by, among others, Mahidhara & Lasdon (1990) and Hernandez (1993). Sparsity in the Jacobian matrix of the constraints can be exploited quite easilly (e.g. Bartholomew-Biggs & Hernandez, 1993). The more difficult issue is to handle the Hessian efficiently.

The approach of Mahidhara & Lasdon centres on using *limited memory* updates (Nocedal 1980) in the framework of SQP Algorithm 1. A limited memory updating procedure holds the approximate Hessian H (or possibly H^{-1}) in the form of a set of vectors characterizing a sequence of l updates like (24) (typically $l \ll n$). The formation of matrices such as Z^tHZ and $AH^{-1}A^t$ is thereby made more efficient.

Hernandez (1993), however, reports little overall benefit from limited memory updates in SQP Algorithm 3. Instead she recommends the use of exact second derivatives because this allows sparsity in the objective function and constraints to be exploited in the solution of equations like (10), (11) from EQP Algorithm 1 which appear at the centre of the QP subproblem.

The main drawback of using exact second derivatives has already been mentioned. Away from the solution the matrix G $(= \nabla^2 L)$ in (10) may not have positive curvature in the tangent space of the active constraints. Therefore we may need to modify G on some iterations to ensure that (10), (11) give a suitable search direction. One approach would be to use Gershgorin disks to estimate the eigenvalues of $\nabla^2 L$ and so obtain modifications to its diagonal elements to ensure that G is positive definite. Although this is a rather crude modification strategy it has proved quite successful and is typically much faster than a comparable quasi-Newton technique. An alternative would be for the solution of (10), (11) to involve a matrix factorization - similar to the modified Cholesky technique in section 2 - which detects and corrects negative pivots.

Both the strategies we have mentioned in the preceding paragraph can cause unneccessary distortion of second derivative information in the space of the constraint normals. Forsgren & Murray (1990) suggest *pivoting rules* to be used during an LBL^t factorization of the coefficient matrix in (10), (11) which enable the identification and correction of only those elements corresponding to second derivative information in the tangent space.

7. Further Reading

In this last section we make some suggestions for further study in the area of constrained optimization. The references cited below provide more detail about issues we have merely touched on, such as convergence properties and comparative numerical experience. We also give a brief account of some other, less widely used, methods which have not been covered in the main text. Finally we mention some generally available software which implements methods discussed in this paper.

7.1 METHODS FOR QP

More details about solution methods for the general QP problem are given, for instance, by Gill et. al. (1982, 1984) and Fletcher (1987), who deal with the computation of the orthogonal matrices Y and Z and modifications to Cholesky factorization for handling indefiniteness in the reduced Hessian Z^tGZ appearing in IQP algorithm 2. A conjugate basis approach (EQP algorithm 4) has been described by Goldfarb & Idnani (1983) and developed by Powell (1985).

7.2 REDUCED GRADIENT METHODS

Reduced gradient methods for both small and large-scale problems are discussed by Lasdon et al (1978) and Smith & Lasdon (1992). Coleman and Conn (1982, 1984) also propose algorithms like those described at the end of section 3 using horizontal and vertical corrections to solve problems with nonlinear constraints. Their work includes a convergence analysis. Byrd (1990) has also discussed convergence of a range of methods based on the reduced Hessian. Conjugate basis variants of some of these ideas are described by Nguyen (1989) and by Bartholomew-Biggs & Nguyen (1994).

7.3 PENALTY & BARRIER FUNCTION METHODS

Pioneering work in this area is due to Fiacco & McCormick (1968, 1990). More recent implementations of penalty function methods have been given by Coope & Fletcher (1979) Gould (1986), Broyden & Attia (1984) and by Conn et al (1992) in the LANCELOT package. These give attention both to avoiding ill conditioning and also to discovering how inaccurately (i.e. cheaply!) each penalty function can be minimized while still ensuring convergence to x^*.

Barrier function methods have regained popularity in the wake of interest aroused in the Karmarkar (1984) method for linear programming ever since Gill et al (1986) pointed out its relationship to the function (31). Recent barrier (or, as they are now more usually called, "interior point") methods are given by Polyak (1992), Wright (1992) and Conn et al (1992).

The l_1 exact penalty function is used as a line search function in some SQP methods and also appears in the Coleman-Conn (1982) reduced gradient method mentioned above. The smooth exact penalty function was studied by Fletcher & Lill (1972) and has since been the subject of many refinements and extensions by Di Pillo & Grippo (Di Pillo, 1993).

7.4 SQP METHODS

The basic SQP approach was described by Wilson (1963). Garcia-Palomares & Mangasarian (1976) subsequently proposed a quasi-Newton version while Han (1977) and Powell (1978) produced the first practical implementations, introducing line searches of the kind used in SQP Algorithm 1. Independently however, Murray (1969) and Biggs (1972) published SQP algorithms related to the penalty function (25), motivated by a wish to avoid the difficult unconstrained minimizations of P(x, r). These ideas were developed (as in SQP Algorithms 2, 3), by Gill et al (1986) and by Bartholomew-Biggs (1987). The l_1 penalty function, trust-region approach in SQP Algorithm 4 is due to Fletcher (1987); but Powell & Yuan (1990) have also proposed a trust-region based SQP technique.

Convergence of SQP methods has been studied by, among many others, Powell (1978), Byrd (1990) and Bartholomew-Biggs & Nguyen (1991).

7.5 OTHER CONSTRAINED MINIMIZATION METHODS

The methods we have considered so far may be regarded as representing "mainstream" thinking in constrained optimization. In this section we give a number of approaches which have been shown to be appropriate in special cases but which (fairly or unfairly!) have not yet gained such wide recognition.

First of all we mention recent work by Kanzow(1992) and Kleinmichel et al (1992) which makes explicit use of optimality conditions (6). These have not been very much used in the algorithms considered above, except as part of a convergence test. Kanzow introduces an *NCP function* $\varphi(\lambda, s)$, say, with the property that

$$\varphi(\lambda,s) = 0 \text{ if and only if } \lambda \geq 0, \ \lambda s = 0 \text{ and } s \geq 0.$$

In other words, the zeros of an NCP-function occur at values of λ and s which satisfy complementarity conditions like those in (6). (Kanzow (1992) gives several realizations of φ.) The use of such functions can be explained if we write the optimality conditions for (1) in a form which uses *slack variables*, s_i, associated with the inequality constraints, i.e.

$$\nabla L(x, \lambda) = 0$$

$$c_i(x) = 0, \ i = 1, ..., m_e$$

$$c_i(x) - s_i = 0, \ i = m_e+1, ..., m$$

$$s_i \geq 0, \ \lambda_i s_i = 0, \ \lambda_i \geq 0, \ i = m_e+1, ..., m.$$

It follows that if a triple (x^*, λ^*, s) solves (1) it also satisfies the nonlinear equations

$$\nabla L(x, \lambda) = 0$$

$$c_i(x) = 0, \ i = 1, ..., m_e$$

$$c_i(x) - s_i = 0, \ i = m_e+1, ..., m$$

$$\varphi_i(\lambda_i, s_i) = 0, \ i = m_e+1, ..., m$$

for some choice of NCP functions φ_i. Kanzow and Kleinmichel et al propose a minimization method in which each iteration is a Newton step towards the solution of this system. It is worth mentioning that this approach is one which only needs the solution of a linear system at every iteration as distinct from most SQP methods which require a complete IQP solution to determine each search direction (see the discussion in section 6.3).

In the case when problem (1) involves highly nonlinear constraints it may be suspected that SQP methods, being based on constraint linearization, may not be very successful. Some evidence of this is given by Brown & Bartholomew-Biggs (1989). Motivated by experience on a class of optimal control problems, Maany (1987) proposes the QFQC algorithm which employs quadratic models of both objective function and constraints. This proves much more effective than SQP on the particular problems for which it was designed.

Another quite frequently suggested approach to highly nonlinear problems involves following a curvilinear path to the solution defined by the solution of a system of differential equations. To give a flavour of such *trajectory following* approaches we mention the ODE system suggested by Tanabe (1980) and Evtushenko & Zhadan (1991) for solving the equality constrained problems. This is

$$dx/dt = - \nabla L(x, u(x)) \tag{46}$$

where t is a parameter and $u(x)$ is a continuous Lagrange multiplier approximation given by

$$u(x) = (AA^t)^{-1}(\tau c - Ag) \tag{47}$$

for a constant $\tau > 0$. A consequence of (47) is that the trajectory defined by (46) satisfies

$$dc(x)/dt = -\tau c(x)$$

and so it tends both to reduce constraint violations and to maintain feasibility in constraints that are already satisfied. It also follows from (46) and (47) that

$$dF(x)/dt = -\|\nabla L(x, u(x))\|^2 + \tau u(x)^t c(x)$$

which ensures that the objective function is monotonically decreasing once the solution trajectory of (46) lies in the feasible subspace.

Evtushenko and Zhadan (1991) suggest the extension of the above ideas to inequality constrained problems via a *space transformation* which implicitly confines the search to the interior of the region defined by the inequality constraints.

Implemention of a trajectory following method involves practical decisions about the numerical solution of a system such as (46). On the one hand we would like to follow the solution curve rather closely in order to maintain feasibility: but on the other hand we wish to avoid the computational costs associated with sophisticated and accurate ODE solvers. A discussion of these issues is given by Brown (1986) which shows that a simple technique like implicit Euler can sometimes be very successful.

Finally - and in complete contrast to all the algorithms mentioned so far - we mention a method proposed by Powell (1992) which is based on linearization of both the function and the constraints and which appears suitable for small scale problems. It is a development of the Nelder & Mead simplex technique for unconstrained optimization and therefore its linear approximations are based on interpolating values of F and c_i at $n+1$ points. Hence it does not require gradient information: and this is perceived as a valuable feature by engineers who need to use optimization algorithms but do not wish to compute derivatives of the complicated expressions which arise in their problems. The approach is based on solving a sequence of linear programming subproblems and includes a trust region strategy to ensure that all subproblem solutions are bounded. The method works quite well when the number of binding constraints in (1) is close to n, but is unlikely to be successful when there are relatively few active constraints at x^* because curvature of F and /or c_i will then be important.

7.6 OPTIMIZATION SOFTWARE

A good deal of software is available which implements methods like those which have been described in this survey. Some of this software is part of large mathematical systems such as NAG and IMSL libraries (Fortran 77 and C versions available, although source code may not be provided). The Numerical Optimisation Centre of the University of Hertfordshire supplies the OPTIMA library in Fortran 77 and in Ada (and the Ada version provides automatic differentiation facilities). All these libraries include several different methods for different types of constrained problem as well as for unconstrained and least squares calculations. It is also worth mentioning some codes which are available for the special case of large scale nonlinear programming. MINOS implements the reduced gradient approach of Murtagh & Saunders (1982) while LSRG2 is also a reduced gradient technique due to Smith & Lasdon (1992). The LANCELOT sequential Augmented lagrangian system (Conn et al 1992) has been mentioned already. More details about the contents and availability of these, and many other optimization codes can be found in a useful survey by More and Wright (1992).

8. References

Bartholomew-Biggs, M.C., *Recursive Quadratic Programming Methods for Nonlinear Constraints*, in "Nonlinear Optimization 1981", (ed. M.J.D. Powell), Academic Press, 1982.

Bartholomew-Biggs, M.C., *The Development of Recursive Quadratic Programming Methods based on the Augmented Lagrangian*, Math. Programming Study 31, pp.21-41, 1987.

Bartholomew-Biggs, M.C. & T.T. Nguyen, *A Local Convergence Analysis for REQP using Conjugate Basis Matrices*, J. Optimization Theory & Applications, 71, pp. 31-45, 1991.

Bartholomew-Biggs, M.C. & T.T. Nguyen, *Orthogonal & Conjugate Basis Methods for solving Equality Constrained Minimization Problems*, Computational Optimization and Applications (to appear) 1994.

Bartholomew-Biggs, M.C. & M. Hernandez, *Some Improvements to the Subroutine OPALQP for Dealing with Large Problems*, Journal of Economic Dynamics & Control (to appear), 1993

Biggs, M.C., *Constrained Minimization using Recursive Equality Quadratic Programming*, in "Numerical Methods for Nonlinear Optimization" (ed. F.A. Lootsma), Academic Press, 1972.

Brown A.A., *Optimisation Methods involving the Solution of Ordinary Differential Equations*, Ph. D. Thesis Hatfield Polytechnic, 1986.

Brown, A.A. & M.C. Bartholomew-Biggs, *ODE versus SQP Methods for Constrained Optimization*, J. Optimization Theory & Applications, 62, pp. 211 -224, 1989.

Broyden, C.G. & Attia, *A Smooth Sequential Penalty Function Method for Solving Nonlinear Programming Problems*, in "System Modelling & Optimization" (ed. P. Thoft-Christenson), Lecture Notes in Control & Information Science 59, Springer-Verlag, 1984.

Byrd, R.H., *On the Convergence of Constrained Optimization Methods with Accurate Hessian Information on a Subspace*, SIAM Journal of Numerical Analysis, 27, pp. 141-153, 1990.

Chamberlain, R., C. Lemarechal, H. Pedersen & M.J.D. Powell, *The Watchdog Technique for Forcing Convergence in Algorithms for Constrained Optimization*, Math. Programming Study 16, pp. 1-17, 1982.

Coleman, T.F. & A. Conn, *Nonlinear Programming via an Exact Penalty Function*: *Asymptotic Analysis*, Math. Programming 24, pp. 123-136, 1982.

Coleman, T.F. & A. Conn, *Nonlinear Programming via an Exact Penalty Function*: *Global Analysis*, Math. Programming 24, pp. 137-161, 1982.

Coleman, T.F. & A. Conn, *On the local Convergence of a quasi-Newton Method for the Nonlinear Programming Problem*, SIAM Journal of Numerical Analysis, 21, pp.755-769, 1984.

Conn, A., N.I.M. Gould & Ph. L. Toint, *LANCELOT: A Fortran Package for Large Scale Nonlinear Optimization*, Springer Series in Computational Mathematics No. 17, Springer-Verlag, 1992.

Conn, A., N.I.M. Gould & Ph. L. Toint, *A Globally Convergent Lagrangian Barrier Algorithm for Optimization with General Inequality Constraints and Simple Bounds*, Report RAL-92-067, Rutherford Appleton Laboratory, 1992

Conn, A., *Algorithms for Large-Scale Nonlinear Optimization*, presented at "Algorithms for Continuous Optimization: the state of the art", Il Ciocco, September 1993.

Coope, I.D. & R. Fletcher, *Some Numerical Experience with a Globally Convergent Algorithm for Nonlinearly Constrained Optimization*, Mathematics Department Report NA/30, University of Dundee, 1979.

Di Pillo, G., *Exact Penalty Methods*, presented at "Algorithms for Continuous Optimization: the state of the art", Il Ciocco, September 1993.

Dixon, L.C.W., *Methods for Automatic Differentiation*, presented at "Algorithms for Continuous Optimization: the state of the art", Il Ciocco, September 1993.

Evtushenko, Y.G. & V.G. Zhadan, *The Space Transformation Techniques in Mathematical Programming*, in "System Modelling & Optimization 15th IFIP Conference", (ed. P. Kall), Lecture Notes in Control & Information Sciences, Springer-Verlag, 1992.

Facchinei, F. & S. Lucidi, *Quadratically and Superlinearly Convergent Algorithms for the Solution of Inequality Constrained Minimization Problems*, Technical report in preparation, University of Rome, 1993.

Fiacco, A.V. & G.P. McCormick, *Nonlinear Programming - Sequential Unconstrained Minimization Techniques*, John Wiley, 1968. Re-issued by SIAM Classics in Applied Mathematics, 4, 1990.

Fletcher, R. & S. Lill, *A Class of Methods for Nonlinear Programming, II: Computational Experience*, in "Nonlinear Programming " (eds J.B. Rosen, O.L. Mangasarian & K. Ritter), Academic Press, New York, 1972.

Fletcher, R., *Optimization in Practice*, John Wiley, 1987.

Forsgren, A.L. & W. Murray, *Newton Methods for Large-Scale Linear Equality Constrained Minimization*, Technical Report SOL 90-6, Systems Optimization Laboratory, Stanford University, 1990.

Garcia-Palomares, U.M. & O.L. Mangasarian, *Superlinearly Convergent Quasi-Newton Algorithms for Nonlinearly Constrained Optimization Problems*, Math. Programming 11, pp. 1-13, 1976.

Gill, P.E., W. Murray & M. Wright, *Practical Optimization*, Academic Press, 1982.

Gill, P.E., W. Murray, M.A. Saunders & M. Wright, *Linearly Constrained Optimization*, in "Nonlinear Optimization 1981", (ed. M.J.D. Powell), Academic Press, 1982.

Gill, P.E., W. Murray, M.A. Saunders & M. Wright, *Software & its Relationship to Methods*, Technical Report SOL 84-10, Systems Optimization Laboratory, Stanford University, 1984.

Gill, P.E., W. Murray, M.A. Saunders & M. Wright *Some Theoretical Properties of an Augmented Lagrangian Merit Function*, Technical Report SOL 86-6, Systems Optimization Laboratory, Stanford University, 1986.

Gill, P.E., W. Murray, M.A. Saunders, J.A. Tomlin & M.H. Wright, *On Projective Newton Barrier Methods for Linear Programming and an Equivalence to Karmarkar's Projective Method*, Math. Programming, Vol. 36, pp. 183-209, 1986.

Goldfarb, D. & A. Idnani, *A Numerically Stable Dual Method for Solving Strictly Convex Quadratic Programs*, Math. Programming, 27, pp. 1-33, 1983.

Gould, N.I.M., *On the Accurate Determination of Search Directions for Simple Differentiable Penalty Functions*, IMA J. Numerical Analysis, Vol. 6, pp. 357-372, 1986.

Han, S.P., *A Globally Convergent Method for Nonlinear Programming*, J. Optimization Theory & Applications, Vol. 22, p. 297, 1977.

Hernandez, M., Ph.D. Thesis (in preparation) University of Hertfordshire, 1993

Kanzow, C., *Newton-type Methods for Nonlinearly Constrained Optimization*, Preprint 62, Institut fur Angewandte Mathematik, Universtitaet Hamburg, 1992.

Karmarkar, N., *A New Polynomial-time Algorithm for Linear Programming*, Combinatorica, Vol. 4, pp. 373-395, 1984.

Kleinmichel, H., C. Richter & K. Schonefeld, *On a Class of Hybrid Methods for Smooth Constrained Optimization*, J. Optimization Theory & Applications, 73, pp 465-499, 1992.

Lasdon L. et.al., *Design and Testing of a Generalized Reduced Gradient Code for Nonlinear Programming*, ACM Transactions on Mathematical Software, 4, pp. 34-50, 1978.

Maany, Z.A., *A New Algorithm for Highly Curved Optimization*, Math. Programming Study 31, pp. 139-154, 1987.

Mahidhara, D. & L. Lasdon, *An SQP Algorithm for Large Sparse Nonlinear Problems*, Technical Report, University of Texas School of Business Administration, 1990.

Mayne, D.Q. & E. Polak, *A Superlinearly Convergent Algorithm for Constrained Optimization Problems*, Math. Programming Study 16, pp. 45-61, 1982.

More, J.J. & S. Wright, *Numerical Optimization Algorithms and Software*, SIAM Tutorial, SIAM National Meeting, 1992. (Available from Mathematics and Computer Science Division, Argonne, National Laboratory).

Murray, W., *An Algorithm for Constrained Minimization*, in "Optimization" (ed. R. Fletcher), Academic Press, 1969.

Murtagh, B.A. & M.A. Saunders, *A Projected Lagrangian Algorithm and its Implementation for Sparse, Nonlinear Constraints*, Mathematical programming Study 16, pp. 84-117, 1982.

Nguyen, T.T., *Conjugate Basis Methods for Constrained Optimization*, Ph.D. Thesis, Hatfield Polytechnic, 1989.

Nocedal, J., *Updating quasi-Newton Matrices with Limited Storage*, Maths. of Computation, 35, pp. 773-782, 1980.

Polyak, R. *Modified Barrier Functions (Theory & Methods)*, Math. Programming, 54, pp. 177-222, 1992.

Powell, M.J.D., *A Fast Algorithm for Nonlinearly Constrained Optimization Calculations*, in "Numerical Analysis, Dundee 1977" (ed. G.A. Watson), Lecture Notes in Mathematics 630, Springer-Verlag, 1978.

Powell, M.J.D., *The Convergence of Variable Metric Methods for Nonlinearly Constrained Optimnization Calculations*, in "Nonlinear Programming 3" (eds. O.L. Mangasarian, R.R.Meyer & S.M. Robinson), Academic Press, 1978.

Powell, M.J.D. & Y. Yuan, *A Recursive Quadratic Programming Algorithm that uses Differentiable Exact Penalty Functions*, DAMTP Report 1984/NA9, Cambridge, 1984.

Powell, M.J.D., *On the Quadratic Programming Algorithm of Goldfarb & Idnani*, Math. Programming Study 25, pp. 46-61, North-Holland, 1985.

Powell, M.J.D. & Y. Yuan, *A Trust Region Algorithm for Equality Constrained Optimization*, Math. Programming, 49, pp. 189-212, 1990.

Powell, M.J.D. ,*A Direct Search Optimization Method that Models the Objective and Constraint Functions by Linear Interpolation*, Report DAMTP 1992/NA5, Cambridge, 1992.

Smith, S. & L. Lasdon, *Solving Large Sparse Nonlinear Programs using GRG*, ORSA Journal of Computing, 4, 1992.

Tanabe, K., *A Geometric Method in Nonlinear Programming*, J. Optimization Theory & Applications, 30, pp. 181-210, 1980.

Wilson, R.B., *A Simplicial Algorithm for Concave Programming*, Ph. D. Thesis, Harvard University Graduate School of Business Administration, 1963.

Wright, M.H., *Interior Methods for Constrained Optimization*, Acta Numerica, Vol.1, pp.341-407, 1992.

Exact Penalty Methods

G. DI PILLO

Dipartimento di Informatica e Sistemistica
Università di Roma "La Sapienza"
Via Eudossiana 18
00185 Roma
ITALY

ABSTRACT. Exact penalty methods for the solution of constrained optimization problems are based on the construction of a function whose unconstrained minimizing points are also solution of the constrained problem. In the first part of this paper we recall some definitions concerning exactness properties of penalty functions, of barrier functions, of augmented Lagrangian functions, and discuss under which assumptions on the constrained problem these properties can be ensured. In the second part of the paper we consider algorithmic aspects of exact penalty methods; in particular we show that, by making use of continuously differentiable functions that possess exactness properties, it is possible to define implementable algorithms that are globally convergent with superlinear convergence rate towards KKT points of the constrained problem.

1 Introduction

"It would be a major theoretic breakthrough in nonlinear programming if a simple continuously differentiable function could be exhibited with the property that any unconstrained minimum is a solution to the constrained problem." In: *J.P.Evans, F.J.Gould* and *J.W.Tolle*, Exact Penalty Functions in Nonlinear Programming, *Mathematical Programming*, vol.4, 1973.

This tutorial paper is an attempt to evaluate, twenty years later, to which extent the breakthrough of the above quotation has been achieved.

In the spirit of the quotation, a considerable amount of investigations, both from the theoretical and the computational point of view, has been devoted to methods which attempt to solve nonlinear programming problems by means of a single minimization of an unconstrained function. Methods of this kind are usually termed *exact penalty methods*, as opposed to the *sequential penalty methods*, which include the methods based on quadratic penalty functions and on augmented Lagrangian functions (see, e. g., [1, 2, 3, 6, 9, 15, 37, 46, 53, 72]).

E. Spedicato (ed.), Algorithms for Continuous Optimization, 209–253.
© 1994 *Kluwer Academic Publishers.*

Exact penalty methods are based on the construction of a function depending on a penalty parameter, such that the unconstrained minimum points of this function are also solutions of the constrained problem for *all sufficiently small* values of the parameter. We can subdivide exact penalty methods in two classes: methods based on *exact penalty functions* and methods based on *exact augmented Lagrangian functions*. In our terminology, the term exact penalty function is used when the variables of the unconstrained problem are in the same space as the variables of the original constrained problem, whereas the term exact augmented Lagrangian function is used when the unconstrained problem has is defined on the product space of the problem variables and of the multipliers.

A common feature of exact penalty methods is that the correspondence between the constrained and the unconstrained minimization problems can only be established with reference to an arbitrary compact set $\mathcal{D}$ containing the problem solutions, so that the threshold value of the penalty parameter depends on $\mathcal{D}$. This causes an inherent difficulty in the unconstrained minimization, since the level set corresponding to the penalty parameter and to the initial point, even if compact, may not be contained in $\mathcal{D}$. It follows that the sequence generated by the unconstrained algorithm may not converge to a problem solution, or may be unbounded. This difficulty can be overcome by incorporating in the exact penalty functions, or in the exact augmented Lagrangian functions, barrier terms that avoid unbounded sequences while preserving the exactness properties. We adopt the denomination of *exact barrier functions* and of *exact augmented Lagrangian barrier functions* for the functions that possess both exactness and barrier properties.

In this paper, we start by recalling some formal definitions of exactness for penalty functions, proposed in [31], which attempt to capture the most relevant aspects of the notion of exactness in the context of constrained optimization. This is motivated by the fact that, up to a certain time, the term *exact penalty function* has been used in the literature without a definite agreement on its meaning. In particular, as remarked in [53], most of the early literature on this subject has been concerned with conditions ensuring that the penalty function has a local (global) minimum at a local (global) minimum point of the constrained problem. However, since the penalty approach is an attempt to solve a constrained problem by the minimization of an unconstrained function, it is of greater interest the study of converse properties, which ensure that local (global) minimimum points of the penalty function are local (global) solutions of the constrained problem. Moreover, in the non convex case, a distinction has to be made between properties of exactness pertaining to global solutions and properties pertaining to local solutions.

With reference to these definitions, we show that, for the same penalty function, different kind of exactness can be established under different qualification requirements on the problem constraints. In particular, we consider nondifferentiable exact penalty functions, that have been the first ones for which exactness properties have been recognized, and continuously differentiable exact penalty functions, more recently developed to a wide extent.

Then we introduce the definition of *exact barrier function*, and we demonstrate the usefulness of incorporating a barrier term in the penalty functions. By this tool the shape of the penalty function is modified in such a way that the penalty function has compact level sets contained in the same set $\mathcal{D}$ where its exactness holds. Hence, if the penalty parameter is sufficiently small, a minimizing sequence is guaranteed to be convergent towards Karush-Kuhn-Tucker points of the constrained problem, without an additional assumption of being bounded.

The same line of development is repeated for exact augmented Lagrangian functions: we give some definitions of exactness, that in this case involve also the Karush-Kuhn-Tucker multipliers; then we analyse a function that satisfy the definitions, according to the qualification conditions satisfied by the constraints; then we show that, also in this case, it is possible to incorporate a barrier term, that is effective on the extended space of the problem variables and of the multipliers, and that prevents the occurrence of unbounded level sets.

After the descriptions of the main classes of functions that enjoy exactness properties, a discussion is in order regarding the assumptions employed in the development of the theoretical aspects. Therefore, we comment on the relative strength of the assumptions employed, and we mention alternative assumptions, under which exactness properties can be proved. In particular we mention the very interesting result that no constraint qualifications conditions may be required outside the feasible set if a feasible point of the constrained problem is known.

Finally we consider the algorithmic point of view, and we focus on the fact that, making use of continuously differentiable exact barrier functions, it is possible to devise methods for the solution of smooth constrained problems, that behave like the methods for the unconstrained minimization of continuously differentiable functions, as concerns the convergence and rate of convergence properties. As representative of these methods, we describe in particular a Newton-type approach and a recursive quadratic programming approach, that are globally stabilized by combining the use of an exact barrier function in the line search and of a suitable updating rule of the penalty parameter.

In the paper all propositions are stated without proof. The proofs can be found in the references quoted; in some cases minor modifications are needed, in order to have different material organized in the same framework.

2 Problem statement, basic notation and preliminary results

The problem considered here is the general constrained nonlinear programming problem:

$$\text{minimize } f(x)$$
$$\text{subject to } g(x) \leq 0, \ h(x) = 0, \tag{C}$$

where $f : \mathbb{R}^n \to \mathbb{R}$, $g : \mathbb{R}^n \to \mathbb{R}^m$, $h : \mathbb{R}^n \to \mathbb{R}^p$, $p \leq n$, are continuously differentiable functions.

We denote by $\mathcal{F} := \{x \in \mathbb{R}^n : g(x) \leq 0, \ h(x) = 0\}$ the feasible set that we assume nonempty, by $\mathcal{G}_C$ and $\mathcal{L}_C$ the set of global solutions and the set of local solutions of Problem (C).

We shall make use of the following assumptions on Problem (C):

Assumption 1 *The feasible set $\mathcal{F}$ is compact.*

Assumption 2 *The objective function f takes a finite number of values on $\mathcal{L}_C$.*

Assumption 1 is introduced for the sake of simplicity, in order to avoid the slightly more cumbersome notation and statements required in its absence. Hence it holds everywhere in the sequel. By this assumption, if $\mathcal{F}$ is nonempty, also $\mathcal{G}_C$ is nonempty.

Assumption 2 will be invoked only when needed.

For any $x \in \mathbb{R}^n$ we define the index sets:

$$
\begin{aligned}
I_0(x) &:= \{i : g_i(x) = 0\} \\
I_+(x) &:= \{i : g_i(x) \geq 0\} \\
I_-(x) &:= \{i : g_i(x) < 0\}.
\end{aligned}
$$

We make reference to some constraint qualification conditions, according to the following terminology.

The *linear independence constraint qualification* (LICQ) holds at $x \in \mathbb{R}^n$ if the gradients $\nabla g_i(x)$, $i \in I_0(x)$, $\nabla h_j(x)$, $j = 1, \ldots, p$, are linearly independent.

The *Mangasarian-Fromovitz constraint qualification* (MFCQ) holds at $x \in \mathbb{R}^n$ if $\nabla h_j(x)$, $j = 1, \ldots, p$, are linearly independent and there exists a $z \in \mathbb{R}^n$ such that

$$
\begin{aligned}
\nabla g_i(x)'z &< 0, \quad i \in I_0(x) \\
\nabla h_j(x)'z &= 0, \quad j = 1, \ldots, p.
\end{aligned}
$$

The *extended Mangasarian-Fromovitz constraint qualification* (EMFCQ) holds at $x \in \mathbb{R}^n$ if $\nabla h_j(x)$, $j = 1, \ldots, p$, are linearly independent and there exists a $z \in \mathbb{R}^n$ such that

$$
\begin{aligned}
\nabla g_i(x)'z &< 0, \quad i \in I_+(x) \\
\nabla h_j(x)'z &= 0, \quad j = 1, \ldots, p.
\end{aligned}
$$

It can be noted that the LICQ implies the MFCQ and that the EMFCQ implies the MFCQ.

The Lagrangian function associated with Problem (C) is the function $L : \mathbb{R}^n \times \mathbb{R}^m \times \mathbb{R}^p \to \mathbb{R}$ defined by

$$L(x, \lambda, \mu) := f(x) + \lambda' g(x) + \mu' h(x).$$

A *Karush-Kuhn-Tucker* (KKT) *triplet* for Problem (C) is a triplet $(\overline{x}, \overline{\lambda}, \overline{\mu}) \in \mathbb{R}^n \times \mathbb{R}^m \times \mathbb{R}^p$ such that $\overline{x} \in \mathcal{F}$ and:

$$\nabla_x L(\overline{x}, \overline{\lambda}, \overline{\mu}) = 0,$$
$$\overline{\lambda}' g(\overline{x}) = 0,$$
$$\overline{\lambda} \geq 0.$$

Let us denote by $\mathcal{K}$ the set

$$\mathcal{K} := \{\overline{x} \in \mathbb{R}^n : \text{there exist } \overline{\lambda}, \overline{\mu} \text{ such that}$$
$$(\overline{x}, \overline{\lambda}, \overline{\mu}) \text{ is a KKT triplet for Problem (C)}\};$$

a point $\overline{x} \in \mathcal{K}$ will be called a *KKT point* for Problem (C).

It is known that if $\overline{x}$ is a local solution of Problem (C) and if the MFCQ holds at $\overline{x}$, then $\overline{x} \in \mathcal{K}$, so that there exist KKT multipliers $\overline{\lambda}, \overline{\mu}$ associated with $\overline{x}$.

Let $\overline{x} \in \mathcal{K}$, we say that *strict complementarity* holds at $\overline{x}$ if $\overline{\lambda}_i > 0$ for all $i \in I_0(\overline{x})$.

Let $\overline{x} \in \mathcal{K}$ be a point where strict complementarity holds, assume that the problem functions f, g and h are twice continuously differentiable in a neighbourhood of $\overline{x}$, and that the second order condition

$$z' \nabla_x^2 L(\overline{x}, \overline{\lambda}, \overline{\mu}) z > 0 \forall z \in \mathbb{R}^n \text{ such that } z \neq 0, \begin{cases} \nabla g_i(\overline{x})' z = 0, & i \in I_0(\overline{x}), \\ \nabla h_j(\overline{x})' z = 0, & j = 1, \ldots, p, \end{cases}$$

is satisfied; then $\overline{x}$ is an *isolated* local solution of Problem (C).

We make use of the following notation. Given the set $\mathcal{A}$, we denote by $\overset{\circ}{\mathcal{A}}$, $\partial \mathcal{A}$ and $\overline{\mathcal{A}}$ respectively the interior, the boundary and the closure of $\mathcal{A}$. Given a vector u with components u_i, $i = 1, \ldots, m$ we denote by u' its transpose, by $\|u\|_q$ its ℓ_q-norm ($q = 2$ when missing), by u^+ the vector with components:

$$u_i^+ := \max[0, u_i], \quad i = 1, \ldots, m$$

and by U the diagonal matrix defined by:

$$U := \text{diag}(u_i), \quad i = 1, \ldots, m.$$

Given two vectors u, v we adopt the short notation (u, v) for the composite vector $(u' \, v')'$. Given a function $F : \mathbb{R}^n \to \mathbb{R}$ we denote by $DF(x, d)$ the directional derivative of F at x along the direction d. We say that $\overline{x}$ is a *critical point* of F if $DF(\overline{x}, d) \geq 0$ for all $d \in \mathbb{R}^n$. If $\overline{x}$ is a critical point of F and F is differentiable at $\overline{x}$ we have $\nabla F(\overline{x}) = 0$; in this case we say that $\overline{x}$ is a *stationary point* of F. Given a vector function $\Phi : \mathbb{R}^n \to \mathbb{R}^s$ we denote by $\nabla \Phi(x)$ the $n \times s$ transpose Jacobian matrix of Φ at x. Finally, we denote by $\mathcal{B}(x; \rho)$ the open ball around x with radius $\rho > 0$.

3 Definitions of exactness for penalty functions

Roughly speaking, an *exact penalty function* for Problem (C) is a function $F(x; \varepsilon)$, where $\varepsilon > 0$ is a *penalty parameter*, with the property that there is an appropriate parameter choice such that *a single unconstrained minimization* of $F(x; \varepsilon)$ yields a solution to Problem (C). In particular, we require that there is an easy way for finding correct parameter values, by imposing that exactness is retained for all ε ranging on some set of nonzero measure. More specifically, we take $\varepsilon \in (0, \ \varepsilon^*]$ where $\varepsilon^* > 0$ is a suitable *threshold value*.

In practice, the existence of a threshold value for the parameter ε and hence the possibility of constructing the exact penalty function $F(x; \varepsilon)$ can only be established with reference to an arbitrary compact set $\mathcal{D}$, that we assume to be regular, that is such that $\mathcal{D} = \overset{\circ}{\mathcal{D}}$. Therefore, instead of Problem (C) we must consider the problem:

$$\text{minimize } f(x), \ x \in \mathcal{F} \cap \mathcal{D}, \tag{Č}$$

where $\mathcal{D}$ is a compact subset of $\mathbb{R}^n$ such that $\mathcal{F} \cap \overset{\circ}{\mathcal{D}} \neq \emptyset$.

In particular, recalling that by Assumption 1 $\mathcal{F}$ is compact, it is possible to assume that the reference set $\mathcal{D}$ contains in its interior the feasible set $\mathcal{F}$:

Assumption 3 $\mathcal{D}$ *is such that* $\mathcal{F} \subset \overset{\circ}{\mathcal{D}}$.

Assumption 3 holds everywhere in the sequel. Under Assumption 3 Problem (Č) and Problem (C) become equivalent, so that we can make reference only to the latter one.

For any given $\varepsilon > 0$, let now $F(x; \varepsilon)$ be a continuous real function defined on $\overset{\circ}{\mathcal{D}}$ and consider the problem:

$$\text{minimize } F(x; \varepsilon), \ x \in \overset{\circ}{\mathcal{D}}. \tag{U}$$

Since $\overset{\circ}{\mathcal{D}}$ is an open set, any local solution of Problem (U), provided it exists, is unconstrained and therefore Problem (U) can be considered as an essentially unconstrained problem. The sets of global and local solutions of Problem (U) are denoted, respectively, by $\mathcal{G}_\mathcal{U}(\varepsilon)$ and $\mathcal{L}_\mathcal{U}(\varepsilon)$:

$$\mathcal{G}_\mathcal{U}(\varepsilon) := \{x \in \overset{\circ}{\mathcal{D}}\colon F(x; \varepsilon) \leq F(y; \varepsilon), \text{ for all } y \in \overset{\circ}{\mathcal{D}}\}$$

$$\mathcal{L}_\mathcal{U}(\varepsilon) := \{x \in \overset{\circ}{\mathcal{D}}\colon \text{ for some } \rho > 0 \ F(x; \varepsilon) \leq F(y; \varepsilon), \text{ for all } y \in \overset{\circ}{\mathcal{D}} \cap \mathcal{B}(x; \rho)\}.$$

A local solution of Problem (U) is a critical point of F, and it is a stationary point if F is differentiable.

There are different kinds of relationships between Problem (C) and Problem (U), which can be associated with different notions of exactness. We adopt here the classification introduced in [31].

A first possibility is that of considering a correspondence between global minimizers of Problem (C) and global minimizers of Problem (U). This correspondence is established formally in the following definition.

Definition 1 *The function $F(x; \varepsilon)$ is a* weakly exact penalty function *for Problem (C) with respect to the set $\mathcal{D}$ if there exists an $\varepsilon^* > 0$ such that, for all $\varepsilon \in (0, \varepsilon^*]$, any global solution of Problem (C) is a global minimum point of Problem (U) and conversely; that is, if for some $\varepsilon^* > 0$:*

$$\mathcal{G}_C = \mathcal{G}_U(\varepsilon), \quad \text{for all } \varepsilon \in (0, \varepsilon^*].$$

The property stated above guarantees that the constrained problem can actually be solved by means of the *global unconstrained minimization* of $F(x; \varepsilon)$ for sufficiently small values of the parameter ε.

The notion of exactness expressed by Definition 1 appears to be of limited value for general nonlinear programming problems, since it does not concern local minimizers of the penalty function, while unconstrained minimization algorithms usually determine only local minimizers. Therefore, we introduce a further requirement concerning local minimizers which gives rise to a stronger notion of exactness.

Definition 2 *The function $F(x; \varepsilon)$ is an* exact penalty function *for Problem (C) with respect to the set $\mathcal{D}$ if there exists an $\varepsilon^* > 0$ such that, for all $\varepsilon \in (0, \varepsilon^*]$, $\mathcal{G}_C = \mathcal{G}_U(\varepsilon)$ and, moreover, any local unconstrained minimizer of Problem (U) is a local solution of Problem (C), that is:*

$$\mathcal{L}_U(\varepsilon) \subseteq \mathcal{L}_C, \quad \text{for all } \varepsilon \in (0, \varepsilon^*].$$

It must be remarked that the notion of exactness given in Definition 2 does not require that all local solutions of Problem (C) correspond to local minimizers of the exact penalty function. A one-to-one correspondence of local minimizers does not seem to be required, in practice, to give a meaning to the notion of exactness, since the condition $\mathcal{G}_C = \mathcal{G}_U(\varepsilon)$ ensures that global solutions of Problem (C) are preserved. However, for the classes of exact penalty functions considered in the sequel, it will be shown that also this correspondence can be established, at least under Assumption 2. Thus, we can also consider the following notion.

Definition 3 *The function $F(x; \varepsilon)$ is a* strongly exact penalty function *for Problem (C) with respect to the set $\mathcal{D}$ if there exists an $\varepsilon^* > 0$ such that, for all $\varepsilon \in (0, \varepsilon^*]$, $\mathcal{G}_C = \mathcal{G}_U(\varepsilon)$, $\mathcal{L}_U(\varepsilon) \subseteq \mathcal{L}_C$, and, moreover, any local solution of Problem (C) is a local unconstrained minimizer of $F(x; \varepsilon)$, that is:*

$$\mathcal{L}_C \subseteq \mathcal{L}_U(\varepsilon) \text{ for all } \varepsilon \in (0, \varepsilon^*].$$

The formal definitions mentioned so far constitute the basis for the statement of sufficient conditions for a penalty function to be exact according to some specified notion of exactness. In particular, in [31] we have established sufficient conditions which apply both to the nondifferentiable and to the continuously differentiable case, thus providing a unified framework for the analysis of exact penalty functions. In this framework, we have considered the best known classes of exact penalty functions, recovering known results and establishing new ones.

Additional exactness properties, employed in computational applications, concern the correspondence between critical or stationary points of the unconstrained Problem (U), and KKT triplets of the constrained Problem (C). As we will see, the penalty functions that are exact according to the definitions given before satisfy also these properties, provided that the value of the penalty parameter ε is below a threshold value ε^*.

In the following section we report the exactness results that can be proved for the most interesting nondifferentiable and differentiable penalty functions.

4 Exact penalty functions

4.1 Nondifferentiable exact penalty functions

Nondifferentiable penalty functions have been the first ones for which some exactness properties have been established, by Zangwill in [73]. In this section we collect the exactness properties of the best known class of nondifferentiable penalty functions, proved in [31] by means of the sufficient conditions mentioned before.

We consider the class of nondifferentiable penalty functions defined by:

$$J_q(x;\varepsilon) := f(x) + \frac{1}{\varepsilon}\big\| \, [g^+(x), \, h(x)] \, \big\|_q,$$

where $1 \leq q \leq \infty$. In particular, we have:

$$J_q(x;\varepsilon) = f(x) + \frac{1}{\varepsilon}\left[\sum_{i=1}^{m} (g_i^+(x))^q + \sum_{j=1}^{p} |h_j(x)|^q \right]^{1/q},$$

for $1 \leq q < \infty$, and:

$$J_\infty(x;\varepsilon) = f(x) + \frac{1}{\varepsilon}\max\big[g_1^+(x), \ldots, g_m^+(x), |h_1(x)|, \ldots, |h_p(x)| \big].$$

$J_1(x;\varepsilon)$ and $J_\infty(x;\varepsilon)$ are the nondifferentiable penalty functions most frequently considered in the literature (see, e.g. [14, 16, 44, 47, 52, 61, 62, 66, 67, 73]).

A preliminary result concerns the existence of the directional derivative $DJ_q(x, d; \varepsilon)$:

Proposition 1 *For all* $\varepsilon > 0$ *and all* $d \in \mathbb{R}^n$ *the function* $J_q(x; \varepsilon)$ *admits a directional derivative* $DJ_q(x, d; \varepsilon)$.

The explicit expression of the directional derivative $DJ_q(x, d; \varepsilon)$ for $q = 1$, $1 < q < \infty$ and $q = \infty$, useful for computational purposes, is given in full detail in [30].

We can now state that, under suitable assumptions on Problem (C), the function $J_q(x; \varepsilon)$ satisfies the definitions of exactness given in the preceding section.

Proposition 2 (a) *Assume that the MFCQ is satisfied at every global minimum point of Problem* (C). *Then, the function* $J_q(x; \varepsilon)$ *is a weakly exact penalty function for Problem* (C)*with respect to the set* $\mathcal{D}$.

(b) *Assume that the EMFCQ is satisfied on* $\mathcal{D}$. *Then, the function* $J_q(x; \varepsilon)$ *is an exact penalty function for Problem* (C)*with respect to the set* $\mathcal{D}$; *moreover, if Assumption 2 holds, the function* $J_q(x; \varepsilon)$ *is a strongly exact penalty function for Problem* (C)*with respect to the set* $\mathcal{D}$.

In the next propositions we report additional exactness results that characterize the correspondence between critical points of $J_q(x; \varepsilon)$ and KKT triplets of Problem (C).

Proposition 3 *Assume that the EMFCQ holds on* $\mathcal{D}$; *then there exists an* $\varepsilon^* > 0$ *such that for all* $\varepsilon \in (0, \varepsilon^*]$ *if* $x_\varepsilon \in \overset{\circ}{\mathcal{D}}$ *is a critical point of* $J_q(x; \varepsilon)$, x_ε *is a KKT point for Problem* (C).

Proposition 4 *Let* $(\overline{x}, \overline{\lambda}, \overline{\mu})$ *be a KKT triplet for Problem* (C). *Then:*

(a) $\overline{x}$ *is a critical point of* $J_q(x; \varepsilon)$, $1 \leq q < \infty$ *for all* $\varepsilon > 0$ *such that:*

$$\overline{\lambda}_i \varepsilon \leq (m+p)^{(1-q)/q}, \quad i \in I_0(\overline{x})$$
$$|\overline{\mu}_j| \varepsilon \leq (m+p)^{(1-q)/q}, \quad j = 1, \dots, p.$$

(b) $\overline{x}$ *is a critical point of* $J_\infty(x; \varepsilon)$, *for all* $\varepsilon > 0$ *such that:*

$$\varepsilon \left[\sum_{i \in I_0(\overline{x})} \overline{\lambda}_i + \sum_{j=1}^{p} |\overline{\mu}_j| \right] \leq 1.$$

4.2 Continuously differentiable exact penalty functions

In this subsection we assume that the problem functions f, g, h are *twice* continuously differentiable.

The key idea for the construction of continuously differentiable exact penalty functions is that of replacing the multiplier vectors λ, μ which appear in the augmented Lagrangian function of Hestenes - Powell - Rockafellar [3] with continuously differentiable *multiplier functions* $\lambda(x), \mu(x)$, depending on the problem variables, and characterized by means of the following definition:

Definition 4 *The vector function* $(\lambda(x), \mu(x))$, *with* $\lambda : \mathbb{R}^n \to \mathbb{R}^m$, $\mu : \mathbb{R}^n \to \mathbb{R}^p$, *is a* multiplier function *for Problem* (C) *if for all* $\overline{x} \in \mathcal{K}$ *it results* $(\lambda(\overline{x}), \mu(\overline{x})) = (\overline{\lambda}, \overline{\mu})$.

Continuously differentiable multiplier functions have been introduced by Fletcher in [43] for the equality constrained case, and by Glad and Polak in [51] for the more general Problem (C). Here we adopt multiplier functions which are a generalization of those proposed in [43, 51], and which have been introduced in [65] for the equality, and in [56] for the inequality constrained case. The main property of the latter multiplier functions is that they can be defined without requiring any assumption outside the feasible set $\mathcal{F}$. In the feasible set, the following constraints qualification assumption is needed.

Assumption 4 *The* LICQ *holds on* $\mathcal{F}$.

Assumption 4 ensures that for any $\overline{x} \in \mathcal{K}$ the associated KKT multipliers $\overline{\lambda}, \overline{\mu}$ are unique. Under this assumption, for any $x \in \mathbb{R}^n$, we can consider the multiplier functions $\lambda(x), \mu(x)$ obtained by minimizing over $\mathbb{R}^m \times \mathbb{R}^p$ the quadratic function in (λ, μ) defined by:

$$\Psi(\lambda, \mu; x) := \left\| \nabla_x L(x, \lambda, \mu) \right\|^2 + \gamma_1 \left\| G(x)\lambda \right\|^2 + \gamma_2 s(x)(\left\| \lambda \right\|^2 + \left\| \mu \right\|^2),$$

where $\gamma_1 > 0$, $\gamma_2 > 0$, $G(x) := \text{diag}\,(g_i(x))$ and

$$s(x) := \sum_{i=1}^{m} g_i^+(x)^r + \|h(x)\|^2,$$

with $r \geq 2$.

In the function $\Psi(\lambda, \mu; x)$ the first and second terms can be viewed as a measure of the violation of the KKT necessary conditions $\nabla_x L(x, \lambda, \mu) = 0$, $\quad G(x)\lambda = 0$. As concerns the third term, it vanishes for all $x \in \mathcal{F}$, so that, for all $x \in \mathcal{F}$ the multiplier functions obtained by minimizing Ψ reduce to the multiplier functions employed in [43, 51]; and it has a positive definite Hessian matrix for all $x \notin \mathcal{F}$, so that, in the minimization of Ψ, makes it possible to dispense with the requirement that the LICQ holds also for $x \notin \mathcal{F}$, as happens for the other multiplier functions.

Let $M(x)$ be the $(m + p) \times (m + p)$ matrix defined by:

$$M(x) := \begin{bmatrix} M_{11}(x) & M_{12}(x) \\ M_{21}(x) & M_{22}(x) \end{bmatrix}.$$

with

$$
\begin{aligned}
M_{11}(x) &:= \nabla g(x)'\nabla g(x) + \gamma_1 G^2(x) + \gamma_2 s(x) I_m, \\
M_{12}(x) &:= M_{21}(x)' = \nabla g(x)'\nabla h(x), \\
M_{22}(x) &:= \nabla h(x)'\nabla h(x) + \gamma_2 s(x) I_p,
\end{aligned}
$$

and I_m (I_p) indicates the $m \times m$ $(p \times p)$ identity matrix; in the next proposition we recall some results stated in [56, 57], and, in particular we give the expression of $(\lambda(x), \mu(x))$.

Proposition 5 *For any $x \in \mathbb{R}^n$:*

(a) the matrix $M(x)$ is positive definite;

(b) there exists a unique minimizer $(\lambda(x), \mu(x))$ of the quadratic function in (λ, μ), $\Psi(\lambda, \mu; x)$, given by:

$$\begin{bmatrix} \lambda(x) \\ \mu(x) \end{bmatrix} = -M^{-1}(x) \begin{bmatrix} \nabla g(x)' \\ \nabla h(x)' \end{bmatrix} \nabla f(x);$$

(c) the function $(\lambda(x), \mu(x))$ is continuously differentiable.

Moreover, if $(\overline{x}, \overline{\lambda}, \overline{\mu})$ is a KKT triplet of Problem (C), we have $\lambda(\overline{x}) = \overline{\lambda}$ and $\mu(\overline{x}) = \overline{\mu}$.

It is to be noted that in computations the inverse matrix $M^{-1}(x)$ is not needed; in fact, since $M(x)$ is symmetric and positive definite, the multiplier functions can be evaluated using a Cholesky factorization for solving the system:

$$M(x) \begin{bmatrix} \lambda(x) \\ \mu(x) \end{bmatrix} = - \begin{bmatrix} \nabla g(x)' \\ \nabla h(x)' \end{bmatrix} \nabla f(x).$$

Let us now recall that the augmented Lagrangian of Hestenes-Powell-Rockafellar used in sequential algorithms for the solution of Problem (C) is the function $L_a : \mathbb{R}^n \times \mathbb{R}^m \times \mathbb{R}^p \to \mathbb{R}$ given by:

$$\begin{aligned} L_a(x, \lambda, \mu; \varepsilon) \quad := \quad & f(x) + \lambda'(g(x) + Y(x, \lambda; \varepsilon) y(x, \lambda; \varepsilon)) + \mu' h(x) \\ + \quad & \frac{1}{\varepsilon} \|g(x) + Y(x, \lambda; \varepsilon) y(x, \lambda; \varepsilon)\|^2 + \frac{1}{\varepsilon} \|h(x)\|^2, \end{aligned} \tag{1}$$

where:

$$\begin{aligned} y_i(x, \lambda; \varepsilon) \quad &:= \quad \left\{ -\min\left[0, g_i(x) + \frac{\varepsilon}{2}\lambda_i\right] \right\}^{1/2}, \quad i = 1, \dots, m \\ Y(x, \lambda; \varepsilon) \quad &:= \quad \operatorname{diag}(y_i(x, \lambda; \varepsilon)). \end{aligned}$$

By replacing in L_a the multiplier vectors λ, μ by the multiplier functions $\lambda(x), \mu(x)$ given in Proposition 6, we obtain the penalty function $W(x; \varepsilon)$ defined by:

$$\begin{aligned} W(x; \varepsilon) \quad := \quad & f(x) + \lambda(x)' \left(g(x) + Y(x; \varepsilon) y(x; \varepsilon)\right) + \mu(x)' h(x) \\ + \quad & \frac{1}{\varepsilon} \|g(x) + Y(x; \varepsilon) y(x; \varepsilon)\|^2 + \frac{1}{\varepsilon} \|h(x)\|^2, \end{aligned}$$

where:

$$y_i(x; \varepsilon) := \left\{ - \min \left[0, g_i(x) + \frac{\varepsilon}{2} \lambda_i(x) \right] \right\}^{1/2}, \quad i = 1, \ldots, m$$
$$Y(x; \varepsilon) := \operatorname{diag}(y_i(x; \varepsilon)).$$

It can be verified that the function $W(x; \varepsilon)$ can also be written in the form:

$$
\begin{aligned}
W(x; \varepsilon) = {} & f(x) + \lambda(x)'g(x) + \mu(x)'h(x) + \frac{1}{\varepsilon} \|g(x)\|^2 + \frac{1}{\varepsilon} \|h(x)\|^2 \\
& - \frac{1}{4\varepsilon} \sum_{i=1}^{m} \{ \min [0, \varepsilon \lambda_i(x) + 2g_i(x)] \}^2.
\end{aligned}
$$

From the above expression, the differentiability assumptions on the problem functions, and Assumption 4 it follows that $W(x; \varepsilon)$ is continuously differentiable on $\mathbb{R}^n$. More precisely, we have:

Proposition 6 *The function $W(x; \varepsilon)$ is continuously differentiable for any $x \in \mathbb{R}^n$, and its gradient is given by:*

$$
\begin{aligned}
\nabla W(x; \varepsilon) = {} & \nabla f(x) + \nabla g(x)\lambda(x) + \nabla h(x)\mu(x) \\
& + \nabla \lambda(x) \left(g(x) + Y(x; \varepsilon)y(x; \varepsilon) \right) + \nabla \mu(x)h(x) \\
& + \frac{2}{\varepsilon} \nabla g(x) \left(g(x) + Y(x; \varepsilon)y(x; \varepsilon) \right) + \frac{2}{\varepsilon} \nabla h(x)h(x).
\end{aligned}
$$

The evaluation of $\nabla W(x; \varepsilon)$ requires the evaluation of the second order derivatives of the problem functions f, g, h, that are present in $\nabla \lambda(x)$, $\nabla \mu(x)$. This aspect is common to all continuously differentiable exact penalty functions.

The properties of exactness of the function $W(x; \varepsilon)$ can be deduced as in [31, 56], and are summarized in the following Proposition.

Proposition 7 (a) *Assume that the LICQ is satisfied on $\mathcal{F}$; then the function $W(x; \varepsilon)$ is a weakly exact penalty function for Problem (C) with respect to the set $\mathcal{D}$.*

(b) *Assume that the EMFCQ is satisfied on $\mathcal{D}$. Then, the function $W(x; \varepsilon)$ is an exact penalty function for Problem (C) with respect to the set $\mathcal{D}$; moreover, if Assumption 2 holds, the function $W(x; \varepsilon)$ is a strongly exact penalty function for Problem (C) with respect to the set $\mathcal{D}$.*

As in the case of nondifferentiable penalty functions, we have additional exactness properties pertaining to the correspondence between stationary points of $W(x; \varepsilon)$ and KKT points of Problem (C). Of main interest are the following:

Proposition 8 *Assume that the EMFCQ holds on $\mathcal{D}$. Then, there exists an $\varepsilon^* > 0$ such that, for all $\varepsilon \in (0, \varepsilon^*]$, if $x_\varepsilon \in \overset{\circ}{\mathcal{D}}$ is a stationary point of $W(x; \varepsilon)$, we have that $(x_\varepsilon, \lambda(x_\varepsilon), \mu(x_\varepsilon))$ is a KKT triplet for Problem (C).*

Proposition 9 *Let $(\overline{x}, \overline{\lambda}, \overline{\mu})$ be a KKT triplet for Problem (C). Then, for any $\varepsilon > 0$, we have $\nabla W(\overline{x}; \varepsilon) = 0$.*

If we make the assumption that the problem functions are *three times* continuously differentiable and if we have $r \geq 3$ in the definition of the multiplier functions, we can prove not only first order results, but also second order results.

Let us introduce the notation:

$$
\begin{aligned}
\nabla g_0(x) &:= [\nabla g_i(x)], \ i \in I_0(x), \\
\nabla \lambda_0(x) &:= [\nabla \lambda_i(x)], \ i \in I_0(x), \\
\nabla \lambda_-(x) &:= [\nabla \lambda_i(x)], \ i \in I_-(x);
\end{aligned}
$$

then we can state the following propositions.

Proposition 10 *(a) The function $W(x; \varepsilon)$ is twice continuously differentiable for all $x \in \mathbb{R}^n$ except, possibly, at the points where $g_i(x) + \varepsilon \lambda_i(x)/2 = 0$ for some i.*

(b) Let $(\overline{x}, \overline{\lambda}, \overline{\mu})$ be a KKT triplet for Problem (C) and assume that strict complementary holds at $(\overline{x}, \overline{\lambda}, \overline{\mu})$. Then, for any $\varepsilon > 0$, the function $W(x; \varepsilon)$ is twice continuously differentiable in a neighbourhood of $\overline{x}$, and the Hessian matrix of $W(x; \varepsilon)$ evaluated at $\overline{x}$ is given by:

$$
\begin{aligned}
\nabla^2 W(\overline{x}; \varepsilon) = \ & \nabla_x^2 L(\overline{x}, \lambda(\overline{x}), \mu(\overline{x})) + \nabla g_0(\overline{x}) \nabla \lambda_0(\overline{x})' + \nabla \lambda_0(\overline{x}) \nabla g_0(\overline{x})' \\
& + \nabla h(\overline{x}) \nabla \mu(\overline{x})' + \nabla \mu(\overline{x}) \nabla h(\overline{x})' - \frac{\varepsilon}{2} \nabla \lambda_-(\overline{x}) \nabla \lambda_-(\overline{x})' \\
& + \frac{2}{\varepsilon} \left[\nabla g_0(\overline{x}) \nabla g_0(\overline{x})' + \nabla h(\overline{x}) \nabla h(\overline{x})' \right].
\end{aligned}
$$

We have written before the explicit expression of $\nabla^2 W(\overline{x}; \varepsilon)$ with the aim to remark that all terms containing third order derivatives of the problem functions vanish at a KKT point $\overline{x}$. As we shall see, this property is very important from the computational point of view, since makes it possible to define consistent approximations of the Newton direction of $W(x; \varepsilon)$ employing only the first and second order derivatives.

Proposition 11 *Let $(\overline{x}, \overline{\lambda}, \overline{\mu})$ be a KKT triplet for Problem (C) and assume that*

(i) strict complementarity holds at $(\overline{x}, \overline{\lambda}, \overline{\mu})$;

(ii) $\overline{x}$ *in an isolated local minimum point for Problem* (C) *satisfying the second order sufficiency condition.*

Then, there exists an ε^* *such that for all* $\varepsilon \in (0, \varepsilon^*]$, $\overline{x}$ *is an isolated local minimum point for* $W(x; \varepsilon)$ *and the Hessian matrix* $\nabla^2 W(\overline{x}; \varepsilon)$ *is positive definite.*

Proposition 12 *Assume that*

(i) the EMFCQ holds on $\mathcal{D}$;

(ii) the strict complementarity holds at any KKT triplet $(\overline{x}, \overline{\lambda}, \overline{\mu})$ *of Problem* (C).

Then, there exists an $\varepsilon^* > 0$ *such that, for all* $\varepsilon \in (0, \varepsilon^*]$, *if* $x_\varepsilon \in \overset{\circ}{\mathcal{D}}$ *is a local unconstrained minimum point of* $W(x; \varepsilon)$, *with positive definite Hessian* $\nabla^2 W(x_\varepsilon; \varepsilon)$, x_ε *is an isolated local minimum point of Problem* (C), *satisfying the second order sufficiency condition.*

5   Exact barrier functions

The properties considered in the definitions of Section 3 do not characterize the behaviour of $F(x; \varepsilon)$ on the boundary of $\overset{\circ}{\mathcal{D}}$. Although this may be irrelevant from the conceptual point of view in connection with the notion of exactness, it may assume a considerable interest from the computational point of view, when unconstrained descent methods are employed for the minimization of $F(x; \varepsilon)$. In fact, it may happen that there exist points of $\overset{\circ}{\mathcal{D}}$ such that a descent path for $F(x; \varepsilon)$ which originates at some of these points crosses the boundary of $\overset{\circ}{\mathcal{D}}$. This implies that the sequence of points produced by an unconstrained algorithm may be attracted towards a stationary point of $F(x; \varepsilon)$ outside the set $\mathcal{D}$ where the exactness properties hold, or may not admit a limit point. Therefore, it could be difficult to construct minimizing sequences for $F(x; \varepsilon)$ which are globally convergent on $\overset{\circ}{\mathcal{D}}$ towards the solutions of the constrained problem.

Consider as an example the simple problem:

$$\begin{aligned} \text{minimize } & x^3 \\ \text{subject to } & x = 0. \end{aligned} \tag{E}$$

This is a very well behaved problem, which has the unique solution $x = 0$. If we consider the nondifferentiable function

$$J_1(x; \varepsilon) = x^3 + \frac{1}{\varepsilon}|x|,$$

it is easily seen that, by taking $\mathcal{D} = [-1.5, 1.5]$ $J_1(x; \varepsilon)$ turns out to be an exact penalty function for Problem (E), for all $\varepsilon \in (0,1]$. However a descent path for $J_1(x;1)$ starting at $\tilde{x} = -2/3$ meets the boundary of $\mathcal{D}$ at the point $x = -1$ and leads the function value to $-\infty$ with $x \to -\infty$, as shown in Figure 1 a). Hence, if one employs a descent algorithm applied to $J_1(x;1)$ for solving Problem (E), unbounded sequences can be generated and global convergence towards the solution cannot be ensured.

In order to avoid this difficulty, it is necessary to impose further conditions on $F(x;\varepsilon)$ and we are lead to introduce the notion of *exact barrier function*.

Definition 5 *The function $F(x;\varepsilon)$ is a (weakly, strongly) exact barrier function for Problem (C) with respect to the set $\mathcal{D}$ if it is a (weakly, strongly) exact penalty function and, moreover, $\forall \varepsilon > 0$ and $\forall \tilde{x} \in \overset{\circ}{\mathcal{D}}$ the level set $\{x \in \overset{\circ}{\mathcal{D}}: F(x;\varepsilon) \leq F(\tilde{x};\varepsilon)\}$ is compact.*

The condition given above excludes the existence of minimizing sequences for $F(x;\varepsilon)$ originating in $\overset{\circ}{\mathcal{D}}$ which have limit points on the boundary.

The definition of barrier function given here is slightly stronger then the definition of *globally exact function* given in [31]. We omit here this last definition, due to the fact that the globally exact functions used in practice satisfy the requirement of Definition 5.

A basic tool for the construction of an exact barrier function consists in taking $\mathcal{D}$ as a *suitable* compact perturbation of the feasible set, and then in defining a function $a(x)$ with the property that $a(x) > 0 \quad \forall x \in \overset{\circ}{\mathcal{D}}$, and $a(x) = 0 \quad \forall x \in \partial\mathcal{D}$; hence $1/a(x)$ is positive $\forall x \in \overset{\circ}{\mathcal{D}}$ and goes to $+\infty$ when $x \to \partial\mathcal{D}$. An assumption on the existence of the compact perturbation of the feasible set has to be made explicitly.

We consider here the nondifferentiable penalty function with barrier properties studied in [23]. This function incorporates a barrier term which goes to infinity on the boundary of a compact perturbation of the feasible set $\mathcal{F}$, and can be viewed as a generalization of the "M_2" penalty function introduced in [39].

Let $\alpha > 0$ and let

$$S_\alpha := \left\{ x \in \mathbb{R}^n : \left\| \left[g^+(x), \ h(x) \right] \right\|_q < \alpha \right\}; \tag{2}$$

clearly, S_α is an open perturbation of the feasible set: $S_\alpha \supset \mathcal{F}$ and $\lim_{\alpha \to 0} \overline{S}_\alpha = \mathcal{F}$.
Suppose that the following assumption is satisfied.

Assumption 5 *The closure $\overline{S}_\alpha$ of the set S_α given in (2), is compact.*

By Assumption 5, we can take $\mathcal{D} \equiv \overline{S}_\alpha$.
Let us introduce the function:

$$a(x) := \alpha - \left\| \left[g^+(x), \ h(x) \right] \right\|_q; \tag{3}$$

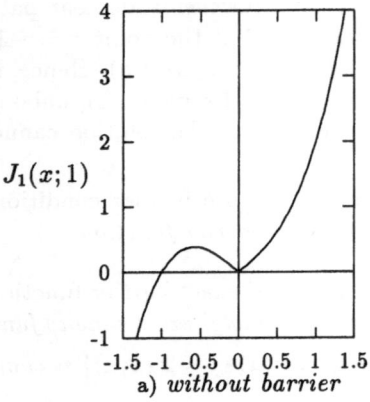

a) *without barrier*

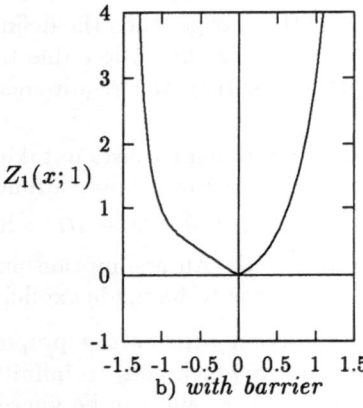

b) *with barrier*

Figure 1: Nondifferentiable exact penalty functions for Problem (E)

we have, obviously, that $a(x) > 0$, for $x \in S_\alpha$, and $a(x) = 0$ for $x \in \partial S_\alpha$.

Then, we consider the following function

$$Z_q(x;\varepsilon) := f(x) + \frac{1}{\varepsilon a(x)} \left\| [g^+(x),\, h(x)] \right\|_q = J_q(x;\varepsilon a(x)) \tag{4}$$

where $\varepsilon > 0$, $1 \leq q \leq \infty$, and $a(x)$ is given by (3).

We can prove that $Z_q(x;\varepsilon)$ enjoys the same exactness properties of $J_q(x;\varepsilon)$; moreover, by construction, we have that the level set $\{x \in S_\alpha : Z_q(x;\varepsilon) \leq Z_q(\tilde{x};\varepsilon)\}$ is compact $\forall \varepsilon > 0$ and $\forall \tilde{x} \in S_\alpha$, as required by Definition 5. Hence we can state the following result:

Proposition 13 *Let S_α be given by (2), and assume that $\overline{S}_\alpha$ is compact. Then the function $Z_q(x;\varepsilon)$ is a (weakly, strongly) exact barrier function with respect to the set $\overline{S}_\alpha$ whenever the function $J_q(x;\varepsilon)$ is a (weakly, strongly) exact penalty function with respect to the set $\overline{S}_\alpha$.*

If we consider again the example stated as Problem (E), we can easily see that the function

$$Z_1(x;\varepsilon) = x^3 + \frac{1}{\varepsilon(1.5 - |x|)}|x|$$

is an exact barrier function with respect to the closure of the set $S = \{x : |x| < 1.5\}$; $Z_1(x;\varepsilon)$ is shown in Figure 1 b) for $\varepsilon = 1$.

The construction of a continuously differentiable exact barrier function can be performed along the same lines followed in the nondifferentiable case, as in [31, 56, 57]. In this case we can consider the open perturbation of the feasible set given by

$$S_\alpha := \{x \in \mathbb{R}^n : \sum_{i=1}^m g_i^+(x)^s + \sum_{j=1}^q h_i(x)^2 < \alpha\}, \tag{5}$$

where $\alpha > 0$ and $s \geq 2$, and the corresponding assumption:

Assumption 6 *The closure $\overline{S}_\alpha$ of the set S_α given in (5), is compact.*

Then we can introduce the function

$$a(x) := \alpha - \sum_{i=1}^m g_i^+(x)^s - \sum_{j=1}^q h_i(x)^2, \tag{6}$$

which takes positive values on S_α and is zero on its boundary, and we can define on the set S_α given by (5) the continuously differentiable function $U(x;\varepsilon)$:

$$U(x;\varepsilon) := W(x;\varepsilon a(x)) \tag{7}$$

where $a(x)$ is given by (6).

The gradient expression $\nabla U(x; \varepsilon)$ for $x \in \mathcal{S}_\alpha$ can be easily evaluated, taking into account (2), (7) and (6).

We can study the exactness properties of $U(x; \varepsilon)$ by taking as reference set $\mathcal{D}$ the set $\overline{\mathcal{S}}_\alpha$. By construction, we have again that the level set $\{x \in \mathcal{S}_\alpha : U(x; \varepsilon) \leq U(\tilde{x}; \varepsilon)\}$ is compact $\forall \varepsilon > 0$ and $\forall \tilde{x} \in \mathcal{S}_\alpha$. Hence we can state the next proposition, which is the analogue of Proposition 13.

Proposition 14 *Let $\mathcal{S}_\alpha$ be given by (5), and assume that $\overline{\mathcal{S}}_\alpha$ is compact. Then the function $U(x; \varepsilon)$ is a (weakly, strongly) exact barrier function with respect to the set $\overline{\mathcal{S}}_\alpha$ whenever the function $W(x; \varepsilon)$ is a (weakly, strongly) exact penalty function with respect to the set $\overline{\mathcal{S}}_\alpha$.*

In Figure 2, a) and b), the functions $W(x; \varepsilon)$ and $U(x; \varepsilon)$ for the example Problem (E) are shown, in the case $\gamma_1 = 1, \gamma_2 = 0.05, \varepsilon = .4, \alpha = 1.5$.

All additional first and second order exactness results that do not depend on the behaviour on the boundary of the reference set $\mathcal{D}$ can be easily recovered for $Z_q(x; \varepsilon)$ and $U(x; \varepsilon)$ from the corresponding results stated for the functions $J_q(x; \varepsilon)$ and $W(x; \varepsilon)$, by taking into account (4) and (7). In particular, second order results require $s \geq 3$ in (5) and (6).

It is to be remarked that, when dealing with exact barrier functions, the barrier is on the boundary of the set $\mathcal{S}_\alpha$, that is on the boundary of the perturbation of the feasible set $\mathcal{F}$, and not on the boundary of the feasible set $\mathcal{F}$ itself, as it happens with the usual sequential barrier functions. This is essential in providing exactness properties. Moreover, it is to be remarked that the barrier must operate for all $\varepsilon > 0$; as it will be seen later, this is essential for the construction of globally convergent algorithms for the solution of Problem (C).

6 Exact augmented Lagrangian functions

An *augmented Lagrangian function* for Problem (C) is a function $A(x, \lambda, \mu; \varepsilon)$: $\mathbb{R}^n \times \mathbb{R}^m \times \mathbb{R}^p \to \mathbb{R}$, of the form:

$$A(x, \lambda, \mu; \varepsilon) = L(x, \lambda, \mu) + V(x, \lambda, \mu; \varepsilon)$$

where $\varepsilon > 0$ is a penalty parameter, and $V(x, \lambda, \mu; \varepsilon)$ is a function added to the Lagrangian L with the aim of providing some desirable property. In the augmented Lagrangian function of Hestenes-Powell-Rockafellar the added term V ensures that if $(\overline{x}, \overline{\lambda}, \overline{\mu})$ is a KKT triplet corresponding to a local solution of Problem (C), then, for sufficiently small values of ε, $\overline{x}$ is also a local solution of the unconstrained problem $\min_x A(x, \overline{\lambda}, \overline{\mu}; \varepsilon)$. The limitation of this approach in the solution of Problem (C) is that the multipliers $(\overline{\lambda}, \overline{\mu})$ are not known, and have to be estimated by sequential procedures.

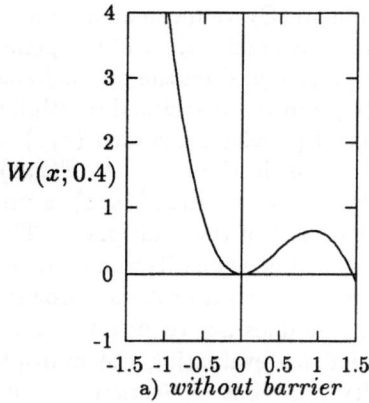

a) *without barrier*

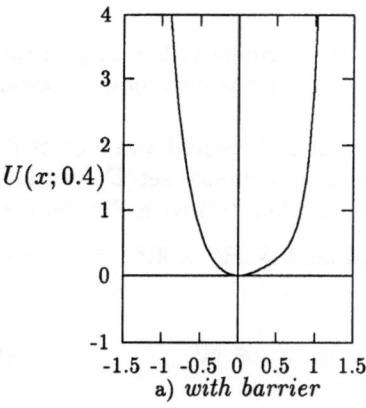

a) *with barrier*

Figure 2: Differentiable exact penalty functions for Problem (E)

By *exact augmented Lagrangian function* for Problem (C) we mean an augmented Lagrangian function $A(x, \lambda, \mu; \varepsilon)$ such that for sufficiently small values of the penalty parameter ε, a single unconstrained minimization of $A(x, \lambda, \mu; \varepsilon)$ on the product space $\mathbb{R}^n \times \mathbb{R}^m \times \mathbb{R}^p$ yields both a solution to Problem (C) and the associated multipliers.

In our terminology, the difference between an exact penalty function $F(x; \varepsilon)$ and an exact augmented Lagrangian function $A(x, \lambda, \mu; \varepsilon)$ lies in the fact that $F(x; \varepsilon)$ is minimized in the same space of the original problem, whereas $A(x, \lambda, \mu; \varepsilon)$ is minimized on the extended space of the problem variables and of the multipliers. Thus, the minimization of $A(x, \lambda, \mu; \varepsilon)$ is a problem of larger dimensionality than the minimization of an exact penalty function for the same constrained problem. However, as shown in the sequel, the computation of $A(x, \lambda, \mu; \varepsilon)$ does not require the matrix inversion, or the solution of a linear system, needed for evaluating the multiplier functions which appear in the differentiable penalty functions of Sections 4 and 5. This may be advantageous in the solution of large dimensional constrained problems and constitutes one of the main motivations for the development of exact augmented Lagrangian functions.

We can introduce different notions of exactness for augmented Lagrangian functions along the same lines followed in Section 3 for defining the exactness of penalty functions.

As in section 3, in order to ensure the existence of a threshold value $\varepsilon^* > 0$ of the penalty parameter we introduce a compact regular reference set $\mathcal{D}$, we define Problem ($\check{C}$), and, under Assumption 3, we identify Problem (C) with Problem ($\check{C}$).

Let $A(x, \lambda, \mu; \varepsilon)$ be a continuous function defined on $\overset{\circ}{\mathcal{D}} \times \mathbb{R}^m \times \mathbb{R}^p$ for any given $\varepsilon > 0$, and consider the problem:

$$\text{minimize} \quad A(x, \lambda, \mu; \varepsilon) \quad x \in \overset{\circ}{\mathcal{D}}, \ \lambda \in \mathbb{R}^m, \ \mu \in \mathbb{R}^p \tag{$\tilde{U}$}$$

which can be viewed as an unconstrained problem due to the fact that the set $\overset{\circ}{\mathcal{D}} \times \mathbb{R}^m \times \mathbb{R}^p$ is open. We denote respectively by $\mathcal{G}_{\tilde{u}}(\varepsilon)$ and $\mathcal{L}_{\tilde{u}}(\varepsilon)$ the set of global solutions and the set of local solutions of Problem ($\tilde{U}$), that is:

$$\mathcal{G}_{\tilde{u}}(\varepsilon) := \{(x, \lambda, \mu) \in \overset{\circ}{\mathcal{D}} \times \mathbb{R}^m \times \mathbb{R}^p : A(x, \lambda, \mu; \varepsilon) \le A(y, \xi, \zeta; \varepsilon),$$
$$\text{for all } (y, \xi, \zeta) \in \overset{\circ}{\mathcal{D}} \times \mathbb{R}^m \times \mathbb{R}^p\}$$

$$\mathcal{L}_{\tilde{u}}(\varepsilon) := \{(x, \lambda, \mu) \in \overset{\circ}{\mathcal{D}} \times \mathbb{R}^m \times \mathbb{R}^p : \text{for some } \rho > 0 \ A(x, \lambda, \mu; \varepsilon) \le A(y, \xi, \zeta; \varepsilon),$$
$$\text{for all } (y, \xi, \zeta) \in \overset{\circ}{\mathcal{D}} \times \mathbb{R}^m \times \mathbb{R}^p \cap \mathcal{B}((x, \lambda, \mu); \rho)\}.$$

The next definition concerns the correspondence between global solutions of Problem (C) and global minimizers of Problem ($\tilde{U}$).

Definition 6 *The function $A(x, \lambda, \mu; \varepsilon)$ is a* weakly exact augmented Lagrangian *function for Problem* (C) *with respect to the set $\mathcal{D}$ if there exists an $\varepsilon^* > 0$ such that, for all $\varepsilon \in (0, \ \varepsilon^*]$:*

(p₁) if $\overline{x}$ is a global solution of Problem (C) *and $\overline{\lambda}, \overline{\mu}$ are associated KKT multipliers, then $(\overline{x}, \overline{\lambda}, \overline{\mu})$ is a global minimum point of Problem $(\tilde{U})$; that is, if $\overline{x} \in \mathcal{G}_C$ and $(\overline{x}, \overline{\lambda}, \overline{\mu})$ is a KKT triplet, we have $(\overline{x}, \overline{\lambda}, \overline{\mu}) \in \mathcal{G}_{\tilde{U}}(\varepsilon)$;*

(p₂) if $(x_\varepsilon, \lambda_\varepsilon, \mu_\varepsilon)$ is a global minimum point of Problem $(\tilde{U})$ then x_ε is a global solution of Problem (C) *and $\lambda_\varepsilon, \mu_\varepsilon$ are associated KKT multipliers; that is, if $(x_\varepsilon, \lambda_\varepsilon, \mu_\varepsilon) \in \mathcal{G}_{\tilde{U}}(\varepsilon)$ we have $x_\varepsilon \in \mathcal{G}_C$ and $(x_\varepsilon, \lambda_\varepsilon, \mu_\varepsilon)$ is a KKT triplet.*

Properties (p₁) and (p₂) ensure that, for sufficiently small values of ε, the global unconstrained minimization of the augmented Lagrangian function $A(x, \lambda, \mu; \varepsilon)$ on the product set $\overset{\circ}{\mathcal{D}} \times \mathbb{R}^m \times \mathbb{R}^p$ yields a global solution of the constrained problem, and the corresponding multipliers.

By imposing a further requirement concerning local minimizers we have the stronger notion of exactness stated below.

Definition 7 *The function $A(x, \lambda, \mu; \varepsilon)$ is an* exact augmented Lagrangian *function for Problem* (C) *with respect to the set $\mathcal{D}$ if there exists an $\varepsilon^* > 0$ such that, for all $\varepsilon \in (0, \ \varepsilon^*]$ properties (p₁) and (p₂) of Definition 6 hold and, moreover:*

(p₃) if $(x_\varepsilon, \lambda_\varepsilon, \mu_\varepsilon)$ is a local minimizer of Problem $(\tilde{U})$, then x_ε is a local solution of Problem (C) *and $\lambda_\varepsilon, \mu_\varepsilon$ are associated KKT multipliers, that is if $(x_\varepsilon, \lambda_\varepsilon, \mu_\varepsilon) \in \mathcal{L}_{\tilde{U}}(\varepsilon)$ we have $x_\varepsilon \in \mathcal{L}_C$ and $(x_\varepsilon, \lambda_\varepsilon, \mu_\varepsilon)$ is a KKT triplet.*

If we impose a one-to-one correspondence between local minimizers of Problem (U) and KKT triplet originating from local solution of Problem (C), we are led to the following definition.

Definition 8 *The function $A(x, \lambda, \mu; \varepsilon)$ is a* strongly exact augmented Lagrangian *function for Problem* (C) *with respect to the set $\mathcal{D}$ if there exists an $\varepsilon^* > 0$ such that, for all $\varepsilon \in (0, \ \varepsilon^*]$ properties (p₁), (p₂) and (p₃) of Definitions 6 and 7 hold and moreover:*

(p₄) if $\overline{x}$ is a local solution of Problem (C) *and $(\overline{\lambda}, \overline{\mu})$ are associated KKT multipliers, then $(\overline{x}, \overline{\lambda}, \overline{\mu})$ is a local unconstrained minimizer of $A(x, \lambda, \mu; \varepsilon)$; that is if $\overline{x} \in \mathcal{L}_C$ and $(\overline{x}, \overline{\lambda}, \overline{\mu})$ is a KKT triplet, we have $(\overline{x}, \overline{\lambda}, \overline{\mu}) \in \mathcal{L}_{\tilde{U}}(\varepsilon)$.*

Augmented Lagrangian functions which are exact according to the definitions given above, can be viewed as exact penalty functions for the nonlinear programming problem defined on the extended space of the problem variables and of the multipliers, and incorporating the KKT optimality conditions as constraints.

More specifically, consider the problem:

$$
\begin{aligned}
\text{minimize} \quad & f(x) \\
\text{subject to} \quad g(x) & \leq 0 \\
h(x) & = 0 \\
\nabla_x L(x, \lambda, \mu) & = 0 \\
\lambda' g(x) & = 0 \\
\lambda & \geq 0;
\end{aligned} \tag{$\hat{\text{C}}$}
$$

then the following proposition is easily verified:

Proposition 15 *Assume that for any local minimum point $\bar{x}$ of Problem (C) there exist KKT multipliers $\bar{\lambda}, \bar{\mu}$. Then, $(\bar{x}, \bar{\lambda}, \bar{\mu})$ is a local (global) solution of Problem ($\hat{\text{C}}$) if and only if $\bar{x}$ is a local (global) solution of Problem (C) and $\bar{\lambda}, \bar{\mu}$ are associated KKT multipliers.*

On the basis of Proposition 15, and arranging in a suitable way for the fact that in Problem (U) only the variable x is restricted to a bounded set, we can study the exactness of augmented Lagrangian functions, along lines similar to those followed in the case of penalty functions.

Augmented Lagrangian functions that enjoy exactness properties have been introduced in [25] for problems with equality constraints, and in [26] for problems with inequality constraints; a general exposition can be found in [3], additional results in [70]. Here we consider, as representative in this class, the continuously differentiable function studied in [54], obtained by adding to the augmented Lagrangian of Hestenes-Powell-Rockafellar, the term

$$
\eta \left\| M(x) \left[\begin{pmatrix} \lambda \\ \mu \end{pmatrix} - \begin{pmatrix} \lambda(x) \\ \mu(x) \end{pmatrix} \right] \right\|^2,
$$

where $\eta > 0$, and $M(x)$, $\lambda(x)$, $\mu(x)$ are the matrix and the multiplier functions introduced in Section 4.2. This term, clearly, weights the difference between the multipliers and their estimates given by the multiplier functions.

In this way, we get the function

$$
S(x, \lambda, \mu; \varepsilon) := L_a(x, \lambda, \mu; \varepsilon) + \eta \left\| M(x) \left[\begin{pmatrix} \lambda \\ \mu \end{pmatrix} - \begin{pmatrix} \lambda(x) \\ \mu(x) \end{pmatrix} \right] \right\|^2,
$$

where $L_a(x, \lambda, \mu; \varepsilon)$ is given by (1).

It is important to realize that, by construction, the evaluation of $S(x, \lambda, \mu; \varepsilon)$ does not require neither matrix inversions nor solutions of linear equations; in fact the matrix $M(x)$ cancels the inverse matrix $M^{-1}(x)$ that appears in the definition of the multiplier function $(\lambda(x), \mu(x))$. Indeed, the augmented Lagrangian function

$S(x, \lambda, \mu; \varepsilon)$ is defined everywhere in $\mathbb{R}^n \times \mathbb{R}^m \times \mathbb{R}^p$, whereas the penalty function $W(x; \varepsilon)$ is defined only at points in $\mathbb{R}^n$ where the multiplier functions are defined.

As concerns the exactness properties of the function $S(x, \lambda, \mu; \varepsilon)$, we can prove the following proposition:

Proposition 16 *(a) Assume that the LICQ is satisfied on $\mathcal{F}$; then the function $S(x, \lambda, \mu; \varepsilon)$ is a weakly exact augmented Lagrangian function for Problem (C) with respect to the set $\mathcal{D}$.*

(b) Assume that the EMFCQ is satisfied on $\mathcal{D}$. Then, the function $S(x, \lambda, \mu; \varepsilon)$ is an exact augmented Lagrangian function for Problem (C) with respect to the set $\mathcal{D}$; moreover, if Assumption 2 holds, the function $S(x, \lambda, \mu; \varepsilon)$ is a strongly exact augmented Lagrangian function for Problem (C) with respect to the set $\mathcal{D}$.

It can be observed that by the definition of the function $S(x, \lambda, \mu; \varepsilon)$, it results

$$S(x, \lambda(x), \mu(x); \varepsilon) = W(x; \varepsilon).$$

This observation helps in understanding that all additional exactness results stated for the continuously differentiable penalty function $W(x; \varepsilon)$ can be restated, with suitable modifications, for the exact augmented Lagrangian $S(x, \lambda, \mu; \varepsilon)$. In particular, as concerns the correspondence between stationary points of $S(x, \lambda, \mu; \varepsilon)$ and KKT points of Problem (C), we have:

Proposition 17 *Assume that the EMFCQ holds on $\mathcal{D}$. Then, there exists an $\varepsilon^* > 0$ such that, for all $\varepsilon \in (0, \varepsilon^*]$, if $(x_\varepsilon, \lambda_\varepsilon, \mu_\varepsilon)$ is a stationary point of $S(x, \lambda, \mu; \varepsilon)$ with $x_\varepsilon \in \overset{\circ}{\mathcal{D}}$, we have that $(x_\varepsilon, \lambda_\varepsilon, \mu_\varepsilon)$ is a KKT triplet for Problem (C).*

Proposition 18 *Let $(\overline{x}, \overline{\lambda}, \overline{\mu})$ be a KKT triplet for Problem (C). Then, for any $\varepsilon > 0$, $(\overline{x}, \overline{\lambda}, \overline{\mu})$ is a stationary point of $S(x, \lambda, \mu; \varepsilon)$.*

If we make the assumption that the problem functions are three times continuously differentiable and if we have $r \geq 3$ in the definition of $M(x)$, we can prove also second order results.

Proposition 19 *(a) The function $S(x, \lambda, \mu; \varepsilon)$ is twice continuously differentiable for all $x \in \mathbb{R}^n$ except, possibly, at points where $g_i(x) + \varepsilon \lambda_i / 2 = 0$ for some i.*

(b) Let $(\overline{x}, \overline{\lambda}, \overline{\mu})$ be a KKT triplet for Problem (C) and assume that strict complementary holds at $(\overline{x}, \overline{\lambda}, \overline{\mu})$. Then, for any $\varepsilon > 0$, the function $S(x, \lambda, \mu; \varepsilon)$ is twice continuously differentiable in a neighbourhood of $(\overline{x}, \overline{\lambda}, \overline{\mu})$.

As in the case of exact penalty functions, third order derivatives do not appear in $\nabla^2 S(\overline{x}, \overline{\lambda}, \overline{\mu}; \varepsilon)$.

Proposition 20 *Let $(\overline{x}, \overline{\lambda}, \overline{\mu})$ be a KKT triplet for Problem (C) and assume that*

(i) strict complementarity holds at $(\overline{x}, \overline{\lambda}, \overline{\mu})$;

(ii) $\overline{x}$ in an isolated local minimum point for Problem (C) satisfying the second order sufficiency condition.

Then, there exists an ε^* such that for all $\varepsilon \in (0, \varepsilon^*]$, $(\overline{x}, \overline{\lambda}, \overline{\mu})$ is an isolated local minimum point for $S(x, \lambda, \mu; \varepsilon)$ and the Hessian matrix $\nabla^2 S(\overline{x}, \overline{\lambda}, \overline{\mu}; \varepsilon)$ is positive definite.

Proposition 21 *Assume that*

(i) the EMFCQ holds on $\mathcal{D}$;

(ii) the strict complementarity holds at any KKT triplet $(\overline{x}, \overline{\lambda}, \overline{\mu})$ of Problem (C).

Then, there exists an $\varepsilon^* > 0$ such that, for all $\varepsilon \in (0, \varepsilon^*]$, if $(x_\varepsilon, \lambda_\varepsilon, \mu_\varepsilon)$ is a local unconstrained minimizer of $S(x, \lambda, \mu; \varepsilon)$ with $x_\varepsilon \in \overset{\circ}{\mathcal{D}}$ and with positive definite Hessian $\nabla^2 S(x_\varepsilon, \lambda_\varepsilon, \mu_\varepsilon; \varepsilon)$, then x_ε is an isolated local minimum point of Problem (C), satisfying the second order sufficiency condition.

On the basis of the definition stated above, the minimization of an exact augmented Lagrangian function allows, in principle, to solve the constrained problem by means of a single unconstrained minimization over the product set $\overset{\circ}{\mathcal{D}} \times \mathbb{R}^m \times \mathbb{R}^p$. However, in order to ensure global convergence of unconstrained algorithms we must exclude the existence of descent paths for $A(x, \lambda, \mu; \varepsilon)$ originating at points $(\tilde{x}, \tilde{\lambda}, \tilde{\mu})$ with $\tilde{x} \in \overset{\circ}{\mathcal{D}}$ such that either x crosses the boundary of $\overset{\circ}{\mathcal{D}}$, or λ or μ become unbounded.

Thus we introduce the notion of *exact augmented Lagrangian barrier function*.

Definition 9 *The function $A(x, \lambda, \mu; \varepsilon)$ is a (weakly, strongly) exact augmented Lagrangian barrier function for Problem (C) with respect to the set $\mathcal{D}$ if it is (weakly, strongly) exact and, moreover, $\forall \varepsilon > 0$ and $\forall (\tilde{x}, \tilde{\lambda}, \tilde{\mu}) \in \overset{\circ}{\mathcal{D}} \times \mathbb{R}^m \times \mathbb{R}^p$ the level set*

$$\{(x, \lambda, \mu) \in \overset{\circ}{\mathcal{D}} \times \mathbb{R}^m \times \mathbb{R}^p : A(x, \lambda, \mu; \varepsilon) \leq A(\tilde{x}, \tilde{\lambda}, \tilde{\mu}; \varepsilon)\}$$

is compact.

The construction of an exact augmented Lagrangian barrier function is more complicated then the construction of an exact barrier function; in fact in the former case the barrier term must operate not only on the variable x, but also on the multipliers λ and μ. However, making reference to the function $S(x, \lambda, \mu; \varepsilon)$, we can see that, in its minimization, if x belongs to $\overset{\circ}{\mathcal{D}}$ and λ is bounded, then also μ is bounded, due to the fact that the matrix $M(x)$ is positive definite. Hence we can obtain an exact augmented Lagrangian barrier function by incorporating into the function $S(x, \lambda, \mu; \varepsilon)$ a barrier term which goes to infinity either when x approaches the boundary of $\mathcal{D}$ or when λ becomes unbounded. To this aim, it is convenient to write S in the form:

$$S(x, \lambda, \mu; \varepsilon) = f(x) + \sum_{i=1}^{m} \left\{ \lambda_i [g_i(x) + y_i(x, \lambda; \varepsilon)^2] + (1/\varepsilon)[g_i(x) + y_i(x, \lambda; \varepsilon)^2]^2 \right\}$$
$$+ \sum_{j=1}^{p} \left\{ \mu_j h_j(x) + (1/\varepsilon) h_j(x)^2 \right\} + \eta \left\| M(x) \left[\begin{pmatrix} \lambda \\ \mu \end{pmatrix} - \begin{pmatrix} \lambda(x) \\ \mu(x) \end{pmatrix} \right] \right\|^2.$$

As proposed in [18], let us consider again the set $\mathcal{S}_\alpha$ and the function $a(x)$ given in (5) and (6), and let us introduce the functions:

$$b_i(x, \lambda; \varepsilon) := \frac{a(x)}{1 + \varepsilon a(x) \lambda_i^2}, \quad i = 1, \ldots, m \tag{8}$$

which take positive values on $\mathcal{S}_\alpha \times \mathbb{R}^m$ and are zero on its boundary. We can define on the set $\mathcal{S}_\alpha \times \mathbb{R}^m \times \mathbb{R}^p$ the augmented Lagrangian function $T(x, \lambda, \mu; \varepsilon)$ obtained by substituting in the function $S(x, \lambda, \mu; \varepsilon)$ the penalty parameter ε that affects each constraint with slack variable $g_i(x) + y_i(x, \lambda; \varepsilon)^2$ by the term $\varepsilon b_i(x, \lambda; \varepsilon)$, $i = 1, \ldots, m$; that is, we can define the function:

$$T(x, \lambda, \mu; \varepsilon) := f(x) + \sum_{i=1}^{m} \left\{ \lambda_i [g_i(x) + v_i(x, \lambda; \varepsilon)^2] + \frac{[g_i(x) + v_i(x, \lambda; \varepsilon)^2]^2}{\varepsilon b_i(x, \lambda; \varepsilon)} \right\}$$
$$+ \sum_{j=1}^{p} \left\{ \mu_j h_j(x) + (1/\varepsilon) h_j(x)^2 \right\} + \eta \left\| M(x) \left[\begin{pmatrix} \lambda \\ \mu \end{pmatrix} - \begin{pmatrix} \lambda(x) \\ \mu(x) \end{pmatrix} \right] \right\|^2,$$

where

$$v_i(x, \lambda; \varepsilon)^2 := y_i(x, \lambda; \varepsilon b_i(x, \lambda; \varepsilon))^2 = -\min \left[0, g_i(x) + \frac{\varepsilon b_i(x, \lambda; \varepsilon)}{2} \lambda_i \right].$$

Under Assumption 6 we can prove that the level set

$$\{(x, \lambda, \mu) \in \mathcal{S}_\alpha \times \mathbb{R}^m \times \mathbb{R}^p : T(x, \lambda, \mu; \varepsilon) \leq T(\tilde{x}, \tilde{\lambda}, \tilde{\mu}; \varepsilon)\}$$

is compact $\forall \varepsilon > 0, \forall (\tilde{x}, \tilde{\lambda}, \tilde{\mu}) \in \mathcal{S}_\alpha \times \mathbb{R}^m \times \mathbb{R}^p$, so that we can state the following result:

Proposition 22 *Let $\mathcal{S}_\alpha$ be given by (5), and assume that $\overline{\mathcal{S}}_\alpha$ is compact. Then the function $T(x, \lambda, \mu; \varepsilon)$ is a (weakly, strongly) exact augmented Lagrangian barrier function with respect to the set $\overline{\mathcal{S}}_\alpha$ whenever the function $S(x, \lambda, \mu; \varepsilon)$ is a (weakly, strongly) exact augmented Lagrangian function with respect to the set $\overline{\mathcal{S}}_\alpha$.*

Clearly, the function $T(x, \lambda, \mu; \varepsilon)$ possesses first and second order additional exactness properties that can be deduced from those of the function $S(x, \lambda, \mu; \varepsilon)$.

7 Discussion of the assumptions

In the preceding sections we have seen that, under suitable assumptions, it is possible to build in several ways a function whose unconstrained minimum points are solution of a constrained problem. Now we discuss to which extent the requirement that these assumptions are satisfied can limit the effectiveness of the approach.

Assumption 1 concerns the compactness of the feasible set $\mathcal{F}$. As already said, this assumption is introduced here mainly for the sake of simplicity. We can develop, as in [31], a theory of exactness making reference to the set $\mathcal{F} \cap \mathcal{D}$.

The compactness of $\mathcal{F}$ is a more substantial requirement when dealing with exact (augmented Lagrangian) barrier functions, because in the cases considered here the compactness of the level sets is due to the barrier that operates on a compact perturbation of the feasible set. However we can develop functions with compact level sets even if $\mathcal{F}$ is unbounded, provided that the objective function $f(x)$ is coercive and a feasible point is known (see [17, 18] for details).

In any case, Assumption 1 is quite reasonable in practice, and then the same is true also for Assumption 2.

Assumption 3 concerns the selection of the set $\mathcal{D}$. This set is used mainly for establishing the *existence result* of the threshold value ε^*, and hence in general its specification is not needed. In the case of exact (augmented Lagrangian) barrier functions a convenient choice is $\mathcal{D} = \overline{\mathcal{S}}_\alpha$, if $\mathcal{S}_\alpha$ is bounded for some α.

Constraint qualifications like the MFCQ or the LICQ on the feasible set $\mathcal{F}$ are required in all methods for the solution of Problem (C). Note that the weak exactness of nondifferentiable penalty function can be established under the weakest assumption that the MFCQ holds on $\mathcal{G}_c$.

The requirement that the EMFCQ holds on $\mathcal{D}$ is more peculiar to the development of exactness results. In the proofs, it is used, together with the compactness of $\mathcal{D}$, in order to show that by an unconstrained minimization we get at least a feasible point, so that $\mathcal{F} \neq \emptyset$.

Indeed, it has been proved in [56] that Problem (C) is feasible if the EMFCQ holds on a perturbation of the feasible set $\mathcal{S}_\alpha$ with compact closure, and that this sufficient condition is also necessary in the case that $g(x)$ is a vector of convex functions and $h(x)$ is a vector of affine functions. This is a nice existence result that has been established in the context of exact penalization. In addition, in [18, 56] it has been shown that, if a feasible point for Problem (C) is known, and if a compact perturbation of the feasible set $\overline{\mathcal{S}}_\alpha$ exists, then the EMFCQ is not required in order to construct an exact (augmented Lagrangian) barrier function; these conditions are frequently meet in practice.

Note that often in other methods for the solution of Problem (C) an EMFCQ assumption is hidden. For instance, the assumption that the system of linearized constraints

$$\begin{cases} \nabla g(x)'d + g(x) \leq 0 \\ \nabla h(x)'d + h(x) = 0 \end{cases}$$

has a solution with respect to d, usual in RQP algorithms, implies that the EMFCQ holds at x.

Finally, we observe that, when dealing with particular penalty functions, the EMFCQ can be weakened making use of a theorem of alternative. This point is illustrated for instance in [28], with reference to the function $W(x; \varepsilon)$.

We will discuss later on what happens from an algorithmic point of view if the EMFCQ does not hold on $\mathcal{D}$.

The last assumption used before is on the existence of a compact perturbation of the feasible set. We can give sufficient conditions ensuring that the set S_α given either by (2) or by (5) has a compact closure *for all values of $\alpha > 0$* [23, 56]. More precisely, we have that $\overline{S}_\alpha$ is compact $\forall \alpha > 0$ if one of the following cases occurs:

i) there exists a function $g_i(x)$ such that $lim_{\|x\| \to \infty} g_i(x) = \infty$

ii) there exists a function $h_j(x)$ such that $lim_{\|x\| \to \infty} |h_j(x)| = \infty$

iii) there exist index sets I, J, not both empty, such that $g_i(x), i \in I$ are convex, $h_j(x), j \in J$ are affine, and the set $\{x : g_i(x) \leq 0, i \in I; h_j(x) = 0, j \in J\}$ is compact.

In particular, case iii) occurs if all problem variables are bounded.

8 Exact penalty algorithms

8.1 Generalities

Up to now we have dealt mainly with the theoretical aspects of exact penalization, concerning the possibility of defining a function whose unconstrained minimum points correspond to the solutions of a constrained problem. Now we consider the algorithmic aspects, with the aim to show that exact penalization makes it possible to define algorithms for the solution of a constrained problem, based on a sequence of the kind

$$x_{k+1} = x_k + \sigma_k d_k, \tag{9}$$

where d_k is a search direction and σ_k is a stepsize, with the same properties that are usual in unconstrained optimization, that is:

- global convergence,

- superlinear convergence rate,

- derivatives up to 2nd order required in computations;

we intend that the convergence is towards points satisfying first order necessary optimality conditions.

For simplicity, we shall consider problems with inequality constraints only; indeed the presence of inequality constraints is the main source of difficulties in the solution of nonlinear programming problems. Hence we particularize Problem (C) as Problem (IC):

$$\begin{array}{ll} \text{minimize} & f(x) \\ \text{subject to} & g(x) \leq 0. \end{array} \tag{IC}$$

For brevity, we shall not make reference to exact augmented Lagrangian functions. Algorithmic applications of these functions are reported, for instance, in [4, 20, 32, 33, 37, 55].

The search direction d_k in the sequence (9) can be obtained:

- from a local model of an exact penalty function $F(x;\varepsilon)$; this approach leads to *descent methods* on $F(x;\varepsilon)$;

- from a local model of Problem (IC); this approach leads to *recursive quadratic programming* (RQP) *algorithms* that use an exact penalty function as a merit function for determining the stepsize.

In both cases, in order to get convergence towards KKT points of Problem (IC), a suitable value $\varepsilon \in (0, \varepsilon^*]$ must be available. Since in general the threshold value ε^* is not known, some *updating rule* must be employed, that drives the penalty parameter into the interval $(0, \varepsilon^*]$.

In both cases, if a superlinear rate of convergence is desired, the use of *nondifferentiable* penalty functions *is not straightforward*. In fact, on the one hand, descent methods for nondifferentiable functions are not superlinearly convergent, and, on the other hand, RQP algorithms that employ nondifferentiable merit function in the line search are inherently affected by the Maratos effect. Therefore, in the sequel, we shall restrict ourselves to continuously differentiable exact penalty functions.

Let $F(x;\varepsilon)$ be a *continuously differentiable* exact penalty function. When using a descent method on $F(x;\varepsilon)$ the main problem is the requirement of higher order derivatives, due to the fact that, in general, the evaluation of $\nabla^2 F(x;\varepsilon)$ requires the evaluation of third order derivatives of the problem functions. When using $F(x;\varepsilon)$ in an RQP algorithm the main problem is to conciliate a superlinear convergence rate with global convergence, so as to avoid the Maratos effect. We shall show that these problems can be overcome.

Local properties of $W(x;\varepsilon)$ and of $U(x;\varepsilon)$ are the same, while global properties of $U(x;\varepsilon)$ are better, due to the compactness of the level sets. Therefore, in the sequel, we shall make reference to the exact barrier function $U(x;\varepsilon)$ given in (7).

Based on function $U(x;\varepsilon)$, we shall illustrate a basic Newton-type approach, and a basic recursive quadratic programming approach, that are typical in their classes.

In both cases, for analytical purposes related to the superlinear rate of convergence, we shall assume that the problem functions f and g are *three times* continuously differentiable, even if, in computations, third order derivatives will never be required.

For convenience, we give here the explicit expressions of the function U and of its gradient ∇U for Problem (IC):

$$
\begin{aligned}
U(x;\varepsilon) &= f(x) + \lambda(x)'\left(g(x) + V(x;\varepsilon)v(x;\varepsilon)\right) \\
&+ \frac{1}{\varepsilon a(x)}\left\|g(x) + V(x;\varepsilon)v(x;\varepsilon)\right\|^2,
\end{aligned}
$$

$$
\begin{aligned}
\nabla U(x;\varepsilon) &= \nabla f(x) + \nabla g(x)\lambda(x) + \nabla\lambda(x)\left(g(x) + V(x;\varepsilon)v(x;\varepsilon)\right) \\
&+ \frac{2}{\varepsilon a(x)}\nabla g(x)\left(g(x) + V(x;\varepsilon)v(x;\varepsilon)\right) \\
&+ \frac{s}{\varepsilon a(x)^2}\left\|g(x) + V(x;\varepsilon)v(x;\varepsilon)\right\|^2 \sum_{i=1}^{m}\nabla g_i(x)g_i^+(x)^{s-1},
\end{aligned}
$$

where:

$$
a(x) := \alpha - \sum_{i=1}^{m} g_i^+(x)^s
$$

$$
v_i(x;\varepsilon) := \left\{-\min\left[0, g_i(x) + \frac{\varepsilon a(x)}{2}\lambda_i(x)\right]\right\}^{1/2}, \quad i = 1,\dots,m
$$

$$
V(x;\varepsilon) := \operatorname{diag}(v_i(x;\varepsilon)).
$$

8.2 A globally convergent Newton-type algorithm

Let $U(x;\varepsilon)$, given by (7), be an exact barrier function for Problem (IC), with respect to a compact set $\overline{S}_\alpha$.

We begin by defining an automatic adjustment rule for the penalty coefficient, that enforces the value of ε to become smaller or equal than the threshold value ε^* that ensures exactness. This rule allows us to describe an implementable algorithm which can be proved to be globally convergent towards KKT points of Problem (IC). Rules of this kind have been analysed in an abstract setting by Polak in [63], and have been used also, e.g., in [29, 34, 51, 60].

The algorithm is based on an additional exactness result that has not been mentioned before, because its interest appears more evident here:

Proposition 23 *Let x_ε be a stationary point of $U(x;\varepsilon)$ and assume that*

$$
g(x_\varepsilon) + V(x_\varepsilon;\varepsilon)v(x_\varepsilon;\varepsilon) = 0;
$$

then $(x_\varepsilon, \lambda(x_\varepsilon))$ is a KKT pair for Problem (C).

In the algorithm we make use of an iteration map $T : \mathcal{S}_\alpha \to 2^{\mathcal{S}_\alpha}$ that satisfies the following assumption:

Assumption 7 *For every fixed value of ε and every starting point $\tilde{x} \in \mathcal{S}_\alpha$, all the points x_k produced by T belong to the level set $\Omega(\tilde{x}; \varepsilon) := \{x \in \mathcal{S}_\alpha : U(x; \varepsilon) \leq U(\tilde{x}; \varepsilon)\}$ and all the limit points of the sequence produced by T are stationary points of $U(x; \varepsilon)$.*

These requirements on the map T can be easily satisfied by every globally convergent algorithm for the unconstrained minimization of U. In fact we can always ensure, by simple devices, that the trial points produced along the search direction remain in $\Omega(\tilde{x}; \varepsilon)$.

Now, we can describe the algorithm:

Algorithm EPS

Data: $\tilde{x} \in \mathbb{R}^n$, $\varepsilon_0 > 0$ and $\delta > 0$.

Step 0: Choose $\alpha > 0$ and $s \geq 2$ such that $\tilde{x} \in \mathcal{S}_\alpha$, set $j = 0$ and $z_0 = \tilde{x}$.

Step 1: Set $k = 0$. If $U(\tilde{x}; \varepsilon_j) \leq U(z_j; \varepsilon_j)$ set $x_0 = \tilde{x}$; else set $x_0 = z_j$.

Step 2: If $\nabla U(x_k; \varepsilon_j) = 0$ go to Step 3; else go to Step 4.

Step 3: If $g(x_k) + V(x_k; \varepsilon_j)v(x_k; \varepsilon_j) = 0$ stop; else go to step 6.

Step 4: If

$$\|\nabla U(x_k; \varepsilon_j)\|^2 + \|\nabla g(x_k)'\nabla U(x_k; \varepsilon_j)\|^2 \geq \delta\|g(x_k) + V(x_k; \varepsilon_j)v(x_k; \varepsilon_j)\|^2$$

go to step 5; else go to Step 6.

Step 5: Compute $x_{k+1} \in T(x_k)$, set $k = k + 1$ and go to Step 2.

Step 6: Choose $\varepsilon_{j+1} \in (0, \varepsilon_j)$, set $z_{j+1} = x_k$, $j = j + 1$ and go to Step 1.

Algorithm EPS enforces the satisfaction of the conditions stated in Proposition 23. It produces an inner sequence $\{x_k\}$, generated by the iteration map T, while the penalty parameter is fixed at the value ε_j, and an outer sequence $\{z_j\}$ of points where an updating of the penalty parameter occurs. At Step 1 we assume as candidate KKT point either the starting point $\tilde{x}$ or the current point z_j, the one that gives the lowest value of the penalty function U for the current value ε_j. At Steps 2 and 3 we test if the conditions of Proposition 23 are satisfied, and stop in the affirmative case. If x_k is a stationary point of U, but $g(x_k) + V(x_k; \varepsilon_j)v(x_k; \varepsilon_j) \neq 0$, we update ε at Step 6; if x_k is not a stationary point of U we may update ε or not, depending on the test at Step 4. The test at Step 4 prevents that $\|\nabla U(x; \varepsilon)\|$ is reduced without a corresponding reduction of $\|g(x) + V(x; \varepsilon)v(x; \varepsilon)\|$, thus avoiding useless iterations of

T. We perform an iteration of T at Step 5, if a suitable reduction of both $\|\nabla U(x;\varepsilon)\|$ and $\|g(x) + V(x;\varepsilon)v(x;\varepsilon)\|$ occurred.

The global convergence of Algorithm EPS can be stated under the same assumptions used to prove the exactness of the barrier function U. Furthermore, a distinguishing feature of Algorithm EPS is its capability to give some information about the original problem even when the EMFCQ does not hold on $\overline{S}_\alpha$ or when the feasible set $\mathcal{F}$ is empty. The convergence properties of Algorithm EPS are analysed in [56], and are described in the following propositions.

If the algorithm produces a finite sequence of points, by applying directly Proposition 23 we obtain a first result.

Proposition 24 *If Algorithm* EPS *terminates at some* $x_\nu \in \mathcal{S}_\alpha$ *then* $(x_\nu, \lambda(x_\nu))$ *is a* KKT *pair for Problem* (IC).

Let us now assume that the algorithm produces an infinite sequence of points $\{x_k\}$. Since the sequence belongs to the compact set $\overline{S}_\alpha$ we have that it admits at least one limit point. On the other hand, by Assumption 7, we have that each limit point is a stationary point of U. As we said before, under the same assumptions required for the exactness of U, we can state that any stationary point of U produced by Algorithm EPS is a KKT point for Problem (IC) and, in particular, we can show that the penalty parameter ε is updated finitely many times.

Proposition 25 *Suppose that the* EMFCQ *holds on* $\overline{S}_\alpha$. *Then the sequence* $\{\varepsilon_j\}$ *produced by Algorithm* EPS *at Step 6 is finite and every limit point* x^* *of the sequence* $\{x_k\} \subseteq \overline{S}_\alpha$ *yields a* KKT *pair* $(x^*, \lambda(x^*))$ *for Problem* (IC).

As already remarked in Section 7, if a feasible point for Problem (IC) is known, then the exactness of U can be established without the requirement that the EMFCQ holds on $\overline{S}_\alpha$. In connection with this result, we have that the convergence result of the preceding proposition can be stated also by replacing the assumption that the EMFCQ holds on $\overline{S}_\alpha$ by the assumption that in the data of Algorithm EPS we take $\tilde{x} \in \mathcal{F}$.

The next proposition follows the line proposed in [7] and [10], of investigating the convergence behaviour when neither the EMFCQ, nor the knowledge of a feasible point, is assumed. In particular this case happens when the feasible set is empty.

Proposition 26 *Let* $\{\varepsilon_j\}$, $\{z_j\} \subseteq \overline{S}_\alpha$ *and* $\{x_k\} \subseteq \overline{S}_\alpha$ *be the sequences produced by Algorithm* EPS. *Then:*

(i) *if the sequence* $\{\varepsilon_j\}$ *is finite, every limit point* x^* *of the sequence* $\{x_k\}$ *yields a* KKT *pair* $(x^*, \lambda(x^*))$ *for Problem* (IC);

(ii) *if the sequence* $\{\varepsilon_j\}$ *is infinite, every limit point* z^* *of the sequence* $\{z_j\}$ *is such that* $z^* \notin \mathcal{F}$ *and the* EMFCQ *does not hold at* z^*.

The points z^* can be further characterized as stationary points of a function that can be interpreted, loosely speaking, as a weighted measure of the violation of the constraints. This last feature is pointed out more clearly by setting $s = 2$ in Algorithm EPS. In particular we have the following result.

Proposition 27 *Let $s = 2$ in Algorithm* EPS. *If the sequences $\{\varepsilon_j\}$ and $\{z_j\} \subseteq \overline{\mathcal{S}}_\alpha$ produced by the algorithm are infinite then every limit point z^* of the sequence $\{z_j\}$ is a stationary point of the distance function*

$$\rho(x) := dist\big[g(x)|\mathbb{R}^m_-\big] := \inf_y\big\{\|g(x) - y\|; y_i \leq 0, i = 1, \ldots, m\big\}.$$

When the feasible set $\mathcal{F}$ is given by convex inequalities, we have the next proposition that follows directly from Proposition 26, Proposition 27 and by the fact, mentioned in Section 7, that the EMFCQ on the set $\mathcal{S}_\alpha$ with compact closure is a necessary condition for $\mathcal{F}$ to be not empty.

Proposition 28 *Let $s = 2$ in Algorithm* EPS. *Assume that $g(x)$ is a vector of convex functions. If the sequences $\{\varepsilon_j\}$ and $\{z_j\}$ produced by the algorithm are infinite then the feasible set $\mathcal{F}$ is empty and every limit point z^* of the sequence $\{z_j\}$ is a minimum point of the distance function $\rho(x)$.*

Therefore, Propositions 26 and 28 ensure that, when the feasible set $\mathcal{F}$ is given by convex inequalities, Algorithm EPS yields a KKT point for Problem (IC) if this problem is feasible, whereas it provides a point which is as close to feasibility as possible if Problem (IC) is nonfeasible.

In order to complete the description of Algorithm EPS we must specify the iteration map T. In principle any method for the unconstrained minimization of the function U can be easily modified to satisfy Assumption 7 and, hence, it can be employed as iteration map T in Algorithm EPS. In particular if the map T is a globally convergent modification of a Newton-type method, and if Proposition 25 applies, then Algorithm EPS turns out to be a globally convergent algorithm, with an ultimate superlinear convergence rate, for solving Problem (IC).

Although the Hessian matrix $\nabla^2 U$, where it exists, requires the third order derivatives of the problem functions, it is possible to define superlinearly convergent Newton-type algorithms based on consistent approximations of the Newton direction of the penalty function U which employ only the first and second order derivatives of the problem functions (see e.g. Proposition 1.15 of [3] for the rate of convergence properties of consistent approximations of the Newton direction). Here, as an example, we describe one of these algorithm and we refer to [27, 34] for other Newton-type algorithms for the function U.

First we need an additional notation. For every $x \in \mathcal{S}_\alpha$ and every $\varepsilon > 0$ we introduce the index set and the subvector of *pseudoactive constraints*:

$$I_A(x; \varepsilon) := \{i : g_i(x) + \frac{\varepsilon}{2}a(x)\lambda_i(x) \geq 0\},$$

$$g_A(x) := [g_i(x)], i \in I_A(x; \varepsilon).$$

Then, we can consider the algorithm:

Algorithm NT.

$$x_{k+1} = x_k + \sigma_k d_k,$$

where the search direction d_k is computed by solving the system:

$$\begin{bmatrix} \nabla^2 L(x_k, \lambda(x_k)) & \nabla g_A(x_k) \\ \nabla g_A(x_k)' & 0 \end{bmatrix} \begin{bmatrix} d_k \\ z_k \end{bmatrix} = - \begin{bmatrix} \nabla f(x_k) \\ g_A(x_k) \end{bmatrix} \tag{10}$$

and the stepsize σ_k is computed by means of some line search procedure applied to the function U.

The following proposition describes the local behaviour of Algorithm NT.

Proposition 29 Let $(\overline{x}, \overline{\lambda})$ be a KKT pair for Problem (IC) satisfying the strict complementarity assumption and let d be the solution of system (10). Then:

(i) for any given $\varepsilon > 0$ there exists a neighbourhood $\mathcal{B}$ of $\overline{x}$ and a continuous matrix $H(x; \varepsilon)$ such that, for all $x \in \mathcal{B}$, we have:

$$H(x; \varepsilon)d = -\nabla U(x; \varepsilon)$$

and in particular we have $H(\overline{x}; \varepsilon) = \nabla^2 U(\overline{x}; \varepsilon)$;

(ii) if Problem (IC) is a quadratic programming problem and the second order sufficiency condition for Problem (IC) holds at $(\overline{x}, \overline{\lambda})$, then, for any given $\varepsilon > 0$, there exists a neighbourhood $\mathcal{B}$ of $\overline{x}$ such that, for any $x \in \mathcal{B}$, we have:

$$\overline{x} = x + d.$$

Part (i) of the preceding proposition and Proposition 11 ensure that the direction d is a consistent approximation of the Newton's direction and that, in a neighbourhood of an isolated local minimum point of Problem (IC), it is a descent direction. By using these results, a globally and superlinearly convergent algorithm for the minimization of $U(x, \varepsilon)$ can be defined by using any stabilization technique (see, e.g., [19, 59]). Part (ii) shows that Algorithm NT converges locally *in one iteration* if the problem considered is a quadratic programming problem and, hence, in this case, it takes advantage of the simple structure of the problem.

Algorithm NT provides a link with Quasi-Newton approaches for the solution of Problem (IC), based on Quasi-Newton approximations of the matrix $\nabla^2 L(x, \lambda(x))$ that appears in (10) [27]. Here, we prefer to conceive Quasi-Newton approaches in the framework of RQP algorithms, considered in the next subsection.

In concluding this subsection, we remark that the global convergence of Algorithm EPS is achieved without the *a priori* assumption, usual in algorithms for constrained optimization, that the sequence $\{x_k\}$ is bounded. Instead, the boundedness of $\{x_k\}$ is a consequence of the barrier property of the function U.

8.3 An RQP algorithm with superlinear convergence rate

An attractive feature of recursive quadratic programming algorithms is that, under standard assumptions, a superlinear convergence rate can be obtained, provided that the unit stepsize is eventually accepted along the direction d_k computed by solving the quadratic programming subproblem. In order to enforce global convergence towards KKT points of the original problem, a standard approach, introduced independently by Pshenichnyj in [66] and by Han in [52], is to define a merit function that measures progress towards the solution, and to choose a stepsize that yields a sufficient decrease along d_k, which is a descent direction for the merit function.

The main difficulty encountered in this kind of approach is that the line search can truncate the stepsize near a solution (Maratos effect, [58]), thus destroying superlinear convergence. Several different techniques have been proposed to overcome this problem, based either on the use of nondifferentiable penalty functions (see e.g. [12, 13, 45, 49, 50]), or on the use of differentiable penalty functions (see e.g. [3, 5, 37, 38, 48, 64, 69]). In particular, in [64] an algorithm using Fletcher's differentiable exact penalty function was proposed for the equality constrained case. The approach of [64] has been extended to the inequality constrained case in [24].

In this subsection, we describe a basic algorithm for the inequality constrained Problem (IC). The algorithm and its properties can be deduced from [24], by assuming as merit function in the line search the exact barrier function U. A more extensive analysis of an RQP algorithm based on U is presented in [41].

In the algorithm the search direction d_k is generated as solution of the quadratic programming subproblem $QP(x_k, H_k)$ defined by:

$$\begin{aligned} \min_d &\ \tfrac{1}{2} d' H_k d + \nabla f(x_k)' d \\ \text{s.t.} &\ \nabla g(x_k)' d + g(x_k) \leq 0, \end{aligned} \tag{QP}$$

where H_k is an $n \times n$ definite positive matrix, obtained by some updating rule. We make the following assumption, needed in the proof of the convergence results:

Assumption 8 *There exist a compact set $\mathcal{H}$ of definite positive $n \times n$ matrices such that:*

(i) the solution $d(x, H)$ of $QP(x, H)$ exists and is continuous on $S_\alpha \times \mathcal{H}$;

(ii) H_k belongs to $\mathcal{H}$ for all k.

It is immediate to verify that the following proposition holds:

Proposition 30 *The pair $(\overline{x}, \overline{\lambda})$ is a KKT pair for Problem (IC) if and only if $\overline{d} = 0$ is a KKT point for Problem $QP(\overline{x}, H_k)$ and $\overline{\lambda}$ is the corresponding multiplier.*

Two relevant features of the algorithm are the use of an approximate directional derivative of U that avoids the need to evaluate the second order derivatives of the problem functions, and an automatic updating procedure for the selection of the penalty parameter.

The directional derivative $\nabla U(x; \varepsilon)'d$ along the direction d is approximated by:

$$
\begin{aligned}
DU(x, d; \varepsilon, t) \quad := \quad & d'\nabla f(x) + d'\nabla g(x)\lambda(x) \\
+ \quad & \frac{1}{t}\Big[\lambda(x + td) - \lambda(x)\Big]'(g(x) + V(x; \varepsilon)v(x; \varepsilon)) \\
+ \quad & d'\frac{2}{\varepsilon a(x)}\nabla g(x)(g(x) + V(x; \varepsilon)v(x; \varepsilon)) \\
+ \quad & d'\frac{s}{\varepsilon a(x)^2}\|g(x) + V(x; \varepsilon)v(x; \varepsilon)\|^2\sum_{i=1}^{m}\nabla g_i(x)g_i^+(x)^{s-1},
\end{aligned}
$$

where t is a positive parameter. We obviously have that:

$$
\lim_{t \to 0} DU(x, d; \varepsilon, t) = \nabla U(x; \varepsilon)'d.
$$

The penalty parameter is updated according to the result of the test at Step 2 in the following Algorithm RQP.

Algorithm RQP

Data: $\tilde{x} \in \mathbb{R}^n$, $\delta_1 \in (0, 1)$, $\delta_2 \in [\delta_1, 1)$, $\mu \in (0, 1/2)$, $H_1 > 0$, $\varepsilon_{1,0} > 0$, $\rho \in (0, 1)$.

Step 0: Choose $\alpha > 0$ and $s \geq 2$ such that $\tilde{x} \in S_\alpha$, set $x_1 = \tilde{x}$, and $k = 1$.

Step 1: Compute d_k as the solution of $QP(x_k, H_k)$.

    If $d_k = 0$ stop; otherwise set $i = 0$ and $\sigma_{k,0} = 1$.

Step 2: If:

$$
DU(x_k, d_k; \varepsilon_{k,i}, \sigma_{k,i}) \leq -\frac{1}{2}d_k'H_kd_k
$$

go to Step 4.

Step 3: Set $\varepsilon_{k,i} = \rho\varepsilon_{k,i}$. If $U(x_k; \varepsilon_{k,i}) > U(x_1; \varepsilon_{k,i})$, set $x_k = x_1$ and go to Step 1, otherwise go to Step 2.

Step 4: If $x_k + \sigma_{k,i}d_k \in S_\alpha$ and:

$$
U(x_k + \sigma_{k,i}d_k; \varepsilon_{k,i}) \leq U(x_k; \varepsilon_{k,i}) + \mu\sigma_{k,i}DU(x_k, d_k; \varepsilon_{k,i}, \sigma_{k,i})
$$

go to Step 5; otherwise set $i = i+1$, $\varepsilon_{k,i} = \varepsilon_{k,i-1}$, choose $\sigma_{k,i} \in [\delta_1\sigma_{k,i-1}, \delta_2\sigma_{k,i-1}]$ and go to Step 2.

Step 5: Set $\varepsilon_{k+1,0} = \varepsilon_{k,i}$, $x_{k+1} = x_k + \sigma_{k,i}d_k$; generate H_{k+1}.

Set $k = k + 1$ and go to Step 1.

Algorithm RQP is motivated by Proposition 30.

At Step 1 we calculate d_k. If $d_k = 0$, by Proposition 30, we have that x_k is a KKT point of Problem (IC). In the case that $\mathrm{QP}(x_k, H_k)$ does not admit a solution, a convenient choice is to make a step along a suitable approximation of the negative gradient direction $-\nabla U(x;\varepsilon)$ [41]. In any case, by Proposition 30, if $\mathrm{QP}(x_k, H_k)$ does not admit a solution, then x_k is not a stationary point of Problem (IC).

At Step 2 we decide whether the value of ε has to be reduced. The reason of the test is that, by using an Armijo-type rule (Step 4), we can force DU, the (approximate) directional derivative of the penalty function along the search direction, to zero, so that the test at step 2 ensures that also d_k tends to zero. Hence we have, by Proposition 30, that x_k converges to a KKT point of Problem (IC). From a slightly different point of view, the test at Step 2 can be viewed as a way to force an angle condition between the gradient of the objective function U and the search direction d_k in the minimization of the merit function U.

If the test at Step 2 is not satisfied we reduce the value of ε (Step 3), and we check whether we are still in the level set of the merit function at the starting point. If this is not the case we restart from x_1, which is a better point.

If the test at Step 2 is satisfied we go to Step 4, where an Armijo-like test is performed using the approximate directional derivative DU. When the Armijo-like test is not satisfied, there can be two reasons. The first one is that the stepsize is too large and we have to reduce it; the second one is that DU is a bad approximation to the directional derivative, and we have to reduce the step used for the finite difference approximation. Since we do not know which is the true cause of the failure, we have to reduce both the stepsize of the line search and the step of the finite difference approximation. This explains why the same parameter $\sigma_{k,i}$ is used for both two quantities. However, once we have reduced the step employed to calculate the finite difference part of DU, we cannot simply perform a new Armijo-like test with the reduce stepsize, first we have to check whether the test at Step 2 is still satisfied.

As in the previous subsection, the global convergence of Algorithm RQP can be stated under the same assumptions used to prove the exactness of the barrier function U. Moreover, again we can get some information about the original problem even if the EMFCQ does not hold on $\overline{S}_\alpha$ or if the feasible set is empty.

Let us assume that Algorithm RQP produces an infinite sequence of points $\{x_k\} \subseteq \overline{S}_\alpha$. We can state that any limit point of $\{x_k\}$ produced by the algorithm is a KKT point for Problem (IC) and, in particular, we can show that the penalty parameter ε is updated finitely many times.

Proposition 31 *Suppose that the EMFCQ holds on S_α. Then there exists an iteration index $\hat{k}$ such that $\varepsilon_{k,i} = \hat{\varepsilon}$, for all $k \geq \hat{k}$ and $i \geq 0$; moreover every limit* *point x^* of the sequence $\{x_k\} \subseteq S_\alpha$ produced by Algorithm RQP yields a KKT pair $(x^*, \lambda(x^*))$ for Problem (IC).*

The convergence result of the preceding proposition can be stated also by replacing the assumption that the EMFCQ holds on $\overline{S}_\alpha$ by the assumption that in the data of Algorithm RQP we take $\tilde{x} \in \mathcal{F}$. In the case that neither the EMFCQ nor the knowledge of a feasible point is assumed, Algorithm RQP behaves mainly like Algorithm EPS; we do not repeat here the conclusion of this kind of analysis, that is performed in [41].

As concerns the rate of convergence, we can state the following result.

Proposition 32 *Assume that the sequence $\{x_k\}$ generated by Algorithm RQP satisfies the following conditions:*

(i) $\lim_{k \to \infty} x_k = x^*$;

(ii) at $(x^*, \lambda(x^*))$ strict complementarity holds;

(iii) $\lim_{k \to \infty} \dfrac{\|x_k + d_k - x^*\|}{\|x_k - x^*\|} = 0.$

Then, there exists an iteration index $\hat{k}$ such that:

$$x_{k+1} = x_k + d_k, \quad \text{for all} \quad k \geq \hat{k}.$$

By the last proposition we have that, if a unit stepsize ensures superlinear convergence, then the unit stepsize is eventually accepted by the algorithm. Hence the Maratos effect cannot occur.

As a final remark, we point out that, as it was the case for Algorithm EPS, also the global convergence of Algorithm RQP is achieved without the *a priori* assumption that the sequence $\{x_k\}$ is bounded, due to the barrier property of the function U. As concerns the sequence $\{H_k\}$, updating rules exploiting the properties of U in such a way to guarantee that $\{H_k\}$ is bounded are currently under investigation.

9 Extensions

Up to now we have considered the exact penalty approach with reference to the general nonlinear programming problem (C) in the continuously differentiable case. In addition we can study, on one hand, the case in which the nonlinear programming problem has some interesting particular structure, and, on the other hand, the case in which the problem functions are not continuously differentiable.

As a problem with particular structure we mention first the *minimax problem*:

$$\text{minimize } f(x) := \max_{i \in I} f_i(x), \quad I = \{1, \ldots, m\},$$

where the functions $f_i(x) : \mathbb{R}^n \to \mathbb{R}$ are continuously differentiable

It is known that the minimax problem, which is nondifferentiable, can be restated as a differentiable constrained minimization problem, by introducing an auxiliary variable $z \in \mathbb{R}$ in the following way:

$$
\begin{aligned}
&\text{minimize } z \\
&\text{subject to } f_i(x) \leq z, \ i = 1, \ldots, m.
\end{aligned} \qquad \text{(MM)}
$$

Problem (MM) is a nonlinear programming problem with unbounded feasible set, so that Assumption 1 is not satisfied. However it has the particularity that a feasible point is easily available. By exploiting this particularity a continuously differentiable exact barrier function can be built, which has the same minimum points of the nondifferentiable function $f(x)$ [35].

From a theoretical point of view, the exact penalty approach adopted in this case allows to prove, in a constructive way, that there exist nondifferentiable minimization problems that have a continuously differentiable fully equivalent counterpart. From an algorithmic point of view, the approach allows us to propose algorithm models that, under suitable smoothness assumptions on the functions f_i, are globally convergent with superlinear convergence rate; a property that other algorithms for the solution of the minimax problem either do not possess, or achieve in a rather cumbersome way.

A second problem of interest with particular structure is the nonlinear programming problem with bounded variables:

$$
\begin{aligned}
&\text{minimize } f(x) \\
&\text{subject to } b_l \leq x \leq b_u,
\end{aligned} \qquad \text{(BV)}
$$

where b_l and b_u are n-vectors, $b_l < b_u$. In this case the evaluation of the multiplier functions does not requires matrix inversions, so that the main drawback of the continuously differentiable exact penalty approach is avoided. Moreover it has been experienced that the index set of pseudoactive constraints $I_A(x; \varepsilon)$ provides a very good estimate of the active set at the solution. Due to these reasons, the algorithms for the solution of Problem (BV) based on the approach of Section 8.2 and described in [42] are most effective, even in the case of a very large dimension n.

A third problem with particular structure arises, for instance, in the numerical solution of optimal control problems. In this case we often deal with a nonlinear programming problem of the form:

$$
\begin{aligned}
&\text{minimize } f(x, u) \\
&\text{subject to } h(x, u) = 0 \\
&\qquad\qquad\; b_l \leq u \leq b_u,
\end{aligned} \qquad \text{(OC)}
$$

where $x \in \mathbb{R}^{n \times N}$ and $u \in \mathbb{R}^{q \times N}$ are the *complete* state and control vectors:

$$
\begin{aligned}
x &:= [x(\Delta), x(2\Delta), \ldots, x(N\Delta)] \\
u &:= [u(0), u(\Delta), \ldots, u((N-1)\Delta)],
\end{aligned}
$$

containing the samples of the state $x(t) \in \mathbb{R}^n$ and of the control $u(t) \in \mathbb{R}^q$ at times $t = k\Delta$, $k = 0, \ldots, N$; $h : \mathbb{R}^{n \times N} \times \mathbb{R}^{q \times N} \to \mathbb{R}^{n \times N}$ is the *complete* state equation vector, b_l and b_u are $q \times N$-vectors of control bounds.

In the numerical solution of optimal control problems, Problem (OC) is usually a very large dimensional one, with the particular property that the Jacobian matrix $\nabla_x h(x, u)$ is, by construction, nonsingular. In this case, it has been proved advantageous a mixed exact augmented Lagrangian - exact penalty function approach, in which the state constraints are handled by means of a simplified exact augmented Lagrangian term that exploits the nonsingularity of $\nabla_x h(x, u)$, and the bounds constraints are handled by exact penalty terms as in Problem (BV). In this way the resulting exact penalty-Lagrangian function depends on only $n \times N$ additional multiplier variables, and its evaluation does not require matrix inversions [36].

As concerns the differentiability properties of the problem functions, the more general case in which f, g and h are Lipschitz continuous has been investigated by many authors (see, e.g. [8, 11, 40, 62, 68, 71]). In particular, in [21, 22, 23] it has been proved that, under suitable constraint qualification assumptions, the nondifferentiable penalty function $J_q(x; \varepsilon)$, as well as the nondifferentiable barrier function $Z_q(x; \varepsilon)$, are exact also in the case of Lipschitz programs; and in [23] some globally convergent algorithms for the solution of Lipschitz programs, based on $Z_q(x; \varepsilon)$, have been analysed.

10 Conclusions

We can now draw some conclusions, with regards to the sentence quoted at the beginning.

We have seen that it is possible in different ways to exhibit functions with the property that their unconstrained minimum points are solutions of a constrained problem. Hence, from an analytical point of view, the exactness theory is today well established.

Some of these functions are very simple, but are not continuously differentiable.

More controversial is the question whether the functions that are continuously differentiable can be considered simple, inasmuch as they require the evaluation of higher order derivatives, and either matrix inversions, or solutions of linear systems of equations, or an increase of the number of variables. However the assessment on the simplicity has to be done in comparison with other approaches for the solution of constrained problems that are able to guarantee both global convergence and superlinear convergence rate. By this comparison we can conclude that the approaches based on exact penalty, or exact augmented Lagrangian functions are at least no more complicated then any other currently available; with the advantages of a built-in preference for minimum points rather then saddle points of the original problem, and of improved convergence properties in the presence of a barrier term.

248

The numerical experiments carried out (see, e.g., [24, 27, 33, 34, 42]) confirm these conclusions.

References

[1] M. BAZARAA AND J. GOODE, *Sufficient conditions for a globally exact penalty function without convexity*, Mathematical Programming Study, 19 (1982), pp. 1–15.

[2] D. BERTSEKAS, *Necessary and sufficient conditions for a penalty method to be exact*, Mathematical Programming, 9 (1975), pp. 87–99.

[3] ——, *Constrained Optimization and Lagrange Multipliers Methods*, Academic Press, New York, 1982.

[4] ——, *Enlarging the region of convergence of Newton's methods for constrained optimization*, Journal of Optimization Theory and Applications, 36 (1982), pp. 221–252.

[5] M. BIGGS, *On the convergence of some constrained minimization algorithms based on recursive quadratic programming*, Journal of the Institute of Mathematics and its Applications, 21 (1978), pp. 67–82.

[6] J. BONNANS, *Théorie de la pénalisation exacte*, Mathematical Modelling and Numerical Analysis, 24 (1990), pp. 197–210.

[7] J. BURKE, *A sequential quadratic programming method for potentially infeasible mathematical programs*, Journal of Mathematical Analysis and Applications, 139 (1989), pp. 319–351.

[8] ——, *Calmness and exact penalization*, SIAM Journal on Control and Optimization, 29 (1991), pp. 493–497.

[9] ——, *An exact penalization viewpoint of constrained optimization*, SIAM Journal on Control and Optimization, 29 (1991), pp. 968–998.

[10] J. BURKE AND S.-H. HAN, *A robust sequential quadratic programming method*, Mathematical Programming, 43 (1989), pp. 277–303.

[11] F. H. CLARKE, *Optimization and Nonsmooth Analysis*, John Wiley & Sons, New York, 1983.

[12] T. COLEMAN AND A. CONN, *Nonlinear programming via an exact penalty method: asymptotic analysis*, Mathematical Programming, 24 (1982), pp. 123–136.

[13] ——, *Nonlinear programming via an exact penalty method: global analysis*, Mathematical Programming, 24 (1982), pp. 137–161.

[14] A. R. CONN, *Constrained optimization using a nondifferentiable penalty function*, SIAM Journal on Numerical Analysis, 10 (1973), pp. 760–784.

[15] ——, *Penalty function methods*, in Nonlinear Optimization 1981, M. J. D. Powell, ed., Academic-Press, New York, 1982, pp. 235–242.

[16] A. R. CONN AND T. PIETRZYKOWSKI, *A penalty function method converging directly to a constrained optimum*, SIAM Journal on Numerical Analysis, 14 (1977), pp. 348–378.

[17] G. CONTALDI, G. DI PILLO, AND S. LUCIDI, *A continuously differentiable exact penalty function for nonlinear programming problems with unbounded feasible set*, Operation Research Letters, 14 (1993), pp. 153–161.

[18] ——, *An exact augmented Lagrangian function for nonlinear programming problems with unbounded feasible set*, Tech. Rep. 01-94, DIS-Università di Roma La Sapienza, Roma, Italy, 1994.

[19] J. DENNIS AND R. SCHNABEL, *Numerical Methods for Unconstrained Optimization and Nonlinear Equations*, Prentice-Hall, Englewood Cliffs, New Jersey, 1983.

[20] M. DEW, *An algorithm for nonlinear equality constrained optimization using the DiPillo-Grippo exact penalty function*, Tech. Rep. 159, Numerical Optimization Centre, Hatfield Polytechnic, Hatfield, UK, 1985.

[21] G. DI PILLO AND F. FACCHINEI, *Exact penalty functions for nondifferentiable programming problems*, in Nonsmooth Optimization and Related Topics, F. Clarke, V. F. Demyanov, and F. Giannessi, eds., Plenum Press, New York, 1989, pp. 89–107.

[22] ——, *Regularity conditions and exact penalty functions in Lipschitz programming problems*, in Nonsmooth Optimization Methods and Applications, F. Giannessi, ed., Gordon and Breach, London, 1992, pp. 107–120.

[23] ——, *Exact barrier function methods for Lipschitz programs*, Applied Mathematics and Optimization, (to appear).

[24] G. DI PILLO, F. FACCHINEI, AND L. GRIPPO, *An RQP algorithm using a differentiable exact penalty function for inequality constrained problems*, Mathematical Programming, 55 (1992), pp. 49–68.

[25] G. Di Pillo and L. Grippo, *A new class of augmented Lagrangians in nonlinear programming*, SIAM Journal on Control and Optimization, 17 (1979), pp. 618–628.

[26] ——, *An augmented Lagrangian for inequality constraints in nonlinear programming problems*, Journal of Optimization Theory and Applications, 36 (1982), pp. 495–519.

[27] G. Di Pillo and L. Grippo, *A class of continuously differentiable exact penalty function algorithms for nonlinear programming problems*, in System Modelling and Optimization, P. Toft-Christensen, ed., Springer-Verlag, Berlin, 1984, pp. 246–256.

[28] G. Di Pillo and L. Grippo, *A continuously differentiable exact penalty function for nonlinear programming problems with inequality constraints*, SIAM Journal on Control and Optimization, 23 (1985), pp. 72–84.

[29] ——, *An exact penalty method with global convergence properties for nonlinear programming problems*, Mathematical Programming, 36 (1986), pp. 1–18.

[30] ——, *On the exactness of a class of nondifferentiable penalty functions*, Journal of Optimization Theory and Applications, 57 (1988), pp. 399–410.

[31] ——, *Exact penalty functions in constrained optimization*, SIAM Journal on Control and Optimization, 27 (1989), pp. 1333–1360.

[32] G. Di Pillo, L. Grippo, and F. Lampariello, *A method for solving equality constrained optimization problems by unconstrained minimization*, in Optimization Techniques, K. Iracki, K. Malanowski, and S. Walukiewicz, eds., Springer-Verlag, Berlin, 1980, pp. 96–105.

[33] ——, *A class of methods for the solution of optimization problems with inequalities*, in System Modelling and Optimization, R. F. Drenick and F. Kozin, eds., Springer-Verlag, Berlin, 1982, pp. 508–519.

[34] G. Di Pillo, L. Grippo, and S. Lucidi, *Globally convergent exact penalty algorithms for constrained optimization*, in System Modelling and Optimization, A. Prekopa, J. Szelezsan, and B. Strazicky, eds., Springer-Verlag, Berlin, 1986, pp. 694–703.

[35] ——, *A smooth method for the minimax problem*, Mathematical Programming, 60 (1993), pp. 187–214.

[36] G. Di Pillo, S. Lucidi, and L. Palagi, *An exact penalty-Lagrangian approach for a class of constrained optimization problems with bounded variables*, Optimization, 28 (1993), pp. 129–148.

[37] L. DIXON, *Exact penalty function methods in nonlinear programming*, Tech. Rep. 103, Numerical Optimization Centre, Hatfield Polytechnic, Hatfield, UK, 1979.

[38] ——, *On the convergence properties of variable metric quadratic programming methods*, Tech. Rep. 110, Numerical Optimization Centre, Hatfield Polytechnic, Hatfield, UK, 1980.

[39] J. EVANS, F. GOULD, AND J. TOLLE, *Exact penalty functions in nonlinear programming*, Mathematical Programming, 4 (1973), pp. 72–97.

[40] F. FACCHINEI, *Exact penalty functions and Lagrange multipliers*, Optimization, 22 (1991), pp. 579–606.

[41] F. FACCHINEI, *A new recursive quadratic programming algorithm with global and superlinear convergence properties*, Tech. Rep. 12-93, DIS-Università di Roma La Sapienza, Roma, Italy, 1993.

[42] F. FACCHINEI AND S. LUCIDI, *A class of penalty functions for optimization problems with bound constraints*, Optimization, 26 (1992), pp. 239–259.

[43] R. FLETCHER, *A class of methods for nonlinear programming with termination and convergence properties*, in Integer and Nonlinear Programming, J. Abadie, ed., North-Holland, Amsterdam, 1970, pp. 157–173.

[44] ——, *Numerical experiments with an exact l_1 penalty function method*, in Nonlinear Programming 4, O. Mangasarian, R. Meyer, and S. Robinson, eds., Academic Press, New York, 1981, pp. 99–129.

[45] ——, *Second order corrections for nondifferentiable optimization*, in Numerical Analysis Dundee 1981, G. Watson, ed., Springer, Berlin, 1982, pp. 144–157.

[46] ——, *Penalty functions*, in Mathematical Programming: the State of the Art, A. Bachem, M. Grötschel, and B. Korte, eds., Springer, Berlin, 1983, pp. 87–114.

[47] R. E. FLETCHER, *An ℓ_1 penalty method for nonlinear constraints*, in Numerical Optimization 1984, P. T. Boggs, R. H. Byrd, and R. B. Schnabel, eds., Siam, Philadelphia, 1984, pp. 26–40.

[48] R. FONTECILLA, T. STEIHAUG, AND R. TAPIA, *A convergence theory for a class of quasi-Newton methods for constrained optimization*, SIAM Journal on Numerical Analysis, 24 (1987), pp. 1133–1151.

[49] M. FUKUSHIMA, *A successive quadratic programming algorithm with global and superlinear convergence properties*, Mathematical Programming, 35 (1986), pp. 253–264.

[50] D. GABAY, *Reduced quasi-Newton methods with feasibility improvement for non-linearly constrained optimization*, Mathematical Programming Study, 16 (1982), pp. 18–44.

[51] T. GLAD AND E. POLAK, *A multiplier method with automatic limitation of penalty growth*, Mathematical Programming, 17 (1979), pp. 140–155.

[52] S. HAN, *A globally convergent method for nonlinear programming*, Journal of Optimization Theory and Applications, 22 (1977), pp. 297–309.

[53] S. HAN AND O. MANGASARIAN, *Exact penalty functions in nonlinear programming*, Mathematical Programming, 17 (1979), pp. 251–269.

[54] S. LUCIDI, *New results on a class of exact augmented Lagrangians*, Journal of Optimization Theory and Application, 58 (1988), pp. 259–282.

[55] ——, *Recursive quadratic programming algorithm that uses an exact augmented Lagrangian function*, Journal of Optimization Theory and Application, 67 (1990), pp. 227–245.

[56] ——, *New results on a continuously differentiable exact penalty function*, SIAM Journal on Optimization, 2 (1992), pp. 558–574.

[57] ——, *On the role of continuously differentiable exact penalty functions in constrained global optimization*, Journal of Global Optimization, (to appear).

[58] N. MARATOS, *Exact penalty function algorithms for finite dimensional and control optimization problems*, PhD Thesis, London University, London, UK, 1978.

[59] J. J. MORÉ AND D. C. SORENSEN, *Newton's method*, in Studies in Numerical Analysis, G. H. Golub, ed., The Mathematical Association of America, Washington, DC, 1984, pp. 29–82.

[60] H. MUKAI AND E. POLAK, *A quadratically convergent primal-dual algorithm with global convergence properties for solving optimization problems with equality constraints*, Mathematical Programming, 9 (1975), pp. 336–349.

[61] T. PIETRZYKOWSKI, *An exact potential method for constrained maxima*, SIAM Journal on Numerical Analysis, 6 (1969), pp. 299–304.

[62] ——, *The potential method for conditional maxima in the locally compact metric spaces*, Numerische Mathematik, 14 (1970), pp. 325–329.

[63] E. POLAK, *On the global stabilization of locally convergent algorithms for optimization and root finding*, Automatica, 12 (1976), pp. 337–342.

[64] M. POWELL AND Y. YUAN, *A recursive quadratic programming algorithm that uses differentiable exact penalty functions*, Mathematical Programming, 35 (1986), pp. 265–278.

[65] ——, *A trust region algorithm for equality constrained optimization*, Mathematical Programming, 49 (1991), pp. 189–211.

[66] B. PSHENICHNYJ, *Algorithms for the general problem of mathematical programming*, Kibernetika (Kiev), 6 (1970), pp. 120–125. In Russian.

[67] B. PSHENICHNYJ AND Y. DANILIN, *Numerical Methods in Extremal Problems*, MIR Publisher, Moscow, 1975. Engl. transl 1978.

[68] E. ROSENBERG, *Exact penalty functions and stability in locally Lipschitz programming*, Mathematical Programming, 30 (1984), pp. 340–356.

[69] K. SCHITTKOWSKI, *The nonlinear programming method of Wilson, Han and Powell with augmented Lagrangian type line search function*, Numerische Mathematik, 38 (1981), pp. 83–114.

[70] C. VINANTE AND S. PINTOS, *On differentiable exact penalty functions*, Journal of Optimization Theory and Applications, 50 (1986), pp. 479–493.

[71] D. WARD, *Exact penalties and sufficient conditions for optimality in nonsmooth optimization*, Journal of Optimization Theory and Applications, 57 (1988), pp. 485–499.

[72] Y. G. YEVTUSHENKO AND V. G. ZHADAN, *Exact auxiliary functions in optimization problems*, U.S.S.R. Computational Mathematics and Mathematical Physics, 30 (1990), pp. 31–42.

[73] W. ZANGWILL, *Non-linear programming via penalty functions*, Management Science, 13 (1967), pp. 344–358.

Stable Barrier-Projection and Barrier-Newton Methods for Linear and Nonlinear Programming[1]

Yu.G. EVTUSHENKO,   V. G. ZHADAN

Computing Center
40 Vavilov Str.
117967 Moscow GSP-1
Russia

ABSTRACT: A space transformation technique is used for the reduction of constrained minimization problems to minimization problems without inequality constraints. The continuous and discrete versions of stable barrier-projection method and Newton's method are applied for solving such reduced LP and NLP problems. The space transformation modifies these methods and introduces additional matrices which play the role of a multiplicative barrier, preventing the trajectories from crossing the boundary of the feasible set. The proposed algorithms are based on the numerical integration of systems of ordinary differential equations. These algorithms do not require feasibility of starting and current points, but they preserve feasibility. Some results about convergence rate analysis for continuous and discrete versions of the methods are presented. We describe primal barrier-projection methods, primal barrier-Newton methods and primal-dual barrier-Newton methods. For LP we develop dual barrier-projection and barrier-Newton methods.

1. Introduction

The purpose of this paper is to indicate the advantages and capabilities of the space transformation techniques for solving linear programming (LP) and nonlinear programming (NLP) problems. Here we give a survey of the most important results and numerical methods which were obtained and published mainly in Russian. We describe the primal barrier-projection and barrier-Newton methods (sections 2,3), primal-dual barrier-Newton method (section 5). For linear programming we developed dual barrier-projection and barrier-Newton methods (section 4).

The analysis of the methods is made on the basis of the qualitative theory of dynamical systems [17], [18] and the stability theory of the solutions of ordinary differential equations [3], [8]. Numerical algorithms are obtained as discretization of dynamical systems. For local convergence analysis we use the Lyapunov linearization principle of determining the stability from the equation of the first approximation

---

[1]Research supported by the grant N 93-012-450 from Russian Scientific fund.

255

E. Spedicato (ed.), Algorithms for Continuous Optimization, 255–285.
© 1994 *Kluwer Academic Publishers.*

about an equilibrium state which is valid for continuous and discrete systems. Non-local convergence is investigated by using the second (direct) method of Lyapunov. This theory is quite general, and, although already published in 1892, it remains one of the most powerful techniques for detecting stability and convergence [3], [8], [10].

Starting from 1973, we have developed a family of numerical methods based on the space transformation techniques [7], [9], [11], [13], [14]. Such an approach enabled us to reduce the original NLP problem to a problem without inequality constraints. The stable versions of the gradient-projection and Newton's methods are used for solving this reduced problem. The numerical methods are found after performing an inverse transformation. These methods are described by systems of ordinary differential equations and stated as initial-value problems.

As a result of space transformation the right-hand side of the differential equation is multiplied by some matrices which prevent the trajectories from crossing the boundary of the feasible set. Therefore these matrices play the role of barrier and we term these methods "barrier-projection" and "barrier-Newton" methods . The space transformation is carried out without using penalty functions and this feature provides a high rate of convergence. The term "barrier" has often been misunderstood. We do not utilize the barriers which are described, for example, in the book [19] written by A. Fiacco and G.Mc Cormick. Numerical methods which are based on penalty functions are inherently unstable, since one has to increase the penalty parameter without bound in order to obtain convergence. In contrast, the unconventional multiplicative barriers which we use do not tend to infinity when a current point approaches the boundary of a feasible set. In our algorithms the barrier functions are continuous and equal to zero on a boundary. These barriers imply the feasibility of the trajectories. Therefore we need not introduce any penalty coefficients.

In the second section we consider a family of barrier-projection methods. In the linear programming case after simplifications and choosing an exponential space-transformation function we obtain Dikin's algorithm [6], sometimes called the "variation on Karmarkar's algorithm". The analysis of this algorithm was given in numerous papers (see, for example, [2], [4], [21], [23], [24], [34], [35]). However there are four main differences with our approaches:

1. We considered LP and NLP problems.

2. From 1983 we developed a stable version of the projection method. Therefore we did not restrict ourselves to the interior point techniques. In our methods the current points are often infeasible, but if the starting points or the current points are feasible, then the subsequent trajectory remains in the feasible set.

3. In all proposed methods we did not resort to a penalty-type approach.

4. We considered the steepest descent variants of our methods where the trajectory could move along the boundary of the feasible set.

The primal-dual barrier-Newton method is obtained after applying Newton's method to solve the nonlinear system of equations that is derived from the generalized Kuhn-Tucker stationary condition. We present some results about convergence rate analysis for continuous and discrete versions and show that superlinear and quadratic convergence can be attained. In a particular case our algorithm bears a resemblance to the algorithm of [38], which was developed for LP. But there are also significant differences:

1. Our algorithms can start the computation from the infeasible region, although they preserve feasibility.

2. Our algorithms enable us to take different stepsizes in the primal space and in the dual space, which is proved to be important for computation.

3. We do not use any penalty or usual barrier functions.

4. There exists a set of initial points such that the method solves a linear programming problem in a finite number of iterates.

The numerical algorithms which we propose for linear and nonlinear programming problems have been implemented, tested and included in the library of algorithms at the Computing Center of the Russian Academy of Sciences. Computer codes were used for solving various practical problems.

2. Space Transformation Technique and Primal Barrier-Projection Methods

Consider the following NLP problem:

$$\text{minimize } f(x) \text{ subject to } x \in X = \{x \in R^n : g(x) = 0_m, x \in P\}, \quad (1)$$

where the functions f and g are twice continuously differentiable, $f(x)$ maps R^n onto R^1 and $g(x)$ maps R^n onto R^m , P is a convex set with nonempty interior, 0_s is the s-dimensional null vector, 0_{sk} is the $s \times k$ rectangular null matrix.

We assume differentiability whenever it is helpful to do so. Subscripts will be used to distinguish values of quantities at a particular iteration and superscripts will indicate components of vectors.

Primal and dual linear programming problems are stated in the standard form:

$$\text{minimize } c^T x \text{ subject to } x \in X = \{x \in R^n : g(x) = b - Ax = 0_m, x \geq 0_n\}, \quad (2)$$

$$\text{maximize } b^T u \text{ subject to } u \in U = \{u \in R^m : v = L_x(x,u) = c - A^T u \geq 0_n\}, \quad (3)$$

258

where c is a n-vector, b is a m-vector, A is a full rank $m \times n$ matrix, and $v \in R^n$ is the vector of dual slack variables.

Define

$$R_{++}^n = \{x \in R^n : x > 0_n\}, \quad \text{int}U = \{u \in R^m : v = c - A^T u > 0_n\},$$

$$\text{ri}X = \{x \in R_{++}^n : Ax = b\},$$

We assume that $\text{ri}X$ and $\text{int}U$ are nonempty and that primal and dual nondegeneracy holds. In this case both problems have unique solutions x_* and u_* respectively.

We introduce a new n-dimensional space with the coordinates $[y^1, \ldots, y^n]$ and make a differentiable transformation from this space to the original one: $x = \xi(y)$. This surjective transformation maps R^n onto P or $\text{int}P$, i.e. $P = \overline{\xi(R^n)}$, where $\bar{B}$ is the closure of B.

Consider the transformed minimization problem

$$\text{minimize } \tilde{f}(y) = f(\xi(y)) \text{ subject to } y \in Y, \tag{4}$$

where $Y = \{y \in R^n : \tilde{g}(y) = g(\xi(y)) = 0_m\}$. The Lagrangian associated with Problem (4) is defined by $\tilde{L}(y, \tilde{u}) = \tilde{f}(y) + \tilde{u}^T \tilde{g}(y)$.

We can use the stable version of the gradient projection method for solving Problem (4). The method is stated as an initial-value problem involving the following system of ordinary differential equations

$$\frac{dy}{dt} = -\tilde{L}_y(y, \tilde{u}(y)), \quad \tilde{L}_y(y, \tilde{u}) = \tilde{f}_y(y) + \tilde{g}_y^T(y)\tilde{u}. \tag{5}$$

The function $\tilde{u}(y)$ is chosen to satisfy the following condition [33]

$$\frac{d\tilde{g}}{dt} = \tilde{g}_y \frac{dy}{dt} = -\tau\tilde{g}(y), \quad \tau > 0. \tag{6}$$

From this condition we obtain the system of linear algebraic equations

$$\tilde{g}_y(y)\tilde{g}_y^T(y)\tilde{u}(y) + \tilde{g}_y(y)\tilde{f}_y(y) = \tau\tilde{g}(y), \quad y_0 \in R^n,$$

where $\tilde{f}_y = \tilde{J}^T f_x$, $\tilde{g}_y = g_x\tilde{J}$, $\tilde{J} = dx/dy$ and $\tilde{J}$ is the Jacobian matrix of the transformation $x = \xi(y)$ with respect to y. If $\tilde{J}$ is a nonsingular matrix, then there exists an inverse transformation $y = \delta(x)$, so it is possible to return from the y-space to the x-space and we obtain in this way a matrix $J(x) = \tilde{J}(\delta(x))$ which is now a function of x. By differentiating $\xi(y)$ with respect to y and taking into account (5), we have

$$\frac{dx}{dt} = \frac{d\xi}{dy}\frac{dy}{dt} = J(x)\frac{dy}{dt} = -G(x)L_x(x, u(x)), \quad x(0, x_0) = x_0 \in P, \tag{7}$$

$$\Gamma(x)u(x) + g_x(x)G(x)f_x(x) = \tau g(x), \tag{8}$$

where we have introduced the two Gram matrices:

$$\Gamma(x) = g_x(x)G(x)g_x^\top(x), \quad G(x) = J(x)J^\top(x).$$

We define the following sets: the nullspace of the matrix $g_x(x)J(x)$ at x:

$$K(x) = \{z \in R^n : \quad g_x(x)J(x)z = 0_m\},$$

the cone of feasible directions at $x \in P$:

$$F(x \mid P) = \{z \in R^n : \quad \exists \lambda(z) > 0 \text{ such that } x + \lambda z \in P, \quad 0 < \lambda \le \lambda(z)\},$$

the conjugate cone to the cone F:

$$F^*(x \mid P) = \{z \in R^n : z^\top y \ge 0 \quad \forall y \in F(x \mid P)\},$$

the linear hull of the cone F^* at $x \in P$:

$$S(x) = \lim F^*(x \mid P) =$$

$$= \{z \in R^n : z = \sum_{i=1}^{s} \lambda^i z_i, \lambda^i \in R; z_i \in F^*(x \mid P), 1 \le i \le s, s = 1, 2, \ldots\}.$$

Definition 1 *The constraint qualification (CQ) for Problem (1) holds at a point $x \in P$ if all vectors $g_x^i(x)$, $1 \le i \le m$, and any nonzero vector $p \in S(x)$ are linearly independent. We say that x is a regular point for Problem (1) if the CQ holds at x.*

We impose the following condition on the transformation $\xi(y)$:

$\mathbf{C_1}$. *At each point $x \in P$ the Jacobian $J(x)$ is defined and $\ker J^\top(x) = S(x)$.*

From this condition it follows that the Jacobian $J(x)$ is nonsingular in the interior of P, it is singular only on the boundary of P.

Lemma 1 *Let the space transformation $\xi(y)$ satisfy C_1, and let the CQ for Problem (1) hold at a point $x \in P$. Then the Gram matrix $\Gamma(x)$ is invertible and positive definite.*

Let W be a $m \times n$ rectangular matrix whose rank is m. We introduce the pseudoinverse matrix $W^+ = W^\top(WW^\top)^{-1}$ and the orthogonal projector $\pi(W) = I_n - W^+W$, where I_n is the $n \times n$ identity matrix.

If at a point x the conditions of Lemma 1 hold, then we can find from (8) the function $u(x)$, substitute it into the right-hand side of (7) and write (7) in the following projective form

$$\frac{dx}{dt} = -J(x)\left\{\pi[g_x(x)J(x)]J^\top(x)f_x(x) + \tau[g_x(x)J(x)]^+g(x)\right\}. \qquad (9)$$

Let $x(t, x_0)$ denote the solution of the Cauchy Problem (7) with initial condition $x_0 = x(0, x_0)$. For the following, we assume that the initial-value problem under consideration is always uniquely solvable. By differentiating $f(x(t, x_0))$ with respect to t we obtain

$$\frac{df}{dt} = - \parallel J^\top(x)L_x(x, u(x)) \parallel^2 + \tau u^\top(x)g(x). \tag{10}$$

Hence the objective function $f(x(t, x_0))$ monotonically decreases on the feasible set X and if the trajectory is close to X, i.e. if $\parallel g(x(t, x_0)) \parallel$ is sufficiently small.

The system of ordinary differential equations (7), where $u(x)$ is given by (8), has the first integral

$$g(x(t, x_0)) = g(x_0)e^{-\tau t}. \tag{11}$$

This means that if $\tau > 0$, then method (7) has a remarkable property: all its trajectories approach the manifold $g(x) = 0_m$ as t tends to infinity and this manifold is an asymptotically stable attractor of the system (see [8], [15], [33]). If $x_0 \in X$, then the trajectory $x(t, x_0)$ of (7) remains in the feasible set X because $g(x(t, x_0)) \equiv 0_m$ and condition C_1 implies $x(t, x_0) \in P$. Therefore if $x_0 \in X$ or $\tau = 0$, then the trajectories of (9) coincide with the trajectories of the following system

$$\frac{dx}{dt} = -J(x)\pi[g_x(x)J(x)]J^T(x)f_x(x). \tag{12}$$

If the condition $x \in P$ is absent in Problem (1), then the space transformation is trivial: $x = y$ and, taking $J(x) = I_n$ in (12), we obtain the well-known gradient projection method [29].

Definition 2 *The pair $[x_*, u_*]$, where $x_* \in P$, is a Kuhn-Tucker pair of Problem (1), if*

$$J^\top(x_*)L_x(x_*, u_*) = 0_n, \quad g(x_*) = 0_m. \tag{13}$$

Lemma 2 *Let conditions of Lemma 1 be satisfied at a point $x_* \in P$. Then x_* is an equilibrium point of system (7) if and only if the pair $[x_*, u_*]$, where $u_* = u(x_*)$, satisfies (13).*

Definition 3 *The strict complementary condition (SCC) holds at a point $[x_*, u_*] \in P \times R^m$, if*

$$L_x(x_*, u_*) \in \ ri \ F^*(x_* \mid P), \tag{14}$$

where ri A is a relative interior of the set A.

The space transformation described above can be used to derive the following second-order sufficient conditions for a point to be an isolated local minimum in Problem (1).

Theorem 1 *Assume that f and g are twice-differentiable functions and the space transformation $\xi(y)$ satisfies C_1. Sufficient conditions that a point $x_* \in P$ be an isolated local minimum of Problem (1) are that there exists a Lagrange multiplier vector u_* satisfying the Kuhn-Tucker conditions (13), that the SCC holds at $[x_*, u_*]$ and that $z^\top J^\top(x_*)L_{xx}(x_*, u_*)J(x_*)z > 0$ for every $z \in K(x_*)$ such that $J(x_*)z \neq 0_n$.*

We denote by $D(z)$ the diagonal matrix containing the components of a vector z. The dimensionality of this matrix is determined by the dimensionality of z.

For the sake of simplicity we consider now the particular case of Problem (1), where the set P is the positive orthant, i.e. $P = R_+^n$. It is convenient for this set P to use a component-wise differentiable space transformation $\xi(y)$

$$x^i = \xi^i(y^i), \quad 1 \leq i \leq n. \tag{15}$$

For such a transformation the corresponding Jacobian matrix is diagonal and

$$\tilde{J}(y) = D(\dot{\xi}(y)), \quad \dot{\xi}(y) = [\dot{\xi}^1(y^1), \dot{\xi}^2(y^2), \ldots, \dot{\xi}^n(y^n)]^\top.$$

Let $\delta(y)$ be the inverse transformation. Denote

$$J(x) = D(\dot{\xi}(y))\mid_{y=\delta(x)}, \quad G(x) = J^2(x) = D(\theta(x))$$

with the vector $\theta(x) = [(\dot{\xi}^1(y^1))^2, (\dot{\xi}^2(y^2))^2, \ldots, (\dot{\xi}^n(y^n))^2]\mid_{y=\delta(x)}$.

We impose on a space transformation $\xi(y)$ the following conditions:

$\mathbf{C_2}$. *The vector function $\theta(x)$ is defined at each point $x \in R_+^n$, and $\theta^i(x^i) = 0$ if and only if $x^i = 0$, where $1 \leq i \leq n$.*

$\mathbf{C_3}$. *The vector function $\theta(x)$ is differentiable in some neighborhood of R_+^n and $\dot{\theta}^i(0) > 0$, $1 \leq i \leq n$.*

Different numerical methods are obtained by different choices of the space transformations. As a rule we perform the following quadratic and exponential transformations

$$x^i = \xi^i(y^i) = \frac{1}{4}(y^i)^2, \quad J(x) = D^{1/2}(x), \quad G(x) = D(x), \tag{16}$$

$$x^i = \xi^i(y^i) = e^{y^i}, \quad J(x) = D(x), \quad G(x) = D^2(x). \tag{17}$$

In these two cases the Jacobian matrix is singular on the boundary of the set $P = R_+^n$. These transformations satisfy C_1, C_2. Condition C_3 holds only for transformation (16). If the SCC holds at $[x_*, u_*]$, then for these transformations we obtain that $L_{x^i}(x_*, u_*) > 0$ for all i such that $x_*^i = 0$.

From method (7) interesting particular cases are derived. In order to distill the essence of multiplicative barriers we consider the simplest case where $X = P = R_+^n$. If we use the space transformation functions (16) and (17), then we obtain

$$\frac{dx}{dt} = -D^\alpha(x)f_x(x), \tag{18}$$

where $x_0 > 0_n$, $\alpha = 1$ for transformation (16) and $\alpha = 2$ for (17).

Let $x^i(t, x_0) = 0$. Then we have $\theta^i(x^i) = 0$ and $dx^i(t, x_0)/dt = 0$ in (18). From the last equality it follows that the trajectory $x(t, x_0)$ of system (18) cannot cross the boundary $x^i = 0$. Thus the Gram matrix $G(x) = D^\alpha(x)$ plays the role of a "barrier", preventing the trajectory $x(t, x_0)$ from passing through the boundary of P. Therefore we call (7) a "primal barrier-projection method".

In our first publication in this field [9] we used the quadratic space transformation (16) and considered the resource allocation problem where $X = \{x \in R_+^n : \sum_{i=1}^n x^i = 1\}$, $x_0 \in X$. In this case $m = 1$ and system (8) was solved explicitly, while system (9) had the form

$$\frac{dx}{dt} = -D(x)[f_x(x) - d(x)e], \quad x_0 \in X,$$

where $e \in R^n$ is the vector of ones, $d(x) = x^\top f_x(x)$. More general cases were considered in subsequent papers [7], [11], [13].

If $P = R_+^n$, then we can simplify definitions 1, 3 and the statement of Theorem 3. Let e^i denote the n-th order unit vector whose i-th component is equal to one and let

$$K(x) = \left\{ \bar{x} \in R^n : g_x(x)D(\theta(x))\bar{x} = 0_m \right\}.$$

Definition 4 *The constraint qualification (CQ) for Problem (1), where $P = R_+^n$, holds at a point x, if all vectors $g^i(x)$, $1 \leq i \leq m$, and all e^j, such that $x^j = 0$, are linearly independent. We say that x is a regular point for Problem (1), if the CQ holds at x. The strict complementary condition (SCC) holds at a point $[x, u]$, if $L_{x^i}(x, u) > 0$ for all i such that $x^i = 0$.*

Theorem 2 *Assume that f and g are twice-differentiable functions and the space transformation $\xi(y)$ satisfies C_2. Sufficient conditions for a point $x_* \in P = R_+^n$ to be an isolated local minimum of Problem (1) are that there exists a Lagrange multiplier vector u_* such that $[x_*, u_*]$ satisfies the Kuhn-Tucker conditions (13), that the SCC holds and that $z^T D(\theta(x_*))L_{xx}(x_*, u_*)D(\theta(x_*))z > 0$ for all $z \in K(x_*)$, satisfying $D(\theta(x_*))z \neq 0_n$.*

Applying the Euler method for solving system (9), we obtain

$$x_{k+1} = x_k - h_k J(x_k) \left\{ \pi[g_x(x_k)J(x_k)]J^\top(x_k)f_x(x_k) + \tau[g_x(x_k)J(x_k)]^+ g(x_k) \right\}, \quad (19)$$

where a stepsize $h_k > 0$.

Theorem 3 *Let $[x_*, u_*]$ be a Kuhn-Tucker pair of Problem (1), where the CQ and the second-order sufficiency conditions of Theorem 2 hold. Let the space transformation $\xi(y)$ satisfy conditions C_2, C_3 and let $\tau > 0$. Then x_* is an asymptotically stable equilibrium state of system (9); there exists a number h_* such that for any fixed $0 < h_k < h_*$ the sequence $\{x_k\}$, generated by (19), converges locally with a linear rate to x_* while the corresponding sequence $\{u_k\}$ converges to u_*.*

The proof of this theorem is given in [13], [16]. It is based on Lyapunov linearization principle. The conditions for asymptotic stability are expressed by mean of the characteristic multipliers of the matrix of the equations of the first approximation about the equilibrium point x_*. The asymptotic stability of a point x_* implies the local convergence of trajectory $x(t, x_0)$ to point x_*. The corresponding statement about the convergence of discrete variants follows from Theorem 2.3.7 (see [8]).

Applying method (7) for solving Problem (2), we obtain the following continuous and discrete versions

$$\frac{dx}{dt} = -G(x)[c - A^T u(x)], \quad x(0, x_0) = x_0 > 0_n, \tag{20}$$

$$x_{k+1} = x_k - h_k G(x_k)[c - A^T u(x_k)], \quad x_0 > 0_n, \tag{21}$$

where the function $u(x)$ is found from linear equation (8) which can be rewritten as follows

$$AG(x)A^T u(x) - AG(x)c = \tau(b - Ax). \tag{22}$$

By differentiating the objective function with respect to t, we obtain

$$c^T \frac{dx}{dt} = - \parallel J(x)(c - A^T u(x)) \parallel^2 + \tau u^T(x)(b - Ax).$$

The system of ordinary differential equations (20) has the first integral

$$Ax(t, x_0) = b + (Ax_0 - b)e^{-\tau t}. \tag{23}$$

From (23) it follows that, if $Ax_0 = b$, then $Ax(t, x_0) \equiv b$ for all $t \geq 0$ and the trajectory $x(t, x_0)$ of (20) remains in the feasible set X, the objective function monotonically decreases along the trajectories.

Theorem 4 *Let x_*, u_* be unique solutions of Problems (2) and (3) respectively. Let the space transformation $\xi(y)$ satisfy the conditions C_2, C_3 and let $\tau > 0$. Then x_* is an asymptotically stable equilibrium state of system (20). There exists $h_* > 0$ such that for any fixed $0 < h_k < h_*$ the sequence $\{x_k\}$, generated by (21), converges locally with a linear rate to x_* while the corresponding sequence $\{u_k\}$ converges to u_*.*

If we use the quadratic and exponential space transformations (16), (17), then from (20), (22) we obtain respectively

$$\frac{dx}{dt} = D(x)(A^T u(x) - c), \quad AD(x)A^T u(x) = AD(x)c + \tau(b - Ax), \tag{24}$$

$$\frac{dx}{dt} = D^2(x)(A^T u(x) - c), \quad AD^2(x)A^T u(x) = AD^2(x)c + \tau(b - Ax). \tag{25}$$

If $\tau > 0$, then method (24) converges locally to x_*, i.e. for any $\epsilon > 0$ there exists $\delta > 0$ such that $\|x_0 - x_*\| < \delta$ implies

$$\lim_{t\to\infty} \|x(t,x_0) - x_*\| = 0. \tag{26}$$

Method (25) converges locally on R_{++}^n, i.e., if $\|x_0 - x_*\| < \delta$ and $x_0 \in R_{++}^n$, then (26) holds. Theorem 4 cannot be used in this case because condition C_3 is not satisfied for exponential space transformation.

Denote $\mu = \max_{1\le i\le n} v_*^i$. In [14], using Lyapunov's linearization principle, we proved that the discrete version of (24) converges linearly if the stepsize h_k is fixed and is such that $h_k < h_* = 2\min[1/\tau, 1/\mu]$.

If we set $\tau = 0$, then (24), (25) yield

$$\frac{dx}{dt} = D(x)(A^T u(x) - c), \quad AD(x)A^T u(x) = AD(x)c, \tag{27}$$

$$\frac{dx}{dt} = D^2(x)(A^T u(x) - c), \quad AD^2(x)A^T u(x) = AD^2(x)c. \tag{28}$$

Method (27) converges locally on the set $Ax = b$, i.e., if x_0 belongs to this set and $\|x_0 - x_*\| < \delta$, then (26) takes place. Method (27) was proposed in 1978 (see [11], [8]). Nonlocal convergence analysis of this method was made in [14] on the basis of the second Lyapunov method. System (27) was also considered in [21].

The discrete and continuous versions of method (28) were investigated in numerous papers (see, for example, [2], [4], [6], [21], [23], [34], [35]). In [2] the discrete version was called "a variation on Karmarkar's algorithm". We should remark that method (28) does not possess the local convergence property. Here the convergence takes place only if x_0 belongs to the relative interior of X. This result was proved by G.Smirnov on the basis of the vector Lyapunov function. He investigated method (25) and proved that for the vector of nonbasic components x_N the following estimate holds: $\|x_N(t,x_0)\| \approx O(t^{-1})$ as $t \to \infty$. If we use method (24), then $\|x_N(t,x_0)\| \approx O(e^{-\lambda_* t})$, where $\lambda_* > 0$. Hence the trajectories of system (24) with the quadratic transformation converge locally faster than the trajectories of system (25) with the exponential transformation. Therefore in our papers and codes we used mainly the quadratic space transformation (16).

There is another interesting case, where P is a n-dimensional box, i.e. $P = \{x \in R^n : a \le x \le b\}$. Here we use the following transformation

$$x^i = [a^i + b^i + (b^i - a^i)\sin y^i]/2, \quad G(x) = D(x - a)D(b - x)$$

and method (7) is written as follows

$$\frac{dx}{dt} == -D(x - a)D(b - x)L_x(x, u(x)), \quad x(0, x_0) = x_0 \in P,$$

$$g_x(x)D(x - a)D(b - x)\left[f_x(x) + g_x^T u(x)\right] = \tau g(x).$$

The statements of Theorems 3 and 4 are generalized for this case.

The preceding results and algorithms admit straightforward extensions for problems involving general inequality constraints by using space dilation. Consider Problem

$$\text{minimize } f(x) \text{ subject to } x \in X = \{x \in R^n : g(x) = 0_m, h(x) \le 0_c\}, \tag{29}$$

where $h(x)$ maps R^n into R^c.

In Problem (29) we do not have nonnegativity constraints on the separated variables. Nevertheless our approach can be used in this case by extension of the space and by converting the inequality constraints to equalities. We introduce an additional variable $p \in R^c$, define $q = m + c$, combine primal, dual variables and all constraints:

$$z = \begin{bmatrix} x \\ p \end{bmatrix} \in R^{n+c}, \quad w = \begin{bmatrix} u \\ v \end{bmatrix} \in R^q, \quad \Phi(z) = \begin{bmatrix} g(x) \\ h(x) + p \end{bmatrix}.$$

Then the original Problem (29) is transformed into the equivalent Problem

$$\text{minimize } f(x) \text{ subject to } z \in Z = \{z \in R^{n+c} : \Phi(z) = 0_q, p \in R^c_+\}. \tag{30}$$

This problem is similar to (1). In order to take into account the constraint $p \ge 0_c$ we introduce a surjective differentiable mapping $\varphi : R^c \to R^c_+$ and make the space transformation $p = \varphi(y)$, where $y \in R^c$, $\overline{\varphi(R^c)} = R^c_+$. Let φ_y be the square $c \times c$ Jacobian matrix of the mapping $\varphi(y)$ with respect to y. We assume that it is possible to define the inverse transformation $y = \psi(p)$ and hence we obtain the $c \times c$ Jacobian and Gram matrices:

$$J(p) = \varphi_y(y) \mid_{y=\psi(p)}, \quad G(p) = J(p)J^\top(p).$$

Combining variables and constraints for the reduced problem, let us define

$$\hat{z} = \begin{bmatrix} x \\ y \end{bmatrix} \in R^{n+c}, \quad \hat{\Phi}(\hat{z}) = \begin{bmatrix} g(x) \\ h(x) + \varphi(y) \end{bmatrix}, \quad \hat{\Phi}_{\hat{z}} = \begin{bmatrix} g_x & 0_{mc} \\ h_x & \varphi_y \end{bmatrix}.$$

Problems (29) and (30) can be formulated as follows:

$$\text{minimize } f(x) \text{ subject to } \hat{z} \in \hat{Z} = \{\hat{z} \in R^{n+c} : \hat{\Phi}(\hat{z}) = 0_q\}. \tag{31}$$

In the last Problem we have only equality constraints, therefore we can use classical optimality conditions and the numerical method described above. After inverse transformation to the space of x and p we obtain

$$\frac{dz}{dt} = -\tilde{G}(p)L_z(z, w(z)). \tag{32}$$

Here
$$L(z,w) = f(x) + w^\top \Phi(z), \quad L_z(z,w) = f_z(z) + \Phi_z^\top(z)w,$$
$$\Phi_z(z)\tilde{G}(p)L_z(z,w(z)) = \tau\Phi(z), \tag{33}$$
$$\tilde{G}(p) = \begin{bmatrix} I_n & 0_{nc} \\ 0_{cn} & G(p) \end{bmatrix}, \quad \Phi_z = \begin{bmatrix} \Phi_x & 0_{mc} \\ & I_c \end{bmatrix}, \quad \Phi_x = \begin{bmatrix} g_x \\ h_x \end{bmatrix}, \quad f_z = \begin{bmatrix} f_x \\ 0_c \end{bmatrix}.$$

System (32) can be rewritten in the more detailed form

$$\frac{dx}{dt} = -L_x(x, w(z)), \quad \frac{dp}{dt} = -G(p)v, \tag{34}$$

where the function $w(z)$ is found from the following linear system of q equations

$$\Gamma(z)w(z) + \Phi_x(x)f_x(x) = \tau\Phi(z), \quad \Gamma(z) = \Phi_x(x)\Phi_x^\top(x) + \begin{bmatrix} 0_{mm} & 0_{mc} \\ 0_{cm} & G(p) \end{bmatrix}.$$

Condition (6) can be written as

$$\frac{dg(x)}{dt} = -\tau g(x), \quad \frac{d(h(x) + p)}{dt} = -\tau(h(x) + p).$$

Therefore system (34) has two first integrals:

$$g(x(t, z_0)) = g(x_0)e^{-\tau t},$$
$$h(x(t, z_0)) + p(t, z_0) = (h(x_0) + p_0)e^{-\tau t}, \quad z_0^\top = [x_0^\top, p_0^\top]. \tag{35}$$

Similarly to (10) we obtain

$$\frac{df}{dt} = -\| L_x \|^2 - \| J^\top(p)v \|^2 + \tau[u^\top g + v^\top(h + p)]. \tag{36}$$

Let us introduce the index set $\sigma(p) = \{i \in [1 : c] : p^i = 0\}$. The set P in Problem (30) has the form $P = \{[x, p] \in R^{n+c} : \quad p \geq 0_c\}$. Therefore definition 1 of CQ for this problem can be reformulated as follows.

Definition 5 *The constraint qualification (CQ) for Problem (30) holds at a point $z \in P$, if all vectors $g_x^i(x)$, $1 \leq i \leq m$ and vectors $h_x^i(x), i \in \sigma(p)$, are linearly independent.*

Let us impose on the mapping $\varphi(y)$ a new condition which is similar to C_1:

C_4. *At each point $p \in R_+^c$ the Jacobian $J(p)$ is defined and*

$$\ker J^\top(p) = \lim F^*(p \mid R_+^c) = \{b \in R^c : \text{ if } i \in \sigma(p), \text{ then } b^i = 0\}. \tag{37}$$

Lemma 3 *Let the mapping $\varphi(y)$ satisfy C_4 and let the CQ for Problem (30) hold at a point $z \in P$. Then the Gram matrix $\Gamma(z)$ is invertible and positive definite.*

We can use all results given in this section for Problem (30) and for the corresponding numerical method (34). It follows from CQ for Problem (30) that at a point z the vectors $\Phi_z^i(z), 1 \leq i \leq q$, are linearly independent. Consequently z is a regular point for Problem (30). The pair $[z_*, w_*]$ is a Kuhn-Tucker pair of Problem (30), if

$$L_x(z_*, w_*) = 0_n, \quad \Phi(z_*) = 0_q, \quad D(p_*)v_* = 0_c. \tag{38}$$

If $v_*^i > 0$ for i such that $p_*^i = 0$, then the strict complementarity condition (SCC) is fulfilled at the point z_*.

Lemma 4 Let conditions of Lemma 3 be satisfied at a point $z_* \in R^n \times R_+^c$. Then z_* is an equilibrium point of system (34) if and only if the pair $[z_*, w_*]$ satisfies (38).

It is convenient to use for Problem (30) a component-wise mapping $p^i = \varphi^i(y^i), 1 \leq i \leq c$. Therefore $\varphi_y = D(\dot\varphi)$, where $\dot\varphi(y)$ is a c-dimensional vector. We denote $\gamma^i(p^i) = \dot\varphi^i(\psi^i(p^i))$, $\theta(p) = [\theta^1(p^1), \ldots, \theta^c(p^c)]$, $\theta^i(p^i) = [\gamma^i(p^i)]^2, 1 \leq i \leq c$. For this mapping the matrix $G(p)$ has the diagonal form $G(p) = D(\theta(p))$. Introduce the following conditions:

C_5. The vector function $\theta(p)$ is defined at each point $p \in R_+^c$, and $\theta^i(p^i) = 0$ if and only if $p^i = 0$, where $1 \leq i \leq c$.

C_6. The vector function $\theta(p)$ is differentiable in some neighborhood of R_+^c and $\dot\theta^i(0) > 0, 1 \leq i \leq c$.

Let us consider the tangent cone

$$\tilde{K}(x) = \{\bar{x} \in R^n : g_x(x)\bar{x} = 0_m; \ \bar{x}^T h_x^i(x) = 0, \ i \in \sigma(h(x))\}.$$

Theorem 5 Let $[z_*, w_*]$ be a Kuhn-Tucker pair of Problem (30), where the CQ and the SCC hold. Let the space transformation $\varphi(y)$ satisfy conditions C_5, C_6 and assume that for any nonzero vector $\bar{x} \in \tilde{K}(x_*)$ the inequality $\bar{x}^T L_{xx}(x_*, w_*)\bar{x} > 0$ holds. Then system (34) with $\tau > 0$ is asymptotically stable at the equilibrium point z_*. The discrete version of the method converges locally with a linear rate, if the stepsize is fixed and sufficiently small.

Consider the simplified version of method (34). Suppose that along the trajectories of system (34) the following condition holds

$$h(x(t, z_0)) + p(t, z_0) \equiv 0_c.$$

From this equality we can define p as a function of h. We exclude from system (34) the additional vector p and integrate the system which does not employ this vector:

$$\frac{dx}{dt} = -L_x(x, w(x)), \tag{39}$$

where

$$\Gamma(x)w(x) + \Phi_x(x)f_x(x) = \tau \begin{bmatrix} g(x) \\ 0_c \end{bmatrix}, \tag{40}$$

$$\Gamma(x) = \Phi_x(x)\Phi_x^\top(x) + \begin{bmatrix} 0_{mm} & 0_{mc} \\ 0_{cm} & G(-h(x)) \end{bmatrix}.$$

Along the trajectories of (39) we have

$$\frac{dg}{dt} = -\tau g(x), \quad \frac{dh}{dt} = -G(-h(x))v(x), \tag{41}$$

$$\frac{df}{dt} = -\parallel L_x(x, w(x)) \parallel^2 - \parallel J^\top(-h(x))v(x) \parallel^2 + \tau u^\top(x)g(x).$$

Let us show that the solution $x(t, x_0)$ does not leave the set X for any $t > 0$, if $x_0 \in X$. Suppose this is not true and let $h^i(x(t, x_0)) > 0$ for some $t > 0$. Then there is a time $0 < t_1 < t$ such that $h^i(x(t_1, x_0)) = 0$ and $\dot{h}^i(x(t_1, x_0)) > 0$. This contradicts (41) since $\theta^i(0) = 0$. Hence $x(t, x_0) \in X$ for all $t \geq 0$. Thus the Gram matrix $G(-h(x))$ plays the role of a "barrier" preventing $x(t, x_0)$ from intersecting the hypersurface $h^i(x) = 0$. The trajectory $x(t, x_0)$ can approach the boundary points only as $t \to +\infty$. If the initial point x_0 is on the boundary, then the entire trajectory of system (39) belongs to the boundary.

Method (39) is closely related to method (7). Let us consider Problem (1), assuming that $P = R_+^n$. We have two alternatives: we can use methods (7) or (39). The main computational work required in any numerical integration method is to evaluate the right-hand sides of the systems for various values of x. This could be done by solving the linear system (8) of m equations or system (40) of $m + n$ equations respectively. One might suspect that the introduction of slack variables p increases the computational work considerably. However, by taking advantage of the simple structure of equation (40), we can reduce the computational time by using the Frobenius formula for an inverse matrix [8]. After some transformations we find that formulas (39), (40) can be written as (7), (8) respectively, if in the last formulas we take

$$G(x) = D(\mu(x)), \quad \mu^i(x^i) = \theta^i(x^i)/[1 + \theta^i(x^i)], \quad 1 \leq i \leq n.$$

Therefore the performances of both seemingly unrelated methods are very similar.

3. Primal Barrier-Newton Method

Let us consider Problem (1) supposing that all functions $f(x)$, $g^i(x)$, $1 \leq i \leq m$, are at least twice continuously differentiable. Assume also for simplicity that $P = R_+^n$ and the transformation $\xi(y)$ has the component-wise form (15).

Equation (8) can be rewritten as

$$g_x(x)D(\theta(x))L_x(x, u(x)) = \tau g(x). \tag{42}$$

Therefore if the space transformation $\xi(y)$ satisfies C_2 and x is a regular point such that

$$D(\theta(x))L_x(x, u(x)) = 0_n,$$

then $[x, u(x)]$ is a Kuhn-Tucker point of Problem (1). In the previous section we used the gradient method for finding a solution x of this equation. Now we will apply Newton's method for this purpose. The continuous version of Newton's method leads to the initial value problem for the following system of ordinary differential equations

$$\Lambda(x)\frac{dx}{dt} = -D(\alpha)D(\theta(x))L_x(x, u(x)), \quad x(t, x_0) = x_0, \tag{43}$$

where $\alpha \in R^n$ is a scaling vector, $\Lambda(x)$ is the Jacobian matrix of the mapping $D(\theta(x))L_x(x, u(x))$ with respect to x:

$$\Lambda(x) = D(\dot{\theta})D(L_x) + D(\theta)L_{xx} + D(\theta)g_x^\mathsf{T}\frac{du}{dx}. \tag{44}$$

Here all matrices and vectors are evaluated at a point x and the function $u(x)$ is defined from (42). By differentiating equality (42) with respect to x, we obtain

$$g_x[D(\dot{\theta})D(L_x) + D(\theta)L_{xx} + D(\theta)g_x^\mathsf{T}\frac{du}{dx}] + E = \tau g_x, \tag{45}$$

where $E(x)$ denotes the $m \times n$ matrix with the elements

$$e_{ij}(x) = \sum_{k=1}^{n} \theta^k(x)\frac{\partial^2 g^i(x)}{\partial x^k \partial x^j}\frac{\partial L(x, u(x))}{\partial x^k}.$$

Let us assume that x is a regular point of Problem (1). From (42) and (45) we find that

$$u = \Gamma^{-1}[\tau g - g_x D(\theta)f_x], \quad \Gamma = g_x D(\theta)g_x^\mathsf{T},$$

$$\frac{du}{dx} = \Gamma^{-1}[\tau g_x - g_x D(\dot{\theta})D(L_x) - g_x D(\theta)L_{xx} - E]. \tag{46}$$

Using the notations $T(x) = g_x^\mathsf{T}(x)\Gamma^{-1}(x)$, $\Omega(x) = T(x)g_x(x)$, we obtain after substitution of (46) into (44)

$$\Lambda = [I_n - D(\theta)\Omega][D(\theta)L_{xx} + D(\dot{\theta})D(L_x)] + D(\theta)T[\tau g_x - E].$$

Lemma 5 *Let conditions of Theorem 3 be satisfied at the point* $[x_*, u_*]$. *Then* $\Lambda(x_*)$ *is a nonsingular matrix.*

Theorem 6 *Let conditions of Theorem 3 be satisfied at the point* $[x_*, u_*]$. *Then for any scaling vector* $\alpha > 0_n$ *and any* $\tau > 0$ *the solution point* x_* *is an asymptotically stable equilibrium point of system (43) and the discrete version*

$$x_{k+1} = x_k - h_k \Lambda^{-1}(x_k)D(\alpha)D(\theta(x_k))L_x(x_k, u_k), \quad u_k = u(x_k) \tag{47}$$

locally converges with at least linear rate to the point x_* if the stepsize h_k is fixed and $h_k < 2/\max_{1 \leq i \leq n} \alpha_i$. If the matrix $\Lambda(x)$ satisfies a Lipschitz condition in a neighborhood of x_*, if $h_k = 1$ and $\alpha = e$, then the sequence $\{x_k\}$ converges quadratically to x_*.

System (43) has the first integral

$$D(\theta(x(t, x_0)))L_x(x(t, x_0), u(x(t, x_0))) = D(e^{-\alpha t})D(\theta(x_0))L_x(x_0, u_0),$$

where $u_0 = u(x_0)$, $D(e^{-\alpha t})$ is a diagonal matrix whose i-th diagonal element is $e^{-\alpha^i t}$. Taking into account (42), we obtain

$$g(x(t, x_0)) = \tau^{-1}g_x(x(t, x_0))D(e^{-\alpha t})D(\theta(x_0))L_x(x_0, u_0).$$

Hence, if the trajectory $x(t, x_0)$ remains in a bounded set, where the CQ holds, then $\| g(x(t, x_0)) \| \to 0$ as $t \to +\infty$.

Suppose that Problem (1) is such that $g_x D(\theta)TE = 0_{mn}$. This condition is satisfied, for example, for a linear programming problem. It is easy to prove that in this case

$$g_x(x)\Lambda(x) = \tau g_x(x), \quad g_x(x) = \tau g_x(x)\Lambda^{-1}(x).$$

Therefore, by differentiating $g(x)$ along the solutions of (43), we have

$$\frac{dg}{dt} = -g_x(x)\Lambda^{-1}(x)D(\alpha)D(\theta(x))L_x(x, u(x)) = -\frac{1}{\tau}g_x(x)D(\alpha)D(\theta(x))L_x(x, u(x)).$$

If $\alpha = e \in R^n$, then using relation (42) we obtain finally:

$$g(x(t, x_0)) = g(x_0)e^{-t}.$$

We come to the conclusion that the feasible manifold $g(x) = 0_m$ is asymptotically stable. The trajectory initiating at a point $x_0 \in X$ does not leave the feasible set. The method (47) was considered in [37] in the case where $P = R^n$ and $D(\theta(x)) \equiv I_n$.

We apply the primal barrier-Newton method (43) for solving the linear programming Problem (2). In this case we have

$$\Lambda(x) = [I_n - H(x)]D(\dot{\theta}(x))D(c - A^T u(x)) + \tau H(x), \tag{48}$$

$$H(x) = D(\theta(x))A^T(AD(\theta(x))A^T)^{-1}A. \tag{49}$$

As in the previous section the vector-function $u(x)$ is found from the linear equation (22).

Introduce a Lebesque level set in R^n

$$\Omega = \{x \in R_+^n : \|Ax - b\| \leq \|Ax_0 - b\|, \ 0_n \leq D(\theta(x))(c - A^T u(x)) \leq D(\theta(x_0))v_0\},$$

where x_0 is an initial point in (43), $v_0 = c - A^T u_0$, $u_0 = u(x_0)$.

Theorem 7 *Suppose that the set Ω is compact and contains a unique stationary point x_*. Assume that the space transformation $\xi(y)$ satisfies C_2 and is such that the matrix $\Lambda(x)$ is nonsingular everywhere on Ω. If the starting point x_0 is such that $x_0 > 0_n$, $v_0 > 0_n$, then*

$$\lim_{t\to\infty} x(t, x_0) = x_*, \quad \lim_{t\to\infty} u(x(t, x_0)) = u_*, \tag{50}$$

where x_*, u_* are the solutions of Problem (2) and (3) respectively.

Integrating (43) using the Euler method, we obtain the following iterative process:

$$x_{k+1} = x_k - h\Lambda^{-1}(x_k)D(\theta(x_k))(c - A^T u(x_k)), \tag{51}$$

where $h > 0$ is a stepsize.

Each equilibrium point x_* of system (43) is a fixed point of iterations (51), i.e. $x_k = x_*$ implies $x_{k+1} = x_*$, and if iterates (51) converge to a regular point x_*, then the pair $[x_*, u(x_*)]$ satisfies the Kuhn-Tucker conditions.

If the conditions of Theorem 2 hold and the space transformation function satisfies conditions C_2, C_3, then the matrix $\Lambda(x_*)$ is nonsingular. Therefore if the stepsize h is fixed and $0 < h < 2$, then the discrete version (51) locally converges to the point x_* with at least linear rate. If matrix $\Lambda(x)$ satisfies the Lipschitz condition in a neighborhood of x_* and $h = 1$, then the sequence $\{x_k\}$ converges quadratically to x_*.

4. Dual Barrier-Projection and Barrier-Newton Methods

For the sake of simplicity we consider in this section only the dual linear programming problem. If we apply method (34) to (3), then we obtain the following dual barrier-projection method

$$\frac{du}{dt} = b - Ax(u, v), \quad \frac{dv}{dt} = -G(v)x(u, v), \tag{52}$$

$$(G(v) + A^T A)x(u, v) = A^T b + \tau(v + A^T u - c),$$

where $u(t, z_0) = u_0, v(t, z_0) = v_0, z_0^T = [u_0^T, v_0^T]$. A very similar calculation as the one in (35), (36) yields the following formulas

$$v(t, z_0) + A^T u(t, z_0) - c = (v_0 + A^T u_0 - c)e^{-\tau t},$$

$$b^T \frac{du}{dt} = \|b - Ax\|^2 + x^T G(v)x + \tau x^T (c - A^T u - v). \tag{53}$$

For the component-wise space transformation $\varphi(y)$ method (52) can be written in the form

$$\frac{du}{dt} = b - Ax(u, v), \quad \frac{dv}{dt} = -D(\theta(v))x(u, v), \tag{54}$$

$$(D(\theta(v)) + A^T A)x(u,v) = A^T b + \tau(v + A^T u - c).$$

If condition C_5 holds, then from (54) we conclude that each component of the vector $v(t, u_0)$ does not change sign. Hence, if $v_0 = c - A^T u_0 > 0_n$, then on the entire trajectory $v(t, z_0) > 0_n$.

If $v_0 = c - A^T u_0$, then $v(t, z_0) = c - A^T u(t, z_0)$ for all $t \geq 0$ and the system (52) can be simplified as

$$\frac{du}{dt} = b - Ax(u), \quad (D(\theta(v)) + A^T A)x(u) = A^T b, \tag{55}$$

where $u(0, u_0) = u_0 \in U$.

Applying the Euler method for numerical integration of systems (54) and (55), we obtain respectively

$$u_{k+1} = u_k + h_k(b - Ax_k), \quad v_{k+1} = v_k + h_k D(\theta(v_k))x_k, \tag{56}$$

$$(D(\theta(v_k)) + A^T A)x_k = A^T b + \tau(v_k + A^T u_k - c),$$

$$u_{k+1} = u_k + h_k(b - Ax_k), \quad (D(\theta(v_k)) + A^T A)x_k = A^T b. \tag{57}$$

These variants of dual barrier-projection method solve simultaneously both the dual and primal linear programming problems. The same property has the primal barrier-projection method (21).

Theorem 8 *Let* x_*, u_* *be unique solutions of nondegenerate Problems (2), (3) respectively and* $v_* = c - A^T u_*$. *Let the space transformation* $\varphi(y)$ *satisfy conditions* C_5, C_6 *and* $\tau > 0$. *Then the following statements are true:*

1. *The pair* $[u_*, v_*]$ *is an asymptotically stable equilibrium state of system (54).*

2. *The solutions* $u(t, z_0), v(t, z_0)$ *of system (54) locally converge to the pair* $[u_*, v_*]$. *The corresponding function* $x(u(t, z_0), v(t, z_0))$ *converges to the optimal solution* x_* *of the primal problem (2).*

3. *There exists* $h_* > 0$ *such that for any fixed* $0 < h_k < h_*$ *the sequence* $\{u_k, v_k\}$ *generated by (56) converges locally with a linear rate to* $[u_*, v_*]$ *while the corresponding sequence* $\{x_k\}$ *converges to* x_*.

4. *The point* u_* *is an asymptotically stable equilibrium state of system (55).*

5. *The solutions* $u(t, u_0)$ *of system (55) locally converge to the optimal solution* u_* *of dual problem (3). The corresponding function* $x(u(t, u_0))$ *converges to the optimal solution* x_* *of primal problem (2).*

6. *There exists $h_* > 0$ such that for any fixed $0 < h_k < h_*$ the sequence $\{u_k\}$ generated by (57) converges locally with a linear rate to u_* while the corresponding sequence $\{x_k\}$ converges to x_*.*

If we use the space transformation $\varphi(y)$ of the forms (16) and (17), then we can rewrite formulas (55) as follows

$$\frac{du}{dt} = b - Ax(u), \quad (D^\alpha(v) + A^T A)x(u) = A^T b. \tag{58}$$

Here $\alpha = 1$ for the quadratic transformation (16) and $\alpha = 2$ for the exponential transformation (17).

From (53) and (58) we obtain that

$$\frac{dv}{dt} = -D^\alpha(v)x, \quad b^T \frac{du}{dt} = \|Ax - b\|^2 + \|D^{\alpha/2}(v)x\|^2 \geq 0.$$

Hence the objective function $b^T u(t, u_0)$ of the dual problem is a monotonically increasing function of t, all components of vector $v(t) = c - A^T u(t, u_0)$ do not change sign.

Denote $\mu = \max_{1 \leq i \leq n} x_*^i$. If $\alpha = 1$, then for algorithm (56) we have $h_k < h_* = 2 \min[1/\tau, 1/\mu]$; for algorithm (57) we obtain $h_k < h_* = 2/\mu$. If $\alpha = 1$, then both algorithms converge linearly.

In order to define the right-hand side of the differential equation (55) we have to solve a system of n linear equations. If $u \in \text{int}U$, then the square positive definite matrix $D^\alpha(v) + A^T A$ is nonsingular, and we can use the Sherman-Morrison-Woodbury formula

$$(D^\alpha(v) + A^T A)^{-1} = D^{-\alpha}(v) \left[I_n - A^T \left(I_m + AD^{-\alpha}(v)A^T \right)^{-1} AD^{-\alpha}(v) \right],$$

where the square matrix $I_m + AD^{-\alpha}(v)A^T$ is also nonsingular. Using this formula, we can transform (55) to the equivalent system

$$\frac{du}{dt} = z(u), \quad (I_m + AD^{-\alpha}(v)A^T)z(u) = b. \tag{59}$$

Here we solve a system of m linear equations. Method (55) was proposed in 1978 (see [11], [8]). If $\alpha = 2$, then method (59) is similar to the dual affine scaling algorithm which was investigated in [1].

We can apply Newton's method for solving the system

$$b - Ax(u) = 0_m,$$

where the function $x(u)$ is defined from the condition

$$(D(\theta(v)) + A^T A)x(u) = A^T b. \tag{60}$$

The continuous version of Newton's method leads to the following system of ordinary differential equations

$$\Lambda(u)\frac{du}{dt} = Ax(u) - b, \quad \Lambda(u) = -A\frac{dx}{du}. \tag{61}$$

By differentiating the relation (60) with respect to u, we obtain

$$-D(\dot\theta(v))D(x)A^T + \left(D(\theta(v)) + A^T A\right)\frac{dx}{du} = 0_{nm}.$$

Therefore

$$\Lambda(u) = -A\left(D(\theta(v)) + A^T A\right)^{-1} D(\dot\theta(v))D(x(u))A^T. \tag{62}$$

Lemma 6 *Let all conditions of Theorem 8 be fulfilled. Then $\Lambda(u_*)$ is a nonsingular matrix.*

Theorem 9 *Let all conditions of Theorem 8 hold. Then u_* is an asymptotically stable equilibrium point of system (61). If the matrix $\Lambda(u)$ satisfies a Lipschitz condition in a neighborhood of u_*, then the discrete version*

$$u_{k+1} = u_k + h_k\Lambda^{-1}(u_k)(Ax_k - b), \quad x_k = x(u_k), \tag{63}$$

locally converges with at least linear rate to the point u_* if the stepsize h_k is fixed and $0 < h_k < 2$. If the matrix $\Lambda(u)$ satisfies a Lipschitz condition in a neighborhood of u_* and $h_k = 1$, then the sequence $\{u_k\}$ converges quadratically to u_*.

If we use the quadratic space transformation (16), then formula (62) is simplified and we can rewrite method (61) as follows

$$\frac{du}{dt} = \left[A\left(D(v) + A^T A\right)^{-1} D(x(u))A^T\right]^{-1} (b - Ax(u)).$$

The discrete version of this method is similar to (63).

5. Primal-Dual Barrier-Newton Methods

For solving Problem (1) it is possible to solve the nonlinear system (13). For simplicity we consider here the case where P is a n-dimensional positive orthant, i.e. $P = R_+^n$ and the component-wise space transformation (15) is used.

Introduce the additional mapping

$$\phi(v) = \left[\phi^1(v^1), \ldots, \phi^n(v^n)\right]$$

and impose on this mapping the following conditions:

C₇.   *The vector function $\phi(v)$ is defined at each point $v \in R_+^n$, and $\phi^i(v^i) = 0$ if and only if $v^i = 0$, where $1 \leq i \leq n$.*

C₈.   *The vector function $\phi(v)$ is differentiable in some neighborhood of R_+^n and $\dot{\phi}^i(0) > 0$, $1 \leq i \leq n$.*

If the functions $\theta(x), \phi(v)$ satisfy conditions C_2 and C_7 respectively, then the necessary conditions (13) can be rewritten in the form

$$D(\theta(x))\phi(L_x(x,u)) = 0_n, \quad g(x) = 0_m, \quad x \in R_+^n. \tag{64}$$

For solving this system we use the continuous version of Newton's method. The computation process is described by the system of ordinary differential equations

$$W(x,u)\begin{pmatrix} \dot{x} \\ \dot{u} \end{pmatrix} = -\begin{pmatrix} \alpha D(\theta(x))\phi(L_x(x,u)) \\ \tau g(x) \end{pmatrix}, \tag{65}$$

where $\alpha > 0, \tau > 0, W$ is a square matrix of order $n + m$,

$$W(x,u) = \begin{pmatrix} M & D(\theta(x))D(\dot{\phi})g_x^T \\ g_x & 0_{mm} \end{pmatrix}, \quad M = D(\dot{\theta})D(\phi) + D(\theta)D(\dot{\phi})L_{xx}. \tag{66}$$

By following the trajectories satisfying (65), we can theoretically obtain a solution of the system of nonlinear equations (64). In practice, we build the iterative procedures using a discretization of dynamical systems.

Lemma 7 *Let $[x_*, u_*]$ be a Kuhn-Tucker pair, where the conditions of Theorem 2 are satisfied. Assume that x_* is a regular point for Problem (1) and the functions $\theta(x), \phi(v)$ satisfy conditions C_2, C_3 and C_7, C_8 respectively. Then the matrix $W(x_*, u_*)$ is nonsingular.*

We combine vectors x, u and define

$$z = \begin{pmatrix} x \\ u \end{pmatrix}, \quad z_0 = \begin{pmatrix} x_0 \\ u_0 \end{pmatrix}, \quad z_* = \begin{pmatrix} x_* \\ u_* \end{pmatrix}, \quad R(z) = \begin{pmatrix} D(\theta(x))\phi(L_x) \\ g(x) \end{pmatrix}.$$

Let $x(t, z_0), u(t, z_0)$ denote the solutions of the Cauchy Problem (65) with initial conditions $x_0 = x(0, z_0), u_0 = u(0, z_0)$. Using this notation, we rewrite the system of equations (65) as

$$W(z)\frac{dz}{dt} = -D(\gamma)R(z), \quad z(0, z_0) = z_0, \tag{67}$$

where γ has the first n components equal to α and all other components equal to τ. We denote $\gamma_* = \min[\alpha, \tau]$.

Theorem 10 *Suppose that the conditions of Lemma 7 hold. Then for any $\alpha > 0, \tau > 0$ the pair $z_*^T = [x_*^T, u_*^T]$ is an asymptotically stable equilibrium point of system (67). If the stepsize h_k is fixed and $0 < h_k < 2/\gamma_*$ then the discrete version*

$$z_{k+1} = z_k - h_k W^{-1}(z_k)D(\gamma)R(z_k) \tag{68}$$

locally converges to the point z_* *with at least linear rate. If* $W(z)$ *satisfies a Lipschitz condition in a neighborhood of* z_* *and* $h_k = \alpha = \tau = 1$, *then the sequence* $\{z_k\}$ *converges quadratically to* z_*.

According to Theorem 10 the convergence of methods (67), (68) is guaranteed only for a very restricted set of starting points. Consider the global behavior of method (67).

Lemma 8 *Let* x *be a regular point, and let the pair* $[x, u]$ *be such that* $L_x^i(x, u) \neq 0, x^i \neq 0,$ *for all* $1 \leq i \leq n,$ *and* $M(x, u)$ *is nonsingular. Then* $W(x, u)$ *is nonsingular.*

Define the nonnegative Lyapunov function

$$F(x, u) = \|D(\theta(x))\phi(L_x(x, u))\| + \|g(x)\|$$

and introduce two sets:

$$\Omega_0 = \left\{ [x, u] : F(x, u) \leq F(x_0, u_0), \quad x \geq 0_n, \quad L_x(x, u) \geq 0_n \right\},$$

$$\tilde{\Omega}_0 = \left\{ [x, u] \in \Omega_0 : \quad x > 0_n, \quad L_x(x, u) > 0_n \right\}.$$

Theorem 11 *Suppose that the set* Ω_0 *is bounded and contains the unique Kuhn-Tucker pair* $[x_*, u_*]$. *Suppose also that for any pair* $[x, u] \in \tilde{\Omega}_0$ *the conditions of Lemma 8 are satisfied. Then all trajectories of (67), starting from a pair* $[x_0, u_0] \in \tilde{\Omega}_0,$ *converge to* $[x_*, u_*]$.

Method (65) is fully applicable to linear programming problems (2) and (3). Formula (66) is simplified in this case:

$$W = \begin{pmatrix} M & -D(\theta)D(\dot{\phi})A^T \\ -A & 0_{mm} \end{pmatrix},$$

where $M = D(\dot{\theta}(x))D(\phi(v))$. If M and the matrix $\Gamma = AM^{-1}D(\theta)D(\dot{\phi})A^T$ are nonsingular, then we can use the Frobenius formula for the inverse matrix:

$$W^{-1} = \begin{pmatrix} M^{-1}[I_n - D(\theta)D(\dot{\phi})A^T\Gamma^{-1}AM^{-1}] & -M^{-1}D(\theta)D(\dot{\phi})A^T\Gamma^{-1} \\ -\Gamma^{-1}AM^{-1} & -\Gamma^{-1} \end{pmatrix}. \qquad (69)$$

The following theorem will guarantee the local convergence of algorithm (67), (68) in a neighborhood of the solution.

Theorem 12 *Let $[x_*, u_*]$ be a nondegenerate optimal pair for the linear programs (2) and (3). Assume that the functions $\theta(x)$ and $\phi(v)$ satisfy conditions C_2, C_3 and C_7, C_8 respectively. Then:*

1. *$W(x_*, u_*)$ is nonsingular;*

2. *$[x_*, u_*]$ is an asymptotically stable equilibrium pair of system (67), and trajectories $[x(t, z_0), u(t, z_0)]$ of (67) converge to $[x_*, u_*]$ on $\tilde{\Omega}_0$;*

3. *the discrete version (68) locally converges with at least a linear rate to the pair $[x_*, u_*]$;*

4. *if $W(z)$ satisfies the Lipschitz condition in a neighborhood of z_*, $h_k = \alpha = \tau = 1$, then the sequence $\{x_k, u_k\}$ converges quadratically to $[x_*, u_*]$.*

The most important feature of the proposed method is that, in contrast to numerous path-following methods, there exists a set of initial points such that the method solves Problems (2) and (3) in a finite number of iterates. It would be extremely interesting to determine the family of sets of starting points which ensures that the solution is reached after a prescribed number of iterations. We will show three such sets. For the sake of simplicity we consider here the case where the stepsizes are fixed and

$$h_k = \alpha = \tau = 1, \quad \theta(x) = x, \quad \phi(v) = v. \tag{70}$$

In this case algorithm (16) is rewritten as follows:

$$D(v_k)x_{k+1} - D(x_k)A^T u_{k+1} = -D(x_k)A^T u_k, \quad Ax_{k+1} = b. \tag{71}$$

Let T_k be a set of pairs $[x_0, u_0]$ such that, if $[x_0, u_0] \in T_k$, then algorithm (68) solves Problems (2) and (3) in k iterations.

Introduce the sets of indexes

$$\sigma(x) = \left\{ 1 \leq i \leq n : x^i = 0 \right\}, \quad \sigma(v) = \left\{ 1 \leq i \leq n : v^i = 0 \right\}.$$

If all components of x and v are non-zero, then we write $\sigma(x) = \sigma(v) = \emptyset$. We define the following three sets of pairs

$$\Omega_1 = \left\{ [x, u] : x = x_*, \ \sigma(x) \cap \sigma(v) = \emptyset \right\}, \quad \Omega_2 = \left\{ [x, u] : u = u_*, \ \sigma(x) \cap \sigma(v) = \emptyset \right\},$$

$$\Omega_3 = \left\{ [x, u] : \sigma(x) = \sigma(x_*), \ \sigma(x) \cap \sigma(v) = \emptyset \right\}.$$

Theorem 13 *Assume that all conditions of Theorem 12 hold. Let the transformation functions be linear. If $[x_0, u_0] \in \Omega_1$ or $[x_0, u_0] \in \Omega_2$, then method (68) yields an optimal solution in a single step. If $[x_0, u_0] \in \Omega_3$, then method (68) yields an optimal solution in at most two steps.*

Now we consider a particular case of the method in which the trajectories $x(t, z_0)$ and $v(t, z_0)$ belong to the positive orthant R_+^n. We define

$$V = \left\{ v \in R^n : \text{exists } u \in R^m \text{such that } v = c - A^T u \right\}. \tag{72}$$

We introduce the vectors $q \in R^n, p \in R^n$, whose i-th components are $\theta^i / \dot{\theta}^i, \phi^i / \dot{\phi}^i$ respectively and define $H = D^{1/2}(q) D^{-1/2}(p)$.

We say that a point $z \in R^{n+m}, z^T = [x^T, u^T]$ is interior, if $x \in R_{++}^n$ and $u \in \text{int} U$. If moreover $x \in \text{ri} X$ then the point z is strictly interior.

If z is an interior point and conditions C_2, C_3, C_7, C_8 are satisfied then H and $AH^2 A^T$ are nonsingular and we can use formula (69) for the inverse matrix. In this case method (67) can be written as

$$\frac{dx}{dt} = H \left(\tau(AH)^+ (b - Ax) - \alpha\pi(AH)Hp \right), \tag{73}$$

$$\frac{du}{dt} = \left(AH^2 A^T \right)^{-1} \left(\alpha Aq + \tau(b - Ax) \right).$$

Introduce a new matrix $\Lambda \in R^{d \times n}$. The columns of Λ^T forms a basis for the null space of A, i.e. $A\Lambda^T = 0_{md}$. Now definition (72) can be written equivalently as $V = \{ v \in R^n : \Lambda(v - c) = 0_d \}$ and from (73) we obtain

$$\frac{dx}{dt} = H \left[\tau(AH)^+ (b - Ax) - \alpha\pi(AH)Hp \right], \tag{74}$$

$$\frac{dv}{dt} = -A^T \frac{du}{dt} = H^{-1}(AH)^+ \left[\tau(Ax - b) - \alpha Aq \right]. \tag{75}$$

Let $x(t, z_0), u(t, z_0), v(t, z_0)$ denote the solutions of systems (73) and (74), (75) with initial conditions $x(0, z_0) = x_0 > 0_n, u(0, z_0) = u_0, v(0, z_0) = c - A^T u_0 = v_0 > 0_n$. In system (74), (75) we used derivatives $\dot{x}$ and $\dot{v}$ instead of $\dot{x}$ and $\dot{u}$ as we did before. We can do this if we are sure that the entire trajectory $v(t, z_0)$ belongs to V, i.e. for all $t \geq 0$ $v(t, z_0) - c$ belongs to the range space of A^T. Differentiating the d-dimensional vector Λv along the solutions of (74) and (75), we obtain $\Lambda \dot{v} = 0_d$. Therefore if $v_0 \in V$, then the trajectory $v(t, z_0)$ belongs to V for all $t \geq 0$ and we can use system (74), (75) for numerical calculations. The trajectories $x(t, z_0), v(t, z_0)$ of (74), (75) start from any positive point and go to the optimum as $t \to \infty$.

For the sake of simplicity we consider the case where $\phi(v) = v$ and $\theta(x)$ is a homogeneous function of x, i.e. $\theta(\beta x) = \beta^\lambda \theta(x), \lambda = \alpha/\tau$. Then according to Euler's formula we have $x = \lambda q(x), \lambda H^2 = D(x)D^{-1}(v)$. Introduce new vectors $\mu \in R^m, \eta \in R^n$ and a scalar function Φ:

$$\mu = \left(AD(x)D^{-1}(v)A^T\right)^{-1}b, \quad \eta = D^{-1}(v)A^T\mu, \quad \Phi(x,v) = x^Tv. \tag{76}$$

Let e be the vector of ones in R^n. Now systems (73) and (74), (75) can be written as

$$\frac{dx}{dt} = \tau D(x)[\eta - e], \quad \frac{du}{dt} = \alpha\mu, \quad \frac{dv}{dt} = -\alpha A^T\mu. \tag{77}$$

The condition $\dot{\theta}(0) > 0$ from C_3 is not necessarily fulfilled for homogeneous functions $\theta(x)$, nevertheless the following theorem is valid.

Theorem 14 *Let $[x_*, u_*]$ be a nondegenerate optimal pair for linear programs (2) and (3). Suppose that $D(\theta(x)) = D^\lambda(x), \phi(v) = v, \alpha > 0, \tau > 0$. Assume that Ω_0 is bounded. Let the starting point z_0 be interior, then the trajectories of (77) are such that:*

1. *the matrix $AD(x(t, z_0))D^{-1}(v(t, z_0))A^T$ is nondegenerate for all $t \geq 0$;*

2. *$z(t, z_0) \in \Omega_0$ and $v(t, z_0) \in V$ for all $t > 0$;*

3. *the objective function $b^Tu(t, z_0)$ of the dual problem increases monotonically;*

4. *the pair $[x(t, z_0), u(t, x_0)]$ is bounded and converges to $[x_*, u_*]$ as $t \to \infty$;*

5. *all components of vectors $D^\lambda(x(t, z_0))v(t, z_0), Ax(t, z_0)$ change monotonically and*

$$D^\lambda(x(t, z_0))v(t, z_0) = e^{-\alpha t}D^\lambda(x_0)v_0,$$
$$Ax(t, z_0) - b = e^{-\tau t}(Ax_0 - b).$$

From (77) we have

$$< \dot{x}, \dot{v} >= \alpha\tau\mu^T(Ax - b). \tag{78}$$

If $Ax_0 = b, x_0 > 0_n$, then $Ax(t, z_0) \equiv b$, and from (78) it follows that the vectors $\dot{x}$ and $\dot{v}$ are orthogonal. The vector $\dot{x}$ belongs to the nullspace of AH and $\dot{v}$ belongs to its orthogonal complement, $x(t, z_0) \in riX$ and $v(t, z_0) \in V$ for all $t \geq 0$. Hence the point $z(t, z_0)$ is strictly interior.

If we set in (77) $\alpha = 1, \tau = 0$ and $x = e$, then we have

$$\frac{dv}{dt} = -A^T(AD^{-1}(v)A^T)^{-1}b, \quad \frac{du}{dt} = (AD^{-1}(v)A^T)^{-1}b.$$

This method was introduced in [14]. It is similar to (59) and to the affine scaling algorithm which was proposed in [1].

Algorithm (77) has one important disadvantage connected with the necessity to know a starting point such that $v_0 \in intV$. It is possible to get rid of this restriction if we use the barrier-Newton method in the extended space of variables x, u, v. The simplest version of the method is described by following system of $2n + m$ differential equations

$$\frac{dx}{dt} = \tau G(A^T \zeta - c), \quad \frac{dv}{dt} = \alpha(c - v - A^T \zeta), \quad \frac{du}{dt} = \alpha(\zeta - u), \qquad (79)$$

where $\zeta = (AGA^T)^{-1}[b - Ax + AGc], G = D(x)D^{-1}(v)$. The essential difference in the requirements on the initial conditions between systems (77) and (79) is that in the latter case we impose only the simplest restrictions: $x_0 > 0_n, v_0 > 0_n$. If $v_0 = c - A^T u_0$, then this system is reduced to (77). More detailed consideration of this approach is given in [14].

If we set in (79) $\alpha = 0, \tau = 1, Ax = b$ and $v = e$ then we obtain (27).

By applying the Euler numerical integration method to system (77) we obtain the simplest discrete version of the method:

$$x_{k+1} = D(x_k)(e + \tau_k(\eta_k - e)), \quad v_{k+1} = D(v_k)(e - \alpha_k \eta_k), \quad u_{k+1} = u_k + \alpha_k \mu_k, \qquad (80)$$

where $x_0 > 0_n, v_0 > 0_n$, and μ_k and η_k are defined by (76).

As before, if $v_0 \in V$, then $v_k \in V$ for all k and the last formula in (80) can be omitted. The objective function $b^T u_k$ also increases monotonically. From (80) we obtain

$$Ax_{k+1} - b = (1 - \tau_k)(Ax_k - b), \qquad (81)$$

$$\Phi_{k+1} = (1 - \tau_k)\Phi_k + (\tau_k - \alpha_k)\mu_k^T Ax_k + \alpha_k \tau_k \mu_k^T (Ax_k - b). \qquad (82)$$

The iterates produced by algorithm (80) are well-defined if vectors x_k, v_k are strictly positive for all k. In order to ensure the positiveness of x_{k+1} and v_{k+1} we have to choose the step lengths α_k and τ_k such that

$$e \geq \alpha_k \eta_k, \quad e \geq \tau_k(e - \eta_k).$$

It is now straightforward to verify that non-negativity conditions hold if α_k and τ_k satisfy

$$0 < \alpha_k \leq \alpha_k^* = 1/[\eta_k^*]_+, \quad 0 < \tau_k \leq \tau_k^* = 1/[1 - \eta_*^k]_+,$$

where $[\alpha]_+ = \max[0, \alpha], \eta_k^*$ and η_*^k are maximal and minimal components of the vector η_k respectively. We will adopt the convention that, if $\eta_*^k \geq 1$, then $\tau_k^* = +\infty$, and, if $\eta_k^* \leq 0$, then $\alpha_k^* = +\infty$.

If we set $\alpha = \tau = 1$ and substitute $b = Ax$ in (80), then formulas (80) coincide with the primal-dual interior point algorithms proposed in [38], if in the latter we ignore the barrier (perturbation) term. In this algorithm the starting point z_0 must be strictly interior. Algorithm (80) does not require feasibility of starting and current points, but according to (81) it preserves feasibility. Another important advantage of algorithm (80) is that it permits us to take different step lengths in the primal space and in the dual space. This property is very useful, especially at the beginning of the computation when the starting point is far from the solution and only one maximal stepsize, either α_k^* or τ_k^*, is very large. This advantage disappears as $k \to \infty$ because the maximal step lengths tend to one.

We specify three classes of procedures for determining the step lengths:

1. step lengths are fixed and small enough, hence the discrete process (80) is close to a continuous one (77);

2. stepsizes are close to one and therefore the discrete process has properties of Newton's method;

3. stepsizes are chosen from steepest descent conditions or from another auxiliary optimization problem.

The convergence and polynomiality of one algorithm from the first class are investigated by G.Smirnov in [32]. Consider the second class. Define the step lengths by the formulas

$$\alpha_k = (1 - \varrho_k)\alpha_k^*, \quad \tau_k = (1 - \varrho_k)\tau_k^*. \tag{83}$$

Here we multiply the maximal stepsizes τ_k^*, α_k^* in primal and dual space by a safety factor $1 - \varrho_k$, $0 < \varrho_k < 1$. The choice $\varrho_k = 0$ corresponds to the case which allows steps to the boundary of the positive orthant and a loss of strict feasibility.

We consider the simplest step length rules, in which the sequence $\{\varrho_k\}$ is generated by:

$$0 < \varrho_k < 1, \quad \lim_{k \to \infty} \varrho_k = 0, \tag{84}$$

$$\varrho_k = \min[\varsigma \Phi_k, 1 - \delta], \quad 0 < \varsigma, \quad 0 < \delta < 1, \tag{85}$$

$$\varrho_k = \varsigma \Phi_k / [1 + \varsigma \Phi_k], \quad 0 < \varsigma. \tag{86}$$

All these choices guarantee that if $x_k > 0_n$ and $v_k > 0_n$, then the new primal and dual slack variables obtained from formulas (80) will remain strictly positive.

Theorem 15 *Let the conditions of Theorem 12 be satisfied. If the step lengths are determined by any rule (83)-(86), and the starting points are such that $x_0 > 0_n, v_0 > 0_n$, then the sequence $\{x_k, u_k\}$ generated by algorithm (80) converges locally and component-wise to the solution $[x_*, u_*]$ with at least superlinear convergence rate and*

$$\overline{\lim_{k \to \infty}} \frac{|x_{k+1}^i - x_*^i|}{|x_k^i - x_*^i|} = 0, \quad \overline{\lim_{k \to \infty}} \frac{|u_{k+1}^j - u_*^j|}{|u_k^j - u_*^j|} = 0, \quad 1 \le i \le n, 1 \le j \le m,$$

$$\lim_{k \to \infty} \alpha_k^* = 1, \quad \lim_{k \to \infty} \tau_k^* = 1, \quad \lim_{k \to \infty} \mu_k = 0_m, \quad \lim_{k \to \infty} \eta_k^* = 1, \quad \lim_{k \to \infty} \eta_*^k = 0.$$

The components of η_k converge to one or to zero. If rules (85) or (86) are used, then $\{x_k, u_k\}$ converge to $[x_*, u_*]$ quadratically.

The proof of this theorem is given in [14].

Consider the third approach to stepsize choice. We may view Φ_{k+1} and $\|Ax_{k+1} - b\|$ as functions of stepsizes α_k and τ_k. It is desirable to choose α_k, τ_k in order to minimize Φ_{k+1} and $\|Ax_{k+1} - b\|$. For solving this bicriterial minimization problem we can apply well-known methods which are developed in multiobjective analysis. Combining both objective functions in a unique criterion with the help of Chebyshev's convolution function, we obtain the following auxiliary minimization Problem

$$\min_{0 \le \alpha_k \le \omega \alpha_k^*} \min_{0 \le \tau_k \le \omega \tau_k^*} F(x_{k+1}, v_{k+1}),$$

where $F(x, v) = \max \{\|Ax - b\|, \Phi(x, v)\}, 0 < \omega < 1$.

This variant of the method is proved to be the most efficient.

6. Conclusion

In this paper we have shown that a great variety of numerical algorithms can be constructed on the basis of space transformation techniques. Generalization of this approach, computational aspects, optimal choice of a stepsize and applications to practical problems are beyond of the scope of the present paper. We aim to publish all these results in English in a near future. We hope that our approach adds new general insight to Karmarkar's algorithm which is so popular in the West.

7. Acknowledgements

The authors would like to express their sincere thanks to C.G.Broyden, M.C.Bartholomew-Biggs, E.Spedicato and Z.Huang for very valuable comments and suggestions. We have also received much helpful input from many other people, too numerous to list here, which we nevertheless gratefully acknowledge.

References

[1] Adler, I., Karmarkar, N., Resende, M. and Veiga, G. (1989) An implementation of Karmarkar's algorithm for linear programming, *Mathematical Programming* **44**, 297-335.

[2] Barnes, E. (1986) A variation on Karmarkar's algorithm for solving linear programming problems, *Mathematical Programming* **36**, 174-182.

[3] Bacciotti, A. (1992) Local stability of nonlinear control systems, Series on Advances in Mathematics for Applied Sciences **8**, World Scientific Publishing Co. Ptc. Ltd., Singapore.

[4] Bayer, D. and Lagarias, J. (1989) The nonlinear geometry of linear programming. Affine and projective scaling trajectories, *Trans. Amer. Math. Soc.* **314**, 499-526.

[5] Bertsekas, D. (1982) Constrained Optimization and Lagrange Multiplier Methods, Academic Press, New York.

[6] Dikin, I. (1967) Iterative solution of problems of linear and quadratic programming, *Sov. Math. Dokl.* **8**, 674-675.

[7] Evtushenko, Yu. (1974) Two numerical methods of solving nonlinear programming problems, *Sov. Math. Dokl.* **15**, 420-423.

[8] Evtushenko, Yu. (1985) Numerical Optimization Techniques. Optimization Software, Inc. Publications Division, New York.

[9] Evtushenko, Yu. and Zhadan, V. (1973) Numerical methods for solving some operations research problems, *U.S.S.R. Comput. Maths. Math. Phys.* **13**, 56-77.

[10] Evtushenko, Yu. and Zhadan, V. (1975) Application of the method of Lyapunov functions to the study of the convergence of numerical methods, *U.S.S.R. Comput. Maths. Math. Phys.* **15**, 96-108.

[11] Evtushenko,Yu. and Zhadan,V. (1978) A relaxation method for solving problems of non-linear programming, *U.S.S.R. Comput. Maths. Math. Phys.* **17**, 73-87.

[12] Evtushenko, Yu. and Zhadan, V. (1989) New approaches in optimization techniques, in H.-J. Sebastian and K. Tammer (eds.), Lecture Notes in Control and Information Sciences, 143, System Modelling and Optimization, Proc. of the 14th IFIP Conference, Springer-Verlag, Leipzig, pp. 23-37, .

[13] Evtushenko, Yu. and Zhadan, V. (1991) Barrier-projective and barrier-Newton numerical methods in optimization (the nonlinear programming case), Computing Center of the USSR Academy of Sciences, Reports at Comput. Math., (in Russian).

[14] Evtushenko, Yu. and Zhadan, V. (1992) Barrier-projective and barrier-Newton numerical methods in optimization (the linear programming case), Computing Center of the Russian Academy of Sciences, Reports in Comput. Math., (in Russian).

[15] Evtushenko, Yu. and Zhadan, V. (1992) The space transformation techniques in mathematical programming, in: P. Kall (ed.), Lecture Notes in Control and Information Sciences, 180, System Modelling and Optimization, Proc. of the 15th IFIP Conference, Springer-Verlag, Zurich, pp. 292-300.

[16] Evtushenko, Yu. and Zhadan, V. (1993) Stable barrier-projection and barrier-Newton methods in nonlinear programming. To appear in *Optimization Methods and Software*.

[17] Faybusovich, L. (1991) Dynamical systems which solve optimization problems with linear constraints, *IMA Journal of Mathematical Control and Information* **8**, 135-149.

[18] Faybusovich, L. (1991) Hamiltonian structure of dynamical systems which solve linear programming problems, *Physica D* **53**, 217-232.

[19] Fiacco, A. and McCormic, G. (1968) Nonlinear programming: Sequential unconstrained minimization techniques, John Wiley & Sons, New York.

[20] Gonzaga, C. (1992) Path following methods for linear programming, *SIAM Review* **34**, 167-224.

[21] Herzel, S., Recchioni, M. and Zirilli, F. (1991) A quadratically convergent method for linear programming, *Linear Algebra and its Applications*, 152, 255-289.

[22] Jarre, F., Sonnevend, G. and Stoer, J. (1988) An implementation of the method of analytic centers, Lect. Notes Control and Information Sci., 111, Springer-Verlag, New York.

[23] Kallio, M. (1986) On gradient projection for linear programming, Working paper 94, Yale School of Organization and Management.

[24] Karmarkar, N. (1984) A new polynomial-time algorithm for linear programming, *Combinatorica*, No. 4, 373-395.

[25] Kojima, M., Mizuno, S. and Yoshise, A. (1989) A primal-dual interior point method for linear programming, in N. Megiddo (ed.), Progress in Mathematical Programming - Interior Point and Related Methods, Springer-Verlag, Berlin, Chap. 2.

[26] Luenberger, D. (1973) Introduction to linear and nonlinear programming, Addison-Wesley Publishing Company, Reading.

[27] Megiddo, N. (1989) Pathways to the optimal set in linear programming, in N. Megiddo (ed.), Progress in Mathematical Programming - Interior Point and Related Methods, Springer-Verlag, Berlin, Chap. 8.

[28] Mehrotra, S. (1990) On the implementation of a (primal-dual) interior point method, Technical Report 90-03, Department of Industrial Engineering and Management Sciences, Northwestern University, Evanston, Illinois.

[29] Rosen, J. (1960) The gradient projection method for nonlinear programming, part 1, linear constraints, *SIAM J. Applied Math.* **8**, 181-217.

[30] McShane, K., Monma, C. and Shanno, D. (1989) An implementation of primal-dual interior point method for linear programming, *ORSA J. Comput.* **1**, 70-89.

[31] Smirnov, G. Convergence of barrier-projection methods of optimization via vector Lyapunov functions, to appear in *Optimization Methods and Software.*

[32] Smirnov, G. Complexity of the barrier-Newton method in linear programming, to appear in *Optimization Methods and Software.*

[33] Tanabe, K. (1980) A geometric method in nonlinear programming, *Journal of Optimization Theory and Applications* **30**, 181- 210.

[34] Vanderbei, R., Meketon, M. and Freedman, B. (1986) A modification of Karmarkar's linear programming algorithm, *Algorithmica* **1**, 395-407.

[35] Wei Zi-luan, (1987) An interior point method for linear programming, *Journal of Computing Mathematics*, Oct., 342-350.

[36] Ye, Y. Tapia, R. and Zhang, Y. (1991) A superlinearly convergent $O(\sqrt{n}L)$-iteration algorithm for linear programming, Technical Report TR91-22, Rice University, Houston, Texas.

[37] Zhadan, V. (1980) On two classes of methods for solving nonlinear programming problems, *Sov. Math. Dokl*, **22**, 388-392.

[38] Zhang, Y. Tapia, R. and Dennis, J. (1990) On the superlinear and quadratic convergence of primal-dual interior point linear programming algorithms, Technical Report TR90-6, Rice University, Houston, Texas.

Large-scale Nonlinear Constrained Optimization: a Current Survey

Andrew R. Conn

IBM T.J. Watson Research Center, Yorktown Heights, USA

Nick Gould

CERFACS, Toulouse, France, EC

Philippe L. Toint

Department of Mathematics, FUNDP, Namur, Belgium, EC

Abstract. Much progress has been made in constrained nonlinear optimization in the past ten years, but most large-scale problems still represent a considerable obstacle.

In this survey paper we will attempt to give an overview of the current approaches, including interior and exterior methods and algorithms based upon trust regions and line searches. In addition, the importance of software, numerical linear algebra and testing will be addressed. We will try to explain why the difficulties arise, how attempts are being made to overcome them and some of the problems that still remain.

Although there will be some emphasis on the LANCELOT and CUTE projects, the intention is to give a broad picture of the state-of-the-art.

Keywords: Large-scale, constraints, nonlinear optimization.

1   Introduction

We shall first state the most general form of the problem that we are addressing, namely

$$\text{minimize}_{x \in \Re^n} \ f(x) \tag{1.1}$$

subject to the general (possibly nonlinear) inequality constraints

$$c_j(x) \leq 0, \quad 1 \leq j \leq l, \tag{1.2}$$

---

[0]This research was supported in part by the Advanced Research Projects Agency of the Department of Defense and was monitored by the Air Force Office of Scientific Research under Contract No F49620-91-C-0079. The United States Government is authorized to reproduce and distribute reprints for governmental purposes notwithstanding any copyright notation hereon.

E. Spedicato (ed.), *Algorithms for Continuous Optimization*, 287–332.

to the (possibly nonlinear) equality constraints

$$c_j(x) = 0, \quad l+1 \le j \le m, \tag{1.3}$$

and the simple bounds

$$l_i \le x_i \le u_i, \quad 1 \le i \le n. \tag{1.4}$$

Here, f and the c_j are all assumed to be twice-continuously differentiable and any of the bounds in (1.4) may be infinite.

We only expect to obtain local minimizers. This presents no problems in convex programming, where all local minima are indeed global (for example, in linear programming), but even for small, general nonlinear programming problems it is usually extremely difficult to verify globality. For large problems, with current techniques it is practically impossible. Fortunately, in many situations, an algorithm that determines local optima suffices.

Our primary interest here is in problems that involve a large number of variables and/or constraints. Consequently, it seems worthwhile to elaborate as to what we mean by large. Firstly, this notion is clearly *computer dependent*. Secondly, the notion of size is *problem dependent*. A highly nonlinear problem in one hundred variables could be considered large, whereas in linear programming it is possible to solve problems in five million variables. Similarly, it also depends upon the *structure of the problem*. Many large-scale nonlinear problems arise from the modeling of very complicated systems that may be subdivided into loosely connected subsystems. This structure may often be reflected in the mathematical formulation of the problem and exploiting it is often crucial if one wants to obtain an answer efficiently. The complexity of the structure is often a key factor in assessing the size of a problem. In addition, the notion of a large problem depends upon the *frequency* with which one expects to solve a particular instance or closely related problem. When one anticipates solving the same class of problems many times, one can afford to expend a significant amount of energy analyzing and exploiting the underlying structure.

Thus, although it is not possible to state categorically that a problem in, say, seven hundred variables is large, suffice it to say that, today, a problem in fifty variables is small and a generally nonlinear problem in five thousand variables and one thousand nonlinear constraints is large.

At the risk of stating the obvious, the world is not linear and *accurate* modeling of physical and scientific phenomena often leads to large-scale nonlinear optimization. In our opinion, the frequent use of linear models is not an indication that nonlinear problems do not abound. Rather, it is a statement of the desire to use an algorithm (the simplex method) that is readily understood and is well-known to be suitable for large problems. We would like to convince you that you should consider solving nonlinear programs when they are more appropriate. It should be emphasized that solutions to large nonlinear problems on moderate workstations in a reasonable amount of time are currently quite possible. Furthermore, in practice one is often only seeking improvement rather than assured optimality (another reason why local solutions may suffice). This fact makes problems that at first sight seem impossible (for example, control problems that one wishes to solve in something like real-time), tractable.

In the past twenty years rather sophisticated and reliable techniques for small-scale problems have been developed (see Chapters 1 and 3 of Nemhauser *et al.*, 1989, and the chapters of Bartholomew-Biggs and Fletcher in this volume). However, efficient algorithms for small-scale problems do not necessarily translate into efficient algorithms for large-scale problems (see, for example, Bartholomew-Biggs and Hernandez, 1994). Thus, it is not adequate to take existing optimization software for small problems and apply it to large ones, hoping that the increased capacity in computing will take care of the growth in problem size. By contrast, we could expect that an efficient method for large-scale problems be at least moderately efficient for small-scale problems. Notwithstanding, it is essential to know and understand the small-scale background.

Without a doubt, the availability of powerful workstations and supercomputers (both parallel and sequential) has encouraged research in algorithms for large-scale problems, but the main reason we can solve very large problems is because we can *exploit structure*. Moreover, the state-of-the-art of large-scale nonlinear programming has progressed so much in the past decade that it is reasonable to ask the question 'Is it worthwhile to design algorithms that are unsuitable for large-scale problems?' At present, the answer is likely to be in the affirmative, for example, for problems where the cost of function evaluations is very high or for problems with extremely nonlinear behaviour and/or difficult scaling.

Furthermore, some of the more mundane tasks, such as the input of problems, are important and non-trivial issues. The evaluation of results is even more important and difficult. The scope of some of the problems tackled by LANCELOT and included in CUTE (see below for more details on these two packages) contains a large number of nonlinear optimization problems of various sizes and difficulty, representing both 'academic' and 'real world' applications. Both constrained and unconstrained examples are included. The problems we have solved to date using LANCELOT range from problems with 20,000 variables and 10,000 nonlinear constraints to small problems with less than 10 variables and constraints. It is worth mentioning that some of the most difficult problems are small (for example, LANCELOT has been unable to solve a problem with 149 variables, a quadratic objective function and 100 nonlinear constraints). It is also worth stating that although LANCELOT was designed with large-scale problems in mind, it is very suitable for solving small-scale problems.

Of course, there are many details that can contribute to the difficulty of a problem. Unfortunately, none of us are good at handling them all. Scaling is a well-known difficulty for which one has methods to try, but it is clear that we would like to be able to do much better. There are approaches that are usually effective in handling indefiniteness, but here again one feels that these are far from ideal. Both primal and dual degeneracy are often perceptible as difficulties. It is not always clear as to how they can best be tackled.

There is a very real difficulty associated with the fact that many practitioners prefer good solutions to bad models rather than less good solutions to more accurate (and thus better and probably more complicated) models. Indeed a related problem that has frequently been an unwelcome accompaniment to nonlinear optimization is that the user of the software needs to be relatively sophisticated.

Finally, there are all the problems related to solving systems of symmetric linear equations, since this is, in many ways, the kernel computation in nonlinear optimization.

2   Basic Background

Although much of the fundamental background is covered in this volume by the contributions of Bartholomew-Biggs (1994) and Fletcher (1994), there are some very basic comments that relate to large-scale optimization that we would like to mention here.

Firstly, the most basic approach to unconstrained optimization is undoubtedly steepest descent. From the point of view of storage, this is a splendid method for large-scale optimization. However, it is intolerably slow since its convergence rate is linear with a rate constant that may be uncomfortably close to one. The other extreme is a safeguarded Newton's method, which has a second-order convergence rate. But in this case, the standard implementation requires too much storage ($O(n^2)$) and too much work per iteration ($O(n^3)$ flops). In fact, much of what we need to concern ourselves with is how to do as little as possible initially (steepest-descent-like) and enough eventually to guarantee an acceptable convergence rate (Newton-like). In effect, this is the standard problem of global versus asymptotic behavior, since the weak behavior of steepest descent is enough to guarantee global convergence (convergence to a stationary point from any starting point).

One effective technique for large structured problems (mentioned in Section 4 of Fletcher, 1994) is intelligent finite differencing (originally due to Curtis *et al.*, 1974). However, the standard steepest descent/Newton's method compromise is quasi-Newton methods. Once again details are given in Fletcher (1994), but essentially the idea is to use low rank updates to an initial approximation to the Hessian matrix (usually a (scaled) identity matrix). These methods possess a sufficiently fast (superlinear) convergence rate. The updates can be posed as minimization problems. For example, PSB (see Powell, 1970) may be determined from

$$\underset{U \in \Re^{n^2}}{\text{minimize}} \ \ \|U\|_F^2 = \sum_{i=1}^{n} \sum_{j=1}^{n} U_{ij}^2 \qquad (2.1)$$

subject to the quasi-Newton equations

$$
\begin{aligned}
U\delta &= \gamma - B\delta \\
U &= U^T,
\end{aligned}
\tag{2.2}
$$

where

$$
\begin{aligned}
B^+ &= B + U \\
\delta &= x^+ - x \\
\gamma &= g^+ - g.
\end{aligned}
\tag{2.3}
$$

Here the superscript $+$ indicates an update, B is the Hessian approximation and g is the gradient of f.

A natural extension to structured problems is to impose sparsity by considering (see Toint, 1981a)

$$
\underset{U \in \Re^{n^2}}{\text{minimize}} \ \|U\|_F^2
\tag{2.4}
$$

subject to the constraints

$$
\begin{aligned}
U\delta &= \gamma - B\delta \\
U &= U^T \\
\text{and } U_{ij} &= 0, ij \in S,
\end{aligned}
\tag{2.5}
$$

where S specifies the sparsity pattern.

Unfortunately this approach has not turned out to be very successful in practise (see Sorensen, 1981). On the other hand, the quasi-Newton approach can be successfully applied to large-scale problems if the partially separable structure of the problem (see below) is taken into account. If quasi-Newton methods are preferred to exact second derivatives[1], it is thus possible to approximate the Hessian of each element function f_i individually, using a secant equation of the type (2.3) for each one of them. This technique is called 'partitioned updating' and was introduced by Griewank and Toint (1982b). This technique is substantially more successful than the sparse updating method just described and is provided as an option within the LANCELOT package.

Another compromise between steepest descent and Newton's method is the method of conjugate directions. In the large-scale case we tend to think of it as closer to steepest descent, but in some contexts (good preconditioners, for example) it may be closer to Newton's method. Steepest descent with the inverse of a positive definite Hessian as preconditioner is indeed Newton's method.

Conjugate direction methods maintain finite Q-convergence (that is, converge for a positive definite quadratic problem in a finite number of iterations; no more than n, the dimension of the space). This is not really very relevant for large-scale problems, where n is large. One can think of conjugacy as a generalization of orthogonality[2].

---

[1]In our experience, this is very seldom necessary.

[2]A set of directions $\{d_i\}_1^k$ are G-conjugate for positive definite G if and only if $d_i^T G d_j = \delta_{ij}$, where δ_{ij} is the Kronecker delta.

Thus it is not that surprising that these directions can be derived via Gram-Schmidt orthogonalization, either as three-term recurrences or using Lanczos orthogonalization, although some care has to be taken to make the process numerically stable (see, for example, Golub and Loan, 1989, Chapter 9). As a consequence of these recurrences, conjugate direction algorithms can be implemented storing only a few vectors (three to five, depending on the precise method used). With exact line searches and exact arithmetic, the method is n-step superlinearly convergent, in general. The proof depends critically upon restarts — otherwise convergence is linear. In practise they nearly always converge linearly, but for large n, n-step superlinear is not much better. Of course, what one wants is a fast linear rate, which preconditioning can achieve.

If we think of Lanczos as

$$Q^T A Q = T, \tag{2.6}$$

where T is tridiagonal and Q is the matrix whose columns are the Lanczos vectors, then the process can also be carried out in block form with T block triangular. One can then work with the blocks separately and exploit a parallel environment (Nash and Sofer, 1991).

Another way to reduce storage is to use limited memory methods. For example Liu and Nocedal (1989) (see also Liu and Nocedal, 1988), use an inverse BFGS update in the form

$$B^+ = V^T B V + \rho \delta \delta^T, \tag{2.7}$$

where $\rho = 1/\gamma^T \delta$ and $V = I - \rho \gamma \delta^T$. The basic idea is to start with a B that can be stored efficiently, for example a scaled version of the identity matrix. One then updates m times, however without storing the updated matrices *explicitly* but instead storing the m pairs γ and δ. Most importantly m is typically very small, say five. The scaling of the initial matrix is also important. Other recent references include Byrd *et al.* (1993) and Zou *et al.* (1993).

However, it is unclear as to whether the relative success of naive preconditioners, limited memory with small m and naive scaling of the identity matrix are mostly a consequence of the not very extensive testing that has been carried out to date. In particular, most problems tested seem to be rather well scaled.

2.1 Solving the Linear System

Typically, the major computational task in optimization is to solve a system of linear equations that arises from the fact that one uses quadratic models and stationary points are characterized by gradients being zero. In addition, optimality conditions and/or reduced methods for constrained problems give rise to (generalized) least squares problems and linear systems involving the Karush-Kuhn-Tucker matrix. Thus progress in solving large linear systems has implications for large-scale optimization (see, for example, the contribution of Björck, 1994, in this volume). If the system is written

$$Bd = -g \qquad (2.8)$$

then ideally, we would like to combine the determination of B with the solution of (2.8). If possible we would choose the matrix of exact second derivatives (maybe in a reduced space) for B. As we will see later, this often can be done if structure is suitably exploited. The linear system can then be solved using direct or iterative techniques.

We first consider direct methods. There are two main approaches, namely multifrontal techniques and sparse Cholesky factorizations. Very briefly, the former approach tries to assemble the required entries in a piecemeal manner. Once a complete column and row are assembled one can do the corresponding elimination, thus building up the corresponding elements of L and U. For details see Conn *et al.* (1993a), Duff *et al.* (1986, Chapter 10), Duff *et al.* (1988), Duff and Reid (1982), Duff and Reid (1983) and Duff and Reid (1993). By contrast, the sparse Cholesky factorization primarily tries to order the rows and columns of B whilst maintaining reasonable stability by including the possibility of adding appropriate quantities to the diagonals of B, if necessary, (Chapter 3 of Conn *et al.*, 1992b, Gill and Murray, 1974, Gill *et al.*, 1992, Schlick, 1993 and Schnabel and Eskow, 1991). For example, Schnabel and Eskow use Gerschgorin bounds to determine the amount to add to the diagonal. They choose diagonal pivots and change the diagonal as little as is reasonable in order to maintain sufficient positive definiteness. All the proposed methods use about $O(n^2)$ additional work as compared with standard Cholesky. It is interesting to remark that these methods are related to l_2 trust region/Levenberg Marquardt algorithms, although the latter are using a rank n update rather than the normally considerably lower rank updates used above, (Hebden, 1973, Levenberg, 1944, Marquardt, 1963 and Moré, 1978).

The iterative method of choice is that of (preconditioned) conjugate gradients. Thus we need to solve (2.8), where B is a (possibly perturbed) approximation to the Hessian matrix $\nabla_{xx} f$. The perturbation may be obtained as the conjugate gradient algorithm proceeds in what we think is an elegant way that preserves conjugacy, see Arioli *et al.* (1993).

3   Some Existing Methods

Let us first consider the most venerable and best known nonlinear optimization algorithm that was designed with large-scale problems in mind. The origins of MINOS (Murtagh and Saunders, 1987) come from Robinson (1972) and Rosen and Kreuser (1972). The method can be considered to be an extension of the simplex method, since both are a reduced gradient technique. Thus the structure exploited is *sparsity* and the essential technology used is closely related to the linear programming technology of the simplex method. In particular, MINOS replaces

$$\text{minimize} \quad F(x) + c^T x + d^T y$$
$$x \in \mathbf{R}^n, y \in \mathbf{R}^m$$
$$\text{subject to} \quad f(x) + A_1 y = b_1$$
$$A_2 x + A_3 y = b_2 \tag{3.1}$$
$$\text{and} \quad l_x \leq x \leq u_x$$
$$l_y \leq y \leq u_y$$

with

$$\text{minimize} \quad F(x) + c^T x + d^T y + \lambda_k^T(f(x) - \tilde{f}(x)) + \tfrac{1}{2}\rho(f(x) - \tilde{f}(x))^T(f(x) - \tilde{f}(x))$$
$$x \in \mathbf{R}^n, y \in \mathbf{R}^m$$

$$\text{subject to} \quad \tilde{f}(x) + A_1 y = b_1$$
$$A_2 x + A_3 y = b_2$$
$$\text{and} \quad l_x \leq x \leq u_x \tag{3.2}$$
$$l_y \leq y \leq u_y,$$
$$\text{where} \quad \tilde{f}(x) = f(x_k) + J_k(x - x_k),$$

and J_k denotes the Jacobian of f evaluated at x_k. We note that the nonlinear contributions to the constraints are linearized. One then formulates a quadratic model for the corresponding augmented Lagrangian objective function (see Fletcher, 1987, Section 12.2). Writing the activities that are determined by the general linear constraints as

$$\hat{A}x = (\ B \quad S \quad N\ )x = b, \tag{3.3}$$

a basis for the null space is given by the columns of the matrix Z, where

$$Z^T = \left(-[B^{-1}S]^T \quad I \quad 0 \right). \tag{3.4}$$

This follows directly from the fact that

$$\hat{A}Z = 0 \text{and} (0 \quad 0 \quad I)Z = 0. \tag{3.5}$$

Since most of the computation in the outline above involves the inverse of the basis matrix, B^{-1}, it is hardly surprising that exploitation of structure in this algorithm mimics exploitation of the same structure in the simplex method.

More recent methods that are closely related to sequential quadratic programming (see Bartholomew-Biggs, 1994, Section 5, for a general description) are what Gill *et al.* (1993b) call transformed Hessian methods (see also Eldersveld, 1992). Thus consider the problem

$$\text{minimize} \quad f(x) \tag{3.6}$$
$$x \in \Re^n$$

subject to

$$c_j(x) \geq 0, \quad 1 \leq j \leq l, \tag{3.7}$$

and the positivity constraints

$$x_i \geq 0 \quad 1 \leq i \leq n. \tag{3.8}$$

They then try to find $(\delta_x, \delta_\lambda)$ by minimizing a *quadratic* model of the Lagrangian subject to a *linear* model of the constraints (3.7). For large problems the efficiency of the linear algebra required to solve the created quadratic program is crucial. One has to repeatedly solve a linear system with the Karush-Kuhn-Tucker matrix

$$\begin{pmatrix} B^{(k)} & A_\omega^T \\ A_\omega & 0 \end{pmatrix}, \text{where} A_\omega = \begin{pmatrix} A \\ I \end{pmatrix}. \tag{3.9}$$

It is worth remarking that solving such systems has general applicability to problems with linear constraints (see, for example, Arioli *et al.*, 1993 and Forsgren and Murray, 1993). Gill, Murray and Saunders use generalized TQ factorizations with

$$A_\omega Q = (0T) \text{ and } Q^T H Q = R^T R. \tag{3.10}$$

Now, the Hessian H required for the gradient of the quadratic program's objective function can be determined from

$$H = Q^{-T} R^T R Q^{-1}. \tag{3.11}$$

The solution to the quadratic program is completely determined by the upper triangular matrix T, the matrix Q and the first $n - t$ rows of the upper trapezoidal matrix R. If we let Z denote the first $n - t$ columns of Q and call the remaining columns of Q, Y, then $Z^T H Z$ is the usual *reduced Hessian* and $Q^T H Q$ is the *transformed Hessian*. Furthermore, Y spans the range space associated with A_ω.

In order to avoid changing A_ω, one adds slacks explicitly and the trick is to choose Q's that are relatively easily invertible, because of the need for (3.11). Moreover, only a part of R need be stored and one can arrange not to lose the structure in H that results from the additional slacks by permuting A_ω appropriately. One can think of this as being a non-orthogonal (and thus appropriate for large-scale) version of NPSOL (Gill *et al.*, 1986).

The above approaches are line-search based. There are also excellent algorithms that are trust-region based. Once again these are mentioned in Fletcher (1994, Section 1) and further details and references are given in Moré (1983). Consider first the unconstrained problem.

The salient features we wish to recall here is that one uses a suitable *model* for the objective[3] that one *trusts* over a suitable region[4]. One then compares the actual reduction with the predicted reduction. If the comparison is sufficiently favorable, the trust region is expanded and the current point is updated. If it is sufficiently unfavorable, the trust region is reduced and the current point is unchanged. Otherwise,

---

[3]e.g., a quadratic model given by the second-order Taylor's expansion about the current point
[4]e.g., a sphere or box

only the current point is updated. Continuity guarantees that eventually reduction of the trust region must ensure that the predicted reduction is close enough to the actual reduction, which in turn guarantees that the trust region is bounded away from zero. Global convergence is assured as long as we do as well as minimizing the model, within the trust region, along the steepest descent direction (which defines the Cauchy point). Eventually the trust region is irrelevant, which guarantees a fast asymptotic convergence rate as long as the underlying model optimization is suitably chosen (for example, a safe-guarded Newton-like method).

The generalization to simple bounds is straightforward. For example, if one uses the l_∞ norm, then the trust region is a box. The feasible region corresponding to simple bounds is also a box. The intersection of two boxes is a box. One now defines a generalized Cauchy point as the minimum along the *projected* gradient path within the trust region, where the projection is with respect to the simple bounds. Since we are dealing with boxes the projection is trivial. Such a projected gradient approach was proposed by McCormick (1969), and independently in Bertsekas (1982) and Levitin and Polyak (1966). More recently it has been exploited extensively in the context of large-scale optimization by many authors, see for example Conn *et al.* (1988b), Dembo and Tulowitski (1983), Moré and Toraldo (1989), and Moré and Toraldo (1991). As in the unconstrained case, global convergence can be guaranteed, provided one does at least as well as the generalized Cauchy point. One obtains better convergence, and ultimately a satisfactory asymptotic convergence rate, by further reducing the model function. This is the trust region basis for the kernel algorithm SBMIN (Conn *et al.*, 1988a) of LANCELOT (Conn *et al.*, 1992b). It can be summarized as follows:

- Find the generalized Cauchy point based upon a local (quadratic) model.

- Fix activities to those at the generalized Cauchy point.

- (Approximately) solve the resulting reduced problem whilst maintaining account of the trust region and bounds.

- Determine whether the current point is acceptable and update the trust region radius accordingly.

The supporting theory in Conn *et al.* (1988a) verifies that the algorithm converges to a first-order stationary point, provided the quadratic model is reasonable. Moreover, the correct activities are identified after a finite number of iterations if strict complementarity[5] is satisfied and the activities determined by the generalized Cauchy point are kept active when the model is further reduced in the inner iteration.

What makes this approach particularly attractive for large-scale problems is that the determination of the generalized Cauchy point is easy (and need not be exact) and one can use suitable *unconstrained* large-scale techniques. An example would

---

[5]The case where strict complementarity fails to hold is considered by Lescrenier (1991).

be truncated, preconditioned conjugate gradients (see, for example, Steihaug, 1983a, Steihaug, 1983b and Toint, 1981b). Furthermore, often one is able to exploit the structure in order to use exact second derivatives (see below). Usually one never needs the Hessian matrix (or its approximation) but rather the corresponding matrix-vector products. Here again it is possible to exploit structure. The standard structure to exploit is sparsity and this is basic to large-scale numerical linear algebra, see for example Duff *et al.* (1986) and George and Liu (1981). In addition, most of the improvements in the simplex method have depended upon such exploitation. LANCELOT exploits a more general form of structure. The basic idea was first introduced in Griewank and Toint (1982a). We introduced a slight generalization, exploiting this very pervasive type of structure, which we call *group partial separability*. Consider two different functions, $f_1(x) = x_{50}^4$ and $f_2(x) = \left[\sum_{i=1}^{5,000,000} x_i\right]^4$, where $x \in \Re^{5,000,000}$. We first note that $\nabla_{xx} f_1$ is very sparse[6] and $\nabla_{xx} f_2$ is completely dense. However, the important structure to note is that both functions have an invariant subspace of dimension $4,999,999$. If we use the linear transformation $w = e^T x$, where e is the vector of ones, then $f_2(x)$ is transformed to w^4. Imagine having sums of such functions, not necessarily independent. Then you have the fundamental idea. Moreover, it is not unusual to have many similar f_i's with just different labellings. In fact the economies of storage are such that often one is able to solve quite large problems on small machines.

A function $f(x)$ is said to be *group partially separable* if:

1. the function can be expressed in the form

$$f(x) = \sum_{i=1}^{n_g} g_i(\alpha_i(x)); \qquad (3.12)$$

2. each of the *group functions* $g_i(\alpha)$ is a twice continuously differentiable function of the single variable α;

3. the function

$$\alpha_i(x) = \sum_{j \in \mathcal{J}_i} w_{i,j} f_j(x^{[j]}) + a_i^T x - b_i \qquad (3.13)$$

is known as the i-th *group*;

4. each of the index sets $\mathcal{J}_i$ is a subset of $\{1, \ldots, n_e\}$;

5. each of the *nonlinear element functions* f_j is a twice continuously differentiable function of a subset $x^{[j]}$ of the variables x. Each function is assumed to have a large invariant subspace. Usually, this is manifested by $x^{[j]}$ comprising a small fraction of the variables x;

---

[6]It has only one non-zero entry.

6. the gradient a_i of each of the *linear element functions* $a_i^T x - b_i$ is, in general, sparse; and

7. the $w_{i,j}$ are known as element *weights*.

An additional degree of freedom may be present in a partially separable structure. Often a distinction can be made between the *elemental* variables (the problem's variables that effectively occur in the expression of the considered element) and *internal* variables associated with a given element[7]. A more thorough introduction to group partial separability is given by Conn *et al.* (1990a). SBMIN assumes that the objective function $f(x)$ is of this form.

To summarize, we now know that LANCELOT is trust-region based, uses SBMIN as its kernel algorithm and exploits structure via group partial separability. We now explain how it is extended to handle general equality constraints. Inequalities are changed to equalities by the addition of slacks. Like MINOS it uses the augmented Lagrangian, which we can think of as a Lagrangian with additional quadratic (exterior) penalty terms.

The objective function and general constraints are combined into a composite function, the *augmented Lagrangian function*,

$$\Phi(x, \lambda, \mu) = f(x) + \sum_{i=1}^{m} \lambda_i c_i(x) + \frac{1}{2\mu} \sum_{i=1}^{m} c_i(x)^2, \qquad (3.14)$$

where the components λ_i of the vector λ are known as *Lagrange multiplier estimates*, and μ is known as the *penalty parameter*.

The constrained minimization problem (1.1), (1.3) and (1.4) is now solved by finding approximate minimizers of Φ for a carefully constructed sequence of Lagrange multiplier estimates, constraint scaling factors and penalty parameters.

The approach can be summarized as

- Test for convergence.
  Convergence occurs when the iterate is sufficiently stationary (i.e., the projected gradient of the augmented Lagrangian with respect to the simple bounds is small enough) and the current approximate minimizer of Φ is sufficiently feasible.

- Major iteration.
  Use the simple bounds algorithm SBMIN to find a sufficiently stationary approximate minimizer of Φ considered as a function of x and constrained explicitly by the simple bounds.

- Choice of update.
  If sufficiently feasible, *update multipliers* and *decrease* the tolerances for feasibility and stationarity.

---

[7]For example, in f_2 above, one could consider w to be an internal variable and the x's to be the elemental variables.

Otherwise, *decrease the penalty parameter* and *reset* the tolerances for feasibility and stationarity.

We use first-order updates for the multipliers, namely

$$\lambda_i^+ = \lambda_i + c_i(x^k)/\mu. \tag{3.15}$$

Reset and update rules for the multipliers, stationarity, feasibility and the penalty parameter are all analyzed in the theory of Conn *et al.* (1991) and Conn *et al.* (1992d). There we are able to show that under suitable conditions there is convergence to a first-order stationary point for the nonlinear programming problem. Furthermore, if we have a single limit point, the algorithm eventually stops reducing the penalty parameter, μ. Finally, under somewhat stronger conditions, one ultimately requires only a single iteration of the simple bounds algorithm to satisfy stationarity for the outer iteration. This, plus many options, is the state-of-the-art of LANCELOT A.

4   A Testing Environment

It is not that astonishing that during our research we were soon led to the frustrating question of testing and evaluating algorithms for large-scale nonlinear optimization. Moreover, there is a rapid appreciation of how difficult this task is — hence the dearth of published nonlinear results obtained with MINOS, even though it has been available for over fifteen years.

The origin of our so-called standard input format (SIF) in LANCELOT was that the setting up of test problems that accounted for the group partially separable structure was tremendously tiresome. Group partial separability simplifies the optimization but complicates the input. Conn *et al.* (1992b, Chapter 2) provide an introduction to the SIF, including the considerations given to its design. Chapter 7 of the same reference serves as a detailed manual on the format.

Additional requirements in a suitable testing environment include

- a large database of test problems and a means of managing it,

- the ability to compare results with the best of the existing optimization packages,

- facilities to test algorithmic ideas on the collection of problems, and finally

- making this all freely available to the community.

Hence the Constrained and Unconstrained Testing Environment of CUTE (Bongartz *et al.*, 1993). This offers a large growing database of test problems written in SIF. The test set covers, amongst others,

- the 'Argonne test set' (Moré et al., 1981), the Testpack report (Buckley, 1989), the Hock and Schittkowski collection (Hock and Schittkowski, 1981), the Dembo network problems (Dembo, 1984), the Moré-Toraldo quadratic problems (Moré and Toraldo, 1991), the Boggs-Tolle problems (Boggs and Tolle, 1989), the Toint-Tuyttens network model problems (Toint and Tuyttens, 1990), and Gould's quadratic programming problems (Gould, 1991),

- most problems from the PSPMIN collection (Toint, 1983),

- problems inspired by the orthogonal regression report by Gulliksson (Gulliksson, 1990),

- some problems from the Minpack-2 test problem collection (Averick et al., 1991, Averick and Moré, 1991) and from the second Schittkowski collection (Schittkowski, 1987) and

- a large number of original problems from a variety of application areas.

Each problem comes with a classification listing the type of problem, degree of available derivatives, origin and size. There are tools provided to create, maintain and update the classification database and also to select problem SIF files on the basis of the classifications. Furthermore, we realize that not everyone, especially non-users of **LANCELOT**, is equally enthusiastic about using partial separability and the SIF. However, the database of test problems provided by **CUTE** is clearly very useful. Thus **CUTE** provides tools to allow an interface between problems, specified using the SIF, and other existing nonlinear programming packages, in addition to providing a relatively easy means of building interfaces with new algorithms. When applicable these tools are provided in sparse and dense formats.

At the present time, interfaces are available for the following:

- **MINOS (see above)**
  We currently have interfaces for MINOS 5.3 and MINOS 5.4.

- **NPSOL of Gill et al. (1986)**
  This package is designed to minimize smooth functions subject to constraints, which may include simple bounds, linear constraints, and smooth nonlinear constraints. The software uses a sequential quadratic programming algorithm, where bounds, linear constraints and nonlinear constraints are treated separately. Unlike MINOS, NPSOL stores all matrices in dense format, and is therefore not intended for large sparse problems.

- **OSL of International Business Machines Corporation (1990)**
  This package obtains solutions to quadratic programming problems where the Hessian matrix is assumed positive semidefinite. It is intended to be suitable for large-scale problems.

- **TENMIN of Schnabel and Chow (1991)**
  This package is intended for problems where the cost of storing one n by n matrix (where n is the number of variables), and factoring it at each iteration, is acceptable. The software allows the user to choose between a tensor method for unconstrained optimization, and an analogous standard method based upon a quadratic model. The tensor method bases each iteration upon a specially constructed fourth-order model of the objective function that is not significantly more expensive to form, store, or solve than the standard quadratic model.

- **UNCMIN of Koontz *et al.* (1985) that corresponds closely to the pseudocode in Dennis and Schnabel (1983)**
  This package is designed for unconstrained minimization and has options that include both line search and trust region approaches. The provided options include analytic gradients or difference approximations with analytic Hessians or finite difference Hessians (from analytic or finite difference gradients) or secant methods (BFGS).

- **VA15 of Liu and Nocedal (1989)**
  This package solves general nonlinear unconstrained problems using a limited memory BFGS method. It is intended for large-scale problems.

- **VE09 of Gould (1991)**
  This package obtains local solutions to general, non-convex quadratic programming problems, using an active set method, and is intended to be suitable for large-scale problems.

- **VE14 of Conn *et al.* (1993g)**
  This package solves bound-constrained quadratic programming problems using a barrier function method and is again intended to be suitable for large-scale problems.

- **VF13 of Powell (1982)**
  This package solves general nonlinearly constrained problems using a sequential quadratic programming technique.

VA15, VE09, VE14 and VF13 are part of the Harwell Subroutine Library (1993).

5 Further Developments

Having described LANCELOT A, we now consider future developments. Firstly it is obvious that we would like to learn from our experiences with LANCELOT A, but this is not necessarily easy. Unfortunately one soon discovers that one should do a great deal of testing, including experience with the best competitive algorithms on the same non-trivial problems. However one also discovers that (fortunately, occasionally) relatively innocuous seeming changes, like changing the initial trust region size from one to two, may change the solution time by several orders of magnitude. A more detailed example of the difficulties of definitive testing is illustrated by the following tale. Amongst our many applications we have some in structural optimization that give rise to minimax problems which, when posed as nonlinear programming problems, contain very many more inequality constraints than variables (see, for example, Achtziger et al., 1992). Consequently if they are solved via LANCELOT A it is necessary to add very many slack variables. In fact the particular incidence we have in mind involved a discrete plate problem[8] with 343 variables and 8,958 inequality constraints. Thus, with the addition of slacks, one has a problem in 9,301 variables and 8,958 equality constraints. The run we made with the LANCELOT default parameters took 117 hours on an IBM RISC/6000 320 — not particularly encouraging! This provided one motivating factor for us to consider handling inequalities directly via barrier functions.

We now consider barrier functions and their extension in more detail. As is discussed in Bartholomew-Biggs (1994, Section 4), historically a shift was introduced to the quadratic penalty function to avoid updating the penalty parameter more than a finite number of times, thus giving the multiplier methods/augmented Lagrangian functions already used above. It seems reasonable to consider doing the same for logarithmic barrier functions and indeed in recent years there has been a flurry of activity in this area (Breitfeld and Shanno, 1993a, Breitfeld and Shanno, 1993b, Conn et al., 1992a, Freund, 1991, Gill et al., 1988, Jittorntrum and Osborne, 1980, Jensen and Polyak, 1993, Nash et al., 1993, Polyak, 1992 and Powell, 1992).

To see the augmented Lagrangian as a shifted/modified quadratic penalty function we note that (3.14) is equivalent to $\hat{\Phi}(x, \lambda, \mu) - \frac{1}{2\mu} \sum_{i=1}^{m} s_i^2$, where

$$\hat{\Phi}(x, \lambda, \mu) = f(x) + \frac{1}{2\mu} \sum_{i=1}^{m} \left[c_i(x) + s_i \right]^2, \qquad (5.1)$$

and the *shifts* $s_i = \mu \lambda_i$. Note that if we assume that the λ_i are bounded, $\mu \to 0$ implies that $s_i \to 0$. But then we can say that the problem

$$\underset{x \in \Re^n}{\text{minimize}} \ f(x) \qquad (5.2)$$

subject to

$$c_i(x) + s_i = 0, \quad 1 \le i \le m, \qquad (5.3)$$

converges to

$$\underset{x \in \Re^n}{\text{minimize}} \ f(x) \qquad (5.4)$$

---

[8]Known as HAIFAL.SIF in the CUTE distribution.

subject to

$$c_i(x) = 0, \quad 1 \le i \le m, \tag{5.5}$$

as μ tends to zero. But (5.1) is the quadratic penalty function for (5.2) and (5.3), and the problem given by (5.2) and (5.3) is equivalent to

$$\underset{x \in \Re^n}{\text{minimize}} \ f(x) \tag{5.6}$$

subject to

$$\frac{1}{2s_i}[c_i(x) + s_i]^2 = 0, \quad 1 \le i \le m, \tag{5.7}$$

provided that $s_i \ne 0$. Now the classical Lagrangian for this latter formulation is

$$f(x) + \frac{1}{2\mu} \sum_{i=1}^m [c_i(x) + s_i]^2 = \hat{\Phi}(x, \lambda, \mu), \tag{5.8}$$

with $s_i = \mu \lambda_i$. Thus one can think of this as a Lagrangian quadratic penalty function. Let us now consider a similar development for the logarithmic barrier function,

$$\Psi(x, \lambda, s) = f(x) - \mu \sum_{i=1}^m \log\left[c_i(x) + s_i\right], \tag{5.9}$$

corresponding to the problem

$$\underset{x \in \Re^n}{\text{minimize}} \ f(x) \tag{5.10}$$

subject to

$$c_i(x) \ge 0, \quad 1 \le i \le m. \tag{5.11}$$

Taking $\mu = \lambda_i s_i$ we rewrite this as

$$\Psi(x, \lambda, s) = f(x) - \sum_{i=1}^m \lambda_i s_i \log\left[c_i(x) + s_i\right], \tag{5.12}$$

with corresponding first-order update

$$\lambda_i^+ = \lambda_i s_i / \left[c_i + s_i\right]. \tag{5.13}$$

Analogously to the presentation above, Polyak points out that (5.11) is equivalent to

$$s_i \log\left[1 + c_i(x)/s_i\right] \ge 0, \quad 1 \le i \le m, \tag{5.14}$$

and the classical Lagrangian for the problem (5.10), subject to (5.14), is given by $\hat{\Psi}(x, \lambda, s)$, where

$$\hat{\Psi}(x, \lambda, s) = f(x) - \sum_{i=1}^{m} \lambda_i s_i \log\left[1 + c_i(x)/s_i\right]. \tag{5.15}$$

But then $\hat{\Psi} = \Psi - \mu \sum_{i=1}^{m} \log\left[s_i\right]$, and the last term is independent of x.

Gill *et al.* (1988) carried out their analysis for linear programs, chose $s_i = \mu \lambda_i$ and used μ to control the algorithm. Polyak (1992) used $s_i = \mu$ and established convergence under the assumption that the Jacobian is full rank and second-order sufficiency and strict complementarity hold. He and Jensen (Polyak, 1992 and Jensen and Polyak, 1993) were able to prove stronger results for linear, quadratic and convex programs. They use λ_i to control the algorithm asymptotically. In Conn *et al.* (1992a) we use $s_i = \mu \lambda_i^{\alpha}$, where $0 < \alpha \leq 1$, with multiplier updates given by (5.13) when appropriate. We accept or reject the multiplier update after approximate inner minimization based upon the relative degree to which we satisfy the complementary slackness conditions written as $c_i \lambda_i^+ / s_i$. If the multiplier update is rejected then we update the penalty parameter. We include a complete convergence analysis and prove that the penalty parameter is updated only a finite number of times. In addition, asymptotically we require only one inner iteration per outer iteration (see Conn *et al.*, 1992d). Finally, we shift the starting point via an auxiliary problem when necessary (see Conn *et al.*, 1992a, for details)

. Now let us consider the numerical results for this Lagrangian barrier approach — more precisely, we consider the modified barrier approach of Jensen *et al.* (1992) with additional quadratic terms. For the discrete plate problem above, it now takes 31 minutes and 54 seconds to determine the solution, which is clearly much better than running LANCELOT with the default options.[9]

However, to emphasize some of the difficulties inherent in evaluating software for large-scale problems, when we tried different values of the penalty parameter within LANCELOT A (the results obtained with Jensen *et al.*, 1992, already included some *tuning*) we obtained the result in 4 hours, 24 minutes and 28 seconds, which already represents considerable improvement over the time using the default penalty parameter value. This improvement is especially noteworthy when one considers that LANCELOT solves the problem in 9,301 variables as opposed to the 343 of the barrier approach. For the record, MINOS 5.3 took 2 hours, 36 minutes and 30 seconds and MINOS 5.4 took 1 hour and 30 minutes.

But the story is not yet over. With a little more thought one can rearrange the linear algebra so that the *effective size* of the augmented Lagrangian approach is equivalent to that of the Lagrangian barrier approach.

To see this, consider Newton's method for minimizing the augmented Lagrangian,

---

[9] Apparently Ben-Tal and Bendsøe (1993) is an even more successful approach to this class of structural problems.

with slacks y added to the inequalites. Then the corresponding augmented Lagrangian becomes

$$\Phi(x, y, \lambda, \mu) = f(x) + \sum_{i=1}^{m} \lambda_i(c_i(x) - y_i) + \frac{1}{2\mu} \sum_{i=1}^{m} (c_i(x) - y_i)^2, \tag{5.16}$$

The linear system that arises is given by

$$\begin{pmatrix} B + \frac{1}{\mu}A^T A & -A^T/\mu \\ -A/\mu & I/\mu \end{pmatrix} \begin{pmatrix} p_x \\ p_y \end{pmatrix} = -\begin{pmatrix} g + A^T\bar{\lambda} \\ -\bar{\lambda} \end{pmatrix}. \tag{5.17}$$

Noting that the coefficient matrix can be factored as

$$\begin{pmatrix} I & -A^T/\mu \\ 0 & I/\mu \end{pmatrix} \begin{pmatrix} B & 0 \\ -A & I \end{pmatrix} \tag{5.18}$$

one can determine the search direction from[10]

$$\begin{aligned} q_y &= -\mu\bar{\lambda} \\ q_x &= -(g + A^T\bar{\lambda}) + \frac{1}{\mu}A^T q_y \\ Bp_x &= q_x \\ p_y &= q_y - Ap_x. \end{aligned} \tag{5.19}$$

Clearly the only significant work is the third equation with the coefficient matrix B. Details are given in Conn *et al.* (1992h).

At this point it is worth mentioning that eliminating slack variables is not the only motivation for considering Lagrangian barrier techniques. In particular, the success of interior methods in linear programming (see, for example, Shanno, 1994, in this volume) suggests that they may be less sensitive to degeneracy. Moreover there is numerical evidence that the Lagrangian barrier approach is superior to standard barrier function methods when applied to problems with simple bounds (see Conn *et al.*, 1993g and Nash *et al.*, 1993) and preliminary evidence suggests that the same is true for more general constraints (see Breitfeld and Shanno, 1993a, and Breitfeld and Shanno, 1993b).

However, there are difficulties associated with the fact that one needs to remain feasible with respect to the *shifted* constraints, the fact that we lack experience (even for small dense problems) with this approach and finally and perhaps most importantly, the fact that a quadratic model is not a good model for a logarithmic barrier function.

In our attempts to improve the current version of LANCELOT we have continued our research along both theoretical and practical lines. One area we have pursued

---

[10]Equation (5.19) is only true if none of the slacks are active, but similar simplifications are possible when there are active slacks.

is that of using a *structured* trust region, which we motivate here by considering the following example:

$$\text{minimize} \atop {x \in \Re^3} \quad x_1^2 + (x_1 + x_2)^2 + e^{(x_2+x_3)^2}. \qquad (5.20)$$

Suppose we take three element functions x_1^2, $(x_1+x_2)^2$ and $e^{(x_2+x_3)^2}$. Traditional trust region methods will tend to keep the radius unnecessarily small because of the third element, even though the first two elements are perfectly modeled by a quadratic. Thus if x_1 is far from its optimal value, it may be prevented from changing rapidly only because of a global trust region dictated by the third element. It is natural to think of using separate trust regions for *separable problems*. The idea is to generalize this by having a separate trust region for each element. In addition, we need an *overall* model on an *overall* trust region. The trust region for each element constrains only the variables for that element. Details are given in Conn *et al.* (1992f).

Another problem that one might associate with that of group partial separability is determining a suitable partitioning into groups and elements. In general this is a difficult problem to do optimally but there are two simpler versions that we have considered. The first is that of 'blowing up the internal variables' and the second is that of 'merging elements and trivial groups'. Since the main computational cost within the conjugate gradient algorithm is the multiplication of the involved matrix with a vector, we see that the cost is certainly dependent upon the representation of the matrix. In the two cases above the trade-off between computing speed and storage requirements is readily determined and can be motivated by geometrical considerations.

For element merging, say between two elements, one needs to consider the amount of overlap of the element Hessians (see Figure 1). If the area of the overlap box in

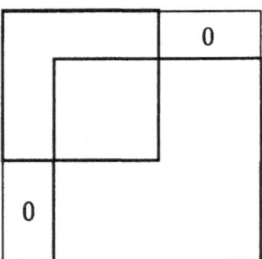

Figure 1: Two elements in the Hessian matrix

the center is greater than the sum of the two areas of the zero blocks then it is worth doing merging. Details are given in Conn *et al.* (1993d).

For blow up consider the following representation (Figure 2) of the blown up Hessian, the block on the left hand side, to that of its internal form, the middle block on the right hand side[11]. In this case the blow up is recommended when the total

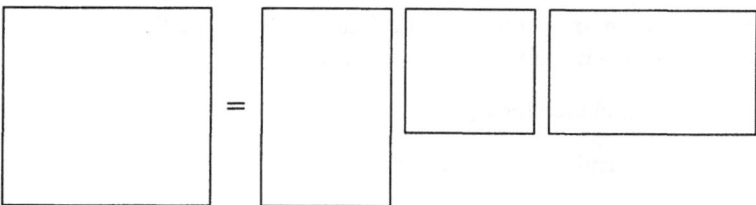

Figure 2: Elemental vs internal Hessian representation

area of the last three blocks is greater than the area of the first block. Once again the reader is referred to Conn *et al.* (1993d) for details, but suffice it to add that in some cases *substantial* improvements can be realized in this simple manner for both types of reformulation (see also Dayde *et al.*, 1994, for an application of similar ideas to preconditioners. In their case it often pays to merge elements even when more zeros are introduced than described here, as better vectorization is possible with bigger blocks.).

Another development for large-scale problems is our work on what we call iterated subspace minimization, which we describe below in the context of unconstrained optimization (1.1). The usual framework is to model f and then do a reasonable amount of work to 'optimize' the model, considering only what happens to the true function *one-dimensionally*. In line search methods the significant work determines the search direction and the true function determines the step-size, whilst in the case of trust region algorithms, the model determines the Cauchy point (and better) and the true function determines the trust region radius. Our motivation is that one does not appear to be using the multidimensional information obtained in the model optimization as well as one might. In addition, we observe that there are high quality algorithms available for solving small-scale problems.

This suggests the following scheme:

1. Determine a full-rank subspace matrix $S_k \in \Re^{n \times s_k}$, where $s_k \ll n$.

2. Approximately solve the s_k-dimensional minimization problem

$$\underset{y \in \Re^{s_k}}{\text{minimize}} \quad f(x_k + S_k y), \tag{5.21}$$

---

[11]The last block represents the transformation matrix from elemental to internal variables and the first block on the right hand side is just its transpose.

and set

$$x_{k+1} = \text{(approx) arg min} \; f(x_k + S_k y), \qquad (5.22)$$
$$y \in \Re^{s_k}$$

where we note that we are using the true function f in (5.22).

This begs the following important questions:

- What is a good choice for s_k?

- How do we determine the matrix S_k?

- What do we mean by "approximate" when solving problem (5.21)?

- Are there methods which are particularly appropriate for solving (5.21)?

- What can we say about the convergence of such a method?

- If we can establish convergence, what can we say about its asymptotic rate?

As we discussed at the beginning of this paper, as long as S_k contains something like the steepest descent direction with a sufficient decrease condition, global convergence is assured. Furthermore, if a Newton-like direction is also represented, we can expect a good asymptotic rate of convergence. Thus we propose to take for our columns of S_k a few directions generated by a preconditioned conjugate gradient algorithm, including the first, and a truncated Newton direction.

As an indication of the usefulness of CUTE, we were able to readily test this idea on thirty-nine unconstrained problems in the database[12]. The average size of the problems tested was around one thousand variables. Compared with the default version of LANCELOT, the new idea was at least twice as fast eleven times, at least ten times as fast twice and twice as slow five times. The remaining problems had comparable times. Details are given in Conn et al. (1994).

In many ways LANCELOT A's major defect is in the way it handles linear constraints. Incorporating them into an augmented Lagrangian function increases their complexity. Thus, in addition to keeping simple bounds explicitly outside the objective function, we wanted to also consider doing the same for linear constraints. The difficulty is that although it is trivial to carry out projections to maintain feasibility with respect to the bounds, it is not so trivial to do the same for linear constraints. In an attempt to improve on this, we first looked at a more general approach that made use of *inexact* projections on convex constraints. We used an approximate generalized Cauchy point and required that Goldstein-like conditions are met. Briefly, we require a feasible step within the trust region, sufficient decrease on the model functions, a sufficiently large step to prevent premature termination and we ask that

---

[12]That is all those available, with the exception of problems which took excessive CPU time (more than 30 minutes) or were variations on the reported problems.

we do a fixed percentage as well as the minimum value of the linearized model in the intersection of the feasible region within a ball of radius greater than or equal to the step. Details are given in Conn *et al.* (1993b). In the case of nonlinear networks, Sartenaer (1993) has obtained some very encouraging numerical results along these lines.

In addition we have extended our previous theory developed for the augmented Lagrangian function to the case where the linear constraints are not incorporated into the objective function. Moreover, as for Karmarkar (1984), we do not exclude the possibility of incorporating the simple bounds into the objective function. The inner iterations are terminated when we are 'sufficiently critical' — based upon identification of the *linear* constraints that are 'dominant'[13]. Details are given in Conn *et al.* (1993e) and Conn *et al.* (1993f) and we are currently preparing similar results for the Lagrangian barrier and mixed cases. We also incorporate the possibility of partitioning the constraints, with separate penalty parameters associated with each partition.

It should be pointed out that these issues are also relevant for the case of near-linear constraints, where, in particular, the idea used in MINOS of considering the deviation from linearity should be a good one.

Some work has been carried out to exploit particular computer architectures. The assumed partially separable form may be exploited in many ways on parallel machines (see, for example, Saludjian, 1993, and Dayde *et al.*, 1994). British Gas are currently experimenting with a parallel LANCELOT-like method for the national gas network.

6 Other Recent Progress

Let us now look at some of the recent work of our colleagues. We first consider the trust region approach of Lalee *et al.* (1993) that is designed for equality constrained problems. The method uses either exact second derivatives or limited memory quasi-Newton. It is intended for large-scale problems and is based on the trust region approach of Omojokun (1991). It uses two trust region problems:

- a *vertical step* that determines the nearest feasible point (measured by the norm of the linearized residuals) in a shrunken trust region, and

- a *horizontal step* that minimizes the model function in the trust region restricted to the null space of the constraint gradients.

This has been implemented as the algorithm ETR and a suitable interface using the CUTE tools has been written.

Defining Z as in reduced gradient methods, B as an approximation to the Hessian of the Lagrangian, the subscript k to denote iteration k, and

$$A_k^T = (\ B \quad N \ ) \tag{6.1}$$

---

[13]$a_i^T x - b_i \leq \kappa_1 \omega$, for some constant κ_1, where ω is the stationarity tolerance.

$$Z^T = -\left(\; B^{-1}N \quad I \;\right),\tag{6.2}$$

the vertical step is given by

$$\underset{v\in\Re^n}{\text{minimize}}\; \|A_k^T v + c_k\|\tag{6.3}$$

subject to

$$\|v\| \le 0.8\Delta_k.\tag{6.4}$$

Writing $d = v_k + Z_k u$, the horizontal step is given by

$$\underset{u\in\Re^n}{\text{minimize}}\; (g_k + B_k v_k)^T Z_k u + \frac{1}{2}u^T Z_k^T B_k Z_k u\tag{6.5}$$

subject to

$$\|Z_k u\| \le \Delta_k.\tag{6.6}$$

The merit function is $f(x) + \mu\|c(x)\|_2$. The implementation uses a modification of MA28 (Duff, 1977) and the limited memory method uses a new compact representation (Byrd et al., 1994), since otherwise the matrix-vector products do not take advantage of sparsity and must work with the Hessian rather than its inverse.

Motivated by real problems in process engineering, Biegler et al. (1993b) and Biegler et al. (1993a) have an implementation of the algorithm of Coleman and Conn (1982) and Coleman and Conn (1984) that is a quasi-Newton algorithm. It is designed for large-scale problems with a small associated reduced space. The reduced Hessian is updated but a correction vector is incorporated to approximate the cross term $Z^T B Y d_y$, where Z is the matrix whose columns span the null space of the activity gradients, Y does the same for the range space and d_y is the component of the step in the range space. This is done with little extra cost and a one-step Q superlinear asymptotic convergence rate is maintained. The fundamental issue is that, for large-scale problems, computing orthogonal bases is expensive. However, with non-orthogonal bases, the vertical or range space component $Y d_y$ can be very large and ignoring the cross term can result in a poor step. Thus Biegler et al. use updates on $Z^T B$ and then compute $Z^T B(Y d_y)$ and $Z^{+T} B^+(Y d_y)$. The former is used in the horizontal or null space step and the latter is used to update $Z^T B Z$. Moreover, these steps are 'juggled' differently, the first being used to guarantee sufficient descent and the second to ensure boundedness of $Z^T B Z$. An interface for CUTE is available. The approach has been extended via limited memory quasi-Newton to the case where the reduced space is not small, again making use of the compact representation of Byrd et al. (1994).

An extension of generalized reduced gradient methods (Abadie and Carpentier, 1966 and Lasdon et al., 1978) to the large-scale case has been incorporated in CONOPT

(Drud, 1985, Drud, 1993) and in the work of Smith and Lasdon (1992), which also makes use of the limited memory approach.

A unique approach is that of reflective Newton methods (Coleman and Li, 1992d and Coleman and Li, 1992c). This relates to the idea of replacing $x_i \geq 0$ by $x_i = |y_i|$ and replacing

$$\underset{x \in \Re^n}{\text{minimize}} \ f(x) \tag{6.7}$$

subject to

$$x \geq 0 \tag{6.8}$$

by

$$\underset{y \in \Re^n}{\text{minimize}} \ \hat{f}(y), \tag{6.9}$$

where $\hat{f}(y) = f(|x|)$. Amongst its advantages is the fact that this transformation does not introduce new minima, one is able to use fixed data structures and strict feasibility is maintained. The method is designed for large problems. The piecewise linear path in x-space that corresponds to a search direction in y is easily determined. They need a generalization of the Armijo-Goldstein criteria and a condition to ensure constraint compatibility. This latter definition ensures that if x is close to a boundary one is able to take a large enough step[14]. In addition, a consistency property guarantees that a first-order step converging to zero implies convergence to a stationary point. The method is implemented for simple bounds and is currently being extended to linear equality constraints. The Newton-like iterations are carried out in a scaled trust-region framework, solved in a dog-leg like method (see Powell, 1975 and Munksgaard and Reid, 1983). Asymptotically a step-size of one is taken and thus second-order convergence is attained.

We have already mentioned that barrier/interior techniques are currently a very active area of research. Nash and Sofer (1993) use a logarithmic barrier function and handle the associated ill-conditioning by using an approximate (explicit) formula for the Newton direction. This formulation (that projects orthogonally to the constraints that cause the ill-conditioning) becomes more accurate as the penalty parameter becomes smaller. In addition they use a special line search as in Murray and Wright (1976) (see also Murray and Wright, 1992), a preconditioned truncated Newton method and extrapolation as in Fiacco and McCormick (1968). Finally they use an expanded form of the Hessian of the barrier function and finite differences to derive efficient matrix-vector products. They report numerical results on simple bound constrained problems as large as 100,000 variables. Nash *et al.* (1993) use a similar implementation of a modified (shifted/Lagrangian) logarithmic barrier function with

---

[14]This, in turn, ensures that the distance to breakpoints remains bounded away from zero. Reflections are likely to be suitable if the angle is reasonable.

additional quadratic terms. More specifically, writing $t = c_i(x)$ and considering a single barrier term Ψ, they use the term

$$
\begin{aligned}
\Psi(\mu^{-1}t + 1) &= \log(\mu^{-1}t + 1) \quad &\text{if } t \geq -\mu/2, \\
&= q(t) \quad &\text{if } t < -\mu/2.
\end{aligned}
\tag{6.10}
$$

Here the quadratic, q, interpolates $q(-\mu/2), q'(-\mu/2)$ and $q''(-\mu/2)$ with the corresponding logarithmic values. Interestingly, in this context they abandoned the special line search of Murray and Wright (1976)[15]. The numerical results reported were better than using just the barrier function.

Breitfeld and Shanno had similar computational results. They used CONMIN (Shanno and Phua, 1980), which is a limited memory BFGS/CG algorithm. In Breitfeld and Shanno (1993a), they suggested replacing equalities by two inequalities which are then shifted. They claim that this is preferable to using an augmented Lagrangian to handle equalities. However, the numerical results to date must still be considered very preliminary.

We now report on some numerical experience and testing in general. Extensive numerical results are available for LANCELOT in Conn *et al.* (1992c) and Conn *et al.* (1993c). These describe tests using all the LANCELOT options on about one thousand problem instances. The basic conclusions are that LANCELOT appears to be very robust and the symmetric rank one update is the best quasi-Newton update in that trust-region context (see also Byrd *et al.*, 1993a, who based upon their convergence analysis, recommend updating even when steps are rejected). From the point of view of general comparisons, there is not a great deal of large-scale experience[16] in the published literature. Eldersveld *et al.* (1993) looked at very sparse problems that have the possibility of having a large reduced space (dimension greater than 700) and where the functions are expensive to evaluate. They considered 109 problems with from 40 to 2,400 variables. They compared NPSOL (Gill *et al.*, 1986), which was not designed for large sparse problems; MINOS, which, although designed for the large-scale case, prefers small reduced subspaces; NLPSPR (Betts and Frank, 1994), which is a sequential quadratic programming method that uses Schur complements on an (increasing) Karush-Kuhn-Tucker matrix; and LSSQP (Gill *et al.*, 1993a and Gill *et al.*, 1993b), which is a transformed Hessian method. Their main conclusions were that NLPSPR was best (although they admit a bias since the code was designed for the class of problems they tested), MINOS was rather disappointing, NPSOL was robust for those problems for which enough storage was available and LSSQP performs well when the reduced space is less than two hundred dimensional. We are currently (Bongartz *et al.*, 1994c), doing an extensive comparison between MINOS and LANCELOT using the CUTE database. We would like to identify, amongst other details, the class of problems for which each is most appropriate and verify if these

---

[15]This suggests that the singularity can be taken care of by adding quadratic terms rather than using a special line search. Breitfeld and Shanno (1993b) made the same observation.

[16]Indeed, there isn't much recent extensive testing for small-scale problems.

findings agree with our preconceptions. As for Eldersveld *et al.* (1993), our preliminary results are that MINOS is not as robust as one would hope, but one should bear in mind that, firstly, we have more expertise with LANCELOT and, secondly, the basis for MINOS is now rather old technology.

Bouaricha and colleagues (Bouaricha and Gould, 1994, Bouaricha and Schnabel, 1994a, Bouaricha and Schnabel, 1994b and Bouaricha and Tuminaro, 1994) are extending the earlier work on tensor methods of Schnabel and Chow (1991) and Schnabel and Frank (1984) to large-scale problems. The basic idea of tensor methods is to base each iteration on a higher order model than standard methods, but in such a way that there is almost no increased cost. The motivation is to improve upon the standard methods when applied to non-singular and (especially) singular problems. As for non-tensor methods, the extension to large problems suggests the use of finite differences, the replacement of orthogonal decompositions and the exploitation of structure. However, because of the nature of the tensor terms, a basic question here is will information in small dimensional subspaces (and in what form) help when the underlying problem is large?

Finally, we should not forget there are methods designed for specially structured large-scale nonlinear programming problems. Some examples follow.

- For nonlinear least-squares problems: Ben Daya and Shetty (1988), Coleman and Plassman (1988), Coleman and Plassman (1992), Golub *et al.* (1986), Gulliksson (1990), Gulliksson (1993), Kaufman and Sylvester (1993), Toint (1987b) and Toint (1987a).

- For minimax, l_p, l_2 and l_∞ problems: Coleman and Li (1992a), Coleman and Li (1992b), Dax (1993), Li (1993b), Li (1993a), Jónasson and Madsen (1992) and Sklar and Armstrong (1993).

- For quadratic programming problems (including those constrained only by simple bounds): Coleman and Hulbert (1993a), Coleman and Hulbert (1993b), Gould (1991), Júdice and Pires (1989), Moré and Toraldo (1989), Moré and Toraldo (1991), Soares *et al.* (1993) and Vanderbei and Carpenter (1993).

- For nonlinear network problems: Ahlfeld *et al.* (1987), Dembo (1986), Sartenaer (1993), Toint and Tuyttens (1990), Toint and Tuyttens (1992), Zenios and Mulvey (1986), Zenios and Mulvey (1988), Zenios and Pinar (1989).

- For location problems: Bongartz *et al.* (1994a), Bongartz *et al.* (1994b) and Calamai and Conn (1987).

- For linear complementarity problems: Júdice and Pires (1993) — see also Júdice (1994), in this volume.

Finally we have said little about automatic differentiation or special architectures. The former still do not seem to have had as much impact in optimization as one might have hoped. Besides the chapter in this volume (Dixon, 1994), we refer the reader to Bischof *et al.* (1991), Bischof and Griewank (1992), Dixon *et al.* (1988), Griewank (1989), Griewank and Corliss (1991) and Griewank *et al.* (1993). For the latter the reader is referred to the chapter of Schnabel in this volume (Schnabel, 1994) and Zenios (1989).

7   In conclusion

We hope we have convinced some of you that it is possible to solve large nonlinear problems in thousands of variables in acceptable time on reasonable workstations. Moreover software packages are available and it is worth pointing out that, although they are designed for large-scale problems, some of them can nevertheless be excellent for the small-scale case. Our hope is that, in the not too distant future, practitioners will be solving nonlinear models rather than linear ones, when the former is the most appropriate one to consider. We also have taken some pains to emphasize the importance of testing. In our opinion, nobody should be publishing papers whose main purpose is to describe an algorithm that is intended to be practically useful, unless they also provide evidence that the algorithm is competitive on significant problems. Even more obvious is the statement that it is meaningless to propose algorithms for large-scale problems and report numerical results only for problems in a few hundred variables.

Besides the relevant chapters in this volume, very good background reading in linear, constrained and unconstrained nonlinear programming is provided in the chapters of Goldfarb and Todd (1989), Dennis and Schnabel (1989), Gill and Murray (1989) and in the book by Nemhauser *et al.* (1989). Recent articles and books devoted primarily to large-scale optimization include Coleman and Li (1990), Coleman (1993), Conn *et al.* (1989), Conn *et al.* (1990b), Conn *et al.* (1992b), Conn *et al.* (1992g) and Wright (1991). The book by Moré and Wright (1993), besides having a useful introduction to the theory, indicates the available software. Some examples of applications are given in Biegler (1992), Chinchalkar and Coleman (1993), Coleman and Liao (1993), Coleman *et al.* (1992), Dunn (1993), Falk and McCormick (1986), Hager (1990), Jones (1967), Kunish and Sachs (1992), Liao (1993), McCormick (1972), McCormick and Sofer (1991), Schrady and Choe (1971), Werbos (1988) and Wu (1993).

Finally, in a subject this complex, a single short article, necessarily, is only able to give an idea of the nature of the main issues in the current research. Moreover we have no doubt that our own particular biases show. Nevertheless we hope that the text and the references will be useful to those interested in what currently is an exciting and vibrant research area.

References

[Abadie and Carpentier, 1966] J. Abadie and J. Carpentier. Généralisation de la méthode du gradient réduit de Wolfe au cas des contraintes non-linéaires. In D.B. Hertz and J. Melese, editors, *Proceedings IFORS Conference*, pages 1041–1053. J. Wiley and Sons, Amsterdam, 1966.

[Achtziger et al., 1992] M. Achtziger, M. P. Bendsøe, A. Ben-Tal, and J. Zowe. Equivalent displacement based formulations for maximum strength topology design. *Impact of Computing in Science and Engineering*, 4:315–345, 1992.

[Ahlfeld et al., 1987] D. P. Ahlfeld, R. S. Dembo, J. M. Mulvey, and S. A. Zenios. Nonlinear programming on generalized networks. *ACM Transactions on Mathematical Software*, 13(3):350–367, 1987.

[Arioli et al., 1993] M. Arioli, T. F. Chan, I. S. Duff, N. I. M. Gould, and J. K. Reid. Computing a search direction for large-scale linearly constrained nonlinear optimization calculations. Technical Report TR/PA/93/34, CERFACS, Toulouse, France, 1993.

[Averick and Moré, 1991] B. M. Averick and J. J. Moré. The Minpack-2 test problem collection. Technical Report ANL/MCS-TM-157, Argonne National Laboratory, Argonne, USA, 1991.

[Averick et al., 1991] B. M. Averick, R. G. Carter, and J. J. Moré. The Minpack-2 test problem collection (preliminary version). Technical Report ANL/MCS-TM-150, Argonne National Laboratory, Argonne, USA, 1991.

[Bartholomew-Biggs and Hernandez, 1994] M. C. Bartholomew-Biggs and M. de F. G. Hernandez. Modifications to the subroutine OPALQP for dealing with large problems. *Journal of Economic Dynamics and Control*, 18:185–204, 1994.

[Bartholomew-Biggs, 1994] M. C. Bartholomew-Biggs. Algorithms for general constrained nonlinear optimization. In E. Spedicato, editor, *Algorithms for continuous optimization: the state of the art*. Kluwer Academic Publishers, Dordrecht, The Netherlands, 1994.

[Ben Daya and Shetty, 1988] M. Ben Daya and C. M. Shetty. Polynomial barrier function algorithm for convex quadratic programming. Research Report J85-5, School of ISE, Georgia Institute of Technology, Atlanta, Georgia, 1988.

[Ben-Tal and Bendsøe, 1993] A. Ben-Tal and M. P. Bendsøe. A new method for optimal truss topology design. *SIAM Journal on Optimization*, 3(2):322–358, 1993.

[Bertsekas, 1982] D. P. Bertsekas. Projected Newton methods for optimization problems with simple constraints. *SIAM Journal on Control and Optimization*, 20(2):221–246, 1982.

[Betts and Frank, 1994] J. T. Betts and P. D. Frank. A sparse nonlinear optimization algorithm. *Journal of Optimization Theory and Applications*, 82(3), 1994, to appear.

[Biegler, 1992] L.T. Biegler. Optimization strategies for complex process models. *Advances in Chemical Engineering*, 18:197–256, 1992.

[Biegler et al., 1993a] L.T. Biegler, J. Nocedal, and C. Schmid. Numerical experience with a reduced Hessian method for large-scale constrained optimization. Research Report (in preparation), EE and CS, Northwestern University, Evanston, USA, 1993.

[Biegler et al., 1993b] L.T. Biegler, J. Nocedal, and C. Schmid. A reduced Hessian method for large-scale constrained optimization. Research Report NAM-03, EE and CS, Northwestern University, Evanston, USA, 1993.

[Bischof and Griewank, 1992] C. Bischof and A. O. Griewank. ADIFOR: A FORTRAN system for portable automatic differentiation. Technical Report MCS-P317-0792, Argonne National Laboratory, Argonne, USA, 1992.

[Bischof et al., 1991] C. Bischof, A. Carle, G. Corliss, P. Hovland, and A. O. Griewank. ADIFOR: Generating derivative codes from Fortran programs. Technical Report MCS-P263-0991, Argonne National Laboratory, Argonne, USA, 1991.

[Björck, 1994] A. Björck. Generalized and sparse least squares problems. In E. Spedicato, editor, *Algorithms for continuous optimization: the state of the art*. Kluwer Academic Publishers, Dordrecht, The Netherlands, 1994.

[Boggs and Tolle, 1989] P. T. Boggs and J. W. Tolle. A strategy for global convergence in a sequential quadratic programming algorithm. *SIAM Journal on Numerical Analysis*, 26(3):600–623, 1989.

[Bongartz et al., 1993] I. Bongartz, A. R. Conn, N. I. M. Gould, and Ph. L. Toint. CUTE: Constrained and Unconstrained Testing Environment. Research Report RC 18860, IBM T. J. Watson Research Center, Yorktown Heights, USA, 1993.

[Bongartz et al., 1994a] I. Bongartz, P. H. Calamai, and A. R. Conn. A projection method for l_p norm location-allocation problems. *Mathematical Programming*, 1994, to appear.

[Bongartz et al., 1994b] I. Bongartz, P. H. Calamai, and A. R. Conn. A second-order algorithm for the continuous capacitated location-allocation problem. Research

Report (in preparation), IBM T. J. Watson Research Center, Yorktown Heights, USA, 1994.

[Bongartz et al., 1994c] I. Bongartz, A. R. Conn, N. I. M. Gould, and Ph. L. Toint. A numerical comparison between the LANCELOT and MINOS packages for large-scale nonlinear optimization. Research Report (in preparation), IBM T. J. Watson Research Center, Yorktown Heights, USA, 1994.

[Bouaricha and Gould, 1994] Ali Bouaricha and N. I. M. Gould. Tensor methods for large sparse unconstrained minimization problems. Technical Report (in preparation), CERFACS, Toulouse, France, 1994.

[Bouaricha and Schnabel, 1994a] Ali Bouaricha and R.B. Schnabel. A software package for large sparse nonlinear equations using tensor methods. Technical Report, (in preparation), CERFACS, Toulouse, France, 1994.

[Bouaricha and Schnabel, 1994b] Ali Bouaricha and R.B. Schnabel. A software package for large sparse nonlinear least squares using tensor methods. Technical Report, (in preparation), CERFACS, Toulouse, France, 1994.

[Bouaricha and Tuminaro, 1994] Ali Bouaricha and R. Tuminaro. Tensor-Krylov methods for large nonlinear equations on sequential and parallel computers. Technical Report, (in preparation), CERFACS, Toulouse, France, 1994.

[Breitfeld and Shanno, 1993a] M. G. Breitfeld and D. Shanno. Preliminary computational experience with modified log-barrier functions for large-scale nonlinear programming. Research Report RRR 08-93, Rutgers Center for Operations Research, New Brunswick, USA, 1993.

[Breitfeld and Shanno, 1993b] M. G. Breitfeld and D. F. Shanno. Computational experience with modified log-barrier methods for nonlinear programming. Research Report RRR 17-93, Rutgers Center for Operations Research, New Brunswick, USA, 1993.

[Buckley, 1989] A. G. Buckley. Test functions for unconstrained minimization. Technical Report CS-3, Computing Science Division, Dalhousie University, Halifax, Canada, 1989.

[Byrd et al., 1993a] R. H. Byrd, H.F. Khalfan, and R. B. Schnabel. Analysis of a symmetric rank-one trust region method. Technical Report CU-CS-657-93, Department of Computer Science, University of Colorado at Boulder, Boulder, USA, 1993.

[Byrd et al., 1993] R. H. Byrd, P. Lu, and J. Nocedal. A limited memory algorithm for bound constrained optimization. Technical Report NAM-08, Department of

318

Electrical Engineering and Computer Science, Northwestern University, Evanston, USA, 1993.

[Byrd et al., 1994] R. H. Byrd, J. Nocedal, and R. B. Schnabel. Representation of quasi-Newton matrices and their use in limited memory methods. *Mathematical Programming, Series A*, 1994, to appear.

[Calamai and Conn, 1987] P. H. Calamai and A. R. Conn. A projected Newton method for l_p norm location problems. *Mathematical Programming*, 38:75–109, 1987.

[Chinchalkar and Coleman, 1993] S. Chinchalkar and T. F. Coleman. Parallel structural optimization applied to bone remodeling on distributed memory machines. Technical Report CTC93TR146, Advanced Computing Research Institute, Cornell Theory Center, Cornell University, Ithaca, USA, 1993.

[Coleman and Conn, 1982] T. F. Coleman and A. R. Conn. Nonlinear programming via an exact penalty function method: Global analysis. *Mathematical Programming*, 24(3):137–161, 1982.

[Coleman and Conn, 1984] T. F. Coleman and A. R. Conn. On the local convergence of a quasi-Newton method for the nonlinear programming problem. *SIAM Journal on Numerical Analysis*, 21(4):755–769, 1984.

[Coleman and Hulbert, 1993a] T. F. Coleman and L. Hulbert. A direct active set method for large sparse quadratic programs with simple bounds. *Mathematical Programming*, 373–406, 1993.

[Coleman and Hulbert, 1993b] T. F. Coleman and L. Hulbert. A globally and superlinearly convergent algorithm for convex quadratic programs with simple bounds. *SIAM Journal on Optimization*, 3:298–321, 1993.

[Coleman and Li, 1990] T. F. Coleman and Y. Li, editors. *Large Scale Numerical Optimization*. SIAM, Philadelphia, USA, 1990.

[Coleman and Li, 1992a] T. F. Coleman and Y. Li. A global and quadratic affine scaling method for linear l_1 problems. *Mathematical Programming*, 56:189–222, 1992.

[Coleman and Li, 1992b] T. F. Coleman and Y. Li. A global and quadratically-convergent method for linear l_∞ problems. *SIAM Journal on Scientific and Statistical Computing*, 29:1166–1186, 1992.

[Coleman and Li, 1992c] T. F. Coleman and Y. Li. On the convergence of reflective Newton methods for large-scale nonlinear minimization subject to bounds. Technical Report CTC 92TR110, Cornell Theory Center, Ithaca, USA, 1992.

[Coleman and Li, 1992d] T. F. Coleman and Y. Li. A reflective Newton method for minimizing a quadratic function subject to bounds on the variables. Technical Report CTC 92TR111, Cornell Theory Center, Ithaca, USA, 1992.

[Coleman and Liao, 1993] T. F. Coleman and A. Liao. An efficient trust region method for unconstrained discrete-time optimal control problems. Technical Report CTC93TR144, Advanced Computing Research Institute, Cornell Theory Center, Cornell University, Ithaca, USA, 1993.

[Coleman and Plassman, 1988] T. F. Coleman and P. E. Plassman. Solution of nonlinear least-squares problems on a multiprocessor. In *Parallel Computing*, pages 44–80. Springer Verlag, Berlin, 1988.

[Coleman and Plassman, 1992] T. F. Coleman and P. E. Plassman. A parallel nonlinear least-squares solver: theoretical analysis and numerical results. *SIAM Journal on Scientific and Statistical Computing*, 13:771–793, 1992.

[Coleman et al., 1992] T. F. Coleman, D. Shalloway, and Z. Wu. Isotropic effective energy simulated annealing searches for low energy molecular cluster states. Technical Report CTC92TR113, Advanced Computing Research Institute, Cornell Theory Center, Cornell University, Ithaca, USA, 1992.

[Coleman, 1993] T. F. Coleman. Large Scale Numerical Optimization: Introduction and overview. In J. Williams and A. Kent, editors, *Encyclopedia of Computer Science and Technology*, Volume 28, supplement 13, pages 167–196. Marcel Dekker, New York, USA, 1993.

[Conn et al., 1988a] A. R. Conn, N. I. M. Gould, and Ph. L. Toint. Global convergence of a class of trust region algorithms for optimization with simple bounds. *SIAM Journal on Numerical Analysis*, 25:433–460, 1988. See also same journal, 26:764–767, 1989.

[Conn et al., 1988b] A. R. Conn, N. I. M. Gould, and Ph. L. Toint. Testing a class of methods for solving minimization problems with simple bounds on the variables. *Mathematics of Computation*, 50:399–430, 1988.

[Conn et al., 1989] A. R. Conn, N. I. M. Gould, and Ph. L. Toint. Large-scale optimization. *Mathematical Programming, Series B*, 45(3), 1989.

[Conn et al., 1990a] A. R. Conn, N. I. M. Gould, and Ph. L. Toint. An introduction to the structure of large scale nonlinear optimization problems and the LANCELOT project. In R. Glowinski and A. Lichnewsky, editors, *Computing Methods in Applied Sciences and Engineering*, pages 42–54. SIAM, Philadelphia, USA, 1990.

[Conn et al., 1990b] A. R. Conn, N. I. M. Gould, and Ph. L. Toint. Large-scale optimization — applications. *Mathematical Programming, Series B*, 48(1), 1990.

[Conn et al., 1991] A. R. Conn, N. I. M. Gould, and Ph. L. Toint. A globally convergent augmented Lagrangian algorithm for optimization with general constraints and simple bounds. *SIAM Journal on Numerical Analysis*, 28(2):545–572, 1991.

[Conn et al., 1992a] A. R. Conn, N. I. M. Gould, and Ph. L. Toint. A globally convergent Lagrangian barrier algorithm for optimization with general inequality constraints and simple bounds. Research Report RC 18049, IBM T. J. Watson Research Center, Yorktown Heights, USA, 1992.

[Conn et al., 1992b] A. R. Conn, N. I. M. Gould, and Ph. L. Toint. LANCELOT: *a Fortran package for large-scale nonlinear optimization (Release A)*, Volume 17 of *Springer Series in Computational Mathematics*. Springer Verlag, Heidelberg, Berlin, New York, 1992.

[Conn et al., 1992c] A. R. Conn, N. I. M. Gould, and Ph. L. Toint. Numerical experiments with the LANCELOT package (Release A) for large-scale nonlinear optimization. Research Report RC 18434, IBM T. J. Watson Research Center, Yorktown Heights, USA, 1992.

[Conn et al., 1992d] A. R. Conn, N. I. M. Gould, and Ph. L. Toint. On the number of inner iterations per outer iteration of a globally convergent algorithm for optimization with general nonlinear equality constraints and simple bounds. In D.F Griffiths and G.A. Watson, editors, *Proceedings of the 14th Biennial Numerical Analysis Conference Dundee 1991*, pages 49–68. Longmans, 1992. (Also as Research Report RC 18382, IBM T. J. Watson Research Center, Yorktown Heights, USA, 1992.)

[Conn et al., 1992f] A. R. Conn, Nick Gould, and Ph. L. Toint. Convergence properties of minimization algorithms for convex constraints using a structured trust region. Research Report RC 18274, IBM T. J. Watson Research Center, Yorktown Heights, USA, 1992.

[Conn et al., 1992g] A. R. Conn, Nick Gould, and Ph. L. Toint. Large-scale nonlinear constrained optimization. In Jr. R. E. O'Malley, editor, *Proceedings of the Second International Conference on Industrial and Applied Mathematics*, pages 51–70. SIAM, Phildelphia, USA, 1992. (Also in M.S. Moonen, G.H. Golub and B.L.R DeMoor, editors, *Linear Algebra for Large-Scale and Real-Time Applications*, Volume 232 of *NATO ASI Series E: Applied Sciences*. Kluwer Academic Publishers, 1993.)

[Conn et al., 1992h] A. R. Conn, Nick Gould, and Ph. L. Toint. A note on exploiting structure when using slack variables. Research Report RC 18435, IBM T. J. Watson Research Center, Yorktown Heights, USA, 1992.

[Conn et al., 1993a] A. R. Conn, N. I. M. Gould, M. Lescrenier, and Ph. L. Toint. Performance of a multifrontal scheme for partially separable optimization. In *Advances in numerical partial differential equations and optimization, Proceedings of the Sixth Mexico-United States Workshop*. Kluwer Academic Publishers, 1993.

[Conn et al., 1993b] A. R. Conn, N. I. M. Gould, A. Sartenaer, and Ph. L. Toint. Global convergence of a class of trust region algorithms for optimization using inexact projections on convex constraints. *SIAM Journal on Optimization*, 3(1):164–221, 1993.

[Conn et al., 1993c] A. R. Conn, N. I. M. Gould, and Ph. L. Toint. Complete numerical results for LANCELOT Release A. Research Report RC 18750, IBM T. J. Watson Research Center, Yorktown Heights, USA, 1993.

[Conn et al., 1993d] A. R. Conn, N. I. M. Gould, and Ph. L. Toint. Improving the decomposition of partially separable functions in the context of large-scale optimization: a first approach. In W. W. Hager, D. W. Hearn, and P.M. Pardalos, editors, *Large Scale Optimization: State of the Art*. Kluwer Academic Publishers, 1993.

[Conn et al., 1993e] A. R. Conn, Nick Gould, A. Sartenaer, and Ph. L. Toint. Local convergence properties of two augmented Lagrangian algorithms for optimization with a combination of general equality and linear constraints. Research Report RC18901, IBM T. J. Watson Research Center, Yorktown Heights, USA, 1993.

[Conn et al., 1993f] A. R. Conn, Nick Gould, A. Sartenaer, and Ph. L. Toint. Local convergence properties of a Lagrangian barrier algorithm for optimization with a combination of general inequality and linear constraints. Research Report (in preparation), IBM T. J. Watson Research Center, Yorktown Heights, USA, 1993.

[Conn et al., 1993g] A. R. Conn, Nick Gould, and Ph. L. Toint. A note on using alternative second-order models for the subproblems arising in barrier function methods for minimization. Research Report RC18898, IBM T. J. Watson Research Center, Yorktown Heights, USA, 1993.

[Conn et al., 1994] A. R. Conn, Nick Gould, A. Sartenaer, and Ph. L. Toint. On iterated-subspace minimization methods for nonlinear optimization. Research Report (in preparation), IBM T. J. Watson Research Center, Yorktown Heights, USA, 1994.

[Curtis et al., 1974] A. Curtis, M. J. D. Powell, and J. Reid. On the estimation of sparse Jacobian matrices. *Journal of the Institute of Mathematics and its Applications*, 13:117–119, 1974.

[Dax, 1993] A. Dax. A row relaxation method for large minmax problems. *BIT*, 33:262–273, 1993.

[Dayde et al., 1994] M. Dayde, J.-Y. L'Excellent, and N. I. M. Gould. On the use of element by element preconditioners to solve large-scale partially separable optimization problems. Technical Report (in preparation), CERFACS, Toulouse, France, 1994.

[Dembo and Tulowitski, 1983] R. S. Dembo and U. Tulowitski. On the minimization of quadratic functions subject to box constraints. School of Organization and Management Working paper series B no. 71, Yale University, 1983.

[Dembo, 1984] R. S. Dembo. A primal truncated-Newton algorithm with application to large-scale nonlinear network optimization. Technical Report 72, Yale School of Management, Yale University, New Haven, USA, 1984.

[Dembo, 1986] R. S. Dembo. The performance of NLPNET, a large scale nonlinear network optimizer. *Mathematical Programming, Series B*, 26:245–249, 1986.

[Dennis and Schnabel, 1983] J. E. Dennis and R. B. Schnabel. *Numerical methods for unconstrained optimization and nonlinear equations*. Prentice-Hall, Englewood Cliffs, USA, 1983.

[Dennis and Schnabel, 1989] J. E. Dennis and R. B. Schnabel. Unconstrained optimization. In G.L. Nemhauser, A.H.G. Rinnooy Kan, and M.J. Todd, editors, *Optimization*, Volume 1 of *Handbooks in Operations Research and Management Science*, pages 73–170. North-Holland, Amsterdam, The Netherlands, 1989.

[Dixon et al., 1988] L. C. W. Dixon, P. Dolan, and R. Price. Finite element optimization: the use of structured automatic differentiation. In A. Osiadacz, editor, *Simulation and Optimization of Large Systems*, pages 117–141. Oxford University Press, Oxford, 1988.

[Dixon, 1994] L. C. W. Dixon. Automatic differentiation and continuous optimization. In E. Spedicato, editor, *Algorithms for continuous optimization: the state of the art*. Kluwer Academic Publishers, Dordrecht, The Netherlands, 1994.

[Drud, 1985] A. Drud. CONOPT: a GRG code for large sparse dynamic nonlinear optimization problems. *Mathematical Programming*, 31(2):153–191, 1985.

[Drud, 1993] A. Drud. CONOPT: a large-scale GRG code. Technical Report, ARKI Consulting and Developing, Bagsvaerd, Denmark, 1993.

[Duff and Reid, 1982] I. S. Duff and J. K. Reid. MA27: A set of Fortran subroutines for solving sparse symmetric sets of linear equations. Report R-10533, AERE Harwell Laboratory, Harwell, UK, 1982.

[Duff and Reid, 1983] I. S. Duff and J. K. Reid. The multifrontal solution of indefinite sparse symmetric linear equations. *ACM Transactions on Mathematical Software*, 9(3):302–325, 1983.

[Duff and Reid, 1993] I. S. Duff and J. K. Reid. MA47: A set of Fortran subroutines for solving sparse symmetric sets of linear equations. Research Report (to appear), Rutherford Appleton Laboratory, Chilton, England, 1993.

[Duff et al., 1986] I. S. Duff, A. M. Erisman, and J. K. Reid. *Direct methods for sparse matrices*. Clarendon Press, Oxford, UK, 1986.

[Duff et al., 1988] I. S. Duff, N. I. M. Gould, M. Lescrenier, and J. K. Reid. The multifrontal method in a parallel environment. In M. G. Cox and S. J. Hammarling, editors, *Reliable Scientific Computation*. Oxford University Press, Oxford, UK, 1988.

[Duff, 1977] I. S. Duff. MA28: A set of Fortran subroutines for sparse unsymmetric linear equations. Report R-8730, AERE Harwell Laboratory, Harwell, UK, 1977.

[Dunn, 1993] J. C. Dunn. Second-order multiplier update calculations for optimal control problems and related large-scale nonlinear programs. *SIAM Journal on Optimization*, 3(3):489–502, 1993.

[Eldersveld et al., 1993] S. K. Eldersveld, J. T. Betts, and W. P. Huffman. A performance comparison of nonlinear programming algorithms for large sparse problems, Presented at the Fourth Stockholm Optimization Days, Royal Institute of Technology, Stockholm, 16–17 August 1993.

[Eldersveld, 1992] S. K. Eldersveld. Large-scale sequential quadratic programming algorithms. Technical Report SOL 92-4, Department of Operations Research, Stanford University, Stanford, USA, 1992.

[Falk and McCormick, 1986] J. E. Falk and G. P. McCormick. Computational aspects of the international coal trade model. In P.T. Harker, editor, *Spacial price equilibrium: Advances in theory, computation and application*, Volume 249 of *Lecture Notes in Economics and Mathematical Systems*. Springer Verlag, Berlin, 1986.

[Fiacco and McCormick, 1968] A. V. Fiacco and G. P. McCormick. *Nonlinear Programming: Sequential Unconstrained Minimization Techniques*. J. Wiley and Sons, New York, 1968. Reprinted as *Classics in Applied Mathematics 4*, SIAM, 1990.

[Fletcher, 1987] R. Fletcher. *Practical Methods of Optimization*. J. Wiley and Sons, Chichester, second edition, 1987.

[Fletcher, 1994] R. Fletcher. Algorithms for unconstrained optimization. In E. Spedicato, editor, *Algorithms for continuous optimization: the state of the art*. Kluwer Academic Publishers, Dordrecht, The Netherlands, 1994.

[Forsgren and Murray, 1993] A. L. Forsgren and W. Murray. Newton methods for large-scale linear equality-constrained minimization. *SIAM Journal on Matrix Analysis and Applications*, 14:560–587, 1993.

[Freund, 1991] R. M. Freund. Theoretical efficiency of a shifted-barrier-function algorithm for linear programming. *Linear Algebra and Applications*, 152:19–41, 1991.

[George and Liu, 1981] A. George and J. W.-H. Liu. *Computer solution of large sparse positive definite systems*. Prentice-Hall, Englewood Cliffs, USA, 1981.

[Gill and Murray, 1974] P. E. Gill and W. Murray. Newton-type methods for unconstrained and linearly constrained optimization. *Mathematical Programming*, 28:311–350, 1974.

[Gill and Murray, 1989] P. E. Gill and W. Murray. Constrained optimization. In G.L. Nemhauser, A.H.G. Rinnooy Kan, and M.J. Todd, editors, *Optimization*, Volume 1 of *Handbooks in Operations Research and Management Science*, pages 73–170. North-Holland, Amsterdam, The Netherlands, 1989.

[Gill et al., 1986] P. E. Gill, W. Murray, M. A. Saunders, and M. H. Wright. User's guide for NPSOL (version 4.0): A Fortran package for nonlinear programming. Technical Report SOL86-2, Department of Operations Research, Stanford University, Stanford, USA, 1986.

[Gill et al., 1988] P. E. Gill, W. Murray, M. A. Saunders, and M. H. Wright. Shifted barrier methods for linear programming. Technical Report SOL88-9, Department of Operations Research, Stanford University, Stanford, USA, 1988.

[Gill et al., 1992] P. E. Gill, W. Murray, D. B. Ponceléon, and M. A. Saunders. Preconditioners for indefinite systems arising in optimization. *SIAM Journal on Matrix Analysis and Applications*, 13:292–311, 1992.

[Gill et al., 1993a] P.E. Gill, W. Murray, and M.A. Saunders. Large-scale SQP methods and their application in trajectory optimization. In R. Bulirsch and D. Kraft, editors, *Control Applications of Optimization*, International Series of Numerical Mathematics. Birkhauser, Basel, Boston, Stuttgart, 1993.

[Gill et al., 1993b] P.E. Gill, W. Murray, and M.A. Saunders. Transformed Hessian methods for large-scale constrained optimization. Presented at the Fourth Stockholm Optimization Days, Royal Institute of Technology, Stockholm, 16–17 August, 1993.

[Goldfarb and Todd, 1989] D. Goldfarb and M. J. Todd. Linear programming. In G.L. Nemhauser, A.H.G. Rinnooy Kan, and M.J. Todd, editors, *Optimization*, Volume 1 of *Handbooks in Operations Research and Management Science*, pages 73–170. North-Holland, Amsterdam, The Netherlands, 1989.

[Golub and Loan, 1989] G. H. Golub and C. F. Van Loan. *Matrix Computations*. Johns Hopkins University Press, Baltimore, second edition, 1989.

[Golub et al., 1986] G. H. Golub, P. E. Manneback, and Ph. L. Toint. A comparison between some direct and iterative methods for large scale geodetic least squares problems. *SIAM Journal on Scientific and Statistical Computing*, 7(3):799–816, 1986.

[Gould, 1991] N. I. M. Gould. An algorithm for large-scale quadratic programming. *IMA Journal of Numerical Analysis*, 11(3):299–324, 1991.

[Griewank and Corliss, 1991] A. Griewank and G. F. Corliss. *Automatic Differentiation of Algorithms: Theory, Implementation, and Application*. SIAM, Philadelphia, USA, 1991.

[Griewank and Toint, 1982a] A. Griewank and Ph. L. Toint. On the unconstrained optimization of partially separable functions. In M. J. D. Powell, editor, *Nonlinear Optimization 1981*, pages 301–312. Academic Press, London and New York, 1982.

[Griewank and Toint, 1982b] A. Griewank and Ph. L. Toint. Partitioned variable metric updates for large structured optimization problems. *Numerische Mathematik*, 39:429–448, 1982.

[Griewank et al., 1993] A. O. Griewank, D. Juedes, J. Srinivasan, and C. Tyner. ADOL-C, a package for the automatic differentiation of algorithms written in C/C++. *ACM Transactions on Mathematical Software*, 1993, to appear.

[Griewank, 1989] A. Griewank. On automatic differentiation. In M. Iri and K. Tanabe, editors, *Mathematical Programming: recent developments and applications*, pages 83–108. Kluwer Academic Publishers, Dordrecht, The Netherlands, 1989.

[Gulliksson, 1990] M. Gulliksson. *Algorithms for Nonlinear Least Squares with Applications to Orthogonal Regression*. PhD Thesis, Institute of Information Processing, University of Umeå, S-901 87 Umeå, Sweden, 1990.

[Gulliksson, 1993] M. Gulliksson. Algorithms for weighted nonlinear least squares problems — especially surface fitting problems, Presented at the Fourth Stockholm Optimization Days, Royal Institute of Technology, Stockholm, 16–17 August 1993.

[Hager, 1990] W. W. Hager. Multipliers methods for nonlinear optimal control. *SIAM Journal on Numerical Analysis*, 27(4):1061–1080, 1990.

[Harwell Subroutine Library, 1993] Harwell Subroutine Library. *A catalogue of subroutines (release 11)*. Advanced Computing Department, Harwell Laboratory, Harwell, UK, 1993.

[Hebden, 1973] M. D. Hebden. An algorithm for minimization using exact second derivatives. Technical Report T.P. 515, AERE Harwell Laboratory, Harwell, UK, 1973.

[Hock and Schittkowski, 1981] W. Hock and K. Schittkowski. *Test Examples for Nonlinear Programming Codes*, Volume 187 of *Lectures Notes in Economics and Mathematical Systems*. Springer Verlag, Berlin, 1981.

[International Business Machines Corporation, 1990] International Business Machines Corporation. *Optimization Subroutine Library: Guide and Reference*, second edition, 1990.

[Jensen and Polyak, 1993] D. Jensen and R. Polyak. On the convergence of a modified barrier method for convex programming. Research Report RC 18570, IBM T. J. Watson Research Center, Yorktown Heights, USA, 1993.

[Jensen et al., 1992] D. Jensen, R. Polyak, and R. Schneur. Numerical experience with modified barrier functions method for linear programming. Research Report RC 18415, IBM T. J. Watson Research Center, Yorktown Heights, USA, 1992.

[Jittorntrum and Osborne, 1980] K. Jittorntrum and M. Osborne. A modified barrier function method with improved rate of convergence for degenerate problems. *Journal of the Australian Mathematical Society (Series B)*, 21:305–329, 1980.

[Jónasson and Madsen, 1992] K. Jónasson and K. Madsen. Corrected sequential linear programming for sparse minimax optimization. Technical Report NI-92-06, Institute for Numerical Analysis, Technical University of Denmark, 2800 Lyngby, Denmark, 1992.

[Jones, 1967] A. P. Jones. The chemical equilibrium problem: An application of SUMT. Technical Report RAC-TP-272, Research Analysis Corporation, Research Analysis Corporation, McLean, USA, 1967.

[Júdice and Pires, 1989] J. J. Júdice and F. M. Pires. Direct methods for convex quadratic programs subject to box constraints. Technical Report, Universidade de Coimbra, Coimbra, Portugal, 1989.

[Júdice and Pires, 1993] J. J. Júdice and F. M. Pires. A block principal pivoting algorithm for large-scale strictly monotone linear complementary problems. Technical Report, Universidade de Coimbra, Coimbra, Portugal, 1993.

[Júdice, 1994] J. J. Júdice. Algorithms for linear complementarity problems. In E. Spedicato, editor, *Algorithms for continuous optimization: the state of the art*. Kluwer Academic Publishers, Dordrecht, The Netherlands, 1994.

[Karmarkar, 1984] N. Karmarkar. A new polynomial-time algorithm for linear programming. *Combinatorica*, 4:373–395, 1984.

[Kaufman and Sylvester, 1993] L. Kaufman and G. Sylvester. Separable nonlinear least-squares with multiple right-hand sides. *SIAM Journal on Matrix Analysis and Applications*, 13:68–89, 1993.

[Koontz et al., 1985] J.E. Koontz, R.B. Schnabel, and B.E. Weiss. A modular system of algorithms for unconstrained minimization. *ACM Transactions on Mathematical Software*, 11:419–440, 1985. Also available as Technical Report CU-CS-240-82, Department of Computer Science, University of Colorado, Boulder, USA.

[Kunish and Sachs, 1992] K. Kunish and E. W. Sachs. Reduced SQP methods for parameter identification problems. *SIAM Journal on Numerical Analysis*, 29(6):1793–1822, 1992.

[Lalee et al., 1993] M. Lalee, J. Nocedal, and T. Plantenga. On the implementation of an algorithm for large-scale equality constrained optimization. EE and CS Technical Report, Northwestern University, Evanston, USA, 1993.

[Lasdon et al., 1978] L. S. Lasdon, A. D. Waren, A. Jain, and M. Ratner. Design and testing of a generalized reduced gradient code for nonlinear programming. *ACM Transactions on Mathematical Software*, 4:34–50, 1978.

[Lescrenier, 1991] M. Lescrenier. Convergence of trust region algorithms for optimization with bounds when strict complementarity does not hold. *SIAM Journal on Numerical Analysis*, 28(2):476–495, 1991.

[Levenberg, 1944] K. Levenberg. A method for the solution of certain problems in least squares. *Quarterly Journal on Applied Mathematics*, 2:164–168, 1944.

[Levitin and Polyak, 1966] E. S. Levitin and B. T. Polyak. Constrained minimization problems. *USSR Comput. Math. and Math. Phys.*, 6:1–50, 1966.

[Li, 1993a] Y. Li. A globally convergent method for l_p problems. *SIAM Journal on Optimization*, 3:609–629, 1993.

[Li, 1993b] Y. Li. Solving l_p-norm problems and applications. Technical Report CTC93TR122, Cornell Theory Center, Cornell University, Ithaca, New York State, 1993.

[Liao, 1993] Aiping Liao. Some efficient algorithms for unconstrained discrete-time optimal control problems. Technical Report CTC93TR159, Advanced Computing Research Institute, Cornell Theory Center, Cornell University, Ithaca, USA, 1993.

[Liu and Nocedal, 1988] D. C. Liu and J. Nocedal. Test results of two limited memory methods for large scale optimization. Technical Report NAM 04, Department. of EE and CS, Northwestern University, Evanston, Illinois 1988.

[Liu and Nocedal, 1989] D. C. Liu and J. Nocedal. On the limited memory BFGS method for large scale optimization. *Mathematical Programming, Series B*, 45:503–528, 1989.

328

[Marquardt, 1963] D. Marquardt. An algorithm for least-squares estimation of non-linear parameters. *SIAM Journal on Applied Mathematics*, 11:431–441, 1963.

[McCormick and Sofer, 1991] G. P. McCormick and A. Sofer. Optimization with unary functions. *Mathematical Programming*, 52:167–179, 1991.

[McCormick, 1969] G. P. McCormick. Anti-zig-zagging by bending. *Management Science*, 15(5):315–320, 1969.

[McCormick, 1972] G. P. McCormick. Computational aspects of nonlinear programming solutions to large-scale inventory problems. Technical Report Technical Memorandum Serial T-63488, Department of Operations Research, George Washington University, Washington DC 20052, 1972.

[Moré and Toraldo, 1989] J. J. Moré and G. Toraldo. Algorithms for bound constrained quadratic programming problems. *Numerische Mathematik*, 14:14–21, 1989.

[Moré and Toraldo, 1991] J. J. Moré and G. Toraldo. On the solution of large quadratic programming problems with bound constraints. *SIAM Journal on Optimization*, 1(1):93–113, 1991.

[Moré and Wright, 1993] J. J. Moré and S. J. Wright. *Optimization Software Guide*. SIAM, Philadelphia, USA, 1993.

[Moré et al., 1981] J. J. Moré, B. S. Garbow, and K. E. Hillstrom. Testing unconstrained optimization software. *ACM Transactions on Mathematical Software*, 7(1):17–41, 1981.

[Moré, 1978] J. J. Moré. The Levenberg-Marquardt algorithm: implementation and theory. In G. A. Watson, editor, *Proceedings Dundee 1977*. Springer Verlag, Berlin, 1978. Lecture Notes in Mathematics.

[Moré, 1983] J. J. Moré. Recent developments in algorithms and software for trust region methods. In A. Bachem, M. Grötschel, and B. Korte, editors, *Mathematical Programming: The State of the Art*, pages 258–287. Springer Verlag, Berlin, 1983.

[Munksgaard and Reid, 1983] N. A. Munksgaard and J. K. Reid. NS02, a Fortran subroutine for solving sparse sets of non-linear equations by Powell's dog-leg algorithm. Technical Report R11047, AERE Harwell Laboratory, Harwell, UK, 1983.

[Murray and Wright, 1976] W. Murray and M. H. Wright. Efficient line search algorithms for the logarithmic barrier function. Technical Report SOL76-18, Department of Operations Research, Stanford University, Stanford, USA, 1976.

[Murray and Wright, 1992] W. Murray and M. H. Wright. Line search procedures for the logarithmic barrier function. Numerical analysis manuscript 92-01, AT&T Bell Laboratories, Murray Hill, USA, 1992.

[Murtagh and Saunders, 1987] B. A. Murtagh and M. A. Saunders. MINOS 5.1 USER'S GUIDE. Technical Report SOL83-20R, Department of Operations Research, Stanford University, Stanford, USA, 1987.

[Nash and Sofer, 1991] S. G. Nash and A. Sofer. A general-purpose parallel algorithm for unconstrained optimization. *SIAM Journal on Optimization*, 1:530–547, 1991.

[Nash and Sofer, 1993] S. G. Nash and A. Sofer. A barrier method for large-scale constrained optimization. *ORSA Journal on Computing*, 5(1):40–53, 1993.

[Nash et al., 1993] S. G. Nash, R. Polyak, and A. Sofer. A numerical comparison of barrier and modified barrier methods for large-scale constrained optimization. Technical Report 93-02, Department of Operations Research, George Mason University, Fairfax, USA, 1993.

[Nemhauser et al., 1989] G. L. Nemhauser, A. H. G. Rinnooy Kan, and M. J. Todd. *Optimization*, Volume 1 of *Handbooks in Operations Research and Management Science*. North-Holland, Amsterdam, 1989.

[Omojokun, 1991] E. O. Omojokun. *Trust region algortihms for optimization with nonlinear equality and inequality constraints*. PhD Thesis, Department of Computer Sciences, University of Colorado, Boulder, USA, 1991.

[Polyak, 1992] R. Polyak. Modified barrier functions (theory and methods). *Mathematical Programming*, 54(2):177–222, 1992.

[Powell, 1970] M. J. D. Powell. A new algorithm for unconstrained optimization. In J. B. Rose, O. L. Mangasarian, and K. Ritter, editors, *Nonlinear Programming*. Academic Press, New York, 1970.

[Powell, 1975] M. J. D. Powell. Convergence properties of a class of minimization algorithms. In O. L. Mangasarian, R.R. Meyer, and S.M. Robinson, editors, *Nonlinear Programming, 2*. Academic Press, New York, 1975.

[Powell, 1982] M.J.D. Powell. Extensions to subroutine VF02. In R.F. Drenick and F. Kozin, editors, *Systems Modelling and Optimization. Lecture notes in control and Information sciences 38*, pages 529 – 538. Springer-Verlag, Berlin, 1982.

[Powell, 1992] M. J. D. Powell. Some convergence properties of the shifted log barrier method for linear programming. Technical Report DAMTP NA7, Department of Applied Mathematics and Theoretical Physics, Cambridge University, Cambridge, UK, 1992.

[Robinson, 1972] S. M. Robinson. A quadratically convergent algorithm for general nonlinear programming problems. *Mathematical Programming*, 3:145–156, 1972.

[Rosen and Kreuser, 1972] J. B. Rosen and J. Kreuser. A gradient projection algorithm for nonlinear constraints. In F. A. Lootsma, editor, *Numerical Methods for Nonlinear Optimization*, pages 39–43. Academic Press, London, 1972.

[Saludjian, 1993] L. Saludjian. Etude d'une version parallèle de Lancelot dans l'environment de programmation par tranferts de message PVM. Rapport de stage de diplome d'études approfondies, Toulouse, France, 1993.

[Sartenaer, 1993] A. Sartenaer. A class of trust region methods for nonlinear network optimization problems, including numerical experiments. Technical Report 93/21, Department of Mathematics, FUNDP, Namur, Belgium, 1993. To appear in SIAM Journal on Optimization.

[Schittkowski, 1987] K. Schittkowski. *More Test Examples for Nonlinear Programming Codes*, Volume 282 of *Lecture notes in economics and mathematical systems*. Springer Verlag, Berlin, 1987.

[Schlick, 1993] T. Schlick. Modified Cholesky factorizations for sparse preconditioners. *SIAM Journal on Scientific and Statistical Computing*, 14(2):424–445, 1993.

[Schnabel and Chow, 1991] R.B. Schnabel and T.-T. Chow. Tensor methods for unconstrained optimization using second derivatives. *SIAM Journal on Optimization*, 1(3):293–315, 1991.

[Schnabel and Eskow, 1991] R. B. Schnabel and E. Eskow. A new modified Cholesky factorization. *SIAM Journal on Scientific and Statistical Computing*, 11:1136–1158, 1991.

[Schnabel and Frank, 1984] R. B. Schnabel and P. D. Frank. Tensor methods for nonlinear equations. *SIAM Journal on Numerical Analysis*, 21(5):815–843, 1984.

[Schnabel, 1994] R. B. Schnabel. Parallel nonlinear optimization. In E. Spedicato, editor, *Algorithms for continuous optimization: the state of the art*. Kluwer Academic Publishers, Dordrecht, The Netherlands, 1994.

[Schrady and Choe, 1971] D. A. Schrady and U. C. Choe. Models for multi-item continuous review inventory policies subject to constraints. *Naval Research Logistics Quarterly*, 18:451–463, 1971.

[Shanno and Phua, 1980] D. F. Shanno and K. H. Phua. Remark on algorithm 500. *ACM Transactions on Mathematical Software*, 6:618–622, 1980.

[Shanno, 1994] D. F. Shanno. Algorithms for linear programming. In E. Spedicato, editor, *Algorithms for continuous optimization: the state of the art*. Kluwer Academic Publishers, Dordrecht, The Netherlands, 1994.

[Sklar and Armstrong, 1993] M. G. Sklar and R. D. Armstrong. Lagrangian approach for large-scale distance value estimation. *Computers Operations Research*, 20:83–93, 1993.

[Smith and Lasdon, 1992] S. Smith and L. Lasdon. Solving large sparse nonlinear programs using GRG. *ORSA Journal on Computing*, 4:2–15, 1992.

[Soares et al., 1993] J. Soares, J. J. Júdice, and F. Facchinei. An active set Newton's algorithm for large-scale nonlinear programs with box constraints. Technical Report, Universidade de Coimbra, Coimbra, Portugal, 1993.

[Sorensen, 1981] D. C. Sorensen. An example concerning quasi-Newton estimates of a sparse Hessian. *SIGNUM Newsletter*, 16:8–10, 1981.

[Steihaug, 1983a] T. Steihaug. The conjugate gradient method and trust regions in large scale optimization. *SIAM Journal on Numerical Analysis*, 20(3):626–637, 1983.

[Steihaug, 1983b] T. Steihaug. Local and superlinear convergence for truncated iterated projection methods. *Mathematical Programming*, 27:199–223, 1983.

[Toint and Tuyttens, 1990] Ph. L. Toint and D. Tuyttens. On large scale nonlinear network optimization. *Mathematical Programming, Series B*, 48(1):125–159, 1990.

[Toint and Tuyttens, 1992] Ph. L. Toint and D. Tuyttens. LSNNO: a Fortran subroutine for solving large scale nonlinear network optimization problems. *ACM Transactions on Mathematical Software*, 18(3):308–328, 1992.

[Toint, 1981a] Ph. L. Toint. A sparse quasi-Newton update derived variationally with a non-diagonally weighted Frobenius norm. *Mathematics of Computation*, 37(156):425–433, 1981.

[Toint, 1981b] Ph. L. Toint. Towards an efficient sparsity exploiting Newton method for minimization. In I. S. Duff, editor, *Sparse Matrices and Their Uses*. Academic Press, London, 1981.

[Toint, 1983] Ph. L. Toint. Test problems for partially separable optimization and results for the routine PSPMIN. Technical Report 83/4, Department of Mathematics, FUNDP, Namur, Belgium, 1983.

[Toint, 1987a] Ph. L. Toint. On large scale nonlinear least squares calculations. *SIAM Journal on Scientific and Statistical Computing*, 8(3):416–435, 1987.

[Toint, 1987b] Ph. L. Toint. VE10AD, a routine for large scale nonlinear least squares. *Harwell Subroutine Library*, 1987.

[Vanderbei and Carpenter, 1993] R. J. Vanderbei and T. J. Carpenter. Symmetric indefinite systems for interior point methods. *Mathematical Programming, Series A*, 58(1):1–32, 1993.

[Werbos, 1988] P. Werbos. Backpropagation: past and future. In *Proceedings of the 2nd International Conference on Neural Networks*. IEEE, New York, 1988.

[Wright, 1991] M. H. Wright. Optimization and large scale computation. In J.P. Mesirov, editor, *Very Large Scale Computation in the 21st Century*, pages 341–407. SIAM, Philadelphia, 1991.

[Wu, 1993] Z. Wu. The effective energy transformation scheme as a general continuation approach to global optimization with application to molecular conformation. Technical Report CTC93TR143, Advanced Computing Research Institute, Cornell Theory Center, Cornell University, Ithaca, USA, 1993.

[Zenios and Mulvey, 1986] S. A. Zenios and J. M. Mulvey. Nonlinear network programming on vector supercomputers: A study on the CRAY X-MP. *Operations Research*, 34(5):667–682, 1986.

[Zenios and Mulvey, 1988] S. A. Zenios and J. M. Mulvey. Vectorization and multitasking of nonlinear network programming algorithms. *Mathematical Programming, Series A*, 42(2):449–470, 1988.

[Zenios and Pinar, 1989] S. A. Zenios and M. C. Pinar. Parallel block-partitioning of truncated Newton for nonlinear network optimization. Technical Report 89-09-08, Decision Sciences Department, The Wharton School, University of Pennsylvania, Philadelphia, USA, 1989.

[Zenios, 1989] S. A. Zenios. Parallel optimization: current status and an annotated bibliography. *ORSA Journal on Computing*, 1:20–43, 1989.

[Zou et al., 1993] X. Zou, I.M. Navon, M. Berger, K. H. Phua, T. Schlick, and F. X. LeDimet. Numerical experience with limited-memory quasi-Newton and truncated Newton methods. *SIAM Journal on Optimization*, 3(3):582–608, 1993.

ABS Methods for Nonlinear Optimization

EMILIO SPEDICATO
Department of Mathematics
University of Bergamo
Piazza Rosate 2
24100 Bergamo
Italy

ZUNQUAN XIA
Department of Applied Mathematics
Dalian University of Technology
Dalian 116024
P.R. China

ABSTRACT. ABS methods have been introduced by Abaffy, Broyden and Spedicato (1984) initially for solving general linear systems and have been later extended to solving linear least squares and nonlinear systems. Applications to nonlinear optimization have been considered recently mainly by Chinese researchers. In this paper we present the basic results on ABS methods for linear systems and then we consider in more detail a number of applications to nonlinear optimization, particularly the construction of general Quasi-Newton updates, subject possibly to sparsity, and a unified formulation of feasible descent direction methods for linearly constrained optimization.

1 The Basic ABS Class

ABS methods have been introduced by Abaffy, Broyden and Spedicato (1984) initially for solving a determined or underdetermined linear system $Ax = b$, $A \in R^{m,n}$, $x \in R^n$, $b \in R^m$, $m \leq n$, by the so called basic or unscaled ABS class of algorithms defined by the following procedure, where e_i denotes the ith unit vector:

(A) Give $x_1 \in R^n$, an arbitrary estimate of the solution x^+; give $H_1 \in R^{n,n}$, an arbitrary nonsingular matrix; set $i = 1$.

(B) Compute $s_i = H_i a_i$. If $s_i \neq 0$ go to (C). If $s_i = 0$ and $a_i^T x_i - b^T e_i = 0$, set $x_{i+1} = x_i$, $H_{i+1} = H_i$ and go to (F), the ith equation is a linear combination of the previous equations; otherwise stop, the system has no solution.

E. Spedicato (ed.), Algorithms for Continuous Optimization, 333–356.

(C) Compute the search vector p_i by

$$p_i = H_i^T z_i \tag{1.1}$$

where $z_i \in R^n$ is an arbitrary vector satisfying $z_i^T H_i a_i \neq 0$.

(D) Update the estimate of the solution by

$$x_{i+1} = x_i - \alpha_i p_i \tag{1.2}$$

where the stepsize α_i is given by

$$\alpha_i = (a_i^T x_i - b^T e_i)/p_i^T a_i \tag{1.3}$$

(E) Update the (Abaffian) matrix H_i by

$$H_{i+1} = H_i - H_i a_i w_i^T H_i / w_i^T H_i a_i \tag{1.4}$$

where $w_i \in R^n$ is an arbitrary vector satisfying $w_i^T H_i a_i \neq 0$.

(F) If $i = m$, stop, x_{m+1} solves the system, otherwise increment i by one and go to (B).

Basic properties of the above procedure are the following:

(A1) $H_i a_i = 0$ iff a_i is a linear combination of $a_1, \ldots, a_{i-1}$

(A2) the search vectors $p_1, \ldots, p_i$ are linearly independent

(A3) let $A_i = (a_1, \ldots, a_i)$, $P_i = (p_1, \ldots, p_i)$ and define L_i by

$$L_i = A_i^T P_i \tag{1.5}$$

then, if $\text{rank}(A_i) = i$, L_i is nonsingular lower triangular

(A4) the vector x_{i+1} solves the first i equations of the system, i.e. $a_j^T x = b^T e_j$, $j = 1, \ldots, i$

(A5) the general solution of the first i equations has the form

$$x = x_{i+1} + H_{i+1}^T q \tag{1.6}$$

with $q \in R^n$ arbitrary

(A6) $\text{Null}(H_{i+1}) = \text{Range}(A_i)$ implying $H_{i+1} A_i = 0$
$\text{rank}(H_{i+1}) = n - i$ if $\text{rank}(A_i) = i$
$\text{Null}(A_i) = \text{Range}(H_{i+1}^T)$

(A7) $H_i H_1^{-1} H_i = H_i$

The matrices defined by recursion (1.4) have been denominated Abaffian matrices at the first international conference on ABS methods, held in Luoyang, September 1991.

2 The Scaled ABS Class

Let $V = (v_1, \ldots, v_m) \in R^{m,m}$ be an arbitrary nonsingular matrix, called the scaling matrix. Then the scaled system $V^T A x = V^T b$ is equivalent to the system $Ax = b$ in the sense that both systems possess the same set of solutions. Let $r(x) = Ax - b$ be the residual of the original system. Let $r_i = r(x_i)$. Then by applying the basic ABS class procedure to the above scaled system we obtain the following recursions, which define the scaled ABS class procedure:

(A') Same as (A)

(B') Compute $s_i = H_i A^T v_i$. If $s_i \neq 0$ go to (C'). If $s_i = 0$ and $r_i^T v_i = 0$ set $x_{i+1} = x_i$, $H_{i+1} = H_i$ and go to (F'), the ith equation of both the original and the scaled system is a linear combination of the previous equations. Otherwise stop, the system has no solutions.

(C') Compute the search vector as in (C'), with $z_i \in R^{n,n}$ arbitrary save that $z_i^T H_i A^T v_i \neq 0$.

(D') Update the estimate of the solution by (1.2) with α_i given by

$$\alpha_i = r_i^T v_i / p_i^T A^T v_i \tag{2.1}$$

(E') Update the Abaffian by

$$H_{i+1} = H_i - H_i A^T v_i w_i^T H_i / w_i^T H_i A^T v_i$$

where $w_i \in R^n$ is an arbitrary vector satisfying $w_i^T H_i A^T v_i \neq 0$.

(F') Same as (F).

We note that at the ith step of the scaled ABS class procedure only the ith column v_i of V is used, implying that V needs not to be defined initially and that v_i can be considered as a parameter, arbitrary save for linear independence from $v_1, \ldots, v_{i-1}$. We note also that x_{m+1} solves both the scaled and the original systems, hence the scaled ABS class is a generalization, with the extra scaling parameter v_i, of the basic class for solving $Ax = b$. It can be shown that essentially any algorithm of the form $x_{i+1} = x_i - \alpha_i p_i$ which, starting from an arbitrary x_1, solves a linear system in a number of steps no greater than the number of equations is a member of the scaled ABS class, i.e. it corresponds to some parameter choices in that class.

Properties (A1) to (A7) are easily reformulated for the scaled ABS class. In particular the factorization relation (1.5) becomes now, with $V_i = (v_1, \ldots, v_i)$

$$V_i^T A_i^T P_i = L_i \tag{2.2}$$

Property (A5) now says that the residual r_{i+1} is orthogonal to the previous scaling vectors, i.e.

$$r_{i+1}^T v_j = 0, \quad j = 1, \ldots, i \tag{2.3}$$

3 Alternative Implementations

The basic and the scaled ABS class have been given here in the so called standard version, based upon the n by n Abaffian matrix H_i. There are several alternative formulations, some discussed in Abaffy and Spedicato (1989), others in more recent papers, e.g. Bodon and Spedicato (1991), Chen, Deng and Xue (1992), Spedicato and Zhu (1993), essentially based upon the use of non square matrices, resulting generally in a reduction of both storage and overhead. Due to space reasons we will consider only some of these versions and only with reference to the unscaled class.

The first version was derived by analogy with the memoryless Quasi-Newton method and is particularly convenient if one is dealing with underdetermined systems with $m \ll n$. It requires, at step i, the storage of $2i - 2$ vectors. It uses the following recursions, started with $s_1 = H_1 a_1$ and $u_1 = H_1^T w_1$

$$p_i = H_1^T z_i - \sum_{j<i} u_j s_j^T z_i \tag{3.1}$$

$$s_i = H_1 a_i - \sum_{j<i} s_j u_j^T a_i \tag{3.2}$$

$$u_i = H_1^T w_i - \sum_{j<i} u_j s_j^T w_i \tag{3.3}$$

Notice that

$$H_{i+1} = H_1 - \sum_{j<i} s_j u_j^T \tag{3.4}$$

Another approach assumes, without loss of generality, that $w_i = z_i/z_i^T H_i a_i$ and that feasible parameters $z_1, \ldots, z_n$ are given initially. Then $p_i = u_i^i$ where $u_j^1 = H_1^T z_j$, $j = 1, \ldots, n$, and the vectors u_j^i are updated, for $i = 1, \ldots, m$, by

$$u_j^{i+1} = u_j^i - (a_i^T u_j^i / a_i^T u_i^i) u_i^i, \quad j = i+1, \ldots, n \tag{3.5}$$

This approach, related to a class of parallel methods considered by Sloboda (1978) requires the storage of $n - i$ vectors at step i. Note that the linear variety comprising all solutions of the first i equations consists of the vectors x having the form

$$x = x_{i+1} + U_i d \tag{3.6}$$

where $d \in R^{n-i}$ is arbitrary and $U^i = (u_{i+1}^{i+1}, \ldots, u_n^{i+1})$.

The version of Spedicato and Zhu (1993) uses formulas formally similar to those in the basic ABS procedure save that the matrix H_i is not n by n but is $n + 1 - i$ by n. At each step after the update of H_i a row of H_{i+1} is deleted. The deleted row is dependent on the remaining rows. It is proved that if the kth component of w_i is nonzero then the kth row of H_i can be deleted. A difference with the version defined by recursions (3.5) is that the vectors z_i and w_i have now dimension $n + 1 - i$ and must not be specified in advance.

4 Particular Algorithms and Numerical Experiments

We consider now some special choices of the parameters in the ABS class, defining algorithms which are related to well known methods but whose ABS formulation may differ in computational complexity, in storage requirement, numerical stability and degree of parallelization.

4.1 THE HUANG AND THE MODIFIED HUANG ALGORITHMS

The Huang algorithm, originally considered by Huang (1975) in a paper which has been seminal for the development of the ABS class, is based upon the well defined choices $H_1 = I$, $z_i = a_i$, $w_i = a_i$, which provide

$$p_i = H_i a_i \tag{4.1}$$

$$H_{i+1} = H_i - H_i a_i a_i^T H_i / a_i^T H_i a_i \tag{4.2}$$

The search vectors generated by the Huang algorithm are orthogonal and coincide, in exact arithmetic, with the vectors generated by the Gram-Schmidt orthogonalization procedure applied on the rows of A. If $x_1 = 0$ then x_{i+1} is the solution of least Euclidean norm of the first i equations and moreover the solution x^+ (of least Euclidean norm) is approached monotonically from below (in Euclidean norm).

The modified Huang algorithm is a modification of the previous algorithm which, while generating the same iterates in exact arithmetic, is more accurate in presence of roundoff, as shown for instance in the experiments of Spedicato and Vespucci (1992). It is based upon the formulas

$$p_i = H_i(H_i a_i) \tag{4.3}$$

$$H_{i+1} = H_i - p_i p_i^T / p_i^T p_i \tag{4.4}$$

From (4.3) we see that the search vector in the original Huang algorithm is re-projected onto the range of H_i, which coincides with the null space of A_{i-1}. This operation tends to annihilate components of the original Huang search vector in the null space of H_i which are not zero due to roundoff. A theoretical analysis of Broyden (1991) shows that the error growth of the $i - 1$ small eigenvalues of H_i, which should be zero if there were no roundoff, is two orders lower if formulas (4.3) and (4.4) are used, resulting actually, under simplifying assumptions, in no further error growth.

The reprojection technique is applicable to most ABS algorithms. It has been shown numerically to substantially improve the accuracy of several methods in ABS formulation, as the QR, Craig, Hestenes-Stiefel algorithms, see for instance Bodon and Spedicato (1990, 1992).

4.2 THE SYMMETRIC ALGORITHM AND THE ALGORITHM OF XIA

This algorithm generalizes the Huang algorithm by letting the initial matrix to be an arbitrary positive definite matrix. Properties of the iterates are consequently changed, for instance the search vectors are now H_1^{-1}-conjugate and if x_1 is zero then x_{i+1} is the solution of least H_1-weighted Euclidean norm. An important choice for H_1, introduced by Xia (1990a) and giving the Xia algorithm, consists in taking an approximation of the inverse Hessian. The Xia algorithm plays a fundamental role in quadratic programming, as discussed in later sections.

4.3 THE IMPLICIT LU OR GAUSS-CHOLESKI ALGORITHM

This algorithm is obtained by the choices $H_1 = I$, $z_i = w_i = e_i$. It is well defined iff A is strongly nonsingular (all the principal submatrices are nonsingular), in which case the scalar product $e_i^T H_i a_i$ appearing at the denominators of (1.3) and (1.4) is identical with the pivot at the ith stage of Gaussian elimination. The matrix P_i in (1.5) is upper triangular, motivating the name. The formulas for the search vector and the Abaffian update are

$$p_i = H_i^T e_i \tag{4.5}$$

$$H_{i+1} = H_i - H_i a_i p_i^T / p_i^T a_i \tag{4.6}$$

From (4.5) p_i is just the ith row of H_i. From (4.6) it follows easily that H_{i+1} has the following structure

$$H_{i+1} = \begin{bmatrix} 0 & 0 \\ S_i & I_{n-i} \end{bmatrix} \tag{4.7}$$

where $S_i \in R^{n-i,i}$ can be shown to have the following structure

$$S_i = -\hat{A}^i (A^{(i)})^{-1} \tag{4.8}$$

where $A^{(i)}$ is the ith principal submatrix of A^T and $\hat{A}^i \in R^{n-i,i}$ is the matrix comprising the last $n - i$ columns of A_i.

It is easy to verify that the number of multiplications required by the implicit LU algorithm is $n^3/3 + O(n^2)$, as for the classical algorithm, while the maximum storage is $n^2/4 + O(n)$. Some other properties are:

- the search vectors are A-semiconjugate, i.e. $p_j^T A p_i = 0,\ j < i$. Hence A-conjugacy holds if A is symmetric but not necessarily positive definite

- if A is symmetric and positive definite then $e_i^T H_i a_i > 0$ and if we set $z_i = e_i/(e_i^T H_i a_i)^{1/2}$ then $P_i^{-1} = L_i^T$ in (1.5), hence the algorithm implicitly generates the Choleski factorization

- if A is symmetric and positive definite then x_{i+1} minimizes the quadratic function $F(x) = (x - x^+)^T A(x - x^+)$ over the linear variety $x_1 + \text{Range}(P_i)$. It also follows that x^+ is approached monotonically from above in the A-weighted Euclidean norm

4.4 THE CONJUGATE DIRECTION SUBCLASS

This is a subclass of the scaled ABS class where $v_i = p_i$, a choice which is well defined if A is square symmetric and positive definite. From the factorization relation (2.2) we obtain, with $P = P_n$

$$P^T A P = D \qquad (4.9)$$

with D diagonal, implying that the search directions are A-conjugate. This subclass contains the Lanczos and the Hestenes-Stiefel algorithms, the last one corresponding to $H_1 = I,\ z_i = w_i = r_i$. It can be shown that also the implicit LU algorithm is a member of this class.

4.5 THE ORTHOGONALLY SCALED SUBCLASS

This subclass is obtained by the choice $v_i = A p_i$ and is well defined for A with full column rank. From (2.2) we obtain, with $V = V_n$

$$V^T V = P^T A^T A P = D \qquad (4.10)$$

hence the search vectors are $A^T A$-conjugate and the scaling vectors are orthogonal. Two important algorithms in this class are the implicit QR algorithm, given by $H_1 = I,\ z_i = w_i = e_i$, for which P_i in (2.2) is upper triangular, and the algorithm with $H_1 = I,\ z_i = w_i = A^T r_i$, which generates the same iterates x_i as the minimum residual conjugate gradient method. The vector x_{i+1} in this subclass minimizes the function $F(x) = r(x)^T r(x)$ over the linear variety $x_1 + \text{Range}(P_i)$.

4.6 THE OPTIMALLY STABLE SUBCLASS

This subclass, well defined for A square nonsingular, corresponds to the choice $v_i = A^{-T}p_i$; the inverse appearing in this definition can be removed in the actual recursions, which are also well defined for underdetermined systems. Taking without loss of generality $z_i = w_i$ the Abaffian update can be written in the form

$$H_{i+1} = H_i - p_i p_i^T / p_i^T p_i \tag{4.11}$$

Setting $z_i = A^T u_i$ the stepsize can be written as

$$\alpha_i = r_i^T u_i / a_i^T p_i \tag{4.12}$$

The search vectors in this subclass are orthogonal. Setting $u_i = e_i$ we get the Huang algorithm. Setting $u_i = r_i$ we obtain an algorithm generating the same iterates x_i as Craig's conjugate gradient algorithm for nonsymmetric systems.

The name of this class comes from the fact that the scaling matrix satisfies the relation

$$V^T A A^T V = D \tag{4.13}$$

with D diagonal, which was shown by Broyden (1985) to characterize the class of algorithms, in his general class equivalent to the scaled ABS class, such that the error in x_{n+1} due to the introduction of a single error ε in x_i is minimized (being actually not greater than ε). It can be shown that the algorithms in the unscaled and in the orthogonally scaled subclass minimize a similar error with respect to the residual r_{n+1}. Hence the Huang algorithm, being a member of both the unscaled and the optimally stable subclasses, possesses both properties, while the implicit LU and QR algorithms are optimal only with respect to the residual error.

4.7 Numerical Experiments

There is already a large body of numerical results on ABS methods. They have shown that these methods can be implemented in a numerically stable way and can be more accurate and faster, especially on vector/parallel computers, than their corresponding classical versions.

The numerical experience relates to the Huang, implicit LU and QR algorithms and to the ABS versions of the Hestenes-Stiefel, STOD and Craig algorithms. The results depend significantly on which version is used of the several alternative formulations of the ABS algorithms. Of particular importance for full problems are the results of Bodon (1993a), who has tested about fifty implementations of the Huang, modified Huang and implicit LU algorithm on an Alliant computer with 8 processors. Her results show that several ABS methods are more accurate (about one order) and faster (up to three times) than the LAPACK code implementing the LU algorithm. In these implementations use was made also of the so called block ABS formulation, see Abaffy and Spedicato (1989), which allows to deal with blocks of equations via recursions which are the natural extension of those characterizing the non block formulation. The following Table from Bodon (1993a) gives results for the LAPACK code, the standard implicit LU and modified Huang algorithm (indicated by the symbols LU and MHUANG) and two block formulations of the implicit LU (LUBLOCK) and the modified Huang algorithm (MHUANGBLOCK). Symbols ER1, ER2, SEC, MFLOP indicate respectively the relative error in the solution, in the residual (versus the norm of b), the time in seconds and the number of megaflops. The superiority of the ABS algorithms comes from the much higher number of achievable megaflops. Notice the higher accuracy obtained by the modified Huang algorithm. The tested problem has coefficient matrix and solution with random integer entries. Computations were done in single precision.

n	ALGORITHM	ER1	ER2	SEC	MFLOP
	LAPACK	3E$-$5	1E$-$6	0.201	3.39
	LU	5E$-$5	7E$-$7	0.100	10.35
100	LUBLOCK	5E$-$4	9E$-$7	0.080	11.40
	MHUANG	7E$-$7	2E$-$7	0.240	21.08
	MHUANGBLOCK	4E$-$6	2E$-$7	0.155	19.69
	LAPACK	1E$-$4	8E$-$6	6.233	13.43
	LU	8E$-$5	2E$-$6	6.749	18.62
500	LUBLOCK	2E$-$4	3E$-$6	3.701	21.17
	MHUANG	2E$-$5	4E$-$7	14.26	35.06
	MHUANGBLOCK	1E$-$5	3E$-$7	4.19	66.41

ABS methods can be implemented in special way on problems with structured or unstructured sparsity pattern. Bodon (1992a,b) has used a special version of the implicit LU algorithm for banded and upper banded (or block angular) systems. The version for banded systems is faster than the corresponding code in LAPACK on the Alliant FX/80 for band width less than 17; it is about three times faster on tridiagonal systems. Spedicato and Tuma (1993) have considered for general sparse systems a Markowitz type strategy coupled with a threshold strategy to zero elements of H_i in the implicit LU algorithm. They have found classes of problems where the obtained algorithm, albeit still at a preliminary stage of development, outperforms the classic LU algorithm, as implemented in the MA28 routine, in both accuracy and speed. Similar results have been obtained by Benzi and Meyer (1993) using the Sloboda type version of the implicit LU algorithm. Finally, theoretical results of Zhu (1987) and Zhu and Yang (1993) show that the implicit LU algorithm can be efficiently implemented on nested dissection type matrices, with a storage requirement lower, and getting relatively better with growing dimension, than that needed by the Choleski method.

5 Solving Linear Least Squares Problems

Overdetermined linear systems can be solved by several ABS approaches for their generalized solution in the least squares sense, i.e. for the vector x^* that solves the normal equations of Gauss

$$A^T A x = A^T b \tag{5.1}$$

System (5.1) is always compatible. The solution is unique if A has full column rank, otherwise one usually is interested in the solution of least Euclidean norm, formally given by relation

$$x^* = A^* b \tag{5.2}$$

where A^* is the Moore-Penrose pseudoinverse.

There are several ABS approaches for the least squares generalized solution. One possibility is obviously to compute a QR or LQ factorization of A by making explicit the implicit factorization associated with the implicit QR or the Huang (or modified Huang) algorithms. Then such a factorization can be used in the traditional way.

A second possibility is to compute the pseudoinverse A^* by ABS techniques, see Spedicato and Bodon (1989) or Spedicato and Xia (1992c). Another approach is based upon the equivalence of the normal equations with the following extended system in the variables x and y

$$Ax = y \tag{5.3}$$

$$A^T y = A^T b \tag{5.4}$$

In order that (5.3) is solvable we must have $y \in \text{Range}(A)$, which implies that y must be the unique solution of least Euclidean norm of the underdetermined system (5.4). Such a solution can be obtained by the Huang algorithm. Any ABS algorithm can then be applied to (5.3), step (B) removing the $m - q$ dependent equations, where $q = \text{rank}(A) \leq n$. If $q < n$ and the solution of least Euclidean norm is wanted, then (5.3) should be solved using the Huang algorithm.

A final approach is based upon the algorithms in the subclass where $v_i = Au_i$, or $V = AU$, $U = (u_1, \ldots, u_n) \in R^{n,n}$ being arbitrary nonsingular and A has full column rank. After n steps of the algorithm, when $H_{n+1} = 0$ if $V^T A$ is nonsingular, it follows from (2.3) that $V^T r_{n+1} = U^T A^T r_{n+1} = 0$. Since U is nonsingular then $A^T r_{n+1} = A^T A x_{n+1} - A^T b = 0$, hence x_{n+1} solves the normal equations and a generalized least squares solution has been found.

If $U = P$ the considered subclass reduces to the orthogonally scaled subclass, whose algorithms therefore all have the property of solving overdetermined linear systems for the least squares solution. This is then true in particular for the implicit QR algorithm. If A has not full column rank the obtained solution is not one of least Euclidean norm, but, if x_1 is zero, it is a solution of basic type, where the last $n - q$ components are zero, $q = \text{rank}(A)$.

Computational experiments of Spedicato and Bodon (1992a,b) have shown that several ABS algorithms for linear least squares are more accurate than the codes available in the NAG or LINPACK libraries, based on the QR factorization via Householder rotations or on the singular value decomposition, this superiority being particularly evident for ill conditioned or rank deficient problems. The best results have been obtained using a version of the implicit QR algorithm where the reprojection technique is applied both to the search and the scaling vectors. Additional experiments of Bodon (1993b) on the Alliant FX 80 have shown that the implicit QR algorithm, which is slower than the NAG QR based code on a sequential machine, becomes now up to three times faster, thanks to a much higher megaflops count.

6 Direct Application of ABS Methods to Unconstrained Optimization

ABS methods can be applied to the unconstrained minimization of a function $f = f(x)$, $x \in R^n$, with gradient $g = g(x)$ and Hessian $G = G(x)$, in several ways. First we notice that all algorithms of the conjugate direction subclass can be applied to minimize a quadratic function $f = (x - x^+)^T A(x - x^+)$ with positive definite Hessian $G = A$ in no more than n steps, since the ABS stepsize α_i can be shown to exactly minimize $f(x_i - \alpha p_i)$ with respect to α so that the statement follows from general properties of A-conjugate directions. Notice that the Hestenes-Stiefel method, which in the context of minimizing quadratic functions is equivalent to the Fletcher-Reeves method or to the Broyden class of Quasi-Newton methods, can be correspondingly implemented in several stable ABS versions, which might lead to stable optimization methods. Notice also that the implicit LU algorithm is a conjugate direction method of great potential interest.

Another possibility, preliminarily investigated by Abaffy, Galantai and Spedicato (1987a), consists in using the Abaffian matrices in a Quasi-Newton fashion. Consider the iteration

$$x_{i+1} = x_i - \alpha_i H_i^T z_i \tag{6.1}$$

where α_i is chosen by some line search criterion and z_i is free subject to the descent condition $g_i^T H_i^T z_i > 0$, which can always be satisfied unless $H_i g_i = 0$. Let us consider updating H_i as in the scaled ABS class. We can interpret A as an approximation to the Hessian and hence assume $A = A^T$. Then if $x' = x_i - \alpha v_i$, we obtain, exactly on quadratic functions or otherwise at first order, that $g' = g_i - \alpha_i A v_i$, hence setting $y_i = g' - g_i$ we get the update

$$H_{i+1} = H_i - H_i y_i w_i^T H_i / w_i^T H_i y_i \tag{6.2}$$

Algorithms based upon (6.1) and (6.2) are particular cases of a class of so called projection methods considered for instance by Pshenichny and Danilin (1978), who used the following more general update

$$H_{i+1} = H_i + \mu_i d_i u_i^T / u_i^T y_i - H_i y_i v_i^T / v_i^T y_i \tag{6.3}$$

where μ_i is arbitrary, $d_i = x_i - x_{i-1}$, u_i, v_i are arbitrary subject to $u_i^T y_j = 0$, $0 \le j \le i - 1$, $v_i^T y_j = 0$, $0 \le j \le i - 1$, $u_i^T y_i \ne 0$, $y_i^T v_i \ne 0$. Methods in this class have the quadratic termination property, hence they are also Q-quadratically locally convergent under mild conditions. Little investigation has been done however on the practical efficiency of these methods.

Another possibility is to use the nonlinear ABS method to solve the stationarity equation $g(x) = 0$, taking possibly into account the symmetry of the Jacobian. The nonlinear ABS algorithms, first introduced by Abaffy, Galantai and Spedicato (1987b), are a generalization of the methods of Newton, Brown or Brent, enjoying essentially the same local convergence properties of Newton's method albeit requiring in some formulations only about half the number of the Jacobian components. They are obtained by a linearization of the equation $g(x) = 0$ where the Jacobian and the residual are generally not evaluated at the beginning of the major iteration but are evaluated row-componentwise in the course of a minor iteration, allowing the use of more recent information. Open questions are how to use efficiently the symmetry of the Jacobian and whether to use a line search criterion in the definition of the stepsize.

It has been shown by Huang (1992) that the orthogonally scaled subclass, which when applied to overdetermined linear systems provides a least squares solution, can also be naturally applied to overdetermined nonlinear systems $f(x) = 0$, with local convergence to a minimizer of the nonlinear function $F(x) = f(x)^T f(x)$. The rate of convergence of these methods is Q-linear if $F(x^*) > 0$, Q-quadratic if $F(x^*) = 0$. These results apply in particular to a modification of the implicit QR algorithm with reduced computational complexity, see Spedicato and Huang (1992). Results of Giudici (1993) show that this method is extremely fast when it converges, usually outperforming the pure Gauss-Newton method.

7 Application to Generating Quasi-Newton Updates

ABS algorithms can be applied to provide in a rather simple way the general solution, in explicit form, of the Quasi-Newton equation, with possibly additional conditions like sparsity, symmetry, extra linear equations and even certain types of nonlinear constraints. While the general solution of the single Quasi-Newton equation, derived by Spedicato and Xia (1992a), had already been obtained in the literature by different approaches, see for instance Adachi (1971), the explicit form of the solution in presence of sparsity and symmetry, derived by Spedicato (1992) and Spedicato and Zhao (1993), was not known before.

We recall that Quasi-Newton methods are based on the iteration $x_{i+1} = x_i - \alpha_i p_i$, where the search vector p_i can be defined as the solution of the linear system

$$B^T p = g \tag{7.1}$$

where B^T is an approximation to the Hessian of $f(x)$ updated to satisfy the Quasi-Newton equation

$$d^T B' = y^T \tag{7.2}$$

where usually $d = x' - x$, $y = g' - g$. Usually one works with B instead of B^T, the present notation being introduced to simplify the ABS formulation.

Now if we let $B' = (b'_1, \ldots, b'_n)$ equation (7.2) is equivalent to the following n linear systems each one consisting of just one equation and all having the same coefficient matrix

$$d^T b'_i = y_i, \quad i = 1, \ldots, n \tag{7.3}$$

Hence just one step of any ABS algorithm is needed to provide the general solution of each equation in (7.3) and moreover the same Abaffian can be used for all equations. Letting b_i be the initial estimate of b'_i and taking without loss of generality $H_1 = I$, $z_1 = w_1 = Wd$, with $W \in R^{n,n}$ an arbitrary matrix such that $d^T W d \neq 0$, it is easy to show that the general solution has the following form, already obtained by Adachi (1971) via a generalized inverse approach

$$B' = B - Wd(B^T d - y)^T / d^T W d + (I - Wdd^T / d^T W d)Q \tag{7.4}$$

where $Q \in R^{n,n}$ is arbitrary.

All formulas considered in the literature follow from (7.4) by suitable choice of the parameters W and Q. For instance if B is symmetric and positive definite and $y^T d > 0$ then we can take $W = B$ and then setting $Q = yy^T / y^T d + (\gamma - 1)B$ gives $B' = \gamma B - \gamma Bdd^T B / d^T B d + yy^T / y^T d$, which is the self scaling BFGS formula.

We consider now the case where the Quasi-Newton equation is adjoined by special linear conditions of the form

$$e^T_{m^i_k} b'_i = \mu_{i,k} \tag{7.5}$$

where $i = 1, \ldots, n$, $k = 1, \ldots, k_i$, $k_i \leq n$ and $M_i = (m^i_1, \ldots m^i_{k_i})$ is a set of k_i distinct positive integers no greater than n, while e_j is the jth unit vector. If $\mu_{i,k} = 0$ the above conditions are sparsity preserving conditions. If $\mu_{i,k}$ is a constant, depending on i and k, then we have a condition keeping the corresponding element of the approximate Hessian constant. If $M_i = (1, 2, \ldots, i - 1)$ and $\mu_{i,k} = b'^T_k e_i$ then we have a symmetry preserving condition. Sparsity and symmetry can be similarly forced by suitable definitions. Notice also that $\mu_{i,k}$ depending in a nonlinear way from already computed components of B' is acceptable.

The general solution of the Quasi-Newton equation with conditions (7.5) is obtained by Spedicato and Zhao (1993) via the following procedure:

- the equations are solved for the different columns of B' in arbitrary order if only sparsity or constant value preserving conditions are given, otherwise in the order $i = 1, 2, \ldots, n$ (or also $i = n, n-1, \ldots, 1$) if symmetry conditions are also present

- equations (7.5) are first dealt with, the Quasi-Newton equation being treated as the last one

- the Huang algorithm is used, giving the following formula for the Abaffian after the k_i-th update

$$H_{k_i+1} = I_{k_i}^* \tag{7.6}$$

where $I_{k_i}^*$ is the matrix obtained from the identity matrix by setting to zero the k_i diagonal elements in the positions $m_1^i, \ldots, m_{k_i}^i$.

Given a vector u, let u_i^* be the vector whose components are the same as those of u except for the components of index m_j^i, $j = 1, \ldots, k_i$, which are equal to zero (in other words, u_i^* is u sparsified according to the sparsity pattern associated with M_i). Define $\bar{t}_i = (t_1, \ldots, t_n)$ by $t_j = \mu_{i,m_j^i}$ if $m_j^i \in M_i$, otherwise by $t_j = b_i^T e_j$. Then the general formula is given by

$$b_i' = \hat{t}_i + q_i^* + \Phi(q_i^*)d_i^* \tag{7.7}$$

where $\Phi(q_i^*) = 0$ if $q_i^* = 0$ otherwise

$$\Phi(q_i^*) = [y_i - d^T(\hat{t}_i + q_i^*)]/(d_i^*)^T d_i^* \tag{7.8}$$

It can be shown moreover, and without the use of continuity conditions on $f(x)$, that the Quasi-Newton equation and conditions (7.5) are compatible.

Setting $b_i = 0$ and $q_i = 0$ gives, from the properties of the Huang algorithm, the solution of least Euclidean norm. This solution gives B' of minimal Frobenius norm if the equations for the different columns of B' are uncoupled. Notice that usually one considers a solution where some norm of the correction to B is minimal. Such a solution can also be expressed in the form (7.8) after some trivial changes in the definition of the right hand side of the Quasi-Newton equation. Work is presently in progress to investigate properties and performance of particular formulas in the class defined by (7.8).

Let us now consider the question of the existence of positive definite symmetric updates. From the Quasi-Newton condition it follows that $y^T d > 0$ is a necessary condition, which is also sufficient when there are no sparsity conditions if B is positive definite (since in such a case the BFGS update maintains positivity) When there are sparsity conditions then $y^T d > 0$ is not sufficient as can be seen by the example where $n = 4$, $d^T = (1, 1, 1, 4)$, $y^T = (3, 2, -3, 20)$, implying $d^T y = 82$, and zeros are forced under the diagonal (and above by symmetry) in the positions $(3, 1)$, $(4, 1)$, $(3, 2)$, $(4, 2)$, $(4, 3)$. Then the general solution is the matrix

$$B' = \begin{bmatrix} 3-\mu & \mu & 0 & 0 \\ \mu & 2-\mu & 0 & 0 \\ 0 & 0 & -3 & 0 \\ 0 & 0 & 0 & 5 \end{bmatrix} \tag{7.9}$$

with μ arbitrary, which is not positive definite. Now we observe that forcing positive definiteness is equivalent to forcing the diagonal elements to be larger than a threshold value, which can be explicitly computed (and very simply if we additionally force diagonal dominance). Hence if the sparsity pattern does not affect the diagonal elements, as is usually the case, and the number of linear conditions for the ith column of B' is less than n (such a number is the sum of the $i-1$ symmetry conditions plus the Quasi-Newton condition plus the number of zeros under the diagonal forced by sparsity) we can add the extra condition that the diagonal element be equal to a value larger than the threshold. This condition clearly does not change the form of the general solution (7.8) apart from an obvious change in the definition of the ith component of $\hat{t}_i$. We cannot however add this condition to the equation relating to the last column, since the elements of it are determined by the $n-1$ symmetry conditions plus the Quasi-Newton condition. It follows that in general we can force a weaker form of positive definiteness, that we call quasi-positive definiteness, where the first $n-1$ principal submatrices (or the last $n-1$ antiprincipal submatrices) are positive definite. The obtained matrices are such that at most two eigenvalues can be nonpositive. Notice that in the literature either more complex conditions have been given for positive definiteness or this property has been satisfied at the expense of violating the Quasi-Newton condition. Notice also that our approach allows to generate positive definite matrices at the cost of not satisfying only the last (or the first, or any arbitrary one after suitable variables pivoting) component of the Quasi-Newton equation.

Another important case which can be dealt with by the ABS approach is the generation of Quasi-Newton updates satisfying an additional set of $m < n$ general linear equations. These updates, going back to a paper by Goldfarb (1969), have been studied by several authors. Zhang (1992a) has obtained the explicit general formula using the symmetric Abaffian. As special cases he has derived a generalization of the Oren's class with the property that $y^T B' y > 0$ for every y such that $B'y \neq 0$. He has also considered the ABS formulation of the factorized formula given by Powell (1989) and how to modify the update in an efficient way when a linear equation is added or is dropped.

8 An ABS Class of Feasible Descent Direction Methods for Linearly Constrained Optimization

In this section we first consider the problem of minimizing a function $f(x)$ subject to the linear constraint $Ax = b$, $A \in R^{m,n}$, $m < n$, rank$(A) = m$. A feasible direction algorithm is based upon the iteration

$$x_{i+1} = x_i - \alpha_i p_i \tag{8.1}$$

where x_0 is given as a feasible point, i.e. $Ax_0 = b$, and all successive iterates are also feasible, implying that

$$A p_i = 0 \tag{8.2}$$

Using the ABS approach and following Spedicato and Xia (1992b) we now give a general class of feasible descent direction algorithms, which provides in addition to new methods alternative formulations of well known algorithms. Then, following Xia, Liu and Zhang (1992, 1993), we discuss some ways of applying these algorithms to the problem with linear inequalities, using an active set strategy.

Let H_{m+1} be the Abaffian obtained after m steps of any ABS algorithm applied to the matrix A. Then from property (A5) the general solution of equation (8.2) has the form

$$p_i = H_{m+1}^T q \tag{8.3}$$

with $q \in R^n$ arbitrary. Hence the iteration $x_{i+1} = x_i - \alpha_i H_{m+1}^T q$ is the most general feasible direction iteration.

One is usually interested in search vectors having the descent property

$$p_i^T g_i > 0 \tag{8.4}$$

or $q^T H_{m+1} g_i > 0$. This condition can always be satisfied for some q unless $H_{m+1} g_i = 0$, in which case $g_i \in \mathrm{Null}(H_{m+1}) = \mathrm{Range}(A^T)$, implying

$$g_i = A^T \lambda \tag{8.5}$$

for some $\lambda \in R^m$. In such a case the K-T conditions of the first order are satisfied, λ is the vector of Lagrange multipliers and x_i is a feasible stationary point.

There are two natural choices of q which satisfy the descent condition. The first choice is obtained by minimizing the function $F(q) = q^T H_{m+1} g_i$ for fixed norm of q, thereby providing a maximum local decrease. This gives $q = -H_{m+1} g_i$ leading to the following formula for the search vector

$$p_i = -H_{m+1}^T H_{m+1} g_i \tag{8.6}$$

Notice that (8.6) defines a class of feasible descent directions having as parameters H_1 and the vectors z_i, w_i used in the construction of H_{m+1}. Notice also that the several alternative implementations of the Abaffian provide alternative implementations of p_i.

Some special cases of (8.6) are the following:

(a) H_{m+1} is generated by the implicit LU update. Then it can be shown that p_i is identical with the search vector in the reduced gradient method of Wolfe (1967). The corresponding formulas for the Lagrange multipliers, see (8.5), have been considered by Gill and Murray (1974)

(b) H_{m+1} is generated by the Huang algorithm. Then p_i is identical with the search vector in the orthogonal projection method of Rosen (1960)

(c) H_{m+1} is generated by the symmetric algorithm with $H_1 = G_i^{-1}$, G_i being a symmetric positive definite approximation of the Hessian at x_i. Then p_i is identical with the search vector in the method of Goldfarb and Idnani (1983), which can be written in the form

$$p_i = [G_i^{-1} - G_i^{-1} A^T (A G_i^{-1} A^T)^{-1} A G_i^{-1}] g_i \qquad (8.7)$$

A second choice is based upon the formula

$$p_i = H_{m+1}^T Z_i H_{m+1} g_i \qquad (8.8)$$

where Z_i is symmetric positive definite. A natural choice for Z_i is $Z_i = \mu_1 I + \mu_2 G_i + \mu_3 G_i^{-1}$, with G_i again a positive definite approximation of the Hessian and the μ's nonnegative and not all zeros.

A third choice for q comes from considering a quadratic model of f at x_i with symmetric definite Hessian G_i

$$f(x_i + p) = f_i + g_i^T H_{m+1}^T q + (1/2) q^T H_{m+1} G_i H_{m+1}^T q \qquad (8.9)$$

The idea is to select q in order to minimize f. There is no unique such a vector q since H_{m+1} is singular. In order to remove the degrees of freedom in q we use formula (3.4) writing

$$H_{m+1} = H_1 - S_m U_m^T \qquad (8.10)$$

where $S_m = (s_1, \ldots, s_m) \in R^{n,m}$, $U_m = (u_1, \ldots, u_m) \in R^{n,m}$. Now let $A^* \in R^{n-m,n}$ be an arbitrary matrix such that $(A^T, (A^*)^T)$ is nonsingular and extend the sequence of Abaffians from H_{m+1} to $H_{n+1} = 0$. Then (8.10) implies

$$0 = H_1 - S_m U_m^T - S_{n-m}^* (U_{n-m}^*)^T \qquad (8.11)$$

where $S_{n-m}^* = (s_{m+1}, \ldots, s_n)$, $U_{n-m}^* = (u_{m+1}, \ldots, u_n)$. Obtaining H_1 from (8.11) and substituting in (8.10) gives

$$H_{m+1} = S_{n-m}^* (U_{n-m}^*)^T \qquad (8.12)$$

For symmetric Abaffians $s_i = u_i = p_i$, hence (8.12) gives

$$H_{m+1} = P_{n-m}^* (P_{n-m}^*)^T \qquad (8.13)$$

where $P_{n-m}^* = S_{n-m}^* = U_{n-m}^*$ is the matrix whose columns are the search vectors associated with the rows of A^*. Formula (8.12) is due to Xia, Liu and Zhang (1992). Formula (8.13) was derived by Xia (1990a), see also Spedicato (1991) for a different proof.

Define now g^*, q^*, G^* by

$$g^* = (U_{n-m}^*)^T g_i \tag{8.14}$$

$$q^* = (S_{n-m}^*)^T q \tag{8.15}$$

$$G^* = (U_{n-m}^*)^T G_i U_{n-m}^* \tag{8.16}$$

Then the quadratic approximation (8.9) can be written more simply in terms of the $(n-m)$-dimensional projected gradient g^*, the parameter q^* and the projected Hessian $G^* \in R^{n-m,n-m}$

$$f(x_i + d) = f_i + (g^*)^T q^* + (1/2)(q^*)^T G^* q^* \tag{8.17}$$

Noticing that G^* is positive definite we obtain from (8.17) the following expression for the unique minimizer of $f(x_i + d)$

$$q^* = -(G^*)^{-1} g^* \tag{8.18}$$

Using the above formulas we obtain the following expression for the search vector p_i in (8.3) based upon the quadratic model

$$p_i = -U_{n-m}^* (G^*)^{-1} (U_{n-m}^*)^T g_i \tag{8.19}$$

If $H_{m+1} g_i \neq 0$ then $(U_{n-m}^*)^T g_i \neq 0$ as well. Hence if x_i is not a K-T point, formula (8.19) also defines a class of feasible direction methods, since there is freedom in the choice of both the Abaffian and the matrix A^*. The formula for the Lagrange multipliers in this approach can be shown to be

$$\lambda_i = R^{-1} U_m^T g_i \tag{8.20}$$

where $R = U_m^T A^T$ is upper triangular.

An open question is the relation between the classes defined by (8.6) and (8.19). We conjecture that despite their different derivation they generate the same set of search directions.

It can be shown that if the Abaffian is constructed using the Huang update then (8.19) gives the same directions as the Goldfarb-Idnani method, while if the implicit LU Abaffian is used then a method is obtained considered by Fletcher (1987) and others.

When linear inequality constraints are present, active set strategies are often used, which results in having to solve at the ith step a problem with linear equality constraints only, $A^{(i)} x = b^{(i)}$, some of which change from iteration to iteration. This implies, in the ABS formulation, that the current Abaffian H_{m+1} or its equivalent formulation has to be modified in an efficient way following the addition or the deletion of a row in $A^{(i)}$. While this problem has not been dealt with for all available representations of the Abaffian, solutions have been given for the important representations (8.12) and (8.13) by Xia et al. (1993). Adding a constraint is a trivial problem. Deleting a constraint results in more work, the correction being expressed by a rank one matrix, whose defining vectors are the solution of triangular systems with readily available elements. The number of operations is second order in the problem dimensions. While a full comparison with the corresponding classical procedures is yet to be done, the ABS formulation is expected to be at least competitive already in the sequential case.

When formulas (8.12) and (8.13) are used it also frequently happens that G_i is modified by a Quasi-Newton update, e.g. by the BFGS formula. Efficient ways for updating H_{m+1} also in this case, with second order complexity, have been derived by Xia et al. (1993), related to procedures considered by Powell (1988, 1989).

Feasible direction methods require a feasible starting point, whose determination in presence of inequalities may be time consuming. Zhang (1992b) has derived a method for checking the feasibility of linear inequalities and for computing a feasible point.

An open problem is determining conditions on the parameters of the given general feasible descent direction class which guarantee at least local convergence. We just note that Tan and Gao (1992) have been able, using the ABS approach, to relax the conditions given by Du (1985) for local convergence of the gradient projection method.

9 Other Applications

ABS formulations and generalizations of several methods for quadratic programming problems have been considered by Xia (1991a,b), including generalized elimination, orthogonal factorization, least squares, penalty and Lagrange type methods. An essential role in this reformulation is again taken by the symmetric algorithm of Xia. Here we only consider a Lagrangean approach to minimizing the quadratic function

$$f(x) = (1/2)x^T G x + g^T x \tag{9.1}$$

subject to $Ax = b$ and assuming that A is full rank and G is symmetric positive definite. The associated Lagrangean is

$$L(x, \lambda) = (1/2)x^T G x + g^T x - \lambda^T (Ax - b) \tag{9.2}$$

Now a stationary point satisfies the nonsingular linear system

$$\begin{bmatrix} G & -A^T \\ -A & 0 \end{bmatrix} \begin{pmatrix} x \\ \lambda \end{pmatrix} = - \begin{pmatrix} g \\ b \end{pmatrix} \tag{9.3}$$

Writing the inverse of the coefficient matrix as follows

$$\begin{bmatrix} G & -A^T \\ -A & 0 \end{bmatrix}^{-1} = \begin{bmatrix} H & -S^T \\ -S & U \end{bmatrix} \tag{9.4}$$

one obtains the following relations, see Fletcher (1987)

$$H = G^{-1} - G^{-1}A^T(AG^{-1}A^T)^{-1}AG^{-1} \tag{9.5}$$

$$S = (AG^{-1}A^T)^{-1}AG^{-1} \tag{9.6}$$

$$U = -(AG^{-1}A^T)^{-1} \tag{9.7}$$

Using the above formulas it is possible to express the solution vectors x^+ and λ^+ explicitly and separately. If we now consider the symmetric algorithm of Xia the following identity follows from a general property of Abaffians, see Abaffy and Spedicato (1989)

$$H_{m+1} = H \tag{9.8}$$

Let us now apply the Xia algorithm on an arbitrary nonsingular matrix whose first m rows consist of A and let P be the matrix whose columns are the obtained search vectors. Now a property of the symmetric algorithm, see Abaffy and Spedicato (1989), states that the search vectors are H_1^{-1}-conjugate, implying

$$P^TGP = D \tag{9.9}$$

with D nonsingular. Assuming that the columns of P have been scaled so that $P^TGP = I$ it follows that $P^TG = P^{-1}$, or $P^T = P^{-1}G^{-1}$ or

$$G^{-1} = PP^T \tag{9.10}$$

But also we have $AP = [L, 0]$ with $L \in R^{m,m}$, and $AP_m = L$, hence we obtain

$$S = L^{-T}P_m^T \tag{9.11}$$

$$U = -L^{-T}L^{-1} \tag{9.12}$$

Among other applications we quote:

- Zhang, Feng and Jiang (1992) have proposed a class of projection-restoration algorithms which is a modification of the class of Rustem (1981). Here the search vector is defined via the Xia algorithm while the implicit LU algorithm is used to guarantee certain positive definiteness and norm boundedness properties, which imply global convergence

- Shi and Liu (1992) have developed an algorithm using ABS techniques for global minimization of concave functions with linear constraints, which is a modification of a method of Falk and Hoffman (1986). The new algorithm relinquishes the assumptions of nondegenerate constraints and boundedness of the feasible region. It terminates in a finite number of steps without the need of solving an LP problem. It can be applied to linear complementarity problems where either it provides a solution or establishes incompatibility

- ABS algorithms have been used by Xia (1992) to construct polyhedrons approximating the subdifferential of a finite convex function, utilized in algorithms for nonsmooth convex optimization.

References

Adachi, N. (1971) 'On variable metric algorithms', J. Optim. Theory Appl. 7, 391-410.

Abaffy, J., Broyden, C.G. and Spedicato, E. (1984) 'A class of direct methods for linear equations', Numerische Mathematik 45, 361-376.

Abaffy, J. and Spedicato, E. (1989) ABS Projection Algorithms: Mathematical Techniques for Linear and Nonlinear Equations, Ellis Horwood, Chichester.

Abaffy, J., Galantai, A. and Spedicato, E. (1987a) 'Application of the ABS class to unconstrained function minimization', Report DMSIA 14/87, University of Bergamo.

Abaffy, J., Galantai, and Spedicato, E. (1987b) 'The local convergence of ABS methods for nonlinear algebraic equations', Numerische Mathematik 51, 429-439.

Benzi, M. and Meyer, C.D. (1993) 'A direct projection method and its application to sparse linear systems', Report NCSU NA-01051593, North Carolina State University.

Bodon, E. (1992a) 'Numerical experiments with ABS algorithms on upper banded systems of linear equations', Report DMSIA 17/92, University of Bergamo.

Bodon, E. (1992b) 'Numerical experiments with ABS algorithms on banded systems of linear equations', Report DMSIA 18/92, University of Bergamo.

Bodon, E. (1993a) 'Numerical results on the ABS algorithms for linear systems of equations', Report DMSIA 9/93, University of Bergamo.

Bodon, E. (1993b) 'Computational performance of the implicit QR algorithm for linear least squares on the Alliant FX 80', Report DMSIA 17/93, University of Bergamo.

Bodon, E. and Spedicato, E. (1990) 'Numerical evaluation of the implicit LU, LQ and QU algorithm in the ABS class', Report DMSIA 20/90, University of Bergamo.

Bodon, E. and Spedicato, E. (1991) 'Factorized ABS algorithms for linear systems: derivation and numerical results', Report DMSIA 14/91, University of Bergamo.

Bodon, E. and Spedicato, E. (1992) 'On some STOD-ABS algorithms for large linear systems', Report DMSIA 15/92, University of Bergamo.

Broyden, C.G. (1985) 'On the numerical stability of Huang and related methods', J. Optim. Meth. Appl. 47, 7-16.

Broyden, C.G. (1991) 'On the numerical stability of Huang's update', Calcolo 28, 303-311.

Chen, Z., Deng, N. and Xue, Y., (1992) 'A general algorithm for underdetermined linear systems', Proceedings of the First International Conference on ABS Algorithms, University of Bergamo.

Du, D. (1985) 'A method of gradient projection under the conditions of nonlinear constraints', ACTA Mathem. Appl. Sinica 8, 7-16.

Falk, J.E. and Hoffmann, K.L. (1986) 'Concave minimization via collapsing polytopes', Operations Research 34, 919-929.

Fletcher, R. (1987) Practical Methods of Optimization, Wiley, New York.

Gill, P.E. and Murray, W. (1974) Numerical Methods for Nonlinear Optimization, Academic Press, New York.

Giudici, S. (1993) 'ABS algorithms for nonlinear least squares', Dissertation, University of Bergamo.

Goldfarb, D. (1969) 'Extension of Davidon's variable metric methods to maximization under linear inequality and equality constraints', SIAM J. Appl. Math. 17, 739-764.

Goldfarb, D. and Idnani, A. (1983) 'A numerically stable dual method for solving strictly convex quadratic programming', Mathem. Progr. 27, 1-33.

Huang, H.Y. (1975) 'A direct method for the general solution of a system of linear equations', J. Optim. Meth. Appl. 16, 429-445.

Huang, Z. (1992) 'A class of ABS methods for nonlinear least squares problems', Report DMSIA 3/92, University of Bergamo, Ricerca Operativa, in press.

Powell, M.J.D. (1989) 'The updating of matrices of conjugate directions in optimization', Report DAMTP 89/NA7, University of Cambridge.

Pshenichny, B.N. and Danilin, Y.M. (1978) Numerical Methods in Extremal Problems, MIR, Moscow.

Rosen, J.B. (1960) 'The gradient projection method for nonlinear programming, Part I:linear constraints,' SIAM J. Appl. Math. 8, 181-217.

Rustem, B. (1981) Projection Methods in Constrained Optimization and Applications in Optimal Policy Decisions, Springer, Berlin.

Shi, G. and Liu, X. (1992) 'An ABS algorithm for concave minimum problems over a convex polyhedroid', Proceedings of the First International Conference of ABS Algorithms, University of Bergamo.

Sloboda, F. (1978) 'A parallel projection method for linear algebraic systems', Apl. Mat. Ceskosl. Akad. Ved., 23, 185-198.

Spedicato, E. (1991) 'De aliqua methodi Huang proprieţate', Report DMSIA 4/91, University of Bergamo.

Spedicato, E. (1992) 'A class of sparse symmetric Quasi-Newton updates', Ricerca Operativa 22, 63-70.

Spedicato, E. and Bodon, E. (1989) 'Solving linear least squares by orthogonal factorization and pseudoinverse computation via the modified Huang algorithm in the ABS class', Computing 42, 195-205.

Spedicato, E. and Bodon, E. (1992a) 'Solution of linear least squares via the ABS algorithms', Mathem. Progr. 58, 111-136.

Spedicato, E. and Bodon, E. (1992b) 'Numerical behaviour of the implicit QR algorithm in the ABS class for linear least squares', Ricerca Operativa 22, 43-55.

Spedicato, E. and Huang, Z. (1992) 'An orthogonally scaled ABS method for nonlinear least squares', Optim. Meth. Soft. 1, 233-242.

Spedicato, E. and Tuma, M. (1993) 'Solving sparse unsymmetric linear systems by implicit Gauss algorithm: stability', Report DMSIA 4/93, University of Bergamo.

Spedicato, E. and Vespucci, M.T. (1992) 'Variations on the Gram-Schmidt and the Huang algorithms for linear systems: a numerical study', Aplikace Mathematiky 2, 81-100.

Spedicato, E. and Xia, Z. (1992a) 'Finding general solutions of the Quasi-Newton equation via the ABS approach', Optim. Meth. Soft. 1, 243-252.

Spedicato, E. and Xia, Z. (1992b) 'A class of descent algorithms for nonlinear function minimization with linear equality constraints', Optim. Meth. Soft. 1, 265-272.

Spedicato, E. and Xia, Z. (1992c) 'On some ABS methods for the computation of the pseudoinverse', Ricerca Operativa 22, 35-41.

Spedicato, E. and Zhao, J. (1993) 'Explicit solution of the Quasi-Newton equation with sparsity and symmetry', Optim. Meth. Soft., in press.

Spedicato, E. and Zhu, M. (1993) 'A reduced ABS-type algorithm I: basic properties', preprint, University of Bergamo.

Tan, Z. and Gao, Y. (1992) 'An improved gradient projection method for nonlinearly constrained systems', Proceedings of the First International Conference on ABS Algorithms, University of Bergamo.

Xia, Z. (1990a) 'Application of ABS algorithms to constrained optimization II: linear inequality constraints', Report DMSIA 29/90, University of Bergamo.

Xia, Z. (1990b) 'An algorithm for minimizing a class of quasidifferentiable functions', Report DMSIA 12, University of Bergamo.

Xia, Z. (1991a) 'Quadratic programming via the ABS algorithm I: equality constraints', Report DMSIA 1/91, University of Bergamo.

Xia, Z. (1991b) 'The ABS class and quadratic programming II: general constraints', Report DMSIA 5/91, University of Bergamo.

Xia, Z. (1992) 'Finding subgradients or descent directions of convex functions by external polyhedral approximation of subdifferentials', Optim. Meth. Soft. 1, 253-264.

Xia, Z., Liu, Y. and Zhang, L. (1992) 'Application of a representation of ABS updating matrices to linearly constrained optimization I', Northeast Operational Research 7, 1-9.

Xia, Z., Liu,Y. and Zhang, L. (1993) 'Application of a representation of ABS matrices to linearly constrained optimization II', preprint, University of Bergamo.

Wolfe, P. (1967) 'Methods for nonlinear constraints', in J. Abadie, Nonlinear Programming, North Holland, Amsterdam.

Zhang, L.(1992a) 'Application of the ABS algorithm to Quasi-Newton methods for linearly constrained optimization problems', Proceedings of the first international conference on ABS algorithms, University of Bergamo.

Zhang, L. (1992b) 'A method for finding a feasible point of inequalities', Proceedings of the first international conference on ABS algorithms, University of Bergamo.

Zhu, M. (1987) 'The implicit LL^T algorithm for sparse nested dissection linear systems', Report NOC 196, Hatfield Polytechnic.

Zhu, M. and Yang, Z. (1993) 'The practical ABS algorithm for large scale nested dissection linear systems', preprint, University of Bergamo.

A Condensed Introduction to Bundle Methods in Nonsmooth Optimization

C. LEMARÉCHAL

INRIA, Rocquencourt
B. P. 105
78153 LE CHESNAY
FRANCE

J. ZOWE

Friedrich-Schiller-Universität Jena
Institut für Angewandte Mathematik
07740 JENA
GERMANY

ABSTRACT. We give a short introduction to (primal) Bundle Methods in Nonsmooth Optimization. The effectivity of these ideas is shown on some nonsmooth real life applications.

1 Introduction

In real life problems one often has to minimize some functional f with discontinuous derivatives. This is what *Nonsmooth (Nondifferentiable) Optimization* deals with.

It is well-known that classical gradient methods fail in a nonsmooth context. This is due to the increasing shortage of information contained in gradients when approaching a point of nondifferentiability. A way out of this dilemma is to work at each iterate with the bundle of all previously computed gradients. This simple idea of 'bundling the information' together with eventual 'null steps' is the basic ingredient of all so-called *Bundle Methods*. In §1 - §3 we give a condensed introduction to these methods while emphasizing the primal viewpoint. We start from the old cutting plane idea, which uses the bundle of already computed (sub)gradients to build up a piecewise linear approximation of the (convex) function f from below in the space of origin (§1). Analyzing the drawbacks of this idea (instability and extremely slow convergence) we are led to three stabilized and accelerated versions (§2). Standard duality arguments show that these refined cutting plane variants can be interpreted as *primal* adaptations of the classical *dual* Bundle concept, which argues in the dual space of (sub)gradients and which is commonly studied in literature(§3).

E. Spedicato (ed.), Algorithms for Continuous Optimization, 357–382.
© 1994 *Kluwer Academic Publishers.*

We restrict ourselves to convex f since in this framework things are most easily explained. It should be mentioned, however, that a complete theory has been developed for the dual bundle approach, which does not require convexity from f.

No proofs are given in the following, since all the material can be found in the fundamental textbook by Hiriart-Urruty and Lemaréchal (1993), where the subject is developed systematically and presented in full detail.

The final §4 shows that codes, based on primal bundle variants, are powerful tools when dealing with nonsmooth (nonconvex, nontrivial, ...) real life problems.

1.1   The nonsmooth problem

Throughout the following f is a convex functional on the n-dimensional euclidean space and we study the problem

$$\text{minimize } f(x) \text{ on } R^n. \tag{1}$$

There are no constraints but, differently from the standard situation, we do not require f to be smooth. The *subdifferential*

$$\partial f(x) = \{s \in R^n \,|\, \langle s, z - x \rangle \le f(z) - f(x) \text{ for all } z \in R^n\} \tag{2}$$

of the convex f at x will serve as substitute for the gradient. This $\partial f(x)$ is a nonempty, convex and compact set, which shrinks to the gradient whenever f is differentiable at x. We make the general assumption:

$$\text{at every } x, \text{ we know } f(x) \text{ and one (arbitrary) } s \in \partial f(x). \tag{3}$$

This assumption is fairly natural and the applications in §4 will show that such $s \in \partial f(x)$ (a so-called *subgradient*) can often be computed using only standard differential calculus (§4.1 and §4.2) or simple tools from sensitivity analysis (§4.3). In other words: we have a black box (sometimes called an *oracle*) which, given x, answers $f(x)$ and some $s \in \partial f(x)$. Thus the situation is similar to that in ordinary smooth optimization, except that s will not vary continuously with x.

1.2   The cutting-plane idea

A long-known algorithm to minimize our convex f is the *cutting-plane* algorithm due to Cheney and Goldstein (1959) and Kelley (1960) and we briefly recall how it works. At step k, let the iterates $y_1, \ldots, y_k$ have been generated, together with the corresponding function values $f(y_1), \ldots, f(y_k)$ and subgradients $s_1 \in \partial f(y_1), \ldots, s_k \in \partial f(y_k)$. We define the *cutting-plane approximation* of f, associated with the above *bundle* of information to be the piecewise affine function

$$R^n \ni y \mapsto \check{f}_k(y) := \max\{f(y_i) + \langle s_i, y - y_i \rangle \,:\, i = 1, \ldots, k\}. \tag{4}$$

It results immediately from convexity that this function under-estimates f, and coincides with it at each sampling point:

$$\check{f}_k \leq f \quad \text{and} \quad \check{f}_k(y_i) = f(y_i) \quad \text{for } i = 1, \ldots, k.$$

Minimization of this cutting-plane model $\check{f}_k$ provides the next iterate y_{k+1}. There is a technical difficulty, however: $\check{f}_k$ may be unbounded from below (think of the first iteration, for example!). Hence, we assume that some convex compact set C is known, so that the problem

$$y_{k+1} \in \operatorname*{Argmin}_{y \in C} \check{f}_k(y) \tag{5}$$

does have a solution. Further, to make sure that our original problem is really solved, C should contain at least one minimum of f (admitting that there is one).

Algorithm 1.1 (basic cutting-plane algorithm) The convex compact set C and the stopping tolerance $\delta \geq 0$ are given.

STEP 0 (initialization). Choose $y_1 \in C$ and compute $f(y_1), s_1 \in \partial f(y_1)$. Set the iteration counter $k = 1$.

STEP 1 (relaxed problem). Solve (5), that is

$$\left| \begin{array}{ll} \min r & r \in R, y \in C, \\ r \geq f(y_i) + \langle s_i, y - y_i \rangle & \text{for } i = 1, \ldots, k \end{array} \right.$$

to obtain a solution (r_{k+1}, y_{k+1}).

STEP 2 (call of oracle). Compute $f(y_{k+1}), s_{k+1} \in \partial f(y_{k+1})$.

STEP 3 (stopping criterion and loop). If

$$f(y_{k+1}) \leq \check{f}_k(y_{k+1}) + \delta,$$

then stop. Otherwise, replace k by $k + 1$ and loop to Step 1. □

Note that, for polyhedral C, Step 1 is just an ordinary linear programming problem. We recall the main convergence properties of this algorithm:

Theorem 1.2 *Denote by $\bar{f}_C$ the minimal value of f over C. There holds for all $k \leq k'$ and all k'':*

$$[r_{k+1} =] \check{f}_k(y_{k+1}) \leq \check{f}_{k'}(y_{k'+1}) \leq \bar{f}_C \leq f(y_{k''}).$$

Assuming $\delta = 0$,

$$f(y_k) \to \bar{f}_C \quad \text{and} \quad \check{f}_k(y_{k+1}) \uparrow \bar{f}_C \text{ when } k \to +\infty.$$

□

Unfortunately, the numerical behaviour of this basic cutting-plane algorithm is extremely poor. For an illustration, consider a simple example: with $n = 1$, take $f(x) = 1/2 x^2$ and start with two iterates $y_1 = 1, y_2 = -\varepsilon < 0$. Then y_3 is obviously the solution of $\check{f}_1(y) = \check{f}_2(y)$, i.e.

$$\tfrac{1}{2} + (1)(y - 1) = \tfrac{1}{2}(-\varepsilon)^2 + (-\varepsilon)(y + \varepsilon),$$

and $y_3 = 1/2 - 1/2\varepsilon$. If ε gets smaller, y_2 increases, and y_3 increases as well. In algorithmic terms, we say: if the current iterate is better (y_2 comes closer to the solution 0), the next iterate is worse (y_3 goes further from this solution), i. e. the algorithm is *unstable*. On this example, the instability is not too serious, but it can become disastrous in less naive situations. Indeed there exists an example constructed by A.S. Nemirovski, for which multiplying the initial gap $f(y_1) - \bar{f}_C$ by a factor $\alpha < 1$ requires some $(1/\alpha)^{1/2 n-1}$ iterations: with 20 variables, a billion of iterations are needed to gain just one digit accuracy!

A more thoughtful analysis of the cutting-plane algorithm reveals that its instability has the same origin as the need to introduce C in (5). The artificial set C had primarily a theoretical motivation: to give a meaning to the problem of minimizing $\check{f}_k$; but this appears to have a pragmatic supplement: to prevent a wild behaviour of $\{y_k\}$. Accordingly, we can even say that C (which after all is not so artificial) should be "small", which also implies "appropriately located" to catch a minimum of f. Actually, one should rather have a varying set C_k, with diameter shrinking to 0 as the cutting-plane iteration proceeds. The next subsection will summarize the ingredients for such a stabilization mechanism.

We mention another major drawback of the above cutting-plane algorithm: the number of constraints in step 1 grows unboundedly, since it is equal to the (possibly astronomic) number of iterations. This will be cured by the so-called *aggregation* technique, to be described in §3.

1.3 Stabilizing devices: leading principles

We describe abstractly the kth iteration of a stabilized algorithm as follows:

(i) we have a *model*, call it φ_k, supposed to represent f;

(ii) we choose a *stability center*, call it x_k;

(iii) we choose a *norming*, call it $\| \cdot \|_k$;

(iv) then we compute a next iterate y_{k+1} realizing a compromise between diminishing the model:

$$\varphi_k(y_{k+1}) < \varphi_k(x_k),$$

while keeping close to the stability center:

$$\|y_{k+1} - x_k\|_k \text{ small}.$$

We emphasize that this stabilizing trick is able to cope simultaneously with two deficiencies of the pure cutting-plane algorithm: the need for a compact set C is totally eliminated, and the stability question will be addressed by a proper management of x_k and/or $\| \cdot \|_k$.

Let us be more concrete with respect to items (i)–(iv) above.

(i) The model φ_k, (the cutting-plane function $\check{f}_k$ of (4) in our context) can be enriched by one affine piece every time a new iterate y_{k+1} is obtained from the model-problem in (iv).

(ii) The stability center should approximate a minimum of f as reasonably as possible; using objective-values to define the word "reasonably", the idea will be to take for x_k one of the sampling points $y_1, \ldots, y_k$ having the best f-value. Actually, a recursive strategy will be used: at each iteration, the stability center is either left as it is, or moved to the new iterate y_{k+1}, depending on how good $f(y_{k+1})$ is.

(iii) Norming is quite an issue, our study will be limited to multiples of a fixed norm (usually equal to the Euclidean norm $\| \cdot \|$).

(iv) Section 2 will be devoted to the stabilized problem; three conceptually equivalent possibilities will be given. One of our objectives is to show that all these approaches are indeed primal variants of the "dual" bundle method, usually described in the literature; see e.g. Zowe (1985) or Lemaréchal (1989) and further Kiwiel (1988) for extensions to constrained and nonconvex problems.

The whole stabilizing idea is primarily characterized by (ii), which relies on the following crucial technique. In addition to the next iterate y_{k+1}, the stabilized problem yields a "nominal decrease" for f. This is a nonnegative number δ_k, giving an idea of the gain $f(x_k) - f(y_{k+1})$ to be expected by a move from the current stability center x_k to the next iterate y_{k+1}; see Example 1.4 below. If the actual gain in f is at least a fraction of the "ideal" gain δ_k, then x_k is set to y_{k+1}. With emphasis put on the management of the stability center, the resulting algorithm looks as follows:

Algorithm 1.3 (schematic stabilized algorithm)   Start from some $x_1 \in R^n$, choose a descent coefficient $m \in \,]0, 1[$ and initialize $k = 1$.

STEP 1. Choose a convex model-function φ_k (for example $\check{f}_k$) and a norming $\| \cdot \|_k$ (for example a multiple of the Euclidean norm).

STEP 2. Solve the stabilized model-problem, whatever it is, to obtain the next iterate y_{k+1}. Upon observation of $\varphi_k(x_k) - \varphi_k(y_{k+1})$, choose a nominal decrease $\delta_k \geq 0$ for f.

STEP 3. If

$$f(y_{k+1}) \leq f(x_k) - m\delta_k, \tag{6}$$

set $x_{k+1} = y_{k+1}$; otherwise set $x_{k+1} = x_k$.

Replace k by $k + 1$ and loop to Step 1.   □

We will say that, when the stability center is changed in Step 3, we make a *descent-step* (or *serious step*); otherwise we make a *null-step*. This terminology is standard in bundle methods.

The set of descent iterations is denoted by $K \subset N$:

$$[k \in K \iff x_{k+1} = y_{k+1}] \quad \text{and} \quad [k \notin K \iff x_{k+1} = x_k].$$

Classical algorithms for smooth optimization can be revisited in the light of the above approach, and this is not a totally frivolous exercise.

Example 1.4 (line-search for smooth f) At the given x_k, take the first-order approximation (s_k is the gradient of f at x_k)

$$y \mapsto \varphi_k(y) = f(x_k) + \langle s_k, y - x_k \rangle \tag{7}$$

as model and choose a symmetric positive definite operator Q for the norming.
(a) Steepest descent, first order. Let the stabilized problem be

$$\min\{\varphi_k(y) : \tfrac{1}{2} \langle Q(y - x_k), y - x_k \rangle \le \tfrac{1}{2}\kappa\} \tag{8}$$

for some radius $\kappa > 0$; stabilization is thus enforced via an explicit *constraint*. From the minimality conditions, there is a multiplier $\mu_k \ge 0$ such that the solution $y = y_{k+1}$ of (8) is characterized by the equation

$$s_k + \mu_k Q(y - x_k) = 0.$$

From there, we obtain

$$
\begin{aligned}
\langle s_k, y - x_k \rangle &= -\mu_k \langle Q(y - x_k), y - x_k \rangle && \text{[multiply by } y - x_k] \\
&= -\mu_k \kappa && \text{[transversality condition]} \\
&= -\sqrt{\kappa \langle s_k, Q^{-1} s_k \rangle}, && \text{[multiply by } \kappa Q^{-1} s_k]
\end{aligned}
$$

which we interpret as follows. Between the stability center and a solution y_{k+1} of the stabilized problem, the model decreases by

$$\varphi_k(x_k) - \varphi_k(y_{k+1}) = \sqrt{\kappa \langle s_k, Q^{-1} s_k \rangle} = \mu_k \kappa =: \delta_k \ge 0.$$

Knowing that $\varphi_k(x_k) = f(x_k)$ and $\varphi_k \simeq f$ close to x_k, the number δ_k can be viewed as a "nominal decrease" for f. Then the next iterate in Algorithm 1.3 is $y_{k+1} = x_k - Q^{-1} s_k / \mu_k$ (assuming $\delta_k > 0$, hence $\mu_k > 0$). Setting $d_k := -Q^{-1} s_k$ (a direction) and $t_k := 1/\mu_k$ (a stepsize), (6) can be written

$$f(y_{k+1}) \le f(x_k) - m\delta_k = f(x_k) - mt_k \langle s_k, d_k \rangle. \tag{9}$$

We recognize a form of Armijo descent test, universally used for classical line-searches.

(b) Steepest descent, second order. With the same φ_k of (7), take the stabilized problem as

$$\min\{\varphi_k(y) + \tfrac{1}{2}\mu \langle Q(y - x_k), y - x_k \rangle \; : \; y \in R^n\} \tag{10}$$

for some *penalty* coefficient $\mu > 0$. The situation is quite comparable to that in (a) above; a slight difference is that the multiplier μ is now explicitly given, as well as the solution $x_k - Q^{-1}s_k/\mu$.

We note in passing a more subtle difference concerning the choice of δ_k: the rationale for (a) was to approximate f by φ_k, at least on some restricted region around x_k. Here, we are bound to consider that f is approximated *by the minimand* in (10); otherwise, why solve (10) at all? It is therefore more logical to set in (b)

$$\delta_k := \varphi(x_k) - \varphi(y_{k+1}) - \frac{1}{2}\mu \langle Q(y_{k+1} - x_k), y_{k+1} - x_k \rangle = \frac{1}{2\mu} \langle s_k, Q^{-1}s_k \rangle \;.$$

A look at (9) shows that, with respect to (a), the nominal decrease is divided by two.
□

2 Three stabilized versions of cutting planes

We give in this section several algorithms realizing the general scheme of §1.3. They use conceptually equivalent stabilized problems in Step 2 of Algorithm 1.3. The notations are those of §1: $\check{f}_k$ is the cutting-plane function of (4), the Euclidean norm $\|\cdot\|$ is assumed for the stabilization (even though our development is only descriptive, and could accomodate more general situations).

2.1 The trust-region point of view

The first idea that comes to mind is to force the next iterate a priori into a ball associated with the given norming, centered at the given stability center, and having a given radius κ. The sequence of iterates is thus defined by

$$y_{k+1} \in \text{Argmin}\{\check{f}_k(y) \; : \; \|y - x_k\| \leq \kappa\}\,.$$

This approach has the same rationale as Example 1.4(a). The original model $\check{f}_k$ is considered as a good approximation of f in a *trust-region* drawn around the stability center: the κ-ball centered at x_k. Accordingly, $\check{f}_k$ is minimized in this trust-region, any point outside it being disregarded. The resulting algorithm in its crudest form is then as follows.

Algorithm 2.1 (cutting planes + trust region) The initial point x_1 is given. Choose a trust-region radius $\kappa > 0$ and a descent coefficient $m \in]0,1[$. Initialize the descent-set $K = \emptyset$, the iteration-counter $k = 1$ and $y_1 = x_1$; compute $f(y_1)$ and $s_1 \in \partial f(y_1)$.

STEP 1. Define the model

$$y \mapsto \check{f}_k(y) := \max\{f(y_i) + < s_i, y - y_i > : i = 1, \dots, k\}.$$

STEP 2. Compute a solution y_{k+1} of

$$\min\{\check{f}_k(y) : \tfrac{1}{2}\|y - x_k\|^2 \leq \tfrac{1}{2}\kappa^2\} \tag{11}$$

and set

$$\delta_k := f(x_k) - \check{f}_k(y_{k+1}) \quad [\geq 0].$$

STEP 3. Compute $f(y_{k+1})$ and $s_{k+1} \in \partial f(y_{k+1})$. If

$$f(y_{k+1}) \leq f(x_k) - m\delta_k,$$

set $x_{k+1} = y_{k+1}$ and append k to the set K (descent-step); otherwise set $x_{k+1} = x_k$ (null-step).

Replace k by $k + 1$ and loop to Step 1. □

Compare the roles of the stability set $B(x_k, \kappa) := \{y : \|y - x_k\| \leq \kappa\}$ in (11) and of the compact set C of Algorithm 1.1. Whereas C was fixed all the time, $B(x_k, \kappa)$ is now changed at every descent-step. Further note that such a descent-step does not require an exact minimization of f on $B(x_k, \kappa)$, and this is crucial for efficiency: accurate minimizing f over the stability set would be pure waste of time if this set is still far from a minimum of f.

The following convergence result holds.

Theorem 2.2 *Let Algorithm 2.1 be used with fixed $\kappa > 0$ and $m \in]0, 1[$. Then $f(x_k)$ tends to the infimal value of f on R^n.* □

We thus have a convergent stabilization of the cutting-plane algorithm. However, this algorithm is too simple. Obviously the radius of the trust-region should shrink to 0 in the limit; otherwise, we are still too close in spirit to the pure cutting-plane philosophy of §1.2.

2.2 The penalization point of view

The trust-region technique of §2.1 was rather abrupt. The next iterate was controlled by a mere *switch*: "on" inside the trust-region, "off" outside it. Something more flexible is obtained if the distance from the stability center acts as a *weight*.

Here we choose a coefficient $\mu > 0$ (the strength of a spring) and our model is

$$y \mapsto \check{f}_k(y) + \tfrac{1}{2}\mu\|y - x_k\|^2 ;$$

needless to say, pure cutting planes would be obtained with $\mu = 0$. This strategy is in the spirit of Example 1.4(b): it is the model itself that is given a stabilized form; its unconstrained minimization will furnish the next iterate, and its decrease will give the nominal decrease for f.

Just as in the previous section, we give the resulting algorithm in its crudest form:

Algorithm 2.3 (cutting planes + stabilization by penalty) The initial point x_1 is given. Choose a spring-strength $\mu > 0$ and a descent-coefficient $m \in]0,1[$. Initialize the descent-set $K = \emptyset$, the iteration-counter $k = 1$, and $y_1 = x_1$; compute $f(y_1)$ and $s_1 \in \partial f(y_1)$.

STEP 1. With $\check{f}_k$ denoting the cutting-plane function (4), compute the solution y_{k+1} of

$$\min[\check{f}_k(y) + \tfrac{1}{2}\mu\|y - x_k\|^2] \qquad (12)$$

and set

$$\delta_k := f(x_k) - \check{f}_k(y_{k+1}) - \tfrac{1}{2}\mu\|y_{k+1} - x_k\|^2 \quad [\geq 0].$$

STEP 2. Compute $f(y_{k+1})$ and $s_{k+1} \in \partial f(y_{k+1})$. If

$$f(y_{k+1}) \leq f(x_k) - m\delta_k$$

set $x_{k+1} = y_{k+1}$ and append k to the set K (descent-step); otherwise set $x_{k+1} = x_k$ (null-step).

Replace k by $k + 1$ and loop to Step 1. $\qquad\square$

With respect to the trust-region variant, an obvious difference is that the above stabilized problem is easier: it is a problem with quadratic objective and affine constraints

$$\left| \begin{array}{ll} \min[r + \tfrac{1}{2}\mu\|y - x_k\|^2] & (y,r) \in R^n \times R, \\ f(y_i) + \langle s_i, y - x_i \rangle \leq r & \text{for } i = 1,\ldots,k. \end{array} \right.$$

Conceptually, however, the two variants are equivalent in the sense that they produce the same $(k+1)^{st}$ iterate, provided that the parameters κ and μ are properly chosen. Standard duality theory tells that the μ in (i) below should be a Lagrange multiplier of the κ-constraint in (11); conversely, if y_{k+1} solves (12), then it also solves (11) for $\kappa := \|y_{k+1} - x_k\|$.

Proposition 2.4

(i) For any $\kappa > 0$, there is $\mu = \mu(\kappa) \geq 0$ such that any solution of the trust-region problem (11) also solves the penalized problem (12).

(ii) Conversely, for any $\mu \geq 0$, there is $\kappa = \kappa(\mu) \geq 0$ such that any solution of (12) (assumed to exist) also solves (11). $\qquad\square$

Note that the equivalence between (11) and (12) is not of practical nature. There is no explicit relation giving a priori one parameter as a function of the other.

The penalized version will be developed more thoroughly in §3, where among others the parameter μ in (12) will vary with the iteration index; compare the remark following Theorem 2.2. For completeness, we just mention here a convergence result which parallels Theorem 2.2:

Theorem 2.5 *Let Algorithm 2.3 be used with fixed $\mu > 0$ and $m \in]0,1[$. Then $f(x_k)$ tends to the infimal value of f on R^n.* $\qquad\square$

2.3 The level point of view

The point made in the previous section is that the second term $1/2\,\mu\|y - x_k\|^2$ in (12) can be interpreted as the dualization of a certain constraint $\|y - x_k\| \leq \kappa$, whose *right-hand side* $\kappa = \kappa(\mu)$ becomes a function of its *multiplier* μ. Likewise, we can interpret the first term $\check{f}_k(y)$ as the dualization of a constraint $\check{f}_k(y) \leq \ell$, whose right-hand side $\ell = \ell(\mu)$ will be a function of its multiplier $1/\mu$. In other words, a third possible stabilized problem is

$$\left| \begin{array}{l} \min \frac{1}{2}\|y - x_k\|^2\,, \\ \check{f}_k(y) \leq \ell \end{array} \right. \tag{13}$$

for some *level* ℓ. A version of Proposition 2.4 can also be proved for (12) and (13), thereby establishing a conceptual equivalence between all three variants.

The level used in (13) suggests an obvious nominal decrease for f: it is natural to set $x_{k+1} = y_{k+1}$ if this results in a definite objective-decrease from $f(x_k)$ towards ℓ. Then the resulting algorithm has the following form, in which we explicitly let the level depend on the iteration.

Algorithm 2.6 (cutting planes + level-stabilization) The initial point x_1 is given. Choose a descent-coefficient $m \in\,]0,1[$. Initialize the descent-set $K = \emptyset$, the iteration-counter $k = 1$, and $y_1 = x_1$; compute $f(y_1)$ and $s_1 \in \partial f(y_1)$.

STEP 1. Choose a level $\ell = \ell_k$ satisfying $\inf \check{f}_k \leq \ell < f(x_k)$.
STEP 2. Compute the solution y_{k+1} of (13).
STEP 3. Compute $f(y_{k+1})$ and $s_{k+1} \in \partial f(y_{k+1})$. If

$$f(y_{k+1}) \leq f(x_k) - m[f(x_k) - \ell]$$

set $x_{k+1} = y_{k+1}$ and append k to the set K (descent-step); otherwise set $x_{k+1} = x_k$ (null-step).
Replace k by $k + 1$ and loop to Step 1. $\qquad\square$

We just give an instance of convergence result, limited to the case of a known minimal value of f.

Theorem 2.7 *Let f have a minimum point $\bar{x}$. Then Algorithm 2.6 applied with $\ell_k \equiv f(\bar{x})$ generates a minimizing sequence $\{x_k\}$.* $\qquad\square$

3 A class of primal bundle algorithms

In this section, we study more particularly the penalized variants of §2.2. Because they work directly in the primal space, they can handle possible linear constraints rather easily. However, we still assume for simplicity that f must be minimized on the whole of R^n.

3.1 The general method

The model will not be exactly $\check{f}_k$ of (4), so we prefer to denote it abstractly by φ_k, a convex function finite everywhere. Besides, the iteration index k is useless for the moment; the stabilized problem is therefore denoted by

$$\min_y [\varphi(y) + \tfrac{1}{2}\mu\|y - x\|^2] \,, \tag{14}$$

x and μ being the kth stability center and penalty coefficient respectively.

The model φ will incorporate the important *aggregation* technique, which allows a control on the amount of subgradients one has to carry along during the iteration. First, we give a useful characterization of the solution of (14).

Lemma 3.1 *With $\varphi : R^n \to R$ convex and $\mu > 0$, (14) has a unique solution y_+, characterized by the formulae*

$$y_+ = x - \frac{1}{\mu}\hat{s}, \quad \hat{s} \in \partial\varphi(y_+).$$

Furthermore,

$$\varphi(y) \geq f(x) + \langle \hat{s}, y - x \rangle - \hat{e} \quad \text{for all } y \in R^n$$

where

$$\hat{e} := f(x) - \varphi(y_+) - \frac{1}{\mu}\|\hat{s}\|^2 \geq 0. \qquad \square$$

Proposition 3.2 *With the notations of Lemma 3.1, take an arbitrary function $\psi : R^n \to R \cup \{+\infty\}$ satisfying*

$$\psi(y) \geq f(x) - \hat{e} + \langle \hat{s}, y - x \rangle =: \bar{f}^a(y) \quad \text{for all } y \in R^n \,, \tag{15}$$

with equality at $y = y_+$. Then y_+ minimizes the function

$$y \mapsto \tilde{\psi}(y) := \psi(y) + \tfrac{1}{2}\mu\|y - x\|^2. \qquad \square$$

The affine function $\bar{f}^a$ appearing in (15) is the *aggregate* linearization of f. It minorizes φ (Lemma 3.1) and can also be written

$$R^n \ni y \mapsto \bar{f}^a(y) = \varphi(y_+) + \langle \hat{s}, y - y_+ \rangle \,.$$

Proposition 3.2 tells us in particular that the next iterate y_+ would not change if, instead of φ, the model were any convex function ψ sandwiched between $\bar{f}^a$ and φ.

The above two results can be used as follows. When (14) is solved, an additional affine piece $(y_+, f(y_+), s_+)$ will be introduced to give the new model

$$\varphi_+(y) = \max\{\varphi(y), f(y_+) + \langle s_+, y - y_+ \rangle\} \,. \tag{16}$$

Before doing so, we may wish to "simplify" φ in (16) to some other ψ, in order to make room in the computer and/or to simplify the next problem (14). For example, we may wish to discard some old affine pieces; this results in some $\phi \leq \varphi$. Then Proposition 3.2 tells us that we may replace φ in (16) by the "skimmed" function

$$\psi(y) = \max\{\phi(y), f(y_+) + \langle s_+, y - y_+\rangle\}.$$

This *aggregation* operation can be used to bound the number of affine pieces in the model φ, without destroying convergence.

In terms of the objective function f, the aggregate linearization $\bar{f}^a$ is not attached to any $\hat{y}$ such that $\hat{s} \in \partial f(\hat{y})$, so the notation (4) is no longer correct for the model. For the same reason, characterizing the model in terms of triples $(y, f, s)_i$ is clumsy: we need only *couples* $(s, r)_i \in R^n \times R$ characterizing affine functions. In addition to its slope s_i, each such affine function can be characterized by its value at the current stability center x; we choose to call $f(x) - e_i$ this value. Calling ℓ the total number of affine pieces, all the necessary information is then characterized by a *bundle* of couples

$$(s_i, e_i) \in R^n \times [0, +\infty[\quad \text{for } i = 1, \ldots, \ell$$

and the model is

$$R^n \ni y \mapsto \varphi(y) = f(x) + \max_{i=1,\ldots,\ell}[-e_i + \langle s_i, y - x\rangle].$$

Finally, we prefer to work in (14) with the inverse of the penalty parameter; its interpretation as a stepsize is much more suggestive. In summary, re-introducing the iteration index k, our stabilized problem to be solved at the kth iteration is

$$\left| \begin{array}{ll} \min[r + \frac{1}{2t_k}\|y - x_k\|^2] & (y, r) \in R^n \times R, \\ r \geq f(x_k) - e_i + \langle s_i, y - x_k\rangle & \text{for } i = 1, \ldots, \ell, \end{array} \right. \tag{17}$$

a quadratic program; the extra variable r stands for φ_k-values. It can easily be seen that (17) has a unique solution (y_{k+1}, r_{k+1}) with $\varphi_k(y_{k+1}) = r_{k+1}$.

The precise algorithm can now be stated, with notations close to §2.2.

Algorithm 3.3 (primal bundle method with Euclidean penalty)
The initial point x_1 is given, together with a maximal bundle size $\bar{\ell}$. Choose a descent-coefficient $m \in]0, 1[$. Initialize the descent-set $K = \emptyset$, the iteration-counter $k = 1$ and the bundle size $\ell = 1$. Compute $f(x_1)$ and $s_1 \in \partial f(x_1)$; set $e_1 = 0$, corresponding to the initial bundle (s_1, e_1), and the initial model

$$y \mapsto \varphi_1(y) := f(x_1) + \langle s_1, y - x_1\rangle.$$

STEP 1 (main computation). Choose a "stepsize" $t_k > 0$. Solve (17) as stated in Lemma 3.1, to obtain the optimal solution $y_{k+1} = x_k - t_k \hat{s}_k$, with $\hat{s}_k \in \partial \varphi_k(y_{k+1})$. Set

$$\hat{e}_k := f(x_k) - \varphi_k(y_{k+1}) - t_k \|\hat{s}_k\|^2,$$

$$\delta_k := f(x_k) - \varphi_k(y_{k+1}) - 1/2\, t_k \|\hat{s}_k\|^2.$$

STEP 2 (descent test). Compute $f(y_{k+1})$ and $s_{k+1} \in \partial f(y_{k+1})$; if the descent test

$$f(y_{k+1}) \leq f(x_k) - m\delta_k \tag{18}$$

is not satisfied declare "null-step" and go to Step 4.

STEP 3 (descent-step). Set $x_{k+1} = y_{k+1}$. Append k to the set K; for $i = 1, \ldots, \ell$ change e_i to

$$e_i + f(x_{k+1}) - f(x_k) - \langle s_i, x_{k+1} - x_k \rangle \;.$$

Change also $\hat{e}_k$ similarly.

STEP 4 (managing the bundle size). If $\ell = \bar{\ell}$ then: delete at least 2 elements from the bundle and insert the element $(s_k, \hat{e}_k)$.

Call again $(s_i, e_i)_{i=1,\ldots,\ell}$ the new bundle thus obtained (note: $\ell < \bar{\ell}$).

STEP 5 (loop). Append $(s_{\ell+1}, e_{\ell+1})$ to the bundle, where $e_{\ell+1} = 0$ in case of descent-step and, in case of null-step:

$$e_{\ell+1} = f(x_k) - [f(y_{k+1}) + < s_{\ell+1}, x_k - y_{k+1} >].$$

Replace ℓ by $\ell + 1$ and define the model

$$y \mapsto \varphi_{k+1}(y) = f(x_{k+1}) + \max_{i=1,\ldots,\ell} [-e_i + < s_i, y - x_{k+1} >].$$

Replace k by $k + 1$ and loop to Step 1. □

3.2 Convergence

A convergence result was mentioned in §2.2 for a simplified version of Algorithm 3.2 (fixed t_k, no aggregation). A good numerical behaviour, however, requires a proper tuning of the stepsize t_k at each iteration. Then the convergence analysis becomes a lot more delicate. Two observations are helpful to orient the choice of t_k. Indeed:

(i) A small t_k is dangerous in case of a descent-step: it might overemphasize the role of the stabilization, resulting in an unduly small move from x_k to y_{k+1}.

(ii) It is dangerous to make a null-step with large t_k. This might bring little improvement in the model φ_{k+1}.

The next lemma reveals two important quantities for convergence analysis: $\hat{s}_k$ and ε_k, which provide an approximate optimality condition for x_k (use(19) below). The rule of the game will be to drive them simultaneously to 0.

Lemma 3.4 *At each iteration of Algorithm 3.3, $\varphi_k \leq f$ and there holds*

$$f(y) \geq f(x_k) + \langle \hat{s}_k, y - x_k \rangle - \varepsilon_k \quad \text{for all } y,$$

with

$$\varepsilon_k := \hat{e}_k = f(x_k) - \varphi_k(y_{k+1}) - t_k \|\hat{s}_k\|^2. \tag{19}$$

$\square$

Following standard proof patterns in bundle theory, we distinguish two cases. First, if there are infinitely many descent steps, the objective function decreases "sufficiently", thanks to the successive descent tests (18).

Theorem 3.5 *Assume that K in Algorithm 3.3 is an infinite set.*

(i) If

$$\sum_{k \in K} t_k = +\infty, \tag{20}$$

then $\{x_k\}$ is a minimizing sequence.

(ii) If, in addition, $\{t_k\}$ has an upper bound on K, and if f has a minimum point, then the whole sequence $\{x_k\}$ converges to such a point. $\square$

It is interesting to note the similarity between (20) and the popular steplength condition in subgradient optimization, see e. g. Poljak (1977) or Shor (1985).

In a second step one fixes the case where the stability center stops at some x_{k_0}. This is our next result.

Theorem 3.6 *Assume that K in Algorithm 3.3 is a finite set: for some k_0, each iteration $k \geq k_0$ produces a null-step. If*

$$t_k \leq t_{k-1} \quad \text{for all } k > k_0 \quad \text{and} \quad \sum_{k > k_0} \frac{t_k^2}{t_{k-1}} = +\infty, \tag{21}$$

then x_{k_0} minimizes f. $\square$

Note that the rules (20) and (21) hold e. g. for constant $t_k = t > 0$; then we get back Theorem 2.5. From a numerical point of view, large t_k give a variant close to the cutting-plane algorithm, which we want to avoid; small t_k lead to something close to the ordinary subgradient algorithm, also known to be inefficient. A good t_k should lie somewhere in between. We have to admit, however, that an appropriate control is a fairly delicate business and needs much more research; some adaptive strategies can be found in Kiwiel (1990), Schramm and Zowe (1992), Lemaréchal (1993).

3.3   Dual interpretations

Consider again (12), written in expanded form:

$$\left|\begin{array}{l} \min r + \frac{1}{2}\mu\|y - x_k\|^2 \quad (y, r) \in R^n \times R, \\ f(y_i) + \langle s_i, y - y_i \rangle \le r \quad \text{for } i = 1, \dots, k. \end{array}\right. \tag{22}$$

The dual of this quadratic program can be formulated explicitly, yielding very instructive interpretations. In the result below, Δ_k is the unit-simplex of R^k; the coefficients

$$e_i := f(x_k) - f(y_i) - \langle s_i, x_k - y_i \rangle \quad \text{for } i = 1, \dots, k$$

are the same as in §3.2: $f(x_k) - e_i$ is the value at the stability center of the ith affine function making up the model. Using standard duality tools, we can prove the following result, which echoes Lemma 3.1.

Lemma 3.7 *For $\mu > 0$, the unique solution of the penalized problem (12) = (22) is*

$$y_{k+1} = x_k - (1/\mu) \sum_{i=1}^{k} \alpha_i s_i, \tag{23}$$

where $\alpha \in R^k$ solves

$$\min_{\alpha \in \Delta_k} \left[\frac{1}{2}\| \textstyle\sum_{i=1}^{k} \alpha_i s_i \|^2 + \mu \sum_{i=1}^{k} \alpha_i e_i \right]. \tag{24}$$

Furthermore, there holds

$$\check{f}_k(y_{k+1}) = f(x_k) - \sum_{i=1}^{k} \alpha_i e_i - (1/\mu)\| \textstyle\sum_{i=1}^{k} \alpha_i s_i \|^2. \qquad \square$$

With the form (24) of stabilized problem, we can play the same game as in §2.2 and §2.3: the linear term in α can be interpreted as the dualization of a constraint whose right-hand side, say ε, is a function of its multiplier μ. In other words: given $\mu > 0$, there is ε such that the solution of (12) is given by (23), where α solves

$$\left|\begin{array}{l} \min \frac{1}{2}\| \textstyle\sum_{i=1}^{k} \alpha_i s_i \|^2, \quad \alpha \in \Delta_k, \\ \sum_{i=1}^{k} \alpha_i e_i \le \varepsilon. \end{array}\right. \tag{25}$$

Conversely, let the constraint in this last problem have a positive multiplier μ; then the associated y_{k+1} given by (23) is also the unique solution of (12) with this μ (cf. Proposition 2.4). In summary, our detour in the dual space gives birth to a fourth conceptually equivalent stabilized algorithm.

Algorithm 3.8 (cutting planes + dual stabilization) The initial point x_1 is given. Choose a descent-coefficient $m \in]0, 1[$. Initialize the descent-set $K = \emptyset$, the iteration-counter $k = 1$, and $y_1 = x_1$; compute $f(y_1)$ and $s_1 \in \partial f(y_1)$.

STEP 1. Choose $\varepsilon \geq 0$ such that the constraint in (25) has a positive multiplier μ.
STEP 2. Solve (25) to obtain an optimal $\alpha \in \Delta_k$ and a multiplier $\mu > 0$.
Set

$$\hat{s} := \sum_{i=1}^{k} \alpha_i s_i, \quad y_{k+1} = x_k - \hat{s}/\mu .$$

STEP 3. Compute $f(y_{k+1})$ and $s_{k+1} \in \partial f(y_{k+1})$. If

$$f(y_{k+1}) \leq f(x_k) - m(\varepsilon + \tfrac{1}{2} \|\hat{s}\|^2 /\mu)$$

set $x_{k+1} = y_{k+1}$ and append k to the set K (descent-step); otherwise set $x_{k+1} = x_k$ (null-step).
Replace k by $k + 1$ and loop to Step 1. □

We obtain exactly the description of *dual bundle methods* as discussed e.g. in Zowe (1985) and Lemaréchal (1989). Here there is no real line-search, the multiplier of the linear constraint in (25) gives directly the stepsize. Thus, classical bundle methods and cutting-plane stabilizations appear as techniques *dual to each other*.

4   Some Applications

From the numerical point of view the *penalized primal* bundle variants discussed in §2.2 and §3 seem to be the most promising approach at the moment. The codes BTC and BTNC (convex and nonconvex case) are implementations of this concept; for details see Schramm and Zowe (1992). A realization of the dual bundle concept is the code M1FC1 of Lemaréchal. Both implementations proved to be successful tools when dealing with large and difficult applied problems. Some of these applications are shortly sketched below. Since we are not interested in a numerical comparison of the two approaches we present the BTC/BTNC-results only.

The first two applications (§4.1 and §4.2) are convex minimax problems, the third one (§4.3) is a nonconvex optimization problem with a variational inequality as side constraint. The eigenvalue problem §4.1 is a special case in so far as one can compute the full subdifferential at each x. This more informative situation can successfully be exploited. In general, however, the computation of a subgradient itself boils down to the solution of a suboptimization problem and in practice it will be too expensive to do this more than once at each iteration step; §4.2 and §4.3 are typical examples for this situation.

The first two applications are discussed in more detail in Schramm and Zowe (1992), the third one in Kočvara and Outrata (1993). Further successful applications of the code BTC/BTNC are reported in Ben-Tal, Eiger, Outrata and Zowe (1992) [design of water distribution networks], in Haslinger, Hoffmann and Kočvara (1993) [shape optimization] in Ben-Tal, Kočvara and Zowe (1993) [optimization of trusses] and in Outrata (1993) [Stackelberg problems and Nash equilibria].

4.1 Minimizing the maximal eigenvalue

Often an application requires the solution of the subproblem

$$\text{minimize} f(x) := \lambda_{max}(A(x)); \tag{26}$$

here $A(.)$ is a real symmetric $m \times m$-matrix, which depends linearly on $x \in R^n$, and $\lambda_{max}(A(x))$ denotes the maximal eigenvalue of $A(x)$. The following properties hold:

- f is convex;

- f is nonsmooth at x, if the maximal eigenvalue $f(x)$ has multiplicity greater than 1;

- if u is eigenvector of $A(x)$ for the eigenvalue $f(x)$ and $\|u\|_2 = 1$, then a subgradient of f at x can easily be computed from the dyadic product uu^T.

Hence we are in the situation described by (3) and bundle type methods are appropriate tools to deal with (26). We encountered the above eigenvalue problem in connection with

(i) stable sets of graphs,

(ii) experimental design.

Here we consider more closely (i). The theoretical background is discussed in detail in a book by Grötschel, Lovász and Schrijver (1988). Let $G = [V, E]$ be a graph, $w = (w_1, \ldots, w_{|V|})^T$ a vector in $R^{|V|}$ with nonnegative components and put $\bar{w} := (\sqrt{w_1}, \ldots, \sqrt{w_{|V|}})^T$. We want to compute the so-called *theta-function* $\vartheta(G; .)$,

$$\vartheta(G; w) := \min_{A \in M} \lambda_{\max}(A + W),$$

where $W = \bar{w}\bar{w}^T$ and

$$M := \{B \,|B \text{ symmetric } n \times n\text{-matrix},$$
$$b_{ii} = 0 \text{ for } i \in V, \, b_{ij} = 0 \text{ for } i, j \text{ nonadjacent}\}.$$

The theta-function is the support-function of the convex set $\text{TH}(G)$ (a set which contains the convex hull of the incidence vectors of all stable sets of G). Its value is known for the two special cases (let $w = (1, \ldots, 1)^T$):

(a) If G is a circle with an odd number n of knots, then

$$\vartheta(G; w) = \frac{n \cos \frac{\pi}{n}}{1 + \cos \frac{\pi}{n}}; \tag{27}$$

(b) if G is an Erdös-Ko-Rado-graph $K(n, r)$, then

$$\vartheta(K(n, r); w) = \binom{n-1}{r-1}. \tag{28}$$

To compute $\vartheta(G; w)$ numerically we have to solve the nonsmooth convex problem

$$\text{minimize} \quad \lambda_{max} \quad (A + W) \quad \text{subject to} \quad A \in M. \tag{29}$$

Since W is constant and the constraints only require A to be symmetric and some components of A to be zero, (29) can be phrased as an unconstrained minimization problem of form (26), where $A = A(x)$ and $x \in R^m$ corresponds to the free components of A. The dimension is equal to

$$m = \frac{n(n-1)}{2} - |\{(ij) \mid i, j \text{ nonadjacent}\}|. \tag{30}$$

Table 1 shows some results we got for case (a) using the code BTC. Note that for circles the dimension m of the optimization problem equals the number n of knots. The starting point was always $x_1 = (-1, \ldots, -1)^T$. The value ϑ in the second column is the precise value of $\vartheta(G; w)$ from (27), niter reports the number of iterations (descent steps), $\#f/g$ the number of function/subgradient evaluations and f is the computed approximation of the optimal value $\vartheta = \inf f(x)$.

n	ϑ	niter	$\#f/g$	f
17	8.42701	19	23	8.42705
23	11.44619	26	30	11.44619
39	19.46833	42	44	19.46833
55	27.47756	49	49	27.47756
111	55.48889	50	50	55.48889

Table 1: Odd Circles

Table 2 reports the corresponding results for case(b). The dimension m and the optimal value $\vartheta = \inf f(x)$ are known from (30) and (28).

Table 3 discusses the improvement of BTC for odd circles, if we add at x to the bundle all subgradients, which we obtain from a system of orthonormal eigenvectors for $f(x)$. We considered eigenvalues as equal, if they differ in value less than 10^{-9}. The extreme gain is probably due to some hidden structure of the problem, since the same technique applied to random graphs did not lead to a consistent improvement.

n	r	Dim	ϑ	niter	$\#f/g$	f
5	2	15	4	28	28	4.000003
6	2	45	5	33	33	5.000013
10	2	630	9	63	63	9.000008
9	3	840	28	63	65	28.000095
10	4	1575	84	125	128	84.000409

Table 2: Erdös-Ko-Rado-Graphs

	BTC			modified BTC		
n	niter	$\#f/g$	f	niter	$\#f/g$	f
17	19	23	8.42705	3	6	8.42701
23	26	30	11.44619	2	3	11.44619
39	42	44	19.46833	3	3	19.46833
55	49	49	27.47756	3	3	27.47756
111	50	50	55.48889	3	7	55.48888

Table 3: Circles – enlarged information

4.2 Traveling salesman problems

In many practical applications one has to solve a problem which can be phrased as a (symmetric) *traveling salesman problem*: Given a complete graph $K_n = (V, E)$ with n knots and distances c_{ij} for each edge $ij \in E$ (with $c_{ij} = c_{ji}$), find a tour T^* with length $c(T^*)$ as small as possible. Since problems of this type often appear in tremendous size (e.g. drilling problems with several thousands of knots), it is in general not possible to solve them exactly. Widely used tools are therefore heuristic, which compute an approximate solution T rather quickly. To judge the quality of such a tour T, it is important to know a lower bound for the length $c(T^*)$ of an optimal tour. Such bounds can be found via the *1-tree relaxation*. One writes the TSP as a linear problem of the form

$$\min\{\langle c, x \rangle \mid Ax = a,\, Bx \leq b,\, x_i \in \{0,1\}\}.$$

The following weak duality relation holds:

$$c(T^*) = \min\{c(T) \mid T \text{ is a tour}\} \geq \max\{\phi(\lambda) \mid \lambda \in R^n\}, \tag{31}$$

where $\phi : R^n \to R$ is defined by

$$\phi(\lambda) := \min\{\langle c, x \rangle + \langle \lambda, Ax - a \rangle \mid Bx \leq b,\, x_i \in \{0,1\}\}.$$

Without the binary constraints $x_i \in \{0,1\}$ one has even equality in (31). For our TSP's the gap was never greater than 1%–3%. Hence we can find a good lower bound by maximizing the function ϕ (i.e. minimizing $-\phi$) which, as minimum of finitely many linear functions in λ, is nonsmooth concave and piecewise linear.

It is known from combinatorics that $\phi(\lambda)$ can be computed via the length $\tilde{c}$ of a so-called minimum spanning 1-tree $x(\lambda)$ for our graph with the modified distances $\tilde{c}_{ij} := c_{ij} + \lambda_i + \lambda_j$; it holds

$$\phi(\lambda) = \tilde{c} - 2 \sum_{i=1}^{n} \lambda_i.$$

There are efficient algorithms to compute such a minimum tree and thus $\phi(\lambda)$; we used the Prim algorithm. Simple subgradient calculus shows that this tree $x(\lambda)$ provides us for free with a subgradient of ϕ at λ, namely

$$s = s(x(\lambda)) := A\,x(\lambda) - a \in \partial\phi(\lambda).$$

A closer look shows that the components s_i of $s(x(\lambda))$ are the degrees of the knots of the 1-tree minus the constant two:

$$s_i = \text{degree}(i) - 2, \quad i = 1, \ldots, n.$$

Thus we are precisely in the situation (3) and can apply the bundle code BTC.

The following *subgradient* variant

$$x_{k+1} := x_k - M\rho^k(\alpha_k s_k + (1 - \alpha_k)s_{k-1})/\|\alpha_k s_k + (1 - \alpha_k)s_{k-1}\| \qquad (32)$$

with fixed $M > 0$, $0 < \rho < 1$ and $0 < \alpha_k \leq 1$ for all k is commonly used in the TSP context to maximize ϕ. For the choice $\alpha_k = 1$ for all k one can establish convergence of the x_k from (32) with geometric convergence speed (with factor $M\rho$).The limit, however, need not be optimal!

Nr.	Problem	Dim	Tour
1	KROL1	100	21282
2	KROL2	100	22141
3	KROL3	100	20749
4	KROL4	100	21294
5	KROL5	100	22068
6	TSP442	442	5069
7	TSP1173	1173	57323
8	V362	362	1966
9	V614	614	2312
10	V1167	1167	5657
11	V2116	2116	6786

Table 4: List of Traveling Salesman Problems

Table 5 below should convince the reader that more sophisticated methods like BTC or M1FC1 do a better job than the subgradient iteration (32). Presented are the results for a collection of synthetic examples (Krolak1, ..., Krolak5) and for some TSP's which come from drilling problems and which are summarized in Table 4. "Dim" in the third column is the number of knots; this is also the dimension of the variable λ in our optimization problem. The fourth column gives the length of a tour, which is considered as a good one (it is not known whether this tour is optimal).

Table 5 shows the results. Here $\#f/g$ denotes the number of function/subgradient evaluations and "lb" is the lower bound which we obtained from the two methods. For the subgradient method we tried several M's and ρ's and report our best results; the method was stopped when we observed no further progress in the leading digits. Obviously we have convergence to a nonoptimal point e.g. for KROL2. Finally, "%" gives the remaining gap in percentage between the length of the tour from Table 4 and the computed lower bound. Further technical details concerning the parameter choice in BTC, the stopping criterion etc. are given in Schramm and Zowe (1992).

	Subgradient Method			BTC		
Nr.	#f/g	lb	%	#f/g	lb	%
1	194	20929	1.66	58	20938	1.62
2	202	21648	2.23	233	21833	1.39
3	264	20451	1.44	79	20473	1.33
4	116	20951	1.61	118	21142	0.71
5	183	21779	1.31	136	21799	1.22
6	229	5043	0.51	378	5051	0.36
7	78	56351	1.70	399	56386	1.63
8	161	1941	1.27	285	1942	1.22
9	129	2253	2.55	179	2254	2.51
10	141	5579	1.38	506	5580	1.36
11	109	6599	2.76	713	6606	2.65

Table 5: Traveling Salesman Problems

4.3 Design of masonry structures

Optimum design problems in mechanics often lead to constraints in form of monotone variational inequalities. As example take contact problems with a nonpenetration condition or a situation where one works with a nonsmooth Hooke's Law. The application discussed below is of the second type; we took this example from Kočvara and Outrata (1993) where all details can be found.

The standard numerical approach to such optimization problems are penalization (regularization) techniques, which replace the variational inequality by an equation. Modern nonsmooth codes suggest to deal directly with the nonsmoothness caused by the variational inequality. To illustrate this approach we consider the design of a pillar made of *masonry*-like material. Such material has an elastic behaviour under pressure but is extremely weak under tension. Our goal is to find the shape u of a pillar of minimal size (meas Ω), which can safely carry the roof of a cathedral. The stress σ in the pillar will vary with the shape u. More precisely: the tensor $\sigma(u)$ is the solution of a variational inequality:

$$\text{find } \sigma \in E(\Omega(u)) \cap M(\Omega(u)) \text{ such that}$$
$$\langle \beta\sigma, \tau - \sigma \rangle_{\Omega(u)} \geq 0 \text{ for all } \tau \in E(\Omega(u)) \cap M(\Omega(u)); \tag{33}$$

here β represents the inverse linear Hooke's law, $E(\Omega(u))$ is the set of statically admissible stresses and

$$M(\Omega(u)) = \{\tau : \Omega(u) \to R^{2\times2}_{sym} \mid \tau_{22} \geq 0 \text{ a.e. in } \Omega(u)\}.$$

The masonry character of the material requires that the vertical component σ_{22} of σ is nonnegative. We even have to demand from σ_{22} to be strictly positive in an a priori given subdomain Ω_0 of the domain Ω of the pillar. Otherwise horizontal cracks may arise and lead to a collapse of the whole structure. Hence our otimization problem reads

$$\text{minimize meas } \Omega(u)$$

$$\text{s.t. } \sigma_{22}(u)(x) > 0 \quad \text{for} \quad x \in \Omega_0(u) \quad \text{and} \quad u \in U_{ad}. \tag{34}$$

Here $\sigma(u)$ is the solution of the variational inequality and U_{ad} specifies the admissible shapes.

Because of the strict inequality in (35) our problem is not well-defined. Therefore we replace this inequality by the condition

$$\lambda(u)(x) = 0 \quad \text{for} \quad x \in \Omega_0(u), \tag{35}$$

where λ is the Karush-Kuhn-Tucker vector associated with the constraint $\sigma_{22}(u) \geq 0$ in the variational inequality (33). We use a standard finite-element discretization, approximate the stress by piecewise constant elements and add the discretized constraint (35) to the objective function as an exact penalty term. Thus we arrive at

$$\text{minimize } \varphi(u, \lambda) := \text{meas } \Omega(u) + r \sum_{i \in D_0} \lambda_i$$

$$\text{s.t. } \lambda = \Lambda(u) \text{ and } u \in U_{ad}. \tag{36}$$

For convenience we have used the same symbols for the continuous and the discrete variables. The set D_0 contains the indices lying in $\Omega_0(u)$ and $r > 0$ is a suitable penalty parameter. Instead of the discretized variational inequality we prefer to work with the equivalent quadratic programming problem

$$\text{minimize} \langle \sigma, B(u)\sigma \rangle$$

$$\text{s.t. } A(u)\sigma = b(u) \text{ and } \sigma \in M(\subset R^{3p});$$

here p is the number of elements, $B(u)$ is the (symmetric positive definite) flexibility matrix, the matrix $A(u)$ and the right hand side $b(u)$ describe the equilibrium condition and M restricts the stress tensor as described above. In this framework the map Λ in (36) assigns to u the (unique) Karush-Kuhn-Tucker vector λ associated with the constraint $\sigma \in M$ (i.e. $\sigma_{22}(u) \geq 0$). It is well-known that $\lambda = \Lambda(u)$ depends in a nonsmooth (and nonconvex) way on the control u and thus the minimization of the objective function

$$f(u) := \text{meas } \Omega(u) + r \sum_{i \in D_0} \Lambda^i(u)$$

requires the use of a nonsmooth code. The results below were obtained with the help of a variant of the code BTNC, which allows additional linear constraints. For all details we refer to Kočvara and Outrata (1993). Let us only mention that we are precisely in the situation described by (3) and that the call of the oracle in (3) is by far the most time consuming job per iteration.

Figure 1 shows the optimal shape and the vertical stress components of a pillar defined by the following data. $\Omega(u) = \{x \in R^2 | 0 \le x^1 \le u, 0 \le x^2 \le 1\}$, the lower horizontal part of the boundary $\partial\Omega$ has prescribed zero displacements and the constant surface traction $g = (5,0)$ acts on the upper half of the left vertical part of $\partial\Omega$. The body force $F = (0,-10)$ corresponds to the self-weight and $\Omega_0(u)$ is defined as the right half of $\Omega(u)$. The domain has been discretized by triangles, given by the 7×7 grid of nodes. The discretized design variables are the x^1-coordinates of the so-called *principal moving nodes*, i.e. the nodes lying on the moving part of the boundary; in this case we have thus $U_{ad} \subset R^7$. On the other hand, the number of variables in the "lower level" problems (30) is 216 (72 triangles x 3 unknown stress components per element). U_{ad} is given by lower and upper bounds and a constant enforcing the Lipschitz continuity of the boundary. As the initial shape the square $[0,1] \times [0,1]$ has been taken; to get the shape in Fig. 1 we needed 61 iterations of the code BTNCLC, capable to minimize nondifferentiable nonconvex functions with respect to linear inequality constraints. The resulting shape exhibits a surprising compliance with some real-world structures, e.g. the supporting pillars of the cathedral shown in Figure 1.

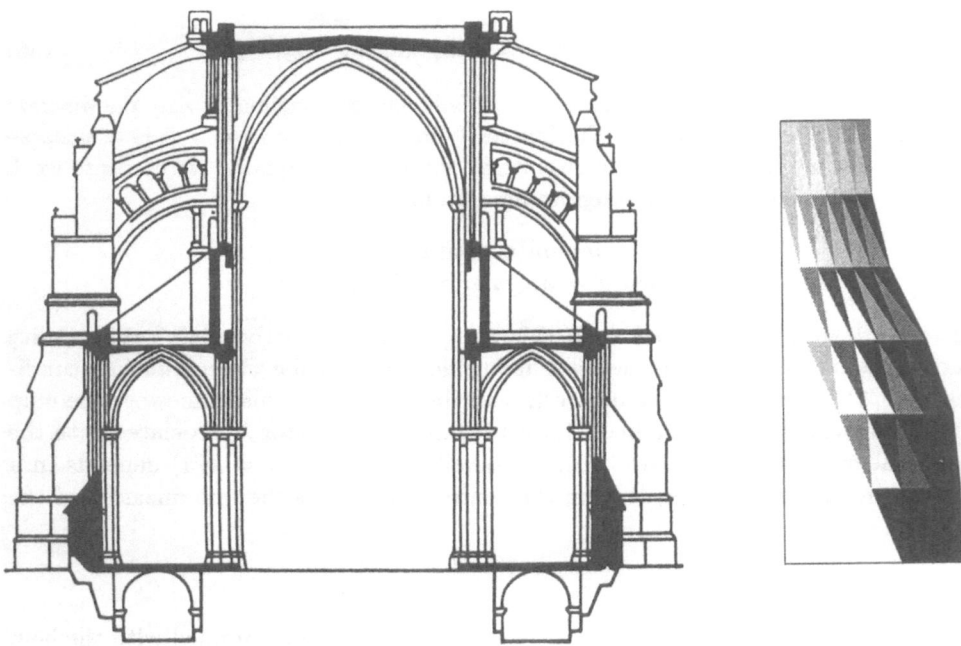

Figure 1

References

Ben-Tal, A., Kočvara, M., Zowe, J. (1993): Two nonsmooth approaches to simultaneous geometry and topology design of trusses. In: *Topology Optimization of Trusses* (M. P. Bendsøe et al. eds.) Kluver, Dordrecht, 31-42.

Ben-Tal, A., Eiger, G., Outrata, J., Zowe, J. (1992): A nondifferentiable approach to decomposable optimization problems with an application to the design of water distribution networks. In: *Proceedings of the 6th Franco-German Conference on Optimization*. Springer.

Cheney, E. W., Goldstein, A. A. (1959): Newton's method for convex programming and Tchebycheff approximation. Numer. Math. 1, 253-268.

Grötschel, M., Lovász, L., Schrijver, A. (1988): *Geometric Algorithms and Combinatorial Optimization*. Springer, Berlin.

Haslinger, J., Hoffmann, K.-H., Kočvara, M.(1993): Control/fictitious domain method for solving optimal shape design problems. M^2AN 27(2), 157-182.

Hiriart-Urruty, J. B., Lemaréchal, C. (1993): *Convex Analysis and Minimization Algorithms* I,II. Springer, Berlin, Heidelberg, New York.

Kelley, J. E.(1960): The cutting plane method for solving convex programs. J. SIAM 8, 703-712.

Kiwiel, K. C.(1988): A survey of bundle methods for nondifferentiable optimization. In: *Proceedings of the XIII. International Symposium on Mathematical Programming*, Tokyo.

Kiwiel, K. C.(1990): Proximity control in bundle methods for convex nondifferentiable minimization. Math. Prog. 46, 1, 105-122.

Kočvara, M., Outrata, J. (1993): A numerical solution of two selected shape optimization problems. In: *Proceedings of the 16th IFIP Conference on System Modelling and Optimization*, Compiègne, 1993.

Lemaréchal, C.(1989): Nondifferentiable optimization. In: *Handbook in OR & MS*, Vol. 1 (G. L. Nemhauser et al., eds.). Elsevier, North-Holland, pp. 529-572.

Lemaréchal, C. (1993): New developments in nonsmooth optimization. In: *Proceedings of the 16th IFIP Conference on System Modelling and Optimization*, Compiègne, 1993.

Outrata(1993): On necessary optimality conditions for Stackelberg problems. J. Optim. Theory Appl., 76,305-320.

Poljak, B. T. (1977): Subgradient methods. A survey of Soviet research. In: *Nonsmooth Optimization* (C. Lemaréchal and R. Mifflin, eds.). Pergamon Press, Oxford, 5-30.

Schramm, H., Zowe, J.(1992): A version of the bundle idea for minimizing a nonsmooth function: conceptual idea, convergence analysis, numerical results. SIAM J. Opt. 2, 121-152.

Shor,N. Z.(1985): *Minimization Methods for Nondifferentiable Functions*. Springer, Berlin Heidelberg.

Zowe, J. (1985): Nondifferentiable optimization. In: *Computational Mathematical Programming* (K. Schittkowski, ed.). Springer, Berlin, 323-356.

COMPUTATIONAL METHODS FOR LINEAR PROGRAMMING

D. F. SHANNO

Rutgers Center for Operations Research
Rutgers University
New Brunswick, NJ 08903
USA

ABSTRACT. The paper examines two methods for the solution of linear programming problems, the simplex method and interior point methods derived from logarithmic barrier methods. Recent improvements to the simplex algorithm, including primal and dual steepest edge algorithms, better degeneracy resolution, better initial bases and improved linear algebra are documented. Logarithmic barrier methods are used to develop primal, dual, and primal-dual interior point methods for linear programming. The primal-dual predictor-corrector algorithm is fully developed. Basis recovery from an optimal interior point is discussed, and computational results are given to document both vast recent improvement in the simplex method and the necessity for both interior point and simplex methods to solve a significant spectrum of large problems.

1   The Simplex Method

Traditional textbook presentation of the simplex method considers the linear programming problem in the form

$$
\begin{aligned}
\text{minimize} \quad & c^T x, \\
\text{subject to} \quad & Ax = b, \\
& x \geq 0,
\end{aligned}
\tag{1}
$$

where $x = (x_1, \ldots x_n)^T$, A is an $m \times n$ matrix, $c = (c_1, \ldots c_n)^T$ and $b = (b_1, \ldots, b_m)^T$ with $b_i \geq 0$, $i = 1, \ldots, m$. A is partitioned as

$$
A = [B|N],
\tag{2}
$$

where B is an $m \times m$ nonsingular basis matrix. Initially enough artificial variables are added to the constraints in (1) to assure that an initial basis matrix B can be found satisfying $x_B = B^{-1}b \geq 0$, $x_N = 0$, $x = (x_B, x_N)^T$. The phase 1 algorithm then drives all artificial variables to zero, and the phase 2 algorithm proceeds to optimize $c^T x$, always satisfying $x_B = B^{-1}b \geq 0$ and $x_N = 0$.

383

E. Spedicato (ed.), Algorithms for Continuous Optimization, 383–413.
© 1994 *Kluwer Academic Publishers.*

For the remainder of this paper, we will denote B and N as the index sets of the basic and nonbasic variables. We now note that several difficulties exist with the numerical efficiency of this form of the simplex method. First, simple upper bound constraints are ostensibly included in the matrix A, while in practice they can be handled much more easily. Also, requiring $x \geq 0$ at all times can lead to large numbers of iterations in which $c^T x$ is not improved on degenerate problems. A significant amount of time can be spend minimizing the sum of the artificial variables, which may lead to a vertex far from optimality. Also, for non-degenerate problems, when a column is selected to enter the basis, a unique column is selected to leave the basis, eliminating choice of the pivot element and sometimes leading to numerical instability. Finally, the traditional choice of entering column, that which minimizes $c_j - c_B^T B^{-1} a_j$, $j \in N$, a_j the j'th column of A, is often not the best neighboring ray along which to proceed in terms of producing the maximum reduction in $c^T x$.

Simplex implementers have been aware of these problems almost since the inception of the simplex algorithm, and have improved performance by addressing these problems over an extended period of time. Progress has accelerated rapidly in the past decade, and with the advent of very fast computer workstations with large memories, the improved simplex methods have recently proved able to solve easily problems that until recently were considered quite intractable. The remainder of this section will document the change in modern simplex methods designed to alleviate the aforementioned problems. Section 4 will provide computational evidence of the dramatic improvement in implemented algorithms.

1.1 THE BOUNDED-VARIABLE SIMPLEX METHOD

The bounded-variable simplex method (Bixby (1992)) considers the problem

$$
\begin{aligned}
\text{minimize} \quad & c^T x, \\
\text{subject to} \quad & Ax = b, \\
& l \leq x \leq u,
\end{aligned}
\tag{3}
$$

where it is possible that some of the $l_i = -\infty$ and some of the $u_i = \infty$. If for some j, $l_j = -\infty$ and $u_j = +\infty$, x_j is said to be free, and if for some j, $l_j = u_j$, x_j is said to be fixed. Again, partitioning A as in (2), we now partition N into N_l corresponding to variables at their lower bounds, N_u corresponding to variables at their upper bounds, N_f corresponding to free variables, and N_s corresponding to fixed variables. A basic solution x corresponding to the basis B is given by

$$
x_{N_l} = l_{N_l}, \tag{4a}
$$
$$
x_{N_u} = u_{N_u}, \tag{4b}
$$
$$
x_{N_s} = l_{N_s} = u_{N_s}, \tag{4c}
$$
$$
x_{N_f} = 0, \tag{4d}
$$
$$
\text{and} \quad x_B = B^{-1}(b - N x_N). \tag{4e}
$$

The basis is called feasible if $l_B \leq x_B \leq u_B$.

Given a feasible basis B with corresponding basic variables x_B, a single step of the bounded variable simplex algorithm is as follows:

(i) Solve $B^T\pi = c_B$ for π.

(ii) Compute all or part of $d_N = c_N - A_N^T\pi$. If $d_j \geq 0$ for all $j \in N_l$, $d_j \leq 0$ for all $j \in N_u$, and $d_j = 0$ for all $j \in N_f$, stop: B is optimal, otherwise select an entering variable x_k, $k \in N$ such that d_k violates these conditions.

(iii) Solve $By = a_k$.

(iv) (Ratio test) If $d_k < 0$, let

$$\theta_i = \begin{cases} \infty & \text{if } y_i = 0, \\ (x_{B_i} - l_{B_i})/y_i & \text{if } y_i > 0, \text{ and} \\ (x_{B_i} - u_{B_i})/y_i & \text{if } y_i < 0, \end{cases}$$

and if $d_k > 0$, let

$$\theta_i = \begin{cases} \infty & \text{if } y_i = 0, \\ (u_{B_i} - x_{B_i})/y_i & \text{if } y_i > 0, \text{ and} \\ (l_{B_i} - x_{B_i})/y_i & \text{if } y_i < 0, \end{cases}$$

for $i = 1, \ldots, m$, let

$$\theta = \min\{\min_i \theta_i, \ u_k - l_k\}.$$

If $\theta = \infty$, stop, (3) is unbounded.

(v) If $d_k < 0$, set $x_B \leftarrow x_B - \theta y$, otherwise $x_B \leftarrow x_B + \theta y$.

If $\theta = u_k - l_k$, update the set of variables at upper and lower bounds.

If $\theta < u_k - l_k$, let j be such that $\theta_j = \theta$ (x_j is the leaving variable). Replace b_j, the j'th column of B with a_k, and set

$$x_{B_j} = \begin{cases} l_k + \theta & \text{if } k \in N_l, \\ u_k - \theta & \text{if } k \in N_u, \\ \theta & \text{if } k \in N_f \text{ and } d_k < 0, \text{ and,} \\ -\theta & \text{if } k \in N_f \text{ and } d_k > 0. \end{cases}$$

Update the index sets B, N_u, N_l, N_f, and N_d, and continue.

This algorithm is a full analogue of the standard simplex algorithm, which will terminate at either an unbounded or optimal solution, reduce the objective function at each step if the problem is not degenerate, and given a feasible solution and proper resolution of degeneracy, is finite. The problem of finding an initial feasible solution will be documented in the next section, and degeneracy, the choice of k, and the solution of the linear equations, in subsequent sections.

1.2 THE INITIAL BASIS

A simple way of finding an initial basic feasible solution to (3) is to set all free variables to zero, all fixed variables to their fixed values, and all other x's at their upper or lower bounds, as in (4a) - (4d). Artificial variables are added to the problem, altering the problem to

$$\text{minimize} \quad e^T z,$$
$$\text{subject to} \quad Ax + Dz \;=\; b, \tag{5}$$
$$l \;\le\; x \;\le\; u,$$
$$z \;\ge\; 0,$$

where z is an m-vector, $e = (1, \ldots, 1)^T$, again an m-vector, and D is a diagonal matrix where $d_k = +1$ if $b_k - a^{k^T} x > 0$ and $d_k = -1$ if $b_k - a^{k^T} x < 0$ with a^k the k'th row of A. Here clearly the initial basis matrix is D, and if there exists a feasible solution to (3), it corresponds to $e^T z = 0$. Note that if $l = 0$, and all nonbasic x's are set to their lower bound once fixed variables have been removed, this corresponds precisely to the standard all artificial basis.

One interesting problem in this regard is whether to set bounded variables at their upper or lower bounds. If one side is infinite, clearly the variables must be set to the finite bound. Free variables, as in the traditional simplex method, are set to zero. Bixby (1992) suggests setting $x_k = l_k$ if $|l_k| \le |u_k|$, and $x_k = u_k$ if $|l_k| > |u_k|$. For some problems, however, other choices may prove better. + Again, as in the traditional simplex method, some artificial variables can be eliminated by taking into account slack variables. Here, as usual, we partition A as

$$A_1 x \;\le\; b_1, \tag{6a}$$
$$A_2 x \;\ge\; b_2, \tag{6b}$$
$$A_3 x \;=\; b_3, \tag{6c}$$

and modify (6) as

$$A_1 x + s_1 \;=\; b_1, \tag{7a}$$
$$A_2 x - s_2 \;=\; b_2, \tag{7b}$$
$$A_3 x + Dz \;=\; b_3, \tag{7c}$$

where $s_1 \ge 0$, $s_2 \ge 0$ and $z \ge 0$. Then s_{1i} and s_{2j} can clearly become members of the initial basis if $b_i - a^{i^T} x \ge 0$ and $b_j - a^{i^T} x \le 0$, respectively.

As one goal in choosing an initial basis is to determine a basis which solves the equations

$$B^T \pi \;=\; c_B \tag{8a}$$
$$\text{and} \quad By \;=\; a_k \tag{8b}$$

efficiently, a diagonal or triangular basis is highly desirable. Further, as numerical accuracy depends on the relative magnitudes of the pivot elements, a triangular matrix with identical diagonal elements is highly desirable. Thus, for numerical purposes, it would be desirable to include all slacks into the basis. Here, the problem is that if $b_i - a^{i^T}x < 0$ or $b_j - a^{j^T}x > 0$, s_i or s_j will violate the bounds $s_i \geq 0$, $s_j \geq 0$. This can be accomodated into the framework of the bounded variable simplex algorithm. Assume we have any basis B, with

$$x_B = B^{-1}(b - Nx_N). \tag{9}$$

Let the objective function of (3) now be defined by $p(x)$ where

$$p_k(x) = \begin{cases} x_k - u_k & x_k > u_k \\ 0 & l_k \leq x_k \leq u_k \\ l_k - x_k & x_k < l_k \end{cases} \tag{10}$$

and

$$p(x) = \sum_{k=1}^{n} p_k(x). \tag{11}$$

(11) defines a piecewise linear objective function, and solving

$$\begin{aligned} \text{minimize} \quad & p(x), \\ \text{subject to} \quad & Ax = b, \\ & l \leq x \leq u, \end{aligned} \tag{12}$$

will produce a feasible solution for any initial basis B. Indeed, artificial variables can easily be incorporated into (12) to obtain an initial basis B simply by considering any $z > 0$ in violation of its bounds. Thus in order to obtain an initial basis, artificials are added to some constraints, with the resultant phase 1 problem

$$\begin{aligned} \text{minimize} \quad & p(x) + e^T z, \\ \text{subject to} \quad & Ax + Dz = b, \\ & l \leq x \leq u, \\ & z \geq 0, \end{aligned} \tag{13}$$

where $p(x)$ is given by (11) and $e = (1, \ldots, 1)^T$ is an s vector corresponding to the s artificials added and D is again a diagonal matrix with $m - s$ diagonal elements 0, the others ± 1.

Basis crashes are attempts to determine an initial matrix B, as close as possible to triangular which has the possibility of containing as many columns as possible of the optimal basis matrix. Bixby (1992) describes one such crash, and the OSL Simplex code (IBM (1991)) contains four different crashes for both the primal and dual simplex algorithms. These initial basis selection algorithms will not be discussed here, other than to note that once structural columns from A have been chosen, together with slack variables, the basis is completed by adding sufficient artificial variables to assure that the initial basis is nonsingular. As a final note on this section, (13) can be modified even further to

$$
\begin{array}{rrcl}
\text{minimize} & p(x) + e^T z & + & \sigma c^T x, \\
\text{subject to} & Ax + Dz & = & b, \\
& l \leq x & \leq & u, \\
& z & \geq & 0,
\end{array}
\tag{14}
$$

where $\sigma \geq 0$ is a parameter which allows movement toward feasibility and optimality simultaneously. Clearly, care must be taken in adjusting σ to assure that an optimal solution to (14) satisfies $z = 0$ and $l \leq x \leq u$.

1.3 THE LINEAR ALGEBRA

From the statement of the bounded variable simplex algorithm, the linear algebra requirements result from having to solve the systems of equations (8a) and (8b). Both can be solved easily assuming we have a factorization of the basis matrix B of the form

$$
B = LU,
\tag{15}
$$

where L is lower triangular and U upper triangular. In any modern simplex code, there are two distinct issues in finding such a factorization.

First, at some iterations, a direct factorization of B as in (15) must be computed. This is certainly true for the initial basis, and is also necessary at selected subsequent iterations to control both stability and efficiency. Suhl and Suhl (1990) describe in detail the issues in computing such a factorization. Ideally, one wishes to maintain the maximum level of sparsity in L and U while assuring that pivot elements chosen in Gauss elimination are sufficiently large to assure numerical stability. To this end, B is first permuted, following principles described by Orchard-Hays (1968), to the form

$$
PBQ = \begin{vmatrix} & & U^1 \\ 0 & & 0 \\ & L^1 & \\ & & N \end{vmatrix},
\tag{16}
$$

where P and Q are permutation matrices. The problem then reduces to that of factoring N.

In order to satisfy the goal of attempting to maintain sparsity, singleton columns and singleton rows of N are identified, and used to determine pivots. When neither singleton rows or columns exist in the remaining matrix to be factored, pivot elements are chosen to satisfy the Markowitz rule which minimizes $(r_i - 1) * (c_j - 1)$ where r_i are the remaining nonzeros in row i and c_j are the remaining nonzeros in column j.

In order to assure stability, only elements of the remaining submatrix which satisfy $|b^s_{i,j}| \geq u * \max\{|b^s_{i,k}|, \quad k \in J^s\}$ are considered. Here the superscripts determine the stage of the elimination and J^s is the index of nonzero elements in the i'th row. The number of rows and columns searched at each stage to find a pivot depends on both the Markowitz count and stability considerations. Full details of the algorithm are in Suhl and Suhl (1990).

The factorization (15) is not stored explicitly, but is stored as a product of respectively upper and lower triangular elementary matrices. The second aspect of computing the factorization (15) is updating the factorization when one column has been dropped from B and one added. The update of choice among contemporary simplex codes is the Forrest-Tomlin update (Forrest and Tomlin (1972)).

Briefly, to understand how updating an LU factorization is accomplished, we first note that changing one column of B gives the updated basis matrix $\hat{B}$ defined by

$$\hat{B} = B + (a_s - a_p)e_p^T, \tag{17}$$

where a_s replaces a_p as the p'th column of B. If $B = LU$, then

$$
\begin{aligned}
L^{-1}\hat{B} &= L^{-1}[a_1 \quad a_{p-1} \quad a_s \quad a_{p+1} \quad a_m] \\
&= [u_1 \quad u_{p-1} \quad L^{-1}a_s \quad u_{p+1} \quad u_m].
\end{aligned} \tag{18}
$$

Thus if the column $L^{-1}a_s$ is shifted to the last column, the remaining matrix has the form

$$
H = \begin{bmatrix}
x & \cdots & \cdots & \cdots & \cdots & x \\
0 & x & \cdots & \cdots & \cdots & \vdots \\
\vdots & 0 & x & \cdots & \cdots & \vdots \\
\vdots & \vdots & x & x & \cdots & \vdots \\
\vdots & \vdots & \vdots & x & \cdots & \vdots \\
0 & 0 & 0 & 0 & x & x
\end{bmatrix},
$$

an upper Hessenberg matrix. Forrest and Tomlin noted that for L and U nonsingular, all the subdiagonal elements in rows p to m are nonzero. Hence rows $p+1, \ldots, p_m$ can be used to eliminate the elements in columns p to $m-1$ of row p, yielding a matrix of the form

$$
\hat{H} = \begin{bmatrix}
x & \cdots & \cdots & \cdots & \cdots & x \\
0 & x & \cdots & \cdots & \cdots & x \\
\vdots & 0 & 0 & \cdots & 0 & x \\
\vdots & \vdots & x & \cdots & \cdots & x \\
0 & 0 & 0 & \cdots & \cdots & \vdots \\
\vdots & \vdots & 0 & \cdots & x & x
\end{bmatrix},
$$

Moving up all lower rows, and putting row p last, gives a matrix in the desired form. This permutation is precisely the inverse of the permutation that moved column p to m.

Algebraically, denoting this permutation by Q, we obtain

$$Q^T \Pi L^{-1} \hat{B} Q = U, \tag{19}$$

or

$$\hat{B} = L\Pi^{-1}(QUQ^T). \tag{20}$$

Here

$$\Pi = (I - e_p r^T), \tag{21}$$

where

$$r^T = [0, \ldots, 0, \quad (h_{p,p}^{(p)}/h_{p+1,p}^{(p)}), \ldots, (h_{p,m-1}^{(m-1)}/h_{m,m-1}^{(m-1)})].$$

From (21)

$$\Pi^{-1} = (I + e_p r^T).$$

In doing the elimination of the row elements in row p, each eliminated element corresponds to an elementary upper triangular matrix

$$\Pi_k = \begin{bmatrix} 1 & & & \vdots & & \\ & \ddots & & -h_{pk}^{(k)}/h_{k+1,k}^{(k)} & & \\ & & \ddots & \vdots & & \\ & & & 1 & & \\ & & & & \ddots & \\ & & & & & 1 \end{bmatrix}, \tag{22}$$

and the matrix $\Pi = \Pi_{m-1} \ldots \Pi_p$, with a similar product form for the inverse. Again, the new factorization is not kept explicitly, but rather in this product form.

As a final note on this section, refactorization occurs when the work of solving the systems using the product form is greater than the cost of refactorizing, or when numerical difficulties occur. Until recently, refactorization has typically been done every 50 iterations, but newer codes can often extend this to 100 iterations.

1.4 DEGENERACY AND PIVOT SELECTION

In the textbook simplex method, the variable leaving the basis is the unique variable (except under degeneracy) which maintains feasibility. This is the variable corresponding to the leaving variable x_j in step (iv) of the bounded simplex algorithm. In practice, the inexact arithmetic of computers requires that bounds be considered satisfied if they are satisfied within a tolerance

$$l - \delta e \le x \le u + \delta e. \tag{23}$$

Harris (1973) suggested using the tolerances to select leaving variables with larger pivots at the cost of perhaps increasing infeasibility, but within the tolerances. The Harris strategy is a two-part strategy. First, a relaxed steplength $\hat{\theta}$ is computed for the problem with the relaxed bounds (23). The second step now chooses the leaving variable from among the set of all variables whose "textbook" step to the exact bound does not exceed $\hat{\theta}$. Of all variables which satisfy this criterion, the leaving variable is chosen as that which corresponds to the largest pivot element. Thus, greater numerical stability is obtained at the cost of allowing a basic variable to violate its bound, but only within the tolerance.

Traditionally, a major problem with the simplex method has been degeneracy. Under degeneracy, a step of $\theta = 0$ occurs, often for a large number of successive iterations. Gill et al.(1989) suggest countering degeneracy by expanding the Harris procedure. To this end, they enlarge the bounds on the variables slightly every iteration, and always assure that a nonnegative step $\alpha_{\min}$ is taken at every iteration, even if this means increasing infeasibility.

Both the CPLEX (CPLEX Optimization (1993)) and the OSL (IBM (1991)) simplex codes attempt to resolve degeneracy by perturbing bounds when degeneracy is encountered. In order to break ties on the leaving variable, each bound is perturbed by a small random amount, thus virtually eliminating the possibility of a tie in choosing the leaving variable. After such a perturbation, the problem is now

$$
\begin{aligned}
\text{minimize} \quad & c^T x, \\
\text{subject to} \quad & Ax = b, \\
& l - \delta r_1 \leq x \leq u + \delta r_2,
\end{aligned}
\tag{24}
$$

where r_1 and r_2 are vectors of random elements in the range $(0, 1)$. While this has proved quite successful in improving the performance of the simplex method on degenerate problems, it creates a new problem, namely that the solution to (24) is not the desired solution to (3). In order to assure that the optimal solution satisfies (3), at least to within the tolerance (23), at some point all nonbasic variables in violation of their bounds must be moved back onto the bounds. When this occurs, it alters the value of x_N in (4e), and thus when x_B is computed by (4e), it is now possible that x_B no longer satisfies

$$
l_B - \delta e \leq x_B \leq u_B + \delta e.
\tag{25}
$$

In this case, a phase 1 algorithm with objective $p(x)$ given by (11) must be reinitiated. To date, computational experience seems to indicate that this is likely to occur if the bounds are tightened quickly after degeneracy has been resolved, but unlikely to occur if (24) is solved to optimality before the bounds are restored to the original bounds.

1.5 CHOOSING THE ENTERING VARIABLE

Step (ii) of the bounded simplex algorithm requires only that a variable be chosen which will reduce the objective function. The textbook simplex method prices every column of N and chooses

$$
\begin{aligned}
d_k^1 &= \max(d_k > 0, \quad k \in N_l), \\
d_k^2 &= \min(d_k < 0, \quad k \in N_u), \\
d_k^3 &= \max(|d_k|, \quad k \in N_f), \\
\text{and} \quad d_k &= \max(d_k^1, \ |d_k^2|, \ d_k^3).
\end{aligned}
$$

Partial pricing, a popular mechanism when the matrix A is so large that it cannot be kept in primary storage, selects a subset of the columns of A to price, optimizes over this subset, finds another subset, and continues until the selected subset cannot improve the objective, at which point full pricing is done, a subset selected, and the algorithm continued until optimality. This option has become less popular with large main memories and especially with vector processing (Forrest and Tomlin (1990)). One recent improvement when partial pricing is used, however, is to select the subset of vectors to be priced randomly, thus minimizing the possibility that problem structure will adversely affect the performance of partial pricing.

Recently, steepest edge algorithms have provided the central focus for algorithms for choosing d_k. While the principal steepest edge algorithm (Goldfarb and Reid (1977)) has been known for some time, recent advances in hardware and data structures for efficient pricing have made the algorithm a much more viable option to partially priced textbook simplex algorithms. The algorithms are simply derived and explained. For simplicity, we will use the development of Forrest and Goldfarb (1992), which assume the linear programming problem is in the form (1). The algorithms apply equally easily to problems in the form (3), but are notationally somewhat more cumbersome.

For problems in the form 1, a feasible solution has the form

$$
x = \begin{bmatrix} B^{-1}b \\ 0 \end{bmatrix}. \tag{26}
$$

A new point $\tilde{x}$ is determined as

$$
\tilde{x} = x + \theta \eta_k, \tag{27}
$$

where η_k is an edge emanating from the current vertex. The edges η_j in this set are

$$
\eta_j = \begin{bmatrix} -B^{-1}N \\ I \end{bmatrix} e_{j-m}, \quad j = m+1, \ldots, n. \tag{28}
$$

The standard simplex method chooses η_k to minimize $c^T \eta_j$, whereas the steepest edge method chooses η_k such that

$$
\frac{c^T \eta_k}{\| \eta_k \|} = \min_{j > m} \left\{ \frac{c^T \eta_j}{\| \eta_j \|} \right\}, \tag{29}
$$

where $\| \cdot \|$ is the Euclidean norm.

Goldfarb and Reid (1977) derived recurrence formulas for updating the quantities $\gamma_j = \| \eta_j \|^2$ for all nonbasic indices j after each pivot. If variable x_k replaces variable x_p in the basis, then

$$\bar{\eta}_p = -\eta_k/\alpha_k, \tag{30a}$$
$$\bar{\eta}_j = \eta_j - \eta_k\bar{\alpha}_j, \quad j > m, \quad j \neq k, \tag{30b}$$
$$\tag{30c}$$

where

$$\bar{\gamma}_p = \gamma_k/\gamma_k^2, \tag{31a}$$
$$\bar{\gamma}_j = \gamma_j - 2\bar{\alpha}_j a_j^T \nu + \bar{\alpha}_j^2 \gamma_k, \quad j > m, \quad j \neq k, \tag{31b}$$

and

$$\alpha_j = \sigma^T a_j, \quad \bar{\alpha}_j = \alpha_j/\alpha_k, \quad Bw = a_k, \quad B^T \nu = w, \quad \text{and } B^T \sigma = e_p.$$

The extra work involved in the primal steepest edge algorithm is the solution of the extra system of equations $B^T \nu = w$, and the inner products of ν with vectors where $\bar{\alpha}_j$ is nonzero. In addition the initial γ_j must all be calculated. Forrest and Goldfarb (1992) report that this additional work is sufficiently little that the greatly improved performance over the traditional simplex method more than compensates.

Forrest and Goldfarb continue to develop three different dual steepest edge algorithms. We will here reproduce only one, the simplest, and in the computational testing of Forrest and Golfarb, the most successful.

This is derived by considering the dual to (1) as

$$\begin{aligned} \text{maximize} \quad & b^T y \\ \text{subject to} \quad & A^T y \leq c. \end{aligned} \tag{32}$$

They show that the edge set for this problem are the rows of $-B^{-1}$ emanating from the vertex $y = B^{-T}c_B$. Here if we denote by ρ_k the solution to

$$-B^T \rho_j = e_j, \tag{33}$$

then we choose the index k by

$$\frac{b^T \rho_k}{\| \rho_k \|} = \min_{i \leq m} \left\{ \frac{b^T \rho_i}{\| \rho_i \|} \right\}.$$

Here we denote $\beta_i = \| \rho_i \|^2$ and note that if variable x_k replaces x_p in the basis, then

$$\bar{B}^{-1} = B^{-1} - \frac{w - e_p}{w_p} e_p^T B^{-1}, \tag{34}$$

where $w = B^{-1}a_k$. Again, it follows that

$$\bar{\rho}_p = (1/w_p)\rho_p, \tag{35a}$$
$$\bar{\rho}_i = \rho_i - (w_i/w_p)\rho_k \tag{35b}$$

and

$$\bar{\beta}_p = (1/w_p)^2\beta_p, \tag{36a}$$
$$\bar{\beta}_i = \beta_i - 2(w_i/w_p)T_i + (w_i/w_p)^2\beta_p, \tag{36b}$$

where

$$B^T\sigma = e_p, \quad B\tau = \sigma, \quad \text{and } \beta_p = \sigma^T\sigma.$$

In general, this is less expensive to implement than in the primal case, as inner products are not required. Other computational efficiencies are discussed by Forrest and Goldfarb (1992). It should be noted that for both the primal and dual steepest edge, full pricing is necessary to preserve all updated norms.

In extensive computational testing, Forrest and Goldfarb found this the overall best of all dual algorithms tested. No clear conclusion could be drawn, however, in comparing this to the primal steepest edge. They did note, however, that the primal software was much more carefully optimized than the dual. In Section 4, computational evidence will be given that the dual steepest edge is perhaps a much better algorithm than one might conclude from the results of Forrest and Goldfarb.

2 Interior Point Methods

Interior point methods for mathematical programming were introduced by Frisch (1955) and developed as a tool for nonlinear programming by Fiacco and McCormick (1968). While Fiacco and McCormick noted that the methods could be applied to linear programming problems, they did not pursue the subject as they did not believe interior point methods could be competitive with the simplex method on linear problems.

Contemporary interest in interior point methods stems from the paper by Karmarkar (1984). Here interior point methods were developed using a projective transformation and a logarithmic potential function. Subsequent work determined that the method developed in this way was a special case of methods derived via logarithmic barrier methods, and hence the development here will dwell entirely on log barrier methods. Readers interested in a full development of both log barrier and projective methods are referred to the surveys by Gonzaga (1992) and Lustig, Marsten, and Shanno (1992c).

2.1 Logarithmic Barrier Methods

The interior point methods introduced by Frisch (1955) and developed by Fiacco and McCormick (1968) were derived from logarithmic barrier methods applied to

$$\begin{aligned}\text{minimize} \quad & f(x), \\ \text{subject to} \quad & g_i(x) \ \geq \ 0, \qquad i = 1,\ldots,m,\end{aligned} \tag{37}$$

where $x = (x_1,\ldots,x_n)^T$. Logarithmic barrier methods transform (37) into a sequence of unconstrained problems of the form

$$\text{minimize} \quad f(x) - \mu_k \sum_{i=1}^{m} \ln g_i(x), \tag{38}$$

where μ_k is a scalar barrier parameter that satisfies $\mu_k > 0$, with $\lim_{k\to\infty} \mu_k = 0$. The algorithm to solve (37) is then

(i) Choose $\mu_0 > 0$ and x^0 such that $g_i(x^0) > 0$ for all $1 \leq i \leq m$.

(ii) Let $x^k = \min\left(f(x) - \mu_k \sum_{i=1}^{m} \ln g_i(x)\right)$.

(iii) If $\mu_k < \epsilon$, then stop. Otherwise, choose $\mu_{k+1} < \mu_k$, set $k = k+1$, and go to (ii).

Fiacco and McCormick show that when $f(x)$ and $g_i(x)$ meet certain general conditions, the sequence $\{x^k\}$ converges to a solution of (37), and that $\lim_{k\to\infty} \mu_k/g_i(x^k) = \lambda_i$, where λ_i is the optimal Lagrange multiplier associated with $g_i(x)$.

2.2 Primal Log Barrier Methods for Linear Programming

The relationship between logarithmic barrier methods and Karmarkar's method was first noted by Gill et al.(1986). They considered the linear programming problem in the standard form (1), transforming the problem to

$$\begin{aligned}\text{minimize} \quad & c^T x \ - \ \mu \sum_{j=1}^{n} \ln x_j, \\ \text{subject to} \quad & Ax \ = \ b.\end{aligned} \tag{39}$$

The Lagrangian for (39) is

$$L(x,y,\mu) = c^T x - \mu \sum_{j=1}^{n} \ln x_j - y^T(Ax - b), \tag{40}$$

and the first order conditions for (40) are

$$\begin{aligned}\nabla_x L \ &= \ c - \mu X^{-1} e - A^T y \ &= \ 0, \\ \nabla_y L \ &= \ -Ax + b \ &= \ 0,\end{aligned} \tag{41}$$

where X is the diagonal matrix whose diagonal elements are the variables x_j, $1 \le j \le n$, and is denoted by X_k when evaluated at the iterate x^k.

With the assumption that there exists a strictly interior feasible point x^k, i.e., $x^k > 0$ and $Ax^k = b$, Newton's method is applied to (41) in an attempt to determine a better estimate to the solution of (1). This yields the search direction

$$\Delta x^k = -\frac{1}{\mu_k} X_k P X_k c + X_k P e, \tag{42}$$

where

$$P = (I - X_k A^T (A X_k^2 A^T)^{-1} A X_k), \tag{43}$$

and the new estimate x^{k+1} to the optimal solution is

$$x^{k+1} = x^k + \alpha_k \Delta x^k \tag{44}$$

for an appropriate step length α_k. The barrier parameter μ_k is then reduced by $\mu_{k+1} = \rho \mu_k$, $0 < \rho < 1$, and the algorithm continues. Note that this is a major departure from the Fiacco-McCormick algorithm in that only one Newton step is taken for each value of μ_k.

Independently of the work on logarithmic barrier methods, Barnes (1986) and Vanderbei, Meketon, and Freedman (1986) were developing what was to become known as the primal affine method. In this method, Δx^k in (42) is replaced by

$$\Delta x^k = -X_k P X_k c, \tag{45}$$

which is the limiting direction in (42) as $\mu_k \to 0$. It was later discovered that this method had been initially proposed by Dikin (1967) more than 15 years before Karmarkar's work.

To understand the relationship between the primal affine and primal logarithmic barrier methods, only the role of the barrier parameter μ_k needs to be considered. It is crucial for interior point methods to remain in the interior of the feasible region, yet from the very beginning, computational experience suggested that choosing α_k in (44) to get very close to the boundary of the region is most efficient. If the problem

$$\begin{array}{ll} \text{minimize} & -\sum_{j=1}^n \ln(x_j), \\ \text{subject to} & Ax = b, \end{array} \tag{46}$$

that tries to find the analytic center of the feasible region is considered, then Newton's method applied to the first order conditions yields

$$\Delta x_k = X_k P e. \tag{47}$$

Thus, the search vector (42) is made up of a centering term to keep away from the boundary, and an affine term which leads toward an optimal solution. As $\mu_k \to 0$, optimality dominates, while for large μ_k, the method proceeds across the interior of the feasible region.

2.3   DUAL LOG BARRIER METHODS

The dual linear programming problem to the standard form primal problem (1) is given by (32). Adding dual slack variables z_j, $1 \leq j \leq n$, to (32) gives the equivalent dual form

$$\text{maximize} \quad b^T y,$$
$$\text{subject to} \quad A^T y + z = c,$$
$$z \geq 0. \tag{48}$$

Dual methods are derived by applying the logarithmic barrier method to (32) by writing the problem

$$\text{maximize} \quad b^T y + \mu \sum_{j=1}^{n} \ln(c_j - a_j^T y), \tag{49}$$

where again a_j is the j'th column of the matrix A. The first order conditions are

$$b - \mu A Z^{-1} e = 0, \tag{50}$$

where Z is the $n \times n$ diagonal matrix with elements $z_j = c_j - a_j^T y$. One step of Newton's method yields

$$\Delta y = \frac{1}{\mu}(AZ^{-2}A^T)^{-1}b - (AZ^{-2}A^T)^{-1}AZ^{-1}e, \tag{51}$$

where the first term in (51) represents a step towards optimality and the second term is a centering step in the dual space.

Again, as in the primal case, the dual affine variant is derived by letting $\mu \to 0$ in (51), yielding the direction

$$\Delta y = (AZ^{-2}A^T)^{-1}b. \tag{52}$$

This is the algorithm proposed independently by Adler et al.(1989).

2.4   PRIMAL-DUAL LOGARITHMIC BARRIER METHODS

The underlying theory of primal-dual interior point methods is due to Megiddo (1989), and was originally developed into a convergent algorithm by Kojima, Mizuno, and Yoshise (1989). The algorithm can be easily derived by considering the first order conditions (41) of the primal problem, or alternatively, by applying the logarithmic barrier method to the dual problem (32) where dual slack variables have been added. Here, the problem is

$$\text{maximize} \quad b^T y + \mu \sum_{j=1}^{n} \ln z_j \tag{53}$$
$$\text{subject to} \quad A^T y + z = c,$$

with the Lagrangian

$$L(x, y, \mu) = b^T y + \mu \sum_{j=1}^{n} \ln z_j - x^T(A^T y + z - c). \tag{54}$$

The first order conditions for (54) are

$$XZe = \mu e, \tag{55a}$$
$$Ax = b, \tag{55b}$$
$$A^T y + z = c, \tag{55c}$$

where X and Z are the previously defined diagonal matrices and e is the n-vector of all ones. Conditions (55b) and (55c) are the usual linear programming optimality conditions of primal and dual feasibility, while (55a) is the usual complementarity condition in the limit as $\mu \to 0$.

As before, Newton's method can be applied to the conditions (55), with resulting steps

$$\Delta y = -(AXZ^{-1}A^T)^{-1} AZ^{-1} \nu(\mu),$$
$$\Delta z = -A^T \Delta y, \tag{56}$$
$$\Delta x = Z^{-1} \nu(\mu) - XZ^{-1} \Delta z,$$

where $\nu(\mu) = \mu e - XZe$. An affine variant of (56) sets $\mu = 0$ at each step.

In comparing primal, dual, and primal-dual methods, it is first instructive to note that all construct a matrix of the form ADA^T, where D is diagonal. The content of D varies, but the computational work does not.

Given this similarity, there are two immediate advantages which appear when examining the primal-dual method. The first is that for primal feasible x and dual feasible y and z, the exact current duality gap $c^T x - b^T y$ is always known. It can easily be shown that, for a feasible point (x, y, z),

$$c^T x - b^T y = x^T z, \tag{57}$$

and thus an excellent measure of how close the given solution is to the optimum is always available. A second advantage of the primal-dual method is that it allows for separate step lengths in the primal and dual spaces, i.e.,

$$x^{k+1} = x^k + \alpha_P^k \Delta x^k,$$
$$y^{k+1} = y^k + \alpha_D^k \Delta y^k, \tag{58}$$
$$z^{k+1} = z^k + \alpha_D^k \Delta z^k.$$

This separate step algorithm was first implemented by McShane, Monma, and Shanno (1989) and has proven highly efficient in practice, significantly reducing the number of iterations to convergence.

Thus far, the choice of the step length parameter has not been addressed. In all implementations which take few iterations, a ratio test is first used to determine the largest steps that can be taken before either some x_j or, respectively, some z_j becomes negative. Let these respective maximum steps be denoted as $\hat{\alpha}_P$ and $\hat{\alpha}_D$. The subsequent step is then chosen to be a constant multiple ρ of the maximum step, i.e.,

$$\alpha_P = \rho\hat{\alpha}_P, \qquad (59)$$
$$\alpha_D = \rho\hat{\alpha}_D.$$

In our computational experience, $\rho = 0.95$ (or even $\rho = 0.9$) appears to be the largest possible safe step for a primal or dual affine variant, while a primal-dual affine variant is not practical for any value of ρ. However for (56), which contains the centering parameter μ, the value $\rho = 0.99995$ works extremely well in practice, with an additional condition that $\alpha_P \le 1$ and $\alpha_D \le 1$. Centering does allow for longer steps, and this largely accounts for the computational superiority of methods using barrier parameters as opposed to affine variants.

2.5 Mehrotra's Predictor-Corrector Method

Mehrotra (1992) introduced a variant to the primal-dual method which computationally has greatly improved efficiency. Mehrotra's method can be derived directly from the first order conditions (55). By substituting $x + \Delta x$, $y + \Delta y$, and $z + \Delta z$ in (55), it is then desired that the new estimate satisfies

$$
\begin{aligned}
(X + \Delta X)(Z + \Delta Z)e &= \mu e, \\
A(x + \Delta x) &= b, \\
A^T(y + \Delta y) + z + \Delta z &= c.
\end{aligned}
\qquad (60)
$$

Collecting terms gives the system

$$
\begin{aligned}
X\Delta z + Z\Delta x &= \mu e - XZe - \Delta X\Delta Ze, & (61a) \\
A\Delta x &= 0, & (61b) \\
A^T\Delta y + \Delta z &= 0, & (61c)
\end{aligned}
$$

where ΔX and ΔZ are $n \times n$ diagonal matrices with elements Δx_j and Δz_j, respectively. This is an implicit system of equations in Δx and Δz. Mehrotra proposed first solving the affine system

$$
\begin{aligned}
X\Delta\hat{z} + Z\Delta\hat{x} &= -XZe, \\
A\Delta\hat{x} &= b - Ax, \\
A^T\Delta\hat{y} + \Delta\hat{z} &= c - A^Ty - z,
\end{aligned}
\qquad (62)
$$

and then substituting the vectors $\Delta\hat{x}$ and $\Delta\hat{z}$ found by solving (62) for the $\Delta X\Delta Ze$ term in the right-hand side of (61). Furthermore, he suggested testing the reduction in complementarity $(x + \alpha_P \Delta \hat{x})(z + \alpha_D \Delta \hat{z})$, where α_P and α_D are again chosen to insure $x > 0$ and $z > 0$. If we let

$$\hat{g} = (x + \alpha_P \Delta \hat{x})^T (z + \alpha_D \Delta \hat{z}), \qquad (63)$$

then Mehrotra's estimate for μ is

$$\mu = (\frac{\hat{g}}{x^T z})^2 (\frac{\hat{g}}{n}). \qquad (64)$$

This chooses μ to be small when the affine direction produces a large decrease in complementarity and chooses μ to be large otherwise. The predictor-corrector algorithm, with a minor variant of Mehrotra's choice of μ, is the current algorithm implemented in OB1 (Lustig, Marsten, and Shanno (1992a)).

2.6 INITIAL FEASIBILITY, BOUNDED VARIABLES, AND FREE VARIABLES

To date, all interior point derivations have assumed the current estimate to the optimizer is feasible. Thus for a primal algorithm, $Ax = b$, for a dual algorithm, $A^T y + z = c$, and for a primal-dual algorithm, both of the above hold for strictly positive x and z. This assumption is easily relaxed. We show here how to relax this for the primal-dual algorithm, but it applies equally to all algorithms developed in this section.

If we do not assume that the point (x, y, z) is feasible, applying Newton's method to (55) yields the system

$$\begin{array}{rcl}
Z\Delta x + X\Delta z & = & \mu e - XZe, \\
A\Delta x & = & b - Ax, \\
A^T\Delta y + \Delta z & = & c - A^T y - z,
\end{array} \qquad (65)$$

which has the solution

$$\begin{array}{rcl}
\Delta y & = & -(AXZ^{-1}A^T)^{-1}(AZ^{-1}\nu(\mu) - AXZ^{-1}r_D - r_P), \\
\Delta z & = & -A^T\Delta y + r_D, \\
\Delta x & = & Z^{-1}\nu(\mu) - Z^{-1}X\Delta z,
\end{array} \qquad (66)$$

where $r_D = c - A^T y - z$ and $r_P = b - Ax$. The directions (66) define a method for handling infeasibility, which is used in the barrier code in the numerical tests reported in Section 4, as well as in OB1.

The bounded variable simplex method of Section 1.1 explicitly handles bounded variables and free variables. These can be handled explicitly by interior point methods as well. We first note that the derivation of interior point methods requires $x_j \geq 0$ for all j. Thus finite lower bounds are used to translate x to a nonnegative vector. Variables x_k with an upper bound but no lower bound are replaced with $-x_k$. When free variables are contained in a problem, some type of translation of the variables is necessary. A standard transformation splits free variables as

$$x_j = x_j^+ - x_j^-,$$ (67)

where $x_j^+ \geq 0$ and $x_j^- \geq 0$. If this transformation is used and the simplex method is applied, it is easy to show that at most one of x_j^+ and x_j^- will be basic at any one time. However, since interior point methods attempt to find points that are as far from the boundary as possible, this can lead to both x_j^+ and x_j^- becoming extremely large. To handle this, first split free variables as in (67). At each iteration, the smaller of the two variables is set to a constant while the difference remains unchanged. To date, this technique has worked well in practice, even for problems with hundreds of free variables.

For upper bounds, the resulting problem is

$$\begin{array}{ll} \text{minimize} & c^T x, \\ \text{subject to} & Ax = b, \\ & 0 \leq x \leq u. \end{array}$$ (68)

Incorporating the upper bounds explicitly gives the first order system

$$\begin{array}{rcl} Ax & = & b, \\ x + s & = & u, \\ A^T y + z - w & = & c, \\ XZe & = & \mu e, \\ SWe & = & \mu e, \end{array}$$ (69)

and the search directions resulting from applying Newton's method to (69) are

$$\begin{array}{rcl} \Delta y & = & (A\Theta A^T)^{-1}[(b - Ax) + A\Theta((c - A^T y - z + w) + \rho(\mu))], \\ \Delta x & = & \Theta[A^T \Delta y - \rho(\mu) - (c - A^T y - z + w)], \\ \Delta z & = & \mu X^{-1}e - Ze - X^{-1}Z\Delta x, \\ \Delta w & = & \mu S^{-1}e - We + S^{-1}W\Delta x, \\ \Delta s & = & -\Delta x, \end{array}$$ (70)

where $\Theta = (X^{-1}Z + S^{-1}W)^{-1}$ and $\rho(\mu) = \mu(S^{-1} - X^{-1})e - (W - Z)e$. Therefore, only the diagonal matrix and the right-hand side change, but the essential computational work remains the same. Note that this derivation assumes that the current iterate always satisfies $x_j + s_j = u_j$ when $u_j < \infty$. In practice, even this simple constraint is relaxed, which causes no difficulties, but makes (70) more notationally complex.

2.7 COMPUTATIONAL ISSUES

An interesting dichotomy between the simplex method and interior point methods is that there is one basic simplex method, and most of the topics discussed in Section 1 are computational issues in improving performance of the simplex method. For interior point methods, there are many possible methods, but once a method has been chosen, there are relatively few choices of implementation issues. Two choices predominate, the choice of the initial point, and the solution of the linear equations. We consider each briefly here.

2.7.1 *Starting Point.*

All interior point methods are sensitive to the initial estimate x^0 to the solution. The starting point currently used by the barrier code tested here is basically that documented in Lustig, Marsten, and Shanno (1992a). Briefly, it computes an estimate $\bar{x}^0 = A^T(AA^T)^{-1}b$, and adjusts all x_j that are below a threshold value to a new value at or above the threshold value. The slack variables for bounds are also set above a threshold value, so that initial bound infeasibility in the constraint $x + s = u$ is allowed. For the dual variables, y^0 is set to 0 and the vectors z^0 and w^0 are set in order to attempt to satisfy the dual feasibility constraints while again staying above a certain threshold. In OB1, the default primal and dual threshold values may be altered by changing them in a control file supplied by the user, called the SPEC file.

Though the rather complex starting point used as a default was determined after extensive computational experimentation, almost any model can be solved in a smaller number of iterations by experimenting to find a problem–specific good starting point. The best choice of a default starting point is still very much an open question, not unlike determining the best initial starting basis for a simplex algorithm.

2.7.2 *The Cholesky Factorization.*

The single largest amount of computation time in solving most linear programming problems using interior point methods is expended on computing the Cholesky factorization of $A\Theta A^T$, where the diagonal matrix Θ is determined by the specific choice of algorithm, as previously explained. The Cholesky factorization computes a lower triangular matrix L such that

$$A\Theta A^T = LDL^T. \tag{71}$$

Since the goal of implementing interior point methods is to solve large problems quickly, it is important that the factorization (71) be done as quickly as possible while minimizing the storage requirement for L. In order to accomplish the latter, both OB1 and the barrier code used in testing for this paper offer the option of two matrix orderings, the multiple minimum degree ordering (see Liu (1985)) and the minimum local fill ordering (see Duff, Erisman, and Reid (1986)). Both are heuristics for permuting the rows of the matrix $A\Theta A^T$ in order to minimize the amount of fill in L. Generally, the minimum local fill algorithm produces a sparser L, but at higher initial cost to obtain the ordering. The multiple minimum degree algorithm produces a denser L, but often at a significant savings in the cost of computing the ordering.

Once the ordering has been completed, the Cholesky factors are computed by (71) using a sparse column Cholesky factorization (see George and Liu (1981)). This again can be implemented in two ways. The first way is called a leftward looking or "pulling" Cholesky, where all previously computed columns that contribute to the factorization are addressed as each new column is computed. The second way is called a rightward looking or "pushing" Cholesky, where after each column is computed, its contribution to all future columns is computed and stored. The latter is more complex to implement in terms of manipulating the data structures of the factorization, but has the advantage of better exploitation of cache memory. OB1 and the barrier code used in computational testing again offer the choice of Cholesky algorithms. The correct choice of algorithm is highly dependent on computer architecture. Further, additional speedups can be obtained by taking advantage of matrix structure. The reader interested in more detail is referred to Lustig, Marsten, and Shanno (1992b).

3   Combining Simplex and Interior Point: The Crossover

As the computational results of the next section will show, the performances of barrier methods and a state-of-the-art simplex method can vary widely. On some problems, even some very large problems, the simplex method is definitely superior. On other problems, interior point methods are every bit as superior. Sometimes, at one stage of a single problem, one method may be preferred, where at another stage the other is preferred (Bixby et al.(1992)).

Further, a class of problems where interior points often dominate are on highly degenerate linear programs arising from integer programming applications. As integer programming codes generally require a vertex solution (and a basis matrix), if interior point methods are to be used in the solution, a vertex solution and basis matrix must be recovered.

Megiddo (1991) proposed a strongly polynomial algorithm for finding an optimal basis, given optimal primal and dual solutions. The paper assumes the primal problem is in the form (1), and the dual in the form (32). While Megiddo's description of his algorithm is theoretical, it can easily be described in the context of a full computational algorithm (Bixby and Lustig (1993)).

To briefly describe the algorithm, we partition x and z into sets where $x > 0$, $z > 0$ and $x = 0$ and $z = 0$, and guess an initial basis matrix B. Now suppose $x_k > 0$, but a_k is not in B. Then by feasibility,

$$B(x_B + x_k B^{-1} a_k) + \tilde{N}\tilde{x}_N = b, \tag{72}$$

where $\tilde{N}$ and $\tilde{x}_N$ omit x_k. Setting $\tilde{x}_B = x_B + x_k B^{-1} a_k$, $x_k = 0$ gives a feasible solution without x_k, provided $\tilde{x}_B \geq 0$. Further, since x is an optimal solution, $z_k = 0$, so

$c^T \tilde{x} = c^T x$. Thus x can be perturbed until some variable reaches 0 without changing either feasibility or optimality. As the variable reaching zero may be basic, however, a pivot may be required by a step of the algorithm.

After all primal variables have been set to zero with the exception of the basic variables, if all m primals are positive, B is the optimal basis. If, however, a pair x_k, z_k with $x_k = 0$, $z_k > 0$ have a_k in B, then the basis is still not optimal.

In this case

$$(B^T(y - y_k B^{-^T} e_k))_k + z_k = 0$$

if and only if $z_k = 0$. Again, y can be adjusted until either $z_k = 0$ or some nonbasic z_k becomes zero. As this corresponds to $x_k = 0$, a_k can be pivoted into the basis, and the process continued until all basic z's are 0.

Clearly, each column is examined at most once. The efficiency of the algorithm will depend in large measure on the accuracy of the first guess to the optimal basis B. The computational experience of Bixby and Lustig (1993) with this algorithm is excellent, as the results of the next section demonstrate.

4   Computational Results

Twenty-eight large models, eight documented in Lustig, Marsten, and Shanno (1992c), and the remainder run by Bob Bixby of Rice University and CPLEX Optimization, were run on a SuperSPARC 10/model 41, using CPLEX 1.0, CPLEX 2.1, and a new currently experimental C-coded barrier code, together with a Megiddo crossover.

Before describing the results, one computational detail has been to date omitted, as it has no real mathematical bearing on the described algorithms, but it does have a significant effect on algorithmic performance. This is preprocessing to reduce problem size. The preprocessor used is standard with CPLEX 2.1 and is described in Lustig, Marsten, and Shanno (1992c).

To first show the measure of improvement in the simplex method over the past few years, the eight LMS models were run with CPLEX 1.0 (primal simplex), CPLEX 1.0 with problem reduction (presolve), and various options of CPLEX 2.2. The model names and sizes before and after presolve are contained in Tables 1 and 2. Table 3 gives the running times for the various options, and in the case of CPLEX 2.2, gives the best times for each model and the algorithm which achieved that time. Finally, the last column gives the time for the barrier code, which is similar to the OB1 code described in Lustig, Marsten, and Shanno (1992a, 1992c), crossover time, and total times. In Table 3, all times are given in SPARC CPU seconds.

The most obvious conclusion to draw from Table 3 is the vast improvement in the CPLEX code from version 1.0 to 2.2. Clearly, presolving is a major benefit, although in one instance, car, it actually hurt.

A second obvious conclusion is that there clearly are models for which barrier methods are distinctly superior, even when it is required to obtain a basic solution, which utilizes barrier plus crossover. This conclusion is reemphasized in Tables 4–6, which give problem statistics for twenty-eight large models, some of which duplicate the models in Tables 1–3. Here simplex was superior on 15, barrier plus crossover on 13. A remaining question of interest is to attempt to determine when one algorithm is better than the other. Of the first eight models, the barrier method relatively only performs very poorly on one model. In this case, AA^T is relatively sparse, but the fill-in from the Cholesky is very bad. Whenever this occurs, it is likely that the simplex is to be preferred. Also, interior points do require more memory, as indicated in the first three "not run" problems of Table 6. The last two "not run" problems are again problems with large fill-in.

Also, for **unknown3**, obviously the primal initial basis was close to optimal. Again, nothing can beat the simplex method when the crash nearly solves the problem.

One case where the advantage is heavily with interior point methods are large, badly degenerate models such as **airfleet**. If the Cholesky is of an acceptable size, barrier methods are generally substantially superior on models exhibiting massive degeneracy.

As a final remark based on these results, the most obvious conclusion is that we are now able to solve routinely models of a size and complexity that just a few years ago were considered completely intractable, and a good solver should contain both simplex and interior point algorithms, together with an efficient crossover algorithm to link them.

REMARKS TO TABLE 6

1. The quantity $|A^T A|/2$ is a lower bound on the size of the Cholesky factors, when they are computed symbolically. In the cases where $|A^T A|$ or $|L|$ is specified in Table 14, it was clear from the size of that number that the barrier method could not compete with the simplex times, and no run was made. In none of these case was this phenomenon due to a manageable number of "dense columns."

2. The models *maintenance, unknown1* and *airfleet* were all suspected in advance to be best solved using steepest-edge. That was the only pricing algorithm used. For the dual, that meant exact initial norms were computed, rather than taking the initial norms of 1.

3. The models *unknown1* and *finland* exceeded the barrier iteration limit of 100, and had to be restarted with a larger iteration limit.

4. The aggregation part of the preprocessor was turned off in solving *stochastic* with the barrier method. That prevented the re-joining of "split columns" and improved the solution time significantly.

5. A total of 26 dense columns were removed from *appliance*. Without doing so, $|L|$ exceeded 3000000.

Name	Rows	Columns	Nonzeros
schedule	23259	29342	75520
energy1	16223	28568	88340
initial	27441	15128	95971
energy2	8335	21200	161160
car	43387	107164	189864
continent	10377	57253	198214
fuel	18401	33905	205789
energy3	27145	31053	268153
TOTAL	174568	323613	1283011

Table 1: Sizes for Sample Models

Name	Rows	Columns	Nonzeros
schedule	5743	12416	40269
energy1	11145	22399	74883
initial	21050	12863	84823
energy2	6845	18307	128948
car	41763	102208	183284
continent	6935	45784	158404
fuel	10163	24854	174703
energy3	9521	28649	188393
TOTAL	113165	267480	1033707
% Remaining*	65%	83%	80%

* Remaining rows in "hardest" models: 30%

Table 2: Presolved Sizes for Sample Models

Name	CPLEX 1.0	CPLEX 1.0 Presolve	CPLEX 2.2	CPLEX 2.2 Option	Barrier	Cross-over	Barrier and Crossover
schedule	174463.5	41268.7	1195.5	Dual SE	2107.3	40.3	2147.6
energy1	8621.8	6970.0	1831.6	Primal	129.8	129.4	259.2
initial	21890.8	4007.3	651.5	Dual Slack SE	39669.3	261.7	39931.0
energy2	80248.2	11533.6	4702.0	Dual	852.3	29.4	881.7
car	9986.3	11927.3	4215.4	Dual SE	539.2	609.5	1148.7
continent	2507.1	1636.5	705.0	Primal	723.3	51.1	774.4
fuel	39451.1	19556.8	5569.8	Primal SE	4831.3	224.6	5055.9
energy3	110788.2	5948.2	1943.8	Dual	796.0	62.0	858.0

Table 3: Running Times for Sample Models

Name	Rows	Columns	Nonzeros
binpacking	42	415953	3526288
distribution	24377	46592	2139096
forestry	1973	60859	2111658
electronics	41366	78750	2110518
maintenance	17619	79790	1681645
finance	26618	38904	1067713
unknown1	43687	164831	722066
finland	56794	139121	658616
crew	4089	121871	602491
gams	8413	56435	435839
roadnet	463	42183	394187
multiperiod	22513	99785	337746
unknown2	44211	37199	321663
appliance	25302	41831	306928
food	27349	97710	288421
airfleet	21971	68136	273539
energy3	27145	31053	268153
unknown5	19168	63488	255504
military	33874	105728	230200
fuel	18800	38540	219880
continent	10377	57253	198214
energy4	25541	19027	194797
car	43387	107164	189864
unknown3	2410	43584	166492
stochastic	28240	55200	161640
energy2	8258	21200	145329
4color	18262	23211	136324
unknown4	1088	2393	110095

Table 4: Sizes for Large Models

Name	Rows	Columns	Nonzeros
binpacking	22	140483	1223824
distribution	22890	22560	1775310
forestry	1888	60754	2109833
electronics	16781	30291	854901
maintenance	17619	79790	1681554
finance	26618	38855	1065664
unknown1	17438	118829	566536
finland	5778	75786	319851
crew	1587	121624	602245
gams	8152	56174	425710
roadnet	462	42182	394186
multiperiod	21022	94541	319798
unknown2	11115	33779	222189
appliance	8142	23638	179402
food	10938	70499	219458
airfleet	18841	65055	225175
energy3	9521	28649	188393
unknown5	14177	57392	322864
military	28221	99915	213210
fuel	10613	24854	174703
continent	69757	45784	158404
energy4	5305	17568	88364
car	41763	102208	183284
unknown3	2410	43584	166492
stochastic	16920	26088	85800
energy2	6845	18307	128948
4color	11831	16835	154905
unknown4	1088	2393	110095

Table 5: Presolved Sizes for Large Models

| Name | Simplex | | Barrier |
	Primal	Dual	+ Crossover
binpacking	**29.5**	62.8	560.6
distribution	**18568.0**	*Not run*	$|AA^T| = 12495446$
forestry	**1354.2**	1911.4	2348.0
electronics	**312.2**	4950.0	$|AA^T| = 4595724$
maintenance	ª57916.3	ª89890.9	**3240.8**
finance	9037.4	**2563.8**	$|AA^T| >$ memory
unknown1	ª59079.1	*Not run*	ª**29177.6**
finland	4594.1	7948.6	ª**4586.6**
crew	7182.6	16172.1	**1264.2**
gams	104278.0	*Not run*	32083.3
roadnet	380.3	1190.8	**318.0**
multiperiod	26771.5	**5023.7**	18362.7
unknown2	2004.6	**961.2**	1131.0
appliance	**861.0**	1599.2	ª1204.6
food	4789.0	1967.6	**974.3**
airfleet	ª71292.5	ª108015.0	**37627.3**
energy3	3091.1	1943.8	**858.0**
unknown5	427.4	516.6	**412.0**
military	25184.0	5606.3	$^b|L| = 6637271$
fuel	7342.3	16452.4	**5055.9**
continent	**705.0**	1517.3	774.4
energy4	**68.7**	83.8	138.2
car	4348.8	5120.4	**1148.7**
unknown3	**46.1**	10212.4	3530.3
stochastic	2670.1	930.0	ª**1700.7**
energy2	5586.2	4702.0	**881.7**
4color	ª**45870.2**	*Not run*	$|L| = 44899242$
unknown4	543.4	**381.0**	2153.4

ª Not run with defaults. See notes below.

^{b} $|L|$ denotes the size of the Cholesky factors.

Table 6: Running Times for Large Models

6. In the cases of *distribution*, *unknown1* and *gams*, dual simplex was run until its running time exceeded the best known solution time, and terminated.

7. *4color* originated from work on a possible new proof of the 4-color theorem. The zero solution was known to be feasible, and the model authors conjectured that this solution was optimal. It was therefore natural to perturb the problem before solving. As it turned out, zero was not optimal.

8. Preprocessing time for the *binpacking* model dominated the solution times and was not included, in order to more clearly show the differences in the algorithms. It is also worth noting that the solve times without preprocessing were not significantly worse than those with preprocessing.

9. One example of the possible effects of tuning is provided by the model *multi-period*. This is a multi-commodity flow problem. Using the network simplex method to get a crash starting basis, followed by the application of the dual simplex method, yields a total solution time of 880.0 seconds.

Acknowledgements

The research is sponsored by the Air Force Office of Scientific Research, Air Force System Command under Grant AFOSR-92-J0046. The United States Government is authorized to reproduce and distribute reprints for governmental purposes notwithstanding any copyright notations thereon.

The author would like to express his sincere thanks to Bob Bixby and Irv Lustig, without whose help this paper could never have been written.

References

[1] Adler, I., Karmarkar, N. K., Resende, M. G. C., and Veiga, G. (1989) An implementation of Karmarkar's algorithm for linear programming , *Mathematical Programming* **44**, 297–335.

[2] Barnes, E. R. (1986) A variation on Karmarkar's algorithm for solving linear programming problems , *Mathematical Programming* **36**, 174–182.

[3] Bixby, R. E. (1992) Implementing the simplex method: The initial basis , *ORSA Journal on Computing* **4**, 267–284.

[4] Bixby, R. E., Gregory, J. W., Lustig, I. J., Marsten, R. E., and Shanno, D. F. (1992) Very large–scale linear programming : A case study in combining interior point and simplex methods , *Operations Research* **40**, 885–897.

[5] Bixby, R. E. and Lustig, I. J. (1993) An implementation of a strongly polynomial time algorithm for basis recovery , In preparation Department of Computational and Applied Mathematics, Rice University Houston, TX, USA.

[6] CPLEX Optimization, Inc. (1993) Using the CPLEX^{TM} Callable Library and CPLEX^{TM} Mixed Integer Library Incline Village, Nevada.

[7] Dikin, I. I. (1967) Iterative solution of problems of linear and quadratic programming , *Doklady Akademii Nauk SSSR* **174**, 747–748 Translated in : *Soviet Mathematics Doklady 8*, 674–675, 1967.

[8] Dikin, I. I. (1974) On the convergence of an iterative process , *Upravlyaemye Sistemi* **12**, 54–60 (In Russian).

[9] Duff, I. S., Erisman, A., and Reid, J. (1986) Direct Methods for Sparse Matrices, Clarendon Press, Oxford, England.

[10] Fiacco, A. V. and McCormick, G. P. (1968) Nonlinear Programming : Sequential Unconstrained Minimization Techniques, John Wiley & Sons, New York.

[11] Forrest, J. J. H. and Tomlin, J. A. (1972) Updating triangular factors of the basis matrix to maintain sparsity in the product form simplex method , *Mathematical Programming* **2**, 263–278.

[12] Forrest, J. J. H. and Tomlin, J. A. (1990) Vector processing in simplex and interior methods for linear programming , *Annals of Operations Research* **22**, 71–100.

[13] Forrest, J. J. H. and Goldfarb, D. (1992) Steepest edge simplex algorithms for linear programming , *Mathematical Programming* **57**, 341–374.

[14] Frisch, K. R. (1955) The logarithmic potential method for convex programming , Unpublished manuscript, Institute of Economics, University of Oslo Oslo, Norway.

[15] George, A. and Liu, J. (1981) Computer Solution of Large Sparse Positive Definite Systems, Prentice Hall, Englewood Cliffs, NJ.

[16] Gill, P. E., Murray, W., Saunders, M. A., Tomlin, J. A., and Wright, M. H. (1986) On projected Newton barrier methods for linear programming and an equivalence to Karmarkar's projective method , *Mathematical Programming* **36**, 183–209.

[17] Gill, P. E., Murray, W., Saunders, M. A., and Wright, M. H. (1989) A practical anticycling procedure for linearly constrained optimization , *Mathematical Programming* **45**, 437–474.

[18] Goldfarb, D. and Reid, J. K. (1977) A practical steepest-edge simplex algorithm , *Mathematical Programming* **12**, 361–371.

[19] Gonzaga, C. C. (1992) Path following methods for linear programming , *SIAM Review* **34(2)**, 167–227.

[20] Harris, P. M. J. (1973) Pivot selection methods of the Devex lp code , *Mathematical Programming* **5**, 1–28.

[21] International Business Machines Corporation (1991) Optimization Subroutine Library Guide and Reference, Release 2.

[22] Karmarkar, N. K. (1984) A new polynomial–time algorithm for linear programming , *Combinatorica* **4**, 373–395.

[23] Kojima, M., Mizuno, S., and Yoshise, A. (1989) A primal–dual interior point algorithm for linear programming , In N. Megiddo, (ed.), Progress in Mathematical Programming : Interior Point and Related Methods, pp. 29–47 Springer Verlag New York.

[24] Liu, J. (1985) Modification of the minimum-degree algorithm by multiple elimination , *ACM Transactions on Mathematical Software* **11**, 141–153.

[25] Lustig, I. J., Marsten, R. E., and Shanno, D. F. (1992) On implementing Mehrotra's predictor–corrector interior point method for linear programming , *SIAM Journal on Optimization* **2**, 435–449.

[26] Lustig, I. J., Marsten, R. E., and Shanno, D. F. (1992) The interaction of algorithms and architectures for interior point methods , In P. M. Pardalos, (ed.), Advances in Optimization and Parallel Computing, pp. 190–205 North–Holland Amsterdam, The Netherlands.

[27] Lustig, I. J., Marsten, R. E., and Shanno, D. F. (1992) Interior point methods : Computational state of the art , Technical Report School of Engineering and Applied Science, Dept. of Civil Engineering and Operations Research, Princeton University Princeton, NJ 08544, USA Also available as RUTCOR Research Report RRR 41–92, RUTCOR, Rutgers University, New Brunswick, NJ, USA. To appear in *ORSA Journal on Computing*.

[28] McShane, K. A., Monma, C. L., and Shanno, D. F. (1989) An implementation of a primal–dual interior point method for linear programming , *ORSA Journal on Computing* **1**, 70–83.

[29] Megiddo, N. (1989) Pathways to the optimal set in linear programming , In N. Megiddo, (ed.), Progress in Mathematical Programming : Interior Point and Related Methods, pp. 131–158 Springer Verlag New York.

[30] Megiddo, N. (1991) On finding primal– and dual–optimal bases , *ORSA Journal on Computing* **3**, 63–65.

[31] Mehrotra, S. (1992) On the implementation of a primal–dual interior point method , *SIAM Journal on Optimization* **2(4)**, 575–601.

[32] Orchard-Hays, W. (1968) Advanced Linear Programming Computing Techniques, McGraw-Hill, New York, NY, USA.

[33] Suhl, U. H. and Suhl, L. M. (1990) Computing sparse LU factorizations for large-scale linear programming bases , *ORSA Journal on Computing* **2**, 325–335.

[34] Vanderbei, R. J., Meketon, M. S., and Freedman, B. A. (1986) A modification of Karmarkar's linear programming algorithm , *Algorithmica* **1(4)**, 395–407.

INFEASIBLE INTERIOR POINT METHODS
FOR SOLVING LINEAR PROGRAMS

J. STOER
Institut für Angewandte Mathematik und Statistik
Universität Würzburg
Am Hubland
D-97074 Würzburg
Germany

ABSTRACT. Interior point methods that follow the primal-dual central path of a dual pair of linear programs (P_0), (D_0) require that these problems are strictly feasible. To get around this difficulty, one technique is to embed (P_0), (D_0) into a family of suitably perturbed strictly feasible linear programs (P_r), (D_r), $r > 0$,

$$(P_r) \quad \begin{aligned} \min\ & (c + r\bar{c})^T x \\ x \in R^n :\ & Ax = b + r\bar{b} \\ & x \geq 0 \end{aligned} \qquad (D_r) \quad \begin{aligned} \max\ & (b + r\bar{b})^T y \\ y \in R^m :\ & A^T y + s = c + r\bar{c} \\ & s \geq 0 \end{aligned}$$

and to follow the path $(x(r), s(r))$, $r > 0$, of all strictly feasible solutions of (P_r), (D_r) with $x_i(r)s_i(r) = r$, $i = 1\ldots,n$ (or, more generally, with $x_i(r)s_i(r) = \rho(r)$ where $\rho(r)$ is a suitable function with $\rho(r) \downarrow 0$ as $r \downarrow 0$). Path following methods compute approximations to this path at parameters $R = r_0 > r_1 > \cdots$, and their complexity is measured by the number $N = N(r, R)$ of steps needed to reduce a given $R > 0$ to a desired $r > 0$. We give an overview on the complexity theory of methods of this type, leading to various estimates on the number $N(r, R)$.

1. Introduction

Consider the pair of dual linear programs

$$(P_0) \qquad \begin{aligned} \lambda_* := \min\ & c^T x \\ x \in \mathbb{R}^n :\ & Ax = b, \quad x \geq 0, \end{aligned}$$

$$(D_0) \qquad \begin{aligned} \lambda_* = \max\ & b^T y \\ y \in \mathbb{R}^m :\ & A^T y \leq c, \end{aligned}$$

where A is an $m \times n$-matrix of rank $A = m$, and b and c are vectors of $\mathbb{R}^m$ and $\mathbb{R}^n$, respectively. Any $y \in \mathbb{R}^m$ is associated with the slack vector $s = s(y) = c - A^T y \in \mathbb{R}^n$ in D_0, and any such s determines y uniquely. Thus we may call a vector $s \in \mathbb{R}^n$ a (strictly) feasible solution of (D_0) if $s \geq 0$ ($s > 0$) and there is a $y \in \mathbb{R}^m$ with

E. Spedicato (ed.), *Algorithms for Continuous Optimization*, 415–434.

$s = c - A^T y$. We will assume that (P_0) and (D_0) have feasible solutions, so that they have optimal solutions and an optimal value $\lambda_* \in R$.

We further use the following customary notation: $e := (1, \ldots, 1)^T \in \mathbb{R}^n$, and

$$[u] := U := \begin{bmatrix} u_1 & & \\ & \ddots & \\ & & u_n \end{bmatrix} \quad \text{if} \quad u = \begin{bmatrix} u_1 \\ \vdots \\ u_n \end{bmatrix}.$$

Then $u \circ v := Uv = Vu = UVe$ denotes the componentwise product of the vectors $u, v \in \mathbb{R}^n$. Powers of a vector u are also defined componentwise, so that $u^2 := u \circ u$, $u^{-1} := U^{-1} e$ etc.

Primal-dual interior point methods for solving (P_0), (D_0) belong to the most efficient interior point methods for solving linear programs. However, they are only applicable to problems (P_0), (D_0) which have strictly feasible solutions $x > 0$ and $s = s(y) > 0$. Then, as is well known, for any $r > 0$, the system

$$Ax = b, \quad x > 0,$$
$$A^T y + s = c, \quad s > 0,$$
$$x \circ s = re,$$

has a unique solution $x = x(r) > 0$, $y = y(r)$, $s = s(r) > 0$ defining the *primal-dual central path* $\{x(r), y(r), s(r)\}_{r>0}$. Primal-dual interior point path following methods try to follow this path for $r \downarrow 0$. In addition, these methods require the knowledge of a strictly feasible pair $x_0 > 0$, $s_0 > 0$ for their start. Therefore, the following major difficulty arises in practical applications of these methods to concrete problems $(P_0), (D_0)$:

• An initial strictly feasible pair $x_0 > 0$, $s_0 > 0$ is not known.

• (P_0), (D_0) may not even have strictly feasible solutions, even though they have feasible solutions.

Among the various proposals to overcome these difficulties the so-called *infeasible interior point methods* seem to be the most promising (see Kojima et al. (1991), Zhang (1992), Mizuno (1992), Potra (1992a,c), Mizuno et al. (1992)), which all generate sequences x_k, (s_k, y_k) of infeasible solutions with $x_k > 0$, $s_k > 0$ of (P_0) and (D_0), respectively, converging to feasibility and optimality simultaneously.

It is the purpose of this paper to study *primal-dual methods* of the latter type. For this purpose it is common to embed (P_0), (D_0) into a family of *perturbed problems* (P_r), (D_r), $r > 0$, converging to (P_0), (D_0) as $r \downarrow 0$,

(P_r)
$$\min \quad (c + r\bar{c})^T x$$
$$x \in \mathbb{R}^n: \quad Ax = b + r\bar{b}, \quad x \geq 0,$$

(D_r)
$$\max \quad (b + r\bar{b})^T y$$
$$y \in \mathbb{R}^m: \quad A^T y + s = c + r\bar{c}, \quad s \geq 0.$$

Here, we choose $\bar{b}, \bar{c}$ so that $(P_r), (D_r)$ have a known feasible solution $x_0 > 0, s_0 > 0$ for some $r = r_0 > 0$, and let $r \downarrow 0$, say by choosing $r_0 = 1$, arbitrary $x_0 > 0$, $s_0 > 0$, $y_0 := 0$ and setting

$$\bar{b} := Ax_0 - b, \quad \bar{c} := s_0 - c + A^T y_0. \tag{1}$$

Then starting with r_0, x_0, s_0, y_0, the vectors $x_k > 0$, $s_k > 0$, y_k are computed as particular strictly feasible solutions of $(P_{r_k}), (D_{r_k})$, which reduce the duality gap

$$(c + r_k \bar{c})^T x_k - (b + r_k \bar{b})^T y_k = x_k^T s_k$$

to zero as $r_k \downarrow 0$.

Such an approach was essentially used with great success by Lustig et al. (1991) in the code OB1 (see also Lustig (1990/91)), who construct these vectors so that $x_k \circ s_k \approx r_k e$. That is, (x_k, s_k, y_k) are an approximation to the point $(x(r_k), s(r_k), y(r_k))$ on the *infeasible central path* of solutions $(x = x(r), y = y(r), s = s(r))$, $r > 0$, to the system

$$Ax = b + r\bar{b}, \quad x > 0,$$
$$A^T y + s = c + r\bar{c}, \quad s > 0,$$
$$x \circ s = re,$$

associated with $(P_r), (D_r)$. Note that on this path

$$(c + r\bar{c})^T x(r) - (b + r\bar{b})^T y(r) = x(r)^T s(r) = rn,$$

so that r measures the duality gap of the perturbed problems $(P_r), (D_r)$ at these central solutions.

Note that if we restrict in (1) the choice of s_0 to $s_0 := x_0^{-1}$, then (1) provides an initial solution x_0, y_0, s_0 on the central path at $r_0 = 1$.

We also note that in our definition of the infeasible central path the parameter r plays a double role, as $r \downarrow 0$ enforces both the reduction of the infeasiblity of $x(r)$, $y(r)$ and $s(r)$, and of the size of the associated duality gap at the same speed (see also Potra (1992a,c), who uses a similar definition). A more general approach is to introduce an extra speed constant $\tau > 0$ and to define the infeasible central path as the solutions $x = x(r) > 0$, $s = s(r) > 0$, $y = y(r)$, $r \downarrow 0$, of

$$Ax = b + r\bar{b}, \quad x > 0$$
$$(PD)_{r,e} \qquad A^T y + s = c + r\bar{c}, \quad s > 0$$
$$x \circ s = r\tau e.$$

belonging to the *speed* $\tau > 0$.

A suitable choice of τ and of the starting point now is (see Mizuno (1992))

$$r_0 := 1, \quad x_0 := s_0 := \rho e, \quad y_0 := 0, \quad \tau := \rho^2$$

with an arbitrary $\rho > 0$ and as before,

$$\bar{b} := Ax_0 - b, \quad \bar{c} := A^T y_0 + s_0 - c.$$

Finally, the vector e may be replaced by any positive *weight vector* $\eta > 0 \in \mathbb{R}^n$, giving rise to an η-*weighted infeasible central path*, or shortly η-*weighted path*, of solutions $x = x(r,\eta)$, $y = y(r,\eta)$, $s = s(r,\eta)$, $r \downarrow 0$, of

$$(PD)_{r,\eta} \qquad \begin{aligned} Ax &= b + r\bar{b}, \quad x > 0 \\ A^T y + s &= c + r\bar{c}, \quad s > 0 \\ x \circ s &= r\tau\eta. \end{aligned}$$

Because of rank $(A) = m$, $s(r,\eta)$ and

$$x(r,\eta) = r\,\tau\,\eta \circ s^{-1}(r,\eta)$$

are uniquely determined by the solution $y = y(r,\eta)$ of the nonlinear system

$$\begin{aligned} A[\tau\,\eta \circ s^{-1}(y)] &= \frac{b}{r} + \bar{b} \\ s = s(y) &= c + r\bar{c} - A^T y > 0. \end{aligned} \tag{2}$$

In this paper we study the complexity of a standard method for following the infeasible central path for $r \downarrow 0$ starting from a known point on the path for $r = r_0 = 1$. The method is a first order method, which uses tangential approximations to the path together with recentering Newton steps to find close approximation to the path at points $r_k, r_k \downarrow 0$.

In Section 2 we will analyze the η-weighted paths and study their limiting behaviour for $r \downarrow 0$. We will see that the paths converge to particular optimal solutions, also if the sets of optimal solutions of (P_0), (D_0) are unbounded. Moreover, also the tangents to these paths converge for $r \downarrow 0$. In Section 3 we describe the path-following method in more detail.

It will turn out that the complexity of the method as measured by the number N of steps to reduce the initial $r_0 = 1$ to $1 > r > 0$ is bounded by

$$N \leq C_0 n \log \frac{1}{r}.$$

Here, C_0 is a universal constant not depending on the linear program to be solved. This result is comparable to the best results of Mizuno (1992) and Potra (1992a,c) for similar but different methods. It is probable (but still an unsolved question) that the complexity bound can be improved to $N = O(\sqrt{n}\log(1/r))$. Moreover, we will provide a very simple proof of the quadratic convergence of the method in the sense that it generates a sequence r_k with $r_{k+1} = O(r_k^2)$, which is based on the existence limit directions of the weighted paths as $r \downarrow 0$.

In the final Section 4 we briefly report on related results of other authors.

2. Analysis of Infeasible Central Paths

We assume for the analysis that the following natural condition is satisfied:

(V) – $rank\ A = m$ (full row rank of A)

      – (P_0) and (D_0) have feasible solutions, say $\bar{x}$ and $(\bar{y}, \bar{s})$

      – (P_r) and (D_r) have for $r = 1$ strictly feasible solutions, say x_0 and (y_0, s_0).

We denote by $\mathcal{LP}$ the set of all linear programs (P_0), (D_0) satisfying (V).

   Under assumption (V), it is elementary to show for all $0 < r \leq 1$ and all weight-vectors $\eta > 0$ and speeds $\tau > 0$

(a) (P_r) and (D_r) are strictly feasible,

(b) (P_r) and (D_r) have compact sets of optimal solutions,

(c) $(PD)_{r,\eta}$ a unique solution $x(r,\eta)$, $y(r,\eta)$, $s(r,\eta)$ (existence of η-weighted infeasible central paths).

Proof: (a) For $0 < r \leq 1$, (P_r) and (D_r) have the strictly feasible solutions $rx_0 + (1-r)\bar{x}$, and $r(y_0, s_0) + (1-r)(\bar{y}, \bar{s})$, respectively.

(b) follows by standard arguments from the strict feasibility of (P_r) and (D_r).

(c) By (2), $y = y(r,\eta)$ maximizes the strictly concave function

$$(b + r\bar{b})^T y + r\,\tau \sum_{i=1}^{n} \eta_i \log s_i(y) \tag{3}$$

on the interior of the compact convex set

$$\{y \mid A^T y \leq c + r\bar{c}\} = \{y \mid s(y) \geq 0\}.$$

    To study the *limiting behavior* for $r \downarrow 0$, we use a crucial Lemma (see Kojima, Megiddo, Mizuno (1991)):

Lemma: *Let $x_0 = s_0 := \rho e$, $\tau = \rho^2$ and $\eta > 0$, and consider any point*

$$x = x(r,\eta), \quad y = y(r,\eta), \quad s = s(r,\eta), \quad 0 < r \leq 1,$$

on the η-weighted central path. Further let x^*, (y^*, s^*) be any optimal solution of (P_0) and (D_0), respectively. Then

$$\|(x, s)\|_1 := e^T x + e^T s \leq r\,\rho\,n + (1-r)\|(x^*, s^*)\|_1 + \rho\,e^T \eta.$$

An immediate Corollary is

Corollary: *The η-weighted central path $(x(r,\eta), y(r,\eta), s(r,\eta))$ stays uniformly bounded for $0 < r \leq 1$ on any compact set of vectors $\eta > 0$.*

Proof of the Lemma: Because of

$$Ax_0 = b + \bar{b}, \qquad\qquad Ax^* = b, \qquad\qquad Ax = b + r\bar{b}$$
$$A^T y_0 + s_0 = c + \bar{c}, \qquad A^T y^* + s^* = c, \qquad A^T y + s = c + r\bar{c},$$

the nonnegative vectors

$$\begin{bmatrix} \hat{x} \\ \hat{y} \\ \hat{s} \end{bmatrix} := r \begin{bmatrix} x_0 \\ y_0 \\ s_0 \end{bmatrix} + (1-r) \begin{bmatrix} x^* \\ y^* \\ s^* \end{bmatrix}$$

satisfy

$$\left. \begin{array}{r} A(\hat{x} - x) = 0 \\ A^T(\hat{y} - y) + (\hat{s} - s) = 0 \end{array} \right\} \implies (\hat{x} - x)^T(\hat{s} - s) = 0.$$

Hence by $0 < rs_0 \le \hat{s}$, $0 < rx_0 \le \hat{x}$, $0 < x$ and $0 < s$,

$$\begin{aligned}
r\rho\|(x,s)\|_1 &= (rs_0)^T x + (rx_0)^T s \\
&\le \hat{s}^T x + \hat{x}^T s = \hat{x}^T \hat{s} + x^T s \\
&= (rx_0 + (1-r)x^*)^T (rs_0 + (1-r)s^*) + x^T s \\
&= r^2 \rho^2 n + r(1-r)\rho\|(x^*,s^*)\|_1 + r\rho^2 e^T \eta
\end{aligned}$$

because of $x^{*T} s^* = 0$ and $x \circ s = r\tau\eta$, which implies $x^T s = r\tau e^T \eta$.

The following Theorem is fundamental:

Theorem 1: (a) *Under assumption* (V) *the following limits exist*

$$\lim_{r\downarrow 0} x(r,\eta) = x_*(= x_*(\eta)), \quad \lim_{r\downarrow 0} \dot{x}(r,\eta),$$

$$\lim_{r\downarrow 0} y(r,\eta) = y_*(= y_*(\eta)), \quad \lim_{r\downarrow 0} \dot{y}(r,\eta),$$

$$\lim_{r\downarrow 0} s(r,\eta) = s_*(= x_*(\eta)), \quad \lim_{r\downarrow 0} \dot{s}(r,\eta).$$

(b) x_*, (y_*, s_*) *are strictly optimal for* (P_0), (D_0), *i.e.* $x^* \circ s^* = 0$ *and* $x^* + s^* > 0$.
(c) *The convergence in* (a) *is uniform on any compact set of vectors* $\eta > 0$.
(d) *For* $\eta = e$, $\tau = 1$, y_* *is the* $\bar{b}$-*center of the (in general noncompact) optimal face* F_0 *of* (D_0), *i.e.*

$$y_* = \arg\max \quad \bar{b}^T y + \sum_{i \in L} \log s_i(y)$$

$$y: \qquad y \in F_0 \,\&\, s_L(y) > 0.$$

Here, the optimal face F_0 of (D_0) is given by

$$F_0 = \{y \mid s(y) = c - A^T y \ge 0 \,\&\, \bar{b}^T y = \lambda_*\}$$

and $L \subset \{1, \ldots, n\}$ is the set of *loose* constraints of F_0,

$$L := \{1 \le i \le n \mid \exists y \in F_0 : s_i(y) = c_i - a_i^T y > 0\}.$$

Its complement is the set

$$T := \{1 \le i \le n \mid i \notin L\}$$

of *tight* constraints of F_0.

We note that results similar to (d) hold for general τ, η and x_*, and (P_0).

We give an outline of the proof, and this only for special values $\eta = e$, $\tau = 1$, and we write briefly $x(r), \ldots$ for $x(r, e), \ldots$. The proof uses arguments introduced by Witzgall, Boggs and Domich (1991) in a related context.

i) By the Lemma, there are convergent sequences, $r_k \downarrow 0$ and

$$\lim_k y(r_k) =: y^*, \quad \lim_k s(r_k) =: s^*, \quad \lim_k x(r_k) =: x^*.$$

ii) By $(PD)_{r,e}$, any such limit is optimal, e.g. $(y^*, s^*) \in F_0$ in the case of (D_0).

iii) Any such limit is *strictly optimal*, say $s_L^* = s_L(y^*) > 0$. This is more difficult to show: For the proof one uses the maximizing property (3) of $y(r, \eta)$, which forces the loose slacks $s_L(y(r))$ to stay bounded away from 0 as $r \downarrow 0$.

iv) By $(PD)_{r_k,e}$, $y_k := y(r_k)$, s_k, x_k satisfy

$$r_k \bar{b} + b = A x_k, \quad x_k \circ s_k = r_k e$$

so that

$$\bar{b} - \sum_{i \in L} \frac{a_i}{s_i(y_k)} = -\frac{b}{r_k} + \sum_{i \in T} \frac{a_i}{s_i(y_k)} \in \text{span}[b, A_T],$$

where span$[B]$ denotes the space spanned by the columns of the matrix B. Taking limits $k \to \infty$ and using $x_L^* = 0$ so that $b = A x^* = A_T x_T^*$, we get

$$\bar{b} - \sum_{i \in L} \frac{a_i}{s_i(y^*)} \in \text{span}[b, A_T] = \text{span}[A_T].$$

Hence, y^* is a $\bar{b}$-center of F_0, which however is *unique* since $\log s_i(y)$ is strictly concave. Therefore the limit $\lim_{r \downarrow 0} y(r)$ exists and is equal to y^*.

v) Let W be a $m \times (m - \text{rk } A_T)$-matrix of full column rank satisfying

$$W^T A_T = 0,$$

so that $W^T b = 0$, because of $b \in \text{span } [A_T]$. Then the solution $y = y(r)$ of (2) also satisfies

$$A[s^{-1}(y)]e - \frac{b}{r} = \bar{b}$$
$$W^T A_L[s_L^{-1}(y)]e_L = W^T \bar{b},$$

$$(4)$$

where $s(y) = c + r\bar{c} - A^T y > 0$.

Differentiation with respect to r gives $(x := x(r), \dot{y} := \dot{y}(r))$

$$A[s^{-2}(y)](A^T \dot{y} - \bar{c}) + \frac{b}{r^2} = 0$$

so that by $x \circ s = re$

$$A[x^2](A^T \dot{y} - \bar{c}) = -b$$
$$W^T A_L [s_L^{-2}(y)](A_L^T \dot{y} - \bar{c}_L) = 0.$$

This is a set of linear equations for $\dot{y} = \dot{y}(r)$. One can show that the limit system of linear equations, $r \to 0$,

$$A_T [x_T^*]^2 (A_T^T \dot{y}^* - \bar{c}_T) = -b$$
$$W^T A_L [s_L^*]^{-2} (A_L^T \dot{y}^* - \bar{c}_L) = 0$$

is uniquely solvable for $\dot{y}^*$. Therefore $\lim_{r \downarrow 0} \dot{y}(r, e)$ exists also and is equal to the solution $\dot{y}^*$ of the limit system.

3. A Typical Path-Following Method

It will be convenient to define for any $z := (x, s)$ in the linear manifold

$$M_r := \{(x, s) \mid \exists y : Ax = b + r\bar{b}, \ A^T y + s = c + r\bar{c}\}$$

distances to the point $z(r, e)$ on the central path, and to the point $z(r, \eta)$ on the η-weighted path by setting

$$d(z, r) := \left\| \frac{x \circ s}{r \tau} - e \right\|_2, \quad d(z, r, \eta) := \left\| \frac{x \circ s}{r \tau} - \eta \right\|_2. \tag{5}$$

We now define a *path-following method* with the aim of constructing sequences r_k and $0 < \tilde{z}_k \in M_{r_k}$ so that r_k converges quickly to 0,

$$r_0 > r_1 > \cdots \to 0,$$

and $d(\tilde{z}_k, r_k) \leq \alpha_0$ for all k, where $\alpha_0 > 0$ is a small constant.

Algorithm I:

Let $0 < \alpha_0 < \alpha_1 < 1/2$.
For given $r_k > 0$, $0 < \tilde{z}_k = (\tilde{x}_k, \tilde{s}_k = s(\tilde{y}_k)) \in M_{r_k}$ with $d(\tilde{z}_k, r_k) \leq \alpha_0$, construct r_{k+1} and $\tilde{z}_{k+1}$ as follows:

1) *Put $\eta_k := \tilde{x}_k \circ \tilde{s}_k/(r_k\tau)$ (then $\|\eta_k - e\|_2 \leq \alpha_0$).*
Consider the path $z(., \eta_k)$ and compute its tangent at r_k:

$$\dot{z}(r_k, \eta_k) =: (\dot{x}_k, \dot{s}_k)$$

$$z_k(r) := \tilde{z}_k + (r - r_k)\dot{z}(r_k, \eta_k).$$

2) *Determine a small r, $0 < r =: r_{k+1} < r_k$ so that*

$$d(z_k(r), r) \leq \alpha_1.$$

3) *Use Newton's method, starting with $z_k(r)$, to find an approximate solution $\tilde{z}_{k+1}$ of $(PD)_{r,e}$ for $r = r_{k+1}$*

$$\tilde{z}_{k+1} \approx z(r_{k+1}, e) \quad \text{with } d(\tilde{z}_{k+1}, r_{k+1}) \leq \alpha_0.$$

One can show (see e.g. Roos and Vial (1988), Zhao and Stoer (1993)) that the number of Newton-Steps is *bounded* and depends *only* on the choice of α_0 and α_1. There are explicit expressions for $r =: r_{k+1} < r_k$ with

$$d(z_k(r), r, \eta_k) = \alpha_1 - \alpha_0,$$

which ensures $d(z_k(r), r) \leq \alpha_1$, namely (see Sonnevend et al. (1991), Zhao and Stoer (1993)) r_{k+1} has to solve the *quadratic equation*

$$f(r_k, \eta_k)(r_k - r_{k+1})\sqrt{\frac{r_k}{r_{k+1}}} = \sqrt{\alpha_1 - \alpha_0}, \tag{6}$$

where

$$f(r, \eta) := \left\| \frac{\dot{x}(r, \eta) \circ \dot{s}(r, \eta)}{r \cdot \tau} \right\|_2^{1/2}. \tag{7}$$

Proof: By differentiation of

$$x(r, \eta) \circ s(r, \eta) = r\tau\eta \implies \dot{x} \circ s + x \circ \dot{s} = \tau\eta$$

follows (with $x = x(r_k, \eta_k)$, $\dot{x} = \dot{x}(r_k, \eta_k)$ etc.)

$$d(z_k(r), r, \eta_k) = \left\| \frac{(x + (r - r_k)\dot{x}) \circ (s + (r - r_k)\dot{s})}{r \cdot \tau} - \eta_k \right\|_2$$

$$= \left\| \frac{\dot{x} \circ \dot{s}}{r \cdot \tau}(r - r_k)^2 \right\|_2 = \alpha_1 - \alpha_0.$$

3.1 COMPLEXITY QUESTIONS

Now it is easy to analyze the complexity of the Algorithm 1. We have to solve the following

Problem: *Given $0 < r < 1$. Find an upper bound for the number $N = N(r)$ of steps the algorithm needs to find sequences $r_k > 0$, $\tilde{z}_k \in M_{r_k}$ with*

$$r_0 = 1 > r_1 > \cdots > r_{N-1} > r \geq r_N \quad and \quad d(\tilde{z}_k, r_k) \leq \alpha_0.$$

To find an answer we consider the *step reduction factor*

$$\beta_k := r_{k+1}/r_k < 1,$$

that solves by (6)

$$1 - \beta_k = \sqrt{\beta_k} \, \frac{\sqrt{\alpha_1 - \alpha_0}}{r_k f(r_k, \eta_k)}. \tag{8}$$

Hence either

$$\beta_k < \frac{1}{4},$$

which gives a fast reduction of r_k, or

$$1 - \beta_k \geq \frac{\sqrt{\alpha_1 - \alpha_0}}{2 r_k f(r_k, \eta_k)} \geq \frac{1}{CD} > 0,$$

where

$$C := 2/\sqrt{\alpha_1 - \alpha_0}, \quad D := \sup_k \, r_k f(r_k, \eta_k). \tag{9}$$

Since

$$\prod_{k=0}^{N} \beta_k \leq \left(1 - \frac{1}{CD}\right)^N \leq r$$

for $N \geq CD \log(1/r)$, we get the *complexity bound*

$$N(r) \leq CD \log \frac{1}{r}. \tag{10}$$

Next, we have to find an upper bound for D by bounding the *complexity function* (see (7), (9)) for general r, η

$$r f(r, \eta) = r \left\| \frac{\dot{x}(r, \eta) \circ \dot{s}(r, \eta)}{r \cdot \tau} \right\|_2^{1/2} = \|\sigma(r, \eta) \circ \xi(r, \eta)\|_2^{1/2}, \tag{11}$$

where $\sigma = \sigma(r, \eta)$, $\xi = \xi(r, \eta)$ are weighted *logarithmic derivatives* of $s(r, \eta)$ and $x(r, \eta)$,

$$\sigma(r, \eta) := r\eta^{1/2} \circ s(r, \eta)^{-1} \circ \dot{s}(r, \eta), \quad \xi(r, \eta) := r\eta^{1/2} \circ x(r, \eta)^{-1} \circ \dot{x}(r, \eta).$$

Differentiation of

$$Ax(r, \eta) = b + r\bar{b}$$
$$A^T y(r, \eta) + s(r, \eta) = c + r\bar{c}$$
$$x(r, \eta) \circ s(r, \eta) = r \tau \eta$$

with respect to r gives equations for $\sigma = \sigma(r, \eta)$ and $\xi = \xi(r, \eta)$ (here $\bar{x}$ is such that $\bar{b} = A\bar{x}$):

$$
\begin{array}{ll}
A\dot{x} = \bar{b} = A\bar{x} & \tilde{A}\xi = \tilde{A}\tilde{u}, \\
A^T \dot{y} + \dot{s} = \bar{c} & \Longleftrightarrow \quad r\tilde{A}^T \dot{y} + \sigma = \tilde{c}, \\
\dot{x} \circ s + x \circ \dot{s} = \tau \eta & \sigma + \xi = \eta^{1/2},
\end{array}
$$

where

$$\tilde{u} := \eta^{-1/2} \circ s \circ \bar{x}/\tau, \quad \tilde{c} := \eta^{-1/2} \circ x \circ \bar{c}/\tau$$

and $\tilde{A} = \tilde{A}(r, \eta) := A[\eta^{1/2} \circ s^{-1}(r, \eta)]$. Their solution σ, ξ is given explicitly by means of the *orthogonal projection* $M = M(r, \eta) := \tilde{A}^T (\tilde{A}\tilde{A}^T)^{-1} \tilde{A}$:

$$
\begin{aligned}
\sigma &= (I - M)\tilde{c} + M\eta^{1/2} - M\tilde{u}, \\
\xi &= \eta^{1/2} - \sigma \\
&= -(I - M)\tilde{c} + (I - M)\eta^{1/2} + M\tilde{u}.
\end{aligned}
\tag{12}
$$

Therefore

$$\|\xi\|_2, \|\sigma\|_2 \le \|\eta^{1/2}\|_2 + \|\tilde{c}\|_2 + \|\tilde{u}\|_2. \tag{13}$$

But by the definitions of $\tilde{c}$ and $\tilde{u}$ and by $\|u \circ v\|_2 \le \|u\|_\infty \|v\|_2$

$$
\begin{aligned}
\|\tilde{c}\|_2 &\le \frac{1}{\tau} \|\eta^{-1/2}\|_\infty \|x\|_\infty \|\bar{c}\|_2, \\
\|\tilde{u}\|_2 &\le \frac{1}{\tau} \|\eta^{-1/2}\|_\infty \|s\|_\infty \|\bar{x}\|_2.
\end{aligned}
\tag{14}
$$

Now, by the algorithm we have for all $\eta = \eta_k$, $k = 0, 1, \ldots$

$$\|\eta - e\|_2 \le \alpha_0 < 1/2 \Rightarrow (1 - \alpha_0)e \le \eta \le (1 + \alpha_0)e$$

so that

$$\|\eta^{\pm 1/2}\|_\infty \le \sqrt{2}, \quad \|\eta^{\pm 1/2}\|_2 \le \sqrt{2n}. \tag{15}$$

Next we choose $\bar{x}$, $A\bar{x} = \bar{b} = Ax_0 - b = A\rho e - A\hat{x}$ so that $\|\hat{x}\|_\infty$ is minimal, and ρ so large that

$$\|\hat{x}\|_\infty, \|c\|_\infty \le \rho. \tag{16}$$

This leads to the bounds

$$
\begin{aligned}
\|\bar{x}\|_\infty &= \|\rho e - \hat{x}\|_\infty \le 2\rho, \quad \|\bar{x}\|_2 \le 2\rho\sqrt{n}, \\
\|\bar{c}\|_\infty &= \|\rho e - c\|_\infty \le 2\rho, \quad \|\bar{c}\|_2 \le 2\rho\sqrt{n}.
\end{aligned}
\tag{17}
$$

Together with (13) – (15) this gives the bound

$$\|\xi\|_2, \|\sigma\|_2 \leq \sqrt{2n} + \frac{2\sqrt{2}\,2\rho\sqrt{n}}{\tau}\|(x,s)\|_\infty$$

$$= \sqrt{2n}\left(1 + 4\frac{\|(x,s)\|_\infty}{\rho}\right). \tag{18}$$

By the Lemma, the number

$$K := \sup_k \|(x(r_k,\eta_k), s(r_k,\eta_k))\|_\infty = \sup_k \|\tilde{z}_k\|_\infty < \infty$$

is finite, which implies by (18) for $x = \tilde{x}_k = x(r_k,\eta_k)$, $s = \tilde{s}_k = s(r_k,\eta_k)$, $\xi = \xi(r_k,\eta_k)$, $\sigma = \sigma(r_k,\eta_k)$,

$$\|\xi(r_k,\eta_k)\|_2, \|\sigma(r_k,\eta_k)\|_2 \leq \sqrt{2n}(1+4K/\rho),$$

so that by (11)

$$r_k f(r_k,\eta_k) = \|\sigma(r_k,\eta_k) \circ \xi(r_k,\eta_k)\|_2^{1/2} \leq \|\sigma(r_k,\eta_k)\|_\infty^{1/2}\|\xi(r_k,\eta_k)\|_2^{1/2}$$

$$\leq \sqrt{2n}(1+4K/\rho).$$

This leads by (9), (10) to a first *complexity estimate*:

$$N(r) \leq \sqrt{2} \cdot C(1 + \frac{4K}{\rho})\sqrt{n}\log(1/r). \tag{19}$$

Therefore, if ρ is large (16) we get a $O(\sqrt{n}L)$-method, *provided that the* $\|\tilde{z}_k\|_\infty$ *do not grow unexpectedly.* (But note in this context that $K \geq \rho$ because of the initial choice $x_0 = s_0 = \rho\,e$.)

We note in passing that the complexity estimate immediately improves for the case of *feasible* interior point methods, where $\bar{b} = 0$ and $\bar{c} = 0$. Then also $\tilde{c} = 0$ and $\tilde{u} = 0$ and we get instead of (18)

$$\|\xi(r_k,\eta_k)\|_2, \|\sigma(r_k,\eta_k)\|_2 \leq \sqrt{2n}$$

$$r_k f(r_k,\eta_k) = \|\sigma(r_k,\eta_k) \circ \xi(r_k,\eta_k)\|_2^{1/2} \leq \sqrt{2n},$$

and instead of (19) the wellknown complexity estimate

$$N(r) \leq \sqrt{2} \cdot C\sqrt{n}\log(1/r)$$

for these methods.

We come back to infeasible interior point methods and the complexity estimate (19). Here, the Lemma can be used to derive a *rigorous* upper bound for K:

Suppose the problem has an optimal solution $z^* = (x^*, s^*)$ with $\|z^*\|_\infty \leq \rho^*$. Since then $\|z^*\|_1 \leq n\|z^*\|_\infty \leq \rho^* n$, and

$$0 < \eta_k \leq (1 + \alpha_0)e \leq (3/2)e$$

implies

$$0 < e^T \eta_k \leq (3/2)n, \tag{20}$$

the Lemma gives the bound

$$\|z_k\|_\infty \leq \|z_k\|_1 \leq r_k \rho n + (1 - r_k)\|z^*\|_1 + \rho e^T \eta_k$$
$$\leq (5/2)n \max\{\rho, \rho^*\},$$

so that $K \leq (5/2)n \max\{\rho, \rho^*\}$ and by (19)

$$N(r) \leq \sqrt{2}C(1 + 10n \max\{1, \rho^*/\rho\})\sqrt{n} \log \frac{1}{r}$$
$$= O\left(n^{1.5} \log \frac{1}{r}\right), \tag{21}$$

which is the same complexity as of an earlier more complicated algorithm proposed by Potra (1992a).

But this estimate can be improved, using ideas of Mizuno (1992) (and also of Potra (1992c)) in the context of other algorithms: We only have to use slightly different estimates for the vectors

$$\tilde{u} = \eta^{-1/2} \circ s \circ \bar{x}/\tau, \quad \tilde{c} = \eta^{-1/2} \circ x \circ \bar{c}/\tau,$$

namely

$$\|\tilde{u}\|_2 \leq \frac{1}{\tau}\|\eta^{-1/2}\|_\infty \|\bar{x}\|_\infty \|s\|_2,$$

$$\|\tilde{c}\|_2 \leq \frac{1}{\tau}\|\eta^{-1/2}\|_\infty \|\bar{c}\|_\infty \|x\|_2,$$

resulting by (13), (15), (17) in

$$\|\xi\|_2, \|\sigma\|_2 \leq \|\eta^{1/2}\|_2 + \frac{\sqrt{2}}{\tau} 2\rho(\|x\|_2 + \|s\|_2).$$

But by (15) $\|\eta^{1/2}\|_2 \leq \sqrt{2n}$, and the Lemma now gives

$$\|x\|_2 + \|s\|_2 \leq \sqrt{2}\|(x, s)\|_1 \leq \sqrt{2}(r \rho n + (1 - r)\rho^* n + (3/2)\rho n)$$
$$\leq 4n \max\{\rho, \rho^*\},$$

so that finally for large $\rho \geq \rho^*$

$$\|\xi\|_2, \|\sigma\|_2, \|\xi \circ \sigma\|_2^{1/2} \leq \sqrt{2n} + 8\sqrt{2n} \max\{1, \frac{\rho^*}{\rho}\} = O(n)$$

resulting in the improved complexity estimate

$$N(r) = O(n \log(1/r)). \tag{22}$$

Thus, the algorithm has *at least* $O(nL)$-complexity, if $(P_0), (D_0)$ are solvable and if ρ is chosen so large that $\|z^*\|_\infty \leq \rho$ and (16) holds.

3.2 STOPPING CRITERIA

Similarly as in Kojima, Megiddo, Mizuno (1991), it is possible to build into the algorithm a *stopping criterion* so that the algorithm stops after *finitely* many steps either with an ε-solution or with the answer that (P_0), (D_0) have no optimal solutions smaller than a prescribed size ρ^*.

Again, the Lemma is important: If (P_0), (D_0) have optimal solutions x^*, s^* with $\|(x^*, s^*)\|_\infty \leq \rho^*$, then $\|(x^*, s^*)\|_1 \leq \rho^* n$ and for all k

$$\|\tilde{z}_k\|_1 \leq r_k \rho n + (1 - r_k)\rho^* n + \rho e^T \eta_k$$
$$\leq r_k \rho n + (1 - r_k)\rho^* n + (3/2)\rho n. \tag{23}$$

This can be used as a necessary condition, and so that the algorithm can be stopped if it is violated. This leads to

Algorithm II:
Let $0 < \alpha_0 < \alpha_1 < 1/2$.
Choose ρ, ρ^*, $\varepsilon > 0$, *put* $r_0 := 1$, $\tau := \rho^2$ *and*

$$(x_0, s_0, y_0) := \rho(e, e, 0), \quad \tilde{z}_0 := (x_0, s_0), \quad \bar{b} := Ax_0 - b, \quad \bar{c} := s_0 - c.$$

1) *Given* $r_k > 0$, $\tilde{z}_k = (\tilde{x}_k, \tilde{s}_k = s(\tilde{y}_k)) \in M_{r_k}$ *with* $\tilde{z}_k > 0$ *so that* $\eta_k := \tilde{x}_k \circ \tilde{s}_k/(r_k\tau)$ *satisfies* $\|\eta_k - e\|_2 \leq \alpha_0$.
   a) *If* $r_k \leq \varepsilon$ *and* $(3/2)\tau r_k \leq \varepsilon$ *stop: An ε-solution has been found.*
   b) *If*

$$\frac{\|\tilde{z}_k\|_1}{\rho n} > r_k + (1 - r_k)\frac{\rho^*}{\rho} + \frac{3}{2},$$

   stop: $(P_0), (D_0)$ *have no optimal solution* z^* *with* $\|z^*\|_\infty \leq \rho^*$.
   Otherwise, continue as before:
2) *Compute the tangent*

$$z_k(r) = \tilde{z}_k + (r - r_k)\dot{z}(r_k, \eta_k)$$

   and determine $0 < r =: r_{k+1} < r_k$ *by* (6) *so that*

$$d(z_k(r_{k+1}), r_{k+1}) \leq \alpha_1.$$

3) *Use Newton's method, starting with* $z_k(r_{k+1})$, *to find a* $\tilde{z}_{k+1} \in M_{r_{k+1}}$ *with*

$$d(\tilde{z}_{k+1}, r_{k+1}) \leq \alpha_0.$$

Since $\tilde{x}_k$, $\tilde{s}_k$, $\tilde{y}_k$ solve $(PD)_{r_k, \eta_k}$ and $\|\eta_k - e\|_2 \leq \alpha_0 < 1/2$, so that $0 < \eta_k \leq (3/2)e$, we get by (22) the following Theorem:

Theorem 2: *For any ρ, ρ^* and $\varepsilon > 0$, algorithm II terminates after at most*

$$O\left(n \log \frac{1}{\varepsilon}\right)$$

steps.
If it stops with step 1a) then the final $\tilde{z}_k = (\tilde{x}_k, \tilde{s}_k) > 0$ and $\tilde{y}_k$ provide an ε-solution,
i.e.

$$\frac{\|A\tilde{x}_k - b\|_\infty}{\|Ax_0 - b\|_\infty} \le \varepsilon, \quad \frac{\|A^T\tilde{y}_k + \tilde{s}_k - c\|_\infty}{\|A^T y_0 + s_0 - c\|_\infty} \le \varepsilon, \ 0 < \tilde{x}_k \circ \tilde{s}_k \le \varepsilon \cdot e.$$

If it stops with step 1b), then (P_0), (D_0) have no optimal solution z^* with
$\|z^*\|_\infty \le \rho^*$.

3.3 SUPERLINEAR CONVERGENCE

It is easy to prove the quadratic convergence of the algorithms considered so far:

Theorem 3: *Under Assumption (V), algorithm I (and algorithm II with $\varepsilon = 0$,*
$\rho^* = \infty$) generates sequences $0 < \tilde{z}_k \in M_{r_k}$ and $r_k \downarrow 0$ with

$$r_{k+1} \le \gamma r_k^2$$

for some $\gamma > 0$ (quadratic convergence of the r_k).

Proof: By Theorem 1, the limits

$$\lim_{r \downarrow 0} \dot{x}(r, \eta), \ \lim_{r \downarrow 0} \dot{y}(r, \eta), \ \lim_{r \downarrow 0} \dot{s}(r, \eta)$$

exist, and the convergence is uniform on the compact subset $\{\eta \mid \|\eta - e\|_2 \le \alpha_0 < 1/2\}$ of $\{\eta > 0 \mid \eta \in \mathbb{R}^n\}$. Now, by the algorithm

$$\|\eta_k - e\|_2 \le \alpha_0, \quad r_k \downarrow 0.$$

Hence, there is an upper bound $\delta > 0$ for

$$\|\dot{x}(r_k, \eta_k)\|_2, \ \|\dot{s}(r_k, \eta_k)\|_2 \le \delta, \quad k \ge 0.$$

Therefore, by (7) for all k

$$0 < r_k f(r_k, \eta_k) = \sqrt{r_k} \|\dot{x}(r_k, \eta_k) \circ \dot{s}(r_k, \eta_k)\|_2^{1/2} / \sqrt{\tau}$$
$$\le \sqrt{r_k} \delta / \sqrt{\tau}.$$

Then (8) gives for $\beta_k = r_{k+1}/r_k$

$$1 \ge 1 - \beta_k = \sqrt{\beta_k} \frac{\sqrt{\alpha_1 - \alpha_0}}{r_k f(r_k, \eta_k)}$$
$$\ge \sqrt{\frac{r_{k+1}}{r_k}} \frac{\sqrt{\tau(\alpha_1 - \alpha_0)}}{\sqrt{r_k}\delta} = \frac{\sqrt{r_{k+1}}}{r_k \delta} \sqrt{\tau(\alpha_1 - \alpha_0)},$$

so that

$$r_{k+1} \le r_k^2 \delta^2 / (\tau(\alpha_1 - \alpha_0)).$$

4. Related Results

In the last two years, many infeasible interior point methods have been proposed, both with the aim to get good complexity estimates and to establish their superlinear or even quadratic convergence. Most of these methods generate points $x_k > 0$, y_k, $s_k > 0$ using Newton's method for solving various linear/nonlinear systems of the type $(PD)_{r,\eta}$. They therefore use Newton directions $\Delta = (\Delta x, \Delta y, \Delta s)$ satisfying

$$\begin{bmatrix} A & 0 & 0 \\ 0 & A^T & I \\ S & 0 & X \end{bmatrix} \begin{bmatrix} \Delta x \\ \Delta y \\ \Delta s \end{bmatrix} = \begin{bmatrix} u \\ v \\ w \end{bmatrix},$$

for suitable $x > 0$, $s > 0$ and right hand sides $R = (u^T, v^T, w^T)^T$ in order to construct the successive "interior points" $x_k > 0$, y_k, $s_k > 0$. We will use the abbreviation

$$B(x, s) := \begin{bmatrix} A & 0 & 0 \\ 0 & A^T & I \\ S & 0 & X \end{bmatrix}$$

for the matrix of the above linear system, which then reads

$$B(x, s) \cdot \Delta = R.$$

We wish to outline some of these recent methods.

1. Mizuno (1992) proposed a *one-step method* with the following characteristics:

Initial Choices: $(x_0, y_0, s_0) := \rho(e, 0, e)$, $r_0 := 1$, $0 < \beta_1 < 1$.
Iteration: Given (x_k, y_k, s_k) with $(x_k, s_k) > 0$,
solve with $\mu_k := \beta_1 x_k^T s_k / n$

$$B(x_k, s_k)\Delta = \begin{bmatrix} b - Ax_k \\ c - s_k - A^T y_k \\ \mu_k e - x_k \circ s_k \end{bmatrix}.$$

Set $(x_{k+1}, y_{k+1}, s_{k+1}) := (x_k, y_k, s_k) + \alpha_k \Delta$ for a suitable $\alpha_k > 0$ guaranteeing

$$Ax_k = b + r_k \bar{b},$$
$$A^T y_k + s_k = c + r_k \bar{c},$$
$$x_k^T s_k \geq r_k x_0^T s_0,$$

and put $r_{k+1} := r_k(1 - \alpha_k)$.
Complexity: $O(n^2 L)$
Superlinear Convergence: unknown.

2. Mizuno (1992) also proposed a *two-step method* (predictor and corrector step) of the improved $O(nL)$ complexity with the following structure:

Initial Choices: $(x_0, y_0, s_0) := \rho(e, 0, e)$, $r_0 := 1$, $0 < \beta_1 < 1$

1. *Predictor step:* As before, given (x_k, y_k, s_k) with $(x_k, s_k) > 0$,
   solve with $\mu_k := \beta_1 x_k^T s_k / n$

$$B(x_k, s_k)\Delta = \begin{bmatrix} b - Ax_k \\ c - s_k - A^T y_k \\ \mu_k e - x_k \circ s_k \end{bmatrix}.$$

Set $(x', y', s') := (x_k, y_k, s_k) + \alpha_k \Delta$ with a suitable choice of $\alpha_k > 0$.

2. *Corrector step:* Solve with $\mu' := x'^T s' / n$

$$B(x', s')\Delta' = \begin{bmatrix} 0 \\ 0 \\ \mu' e - x' \circ s' \end{bmatrix}.$$

Set $(x_{k+1}, y_{k+1}, s_{k+1}) := (x', y', s') + \Delta'$.
Complexity: $O(nL)$
Superlinear Convergence: unknown.

3. Potra (1992b) proposed among others a $O(nL)$ method with *quadratic convergence*. It is however a *three-step method*, each major step consisting of *one* predictor and *two* corrector steps:

Initial Choices: $(x_0, y_0, s_0) := \rho(e, 0, e)$.
Predictor step: Given (x_k, y_k, s_k), solve

$$B(x_k, s_k)\Delta = \begin{bmatrix} 0 \\ 0 \\ -x_k \circ s_k \end{bmatrix}.$$

Set $(\bar{x}, \bar{y}, \bar{s}) := (x_k, y_k, s_k) + \theta \Delta$ with a suitable (rather complicated) choice of $\theta > 0$.

1st corrector step: Solve

$$B(\bar{x}, \bar{s})\bar{\Delta} = \begin{bmatrix} 0 \\ 0 \\ -\Delta x \circ \Delta s \end{bmatrix}.$$

Set $(\hat{x}, \hat{y}, \hat{s}) := (\bar{x}, \bar{y}, \bar{s}) + \theta^2 \bar{\Delta}$.

2nd corrector step: Solve with $\hat{\mu} := \hat{x}^T \hat{s} / n$

$$B(\hat{x}, \hat{s})\hat{\Delta} = \begin{bmatrix} 0 \\ 0 \\ \hat{\mu} e - \hat{x} \circ \hat{s} \end{bmatrix}.$$

Set $(x_{k+1}, y_{k+1}, s_{k+1}) := (\hat{x}, \hat{y}, \hat{s}) + \hat{\Delta}$.

4. Substantially better results have been obtained by Ye, Todd and Mizuno (1992), but they refer to a *different* infeasible path which may be viewed as resulting from $(PD)_{r,e}$ by *homogenization techniques*.

Choosing again any $x_0 > 0, s_0 > 0, y_0$ and setting $\bar{b} := Ax_0 - b, \bar{c} := s_0 - c + A^T y_0$, and in addition $\beta := x_0^T s_0 + 1, \gamma := c^T x_0 - b^T y_0 + 1$, they study the *homogeneous selfdual* linear program in the variables $y \in \mathbb{R}^m, x, s \in \mathbb{R}^n, \tau, \theta, \kappa \in \mathbb{R}$

$$(HLP) \qquad\qquad \min \quad \beta\theta$$

s.t.
$$
\begin{array}{llllll}
 & Ax & -b\tau & -\bar{b}\theta & = 0 & \\
-A^T y & & +c\tau & +\bar{c}\theta & = s & \geq 0 \\
+b^T y & -c^T x & & +\gamma\theta & = \kappa & \geq 0 \\
+\bar{b}^T y & -\bar{c}^T x & -\gamma\tau & & = -\beta & \\
 & x \geq 0 & \tau \geq 0 & & &
\end{array}
$$

Crucial variables: the vector $z = (x, \tau, s, \kappa)$ of nonnegative variables.

Initial strictly feasible solution:

$$y_0, x_0, s_0, \tau_0 = \theta_0 = \kappa_0 = 1.$$

For this homogeneous problem they showed the following nice theoretical properties:

Optimal solution: By selfduality, (HLP) always has an optimal solution

$$(y_*, x_*, \tau_*, \theta_* = 0, s_*, \kappa_*)$$

with $x_* + s_* > 0, \tau_* + \kappa_* > 0$ *(strict selfcomplementarity)*.

Relation to $(P_0), (D_0)$: (P_0) and (D_0) have optimal solutions iff $\tau_* > 0$, and then $x_*/\tau_*, y_*/\tau_*, s_*/\tau_*$ are such optimal solutions.

The *primal-dual central path* belonging to (HLP) is given by the (unique) *feasible solution*

$$z = z(r) = (x(r), \tau(r), s(r), \kappa(r)) \text{ and } y(r), \theta(r)$$

of (HLP) satisfying

$$\begin{bmatrix} x \\ \tau \end{bmatrix} \circ \begin{bmatrix} s \\ \kappa \end{bmatrix} = r \begin{bmatrix} e \\ 1 \end{bmatrix} =: r\bar{e}$$

Replacing the vector $\bar{e} = (e^T, 1)^T$ by any positive weight vector $0 < \bar{\eta} \in \mathbb{R}^{n+1}$, we get a $\bar{\eta}$-*weighted* central path $z(r, \eta), r > 0$.

A primal-dual-algorithm, working again with two constants $0 < \alpha_0 < \alpha_1 < 1/2$, can be formulated as in Section 3. It generates sequences $r_k \downarrow 0$, and strictly feasible points $\tilde{z}_k$ with $d(\tilde{z}_k, r_k) \leq \alpha_0$, where

$$d(z, r) := \left\| \frac{1}{r} \begin{pmatrix} x \\ \tau \end{pmatrix} \circ \begin{pmatrix} s \\ \kappa \end{pmatrix} - \begin{pmatrix} e \\ 1 \end{pmatrix} \right\|_2.$$

Since we pursue a central path of feasible interior solutions of (HLP), the number $N(r)$ of steps to reduce a $r_0 = 1$ to a $0 < r < 1$ can now be bounded by the better estimate

$$N(r) \leq C\sqrt{n}\log\frac{1}{r}$$

with a *universal constant* C.

The method is (presumably) also quadratically convergent.

The original method of Ye, Todd and Mizuno is related but different: It is a two-step predictor-corrector method similar to the method of Mizuno described under 2) above. It uses the parameter

$$\mu_k = (x_k^T s_k + \tau_k \kappa_k)/(n+1)$$

to measure the optimality and the progress along the central path, and not the parameters r_k or η_k. Their method has complexity $O(\sqrt{n}L)$, its superlinear convergence is unknown.

References

Anstreicher, K.M. (1989) "A combined phase I-phase II projective algorithm for linear programming," *Mathematical Programming* **43**, 209–223.

Kojima, M., Megiddo, N., and Mizuno, S. (1991) "A primal-dual exterior point algorithm for linear programming," Research Report RJ 8500, IBM Almaden Research Center (San Jose, CA).

Kojima, M., Mizuno, S., and Yoshise, A. (1989) "A polynomial-time algorithm for a class of linear complementarity problems," *Mathematical Programming* **44**, 1–26.

Kojima, M., Mizuno, S., and Yoshise, A. (1991) "A little theorem of the Big M in interior point algorithms," Technical Report (revised September, Department of Information Sciences, Tokyo Institute of Technology (Tokyo).

Lustig, I.J. (1990/91) "Feasibility issues in a primal-dual interior-point method for linear programming," *Math. Programming* **49**, 145–162.

Lustig, I.J., Marsten, R.E., and Shanno, D.F. (1991) "Computational experience with a primal-dual interior point method for linear programming," *Linear Algebra and Its Applications* **152**, 191–222.

Monteiro, R.C., and Adler, I. (1989) "Interior path following primal-dual algorithms. Part I: Linear programming," *Mathematical Programming* **44**, 27–42.

Mizuno, S. (1992) "Polynomiality of the Kojima-Megiddo-Mizuno algorithm for linear programming," Technical Report 1006, revised Version, School of Operations Research and Industrial Engineering, Cornell University (Ithaca, New York).

Mizuno, S., Kojima, M., and Todd, M.J. (1992) "Infeasible-interior-point primal-dual potential-reduction algorithms for linear programming," Technical Report 1023, School of Operations Research and Industrial Engineering, Cornell University (Ithaca, New York).

Mizuno, S., Todd, M.J., and Ye, Y. (1993) "A surface of analytic centers and infeasible-interior-point algorithms for linear programming,"Technical Report 93-1, School of Operations Research and Industrial Engineering, Cornell University (Ithaca, New York).

Potra, F.A. (1992a) "An infeasible interior-point predictor-corrector method for linear programming," Technical Report 26, revised Version, Department of Mathematics, The University of Iowa (Iowa City, Iowa).

Potra, F. A. (1992b) "On a predictor-corrector method for solving linear programs from infeasible starting points," Technical Report 34, Department of Mathematics, The University of Iowa (Iowa City, Iowa).

Potra, F.A. (1992c) "A quadratically convergent infeasible interior-point algorithm for linear programming," Technical Report 28, revised Version, Department of Mathematics, The University of Iowa (Iowa City, Iowa).

Roos, C., and Vial, I.-Ph. (1988) "A polynomial method of approximate weighted centers for linear programming," Technical Report 88-68, Technical University Delft (Delft).

Sonnevend, G. (1985) "An analytical center for polyhedrons and new classes of global algorithms for linear (smooth, convex) programming," *Lecture Notes in Control and Information Sciences 84* (Springer, New York).

Sonnevend, G., Stoer, J., and Zhao, G. (1991) "On the complexity of following the central path by linear extrapolation in linear programs," *Mathematical Programming* **52**, 527–553.

Stoer, J. (1993) "The complexity of an infeasible interior-point path-following method for the solution of linear programs,"to appear in *Optimization Methods and Software*.

Todd, M.J. (1992) "On Anstreicher's combined phase I-phase II projective algorithm for linear programming," *Mathematical Programming* **55**, 1–16.

Witzgall, C., Boggs, P.T., and Domich, P.D. (1991) "On the convergence behavior of trajectories for linear programming," in: *Mathematical Developments Arising from Linear Programming Algorithms (J.C. Lagarias, M. J. Todd, eds.)*, Contemporary Mathematics, Vol. 114 (American Mathematical Society, Providence, RI) pp. 161–187.

Ye, Y., Todd, M.J., and Mizuno, S. (1992) "An $O(\sqrt{n}L)$-iteration homogeneous and self-dual linear programming algorithm," Technical Report No. 1007, School of Operations Research and Industrial Engineering, Cornell University (Ithaca, New York).

Zhao, G., and Stoer, J. (1993) "Estimating the complexity of a class of path-following methods for solving linear programs by curvature integrals," *Appl. Math. Optim.* **27**, 85–103.

Zhang, Y. (1992) "On the convergence of an infeasible interior-point algorithm for linear programming and other problems," Technical Report, Department of Mathematics and Statistics, University of Maryland (Baltimore, Maryland).

Algorithms for Linear Complementarity Problems

JOAQUIM J. JÚDICE

Departamento de Matemática
Universidade de Coimbra
3000 - Coimbra
PORTUGAL

ABSTRACT. This paper presents a survey of the Linear Complementarity Problem (LCP). The most important existence and complexity results of the LCP are first reviewed. Direct, iterative and enumerative algorithms are then discussed together with their benefits and drawbacks.

Some important global optimization problems can be solved by a sequential technique that is based on the solution of a Generalized Linear Complementarity Problem (GLCP). This problem is also discussed in this paper, namely its complexity and some procedures for its solution.

1. Introduction

The Linear Complementarity Problem (LCP) consists of finding vectors $z \in \mathbb{R}^n$ and $w \in \mathbb{R}^n$ such that

$$w = q + Mz$$
$$z \geq 0, \quad w \geq 0 \tag{1}$$
$$z^T w = 0$$

where q is an n-vector and M is a square matrix of order n. This problem has emerged as one of the most important areas of optimization. Many applications of the LCP have been found in several areas of science, engineering and economics [21, 37, 78, 81, 86]. A number of important optimization problems can also be solved by finding a solution of its associated LCP [21, 78].

It is nowadays well accepted that the class of the matrix M of the LCP plays an important role on its difficulty. If the matrix M is Row Sufficient ($RSuf$) [22] then the LCP can be solved efficiently by a direct or an iterative method. In particular, a LCP with a Positive Definite (PD) or Positive Semi-Definite (PSD) matrix can be solved in polynomial time [61]. Furthermore, the LCP can be processed in strongly polynomial time if M is a Z-matrix [12] and, in particular, if M is a Nonsingular M (NSM) matrix [7]. In general, the LCP is NP-hard [14] and an enumerative algorithm or some global optimization technique should be employed for finding its solution. In

435

E. Spedicato (ed.), Algorithms for Continuous Optimization, 435–474.

this paper we survey the most important theoretical existence and complexity results for the LCP. Direct, iterative and enumerative algorithms are also discussed, together with their benefits and drawbacks.

As discussed in [9, 44, 46], linear complementarity also plays an important role in some global optimization problems that have been referred in the literature. These include linear and linear-quadratic bilevel programming problems, concave quadratic and bilinear programs and $0 - 1$ integer programming problems. All these problems can be solved by a sequential technique that processes in each iteration a Generalized Linear Complementarity Problem (GLCP) of the form

$$
\begin{aligned}
w &= q + Mz + Ny \\
v &= p + Rz + Sy \\
z, w, v, y &\geq 0 \\
z^T w &= 0
\end{aligned}
\tag{2}
$$

where q and p are given vectors, M, N, R and S are matrices of appropriate dimensions and M is a PSD matrix. In this paper we also discuss this GLCP, namely its complexity and the algorithms that have been designed for its solution.

The organization of this paper is as follows. Section 2 addresses the existence and complexity of the LCP. Algorithms for the LCP are discussed in Section 3. Section 4 is devoted to the GLCP and some concluding remarks are presented in the last part of the paper.

Some notation that is used throughout the paper should also be mentioned. Matrices and sets are denoted by capital letters, while small letters are used to represent vectors. $|S|$ denotes the number of elements of the set S. Real numbers are represented by greek letters or as components of vectors. The row i and column j of A are denoted by $A_{i\cdot}$ and $A_{\cdot j}$ respectively. Furthermore A_{RS} represents the submatrix of A containing the rows $i \in R$ and the columns $j \in S$. In particular, A_{rS} (A_{Sr}) is the vector containing the elements of the row (column) r of A whose indices belong to the set S. Finally, transposition is denoted by the symbol T.

2. Existence and complexity of the LCP

2.1. DEFINITION AND FIRST PROPERTIES

As stated in the previous section, given a vector $q \in \mathbb{R}^n$ and a square matrix M of order n, the LCP consists of finding vectors $z \in \mathbb{R}^n$ and $w \in \mathbb{R}^n$ such that

$$
w = q + Mz \tag{3}
$$

$$
z \geq 0, \quad w \geq 0 \tag{4}
$$

$$
z^T w = 0 \tag{5}
$$

We also represent by LCP(q,M) a LCP with vector q and matrix M. So the LCP contains a set of linear constraints defined by (3) and (4). As in linear programming, (z, w) is a feasible solution of the LCP if it satisfies these linear constraints, that is, if it is an element of the so-called feasible set

$$K = \{(z, w) : w = q + Mz, \ z \geq 0, \ w \geq 0\} \tag{6}$$

If z and w are nonnegative vectors then the condition (5) holds if and only if

$$z_i w_i = 0, \ i = 1, \cdots, n \tag{7}$$

A solution (z, w) satisfying (3) and (7) is called complementary and (7) is named the complementarity condition. So a solution of the LCP is feasible and complementary. By looking cautiously at the conditions (7), we realize that they possess a combinatorial nature similar to the $0 - 1$ integer variables. In fact, it follows from (7) that any solution (z, w) of the LCP must satisfy

$$z_i > 0 \Rightarrow w_i = 0$$
$$w_i > 0 \Rightarrow z_i = 0$$

for all $i = 1, \cdots, n$. So the LCP is an optimization problem in which linear, nonlinear and conbinatorial programming are present in its definition. In our opinion, this is the main reason for the large number of algorithms that have been proposed for its solution.

Another interesting feature of the LCP is that the existence and uniqueness of a solution depends on the class of the matrix M. Before presenting some important classes, we give some examples of the four cases that may occur.

(i) The LCP has a unique solution. For instance, if

$$M = \begin{bmatrix} 1 & 0 \\ 0 & 1 \end{bmatrix} \quad \text{and} \quad q = \begin{bmatrix} 1 \\ 1 \end{bmatrix}$$

then $z = (0, 0)$ and $w = (1, 1)$ is the unique solution of the LCP.

(ii) The LCP has multiple solutions. As an example of this type of LCP, consider

$$M = \begin{bmatrix} 1 & 1 \\ 1 & 1 \end{bmatrix} \quad \text{and} \quad q = \begin{bmatrix} -1 \\ -1 \end{bmatrix}$$

Then $z = (\lambda, 1 - \lambda)$ and $w = (0, 0)$ is a solution of the LCP for any $0 \leq \lambda \leq 1$. We note that M is singular in this example, but we can easily find another one in which M is nonsingular.

(iii) The LCP is infeasible, that is, the feasible set is empty ($K = \emptyset$). This occurs for

$$M = \begin{bmatrix} -1 & 0 \\ 1 & 2 \end{bmatrix} \quad \text{and} \quad q = \begin{bmatrix} -1 \\ 2 \end{bmatrix}$$

since there is no $z_1 \geq 0$ such that $w_1 = -1 - z_1 \geq 0$.

(iv) The LCP is feasible, but has no solution. As an example of this case, let

$$M = \begin{bmatrix} 1 & 0 \\ 1 & -2 \end{bmatrix} \quad \text{and} \quad q = \begin{bmatrix} 3 \\ -1 \end{bmatrix}$$

In fact, any $z = (z_1, 0)$ with $z_1 \geq 1$ is feasible, but $w_1 = 3 + z_1 > 0$. Hence $z_1 w_1 > 0$ in any feasible solution of the LCP and the LCP has no solution.

2.2. CLASSES OF MATRICES

In the last section we have shown that the class of the matrix M plays an important role on the existence and uniqueness of a solution to the LCP. The classes of matrices Q and Q_0 have been introduced in this context [21, 78] and satisfy the following equivalences

- $M \in Q \Leftrightarrow$ LCP has a solution for each $q \in \mathbb{R}^n$.

- $M \in Q_0 \Leftrightarrow$ LCP has a solution for each $q \in \mathbb{R}^n$ such that $K \neq \emptyset$.

Unfortunately, there exist no practical necessary and sufficient conditions for a matrix M to belong to each one of these classes. However, there are some interesting necessary or sufficient conditions for this purpose. These conditions are given in terms of other classes of matrices. Next we introduce some of the most relevant classes in this context. We recommend [21, 49, 78] for properties of these matrices and some other classes that have been related with the LCP.

- $M \in PD \Leftrightarrow z^T M z > 0$ for all $z \neq 0$.

- $M \in PSD \Leftrightarrow z^T M z \geq 0$ for all z.

- $M \in P \Leftrightarrow$ all principal minors are positive.

- $M \in P_0 \Leftrightarrow$ all principal minors are nonnegative.

- $M \in CSuf \Leftrightarrow [z_i(Mz_i) \leq 0 \text{ for all } i = 1, \cdots, n] \Rightarrow [z_i(Mz_i) = 0 \text{ for all } i = 1, \cdots, n]$.

- $M \in RSuf \Leftrightarrow M^T \in CSuf$

- $M \in Suf \Leftrightarrow M \in CSuf$ and $M \in RSuf$.

- $M \in Z \Leftrightarrow m_{ij} \leq 0$ for all $i \neq j$.

- $M \in NSM \Leftrightarrow M \in Z$ and $M \in P$.

- $M \in S \Leftrightarrow$ there exists $z \geq 0$ such that $Mz > 0$.

- $M \in NSD \Leftrightarrow z^T M z \leq 0$ for all z.

- $M \in NSD_+ \Leftrightarrow M \in NSD$ and $m_{ij} \geq 0$ for all $i \neq j$.

- $M \in IND \Leftrightarrow M \notin PSD$ and $M \notin NSD$.

The relationships among these classes and the matrices Q and Q_0 are displayed in the diagram of figure 1, in which each arrow represents a strict inclusion.

Furthermore we can state the following properties

- $CSuf \not\subseteq Q_0 \Rightarrow P_0 \not\subseteq Q_0$
  For instance
  $$M = \begin{bmatrix} 0 & 1 \\ 0 & 1 \end{bmatrix} \in CSuf \quad \text{and} \quad M \notin Q_0$$
  In fact, if $q = [-1, 2]^T$, then the LCP(q,M) is feasible but has no solution.

- $NSD \subset Q_0$ for $n = 2$, but $NSD \not\subseteq Q_0$ for $n \geq 3$.

- $IND \cap Q_0 \neq \emptyset$, but $IND \not\subseteq Q_0$.

- $M \in S \Leftrightarrow K \neq \emptyset$ for each q. Hence $S \cap Q_0 = Q$.

- $M \in P \Leftrightarrow$ LCP(q,M) has a unique solution for each q.

- M symmetric $\Rightarrow P = PD$ and $P_0 = PSD$

2.3. REDUCTIONS TO QUADRATIC PROGRAMMING PROBLEMS

The LCP can always be reduced into a quadratic program. If M is symmetric, then the LCP is equivalent to the problem of finding a stationary point of the following quadratic program [21]

$$QP_1: \quad \text{Minimize} \quad q^T z + \frac{1}{2} z^T M z \qquad (8)$$
$$\text{subject to} \quad z \geq 0$$

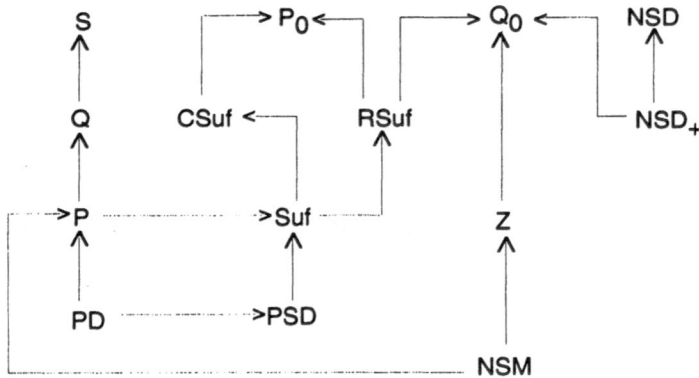

Figure 1: Classes of matrices for the LCP.

Furthermore, if $M \in PSD$ then QP_1 is a convex program and $\bar{z}$ is a solution of the LCP if and only if $\bar{z}$ is a global minimum of QP_1.

If M is an unsymmetric matrix, consider the following quadratic program

$$\text{Minimize} \qquad z^T w$$
$$\text{subject to} \qquad (z, w) \in K$$

which is equivalent to

$$QP_2: \quad \text{Minimize} \quad f(z) = q^T z + \frac{1}{2} z^T (M + M^T) z \qquad (9)$$
$$\text{subject to} \qquad q + Mz \geq 0, \ z \geq 0$$

Let $\bar{K}$ be the constraint set of QP_2. Then the function $f(z)$ is bounded from below on $\bar{K}$ and there are three possible cases:

- $\bar{K} = \emptyset \Rightarrow$ LCP is infeasible.

- $f(\bar{z}) = \min_{z \in \bar{K}} f(z) = 0 \Rightarrow \bar{z}$ is solution of the LCP.

- $\min_{z \in \bar{K}} f(z) > 0 \Rightarrow$ the LCP is feasible but has no solution.

It follows from the definition of the matrices Q and Q_0, that case 3 cannot occur if $M \in Q_0$ and $M \in Q$ if and only if QP_2 has a global minimum with value zero for each $q \in \mathbb{R}^n$.

This latter reduction shows that if M is an unsymmetric matrix, then a solution of the LCP can be obtained by finding a global minimum of the quadratic program QP_2. This is a difficult task in general [89]. However, if $M \in RSuf$ then any stationary point of QP_2 is a solution of the LCP [21]. Then the LCP is much easier in this case. If M is symmetric, then again any stationary point of QP_1 is a solution of the LCP. This means that a LCP with a symmetric matrix is in general less difficult than a LCP in which M is unsymmetric. In our opinion, this fact has not been well recognized in the literature of the LCP [21, 78]. However, it will be exploited in the section devoted to the algorithms for the LCP.

These reductions of the LCP into quadratic programs lead to our first classification of the LCP, in which a LCP is said to be strictly convex, convex, concave or indefinite depending on the status of the equivalent quadratic program. This is in turn related with the class of the Hessian matrix of the quadratic function [89]. Hence

- LCP is strictly convex $\Leftrightarrow M \in PD$

- LCP is convex $\Leftrightarrow M \in PSD$

- LCP is concave $\Leftrightarrow M \in NSD$

- LCP is indefinite $\Leftrightarrow M \in IND$

As in quadratic programming [89], these four types of LCPs possess different complexities. Strictly convex and convex LCPs can be solved in polynomial-time [61], but concave and indefinite LCPs are NP-hard problems [14]. However, there are some classes of concave and indefinite LCPs that can be solved in polynomial-time [5, 70, 71, 85, 103]. These are also the cases of LCPs with Z and NSD_+ matrices [12, 49]. Finally, it is not clear so far what is the complexity of the LCP when its matrix M is P, $RSuf$ or even symmetric. In fact, in these cases the LCP reduces to the problem of finding a stationary point of a quadratic program, whose complexity is not well-known so far.

2.4. REDUCTION TO AN AFFINE VARIATIONAL INEQUALITY PROBLEM

The LCP is equivalent to the following affine variational inequality problem [21]

$$\text{AVI:} \quad \text{Find} \quad \bar{z} \in K_1 = \{z \in \mathbb{R}^n : z \geq 0\} \quad \text{such that} \quad (10)$$
$$(q + M\bar{z})^T(z - \bar{z}) \geq 0 \quad \text{for all } z \in K_1$$

Based on this equivalence, a LCP is said to be strictly monotone or monotone depending on the status of the AVI [42]. Hence

- LCP is strictly monotone $\Leftrightarrow M \in P$

- LCP is monotone $\Leftrightarrow M \in PSD$

So we use the terms monotone and convex for a LCP with a *PSD* matrix. Furthermore, a strictly convex LCP is also strictly monotone, but the converse is only true for symmetric matrices.

In many applications [37] the constraint set of the AVI problem contains finite or infinite lower and upper bounds on the variables, that is, takes the following form

$$K_1 = \{z : l_i \le z_i \le u_i, \ i = 1, \cdots, n\}$$

where $-\infty \le l_i \le z_i \le u_i \le +\infty$ for all $i = 1, \cdots, n$. In this case the AVI problem is equivalent to the following Linear Complementarity Problems with Bounds (BLCP):

$$\left. \begin{aligned} w &= q + Mz \\ l_i &\le z_i \le u_i \\ z_i &= l_i \Rightarrow w_i \ge 0 \\ z_i &= u_i \Rightarrow w_i \le 0 \\ l_i &< z_i < u_i \Rightarrow w_i = 0 \end{aligned} \right\} \ i = 1, \cdots, n \tag{11}$$

It is important to note that the existence of a solution to the BLCP also depends on the bounds l_i and u_i. For instance, if all the bounds are finite, then the BLCP has a solution [42] independently of the matrix M. Furthermore, if $M \in P$ then the BLCP has a unique solution for each $q \in \mathbb{R}^n$ [1]. However, the converse is only true if all the bounds are finite [1].

If all the bounds are infinite, then the BLCP becomes the system of equations $Mz = -q$. Furthermore, the BLCP reduces to the LCP if $l_i = 0$ and $u_i = +\infty$ for all $i = 1, \cdots, n$. In the next section we discuss the main algorithms for the LCP. Some of these procedures may be extended to deal with the BLCP. These extensions are not described in this paper, but we provide some references in which these techniques may be found.

3. Algorithms for the LCP

3.1. BASIC SOLUTIONS AND PIVOT OPERATIONS

Consider the system of linear equations

$$Ax = b \tag{12}$$

where $A \in \mathbb{R}^{n \times m}$, $b \in \mathbb{R}^n$, $x \in \mathbb{R}^m$ and $rank(A) = n < m$. A basic solution for this system assumes a partition of the matrix A in the form

$$A = [\ B \quad N \ ]$$

where B is a nonsingular matrix of order n, called basis. If R and S are the sets of indices corresponding to the columns of A that are in B and N respectively, then the basic solution with basis B is given by

$$x_S = 0, \quad Bx_R = b$$

The variables x_i, $i \in R$ are called basic, while those associated to the indices $i \in S$ are said to be nonbasic. A basic solution is nondegenerate if all the basic variables have nonzero values. Otherwise it is said to be degenerate.

A pivot operation transforms a basic solution with basis B into another basic solution with a different basis $\bar{B}$. Therefore this new basic solution satisfies

$$x_{\bar{S}} = 0, \quad \bar{B}x_{\bar{R}} = b$$

where, as before, $\bar{R} \cup \bar{S} = \{1, \cdots, m\}$ and $\bar{R} \cap \bar{S} = \emptyset$. The pivot operation is said to be single if R and $\bar{R}$ differ exactly in one element. Otherwise is called block.

Consider again the LCP

$$w = q + Mz$$
$$z \geq 0, \quad w \geq 0$$
$$z^T w = 0$$

Hence a basic solution is available $(w, z) = (q, 0)$. Furthermore this solution is complementary, that is, satisfies $z_i w_i = 0$ for all $i = 1, \cdots, n$. Pivoting algorithms are procedures that use complementary basic solutions in each iteration and terminate when feasibility is reached. In the next two subsections we describe the most important pivoting algorithms for the LCP.

3.2. LEMKE'S ALGORITHM

This method [64] is probably the most famous procedure for solving the LCP. In this algorithm a positive vector p and an artificial variable z_0 are added to the LCP in order to get the following Generalized Linear Complementarity Problem (GLCP):

$$w = q + Mz + z_0 p$$
$$z \geq 0, \quad w \geq 0, \quad z_0 \geq 0 \tag{13}$$
$$z^T w = 0$$

We assume that the vector q has at least a negative component, since $z = 0$ is a solution of the LCP otherwise. Then there exists a value $\bar{z}_0 > 0$ such that $q + \bar{z}_0 p \geq 0$. This value $\bar{z}_0$ can be obtained by performing a pivot operation that replaces z_0 with a basic variable w_r, where r satisfies

$$-\frac{q_r}{p_r} = \max\{-\frac{q_i}{p_i} : q_i < 0\} \tag{14}$$

After this operation is performed, a basic solution of the GLCP is at hand. Lemke's algorithm is a pivoting procedure that moves in the set of solutions of the GLCP (13) until finding a solution of this GLCP with $z_0 = 0$ or some indication that it is not possible to get such a vector. In each iteration there are two rules for the choice of the driving nonbasic and blocking basic variables that should be interchanged by the pivot operation. These rules are stated as follows:

Rule 1 The nonbasic driving variable is the complementary of the blocking variable that has been chosen in the previous iteration.

Rule 2 The basic blocking variable is the one that blocks the increase of the driving variable by not allowing the value of any basic variable to become negative.

We note that the rule 2 is a simple minimum quotient rule similar to the one that is incorporated in the simplex method for linear programming. Furthermore the use of these two rules implies that basic solutions of the GLCP are used in each iteration of the algorithm.

Now suppose that x_s is the driving variable chosen by rule 1. If there is no blocking variable, then x_s may increase to infinity and the so-called termination in ray occurs. Suppose that this is not the case and let x_r be the blocking variable. Then a pivot operation is performed by changing x_s and x_r and a new basic solution of the GLCP is obtained. If $z_0 = 0$ then a solution of the LCP is at hand. Otherwise the driving variable of the next iteration is the complementary of the variable x_r and the algorithm proceeds as before. The steps of the algorithm can be stated as follows.

Lemke's algorithm

Let $A = [I \ -M \ -p]$ and $x = [w, z, z_0]$. Find r by (14) and let $R = \{1, \cdots, r-1, r+1, \cdots, n, 2n+1\}$ and $S = \{1, \cdots, 2n+1\} - R$. Let $B = [A_{.j}]_{j \in R}$ be the basis matrix associated with R, and $B\bar{x}_R = q$, $\bar{x}_S = 0$.

Step 1. Choose the driving variable x_s, $s \in S$ by rule 1 ($x_s = z_r$ in the first iteration).

Step 2. Solve $Bv = A_{.s}$.

Step 3. Compute the stepsize α by

$$\alpha = \begin{cases} \min\{\frac{\bar{x}_i}{v_i} : i \in R, v_i > 0\} = \frac{\bar{x}_r}{v_r} \\ +\infty, \ \text{if} \ v_i \leq 0 \ \text{for all} \ i \in R \end{cases}$$

Step 4. If $\alpha = +\infty$, stop: termination in ray.

Step 5. Update the values of the variables by $\bar{x} = \bar{x} + \alpha\beta$, where $\beta \in \mathbb{R}^{2n+1}$ satisfies

$$\beta_i = \begin{cases} 1 \ \text{if} \ i = s \\ v_i \ \text{if} \ i \in R \\ 0 \ \text{if} \ i \in S - \{s\} \end{cases}$$

Set $R = R - \{r\} \cup \{s\}$ and $S = S - \{s\} \cup \{r\}$.

Step 6. if $z_0 = 0$ stop: $(x_R = \bar{x}_R, x_S = 0)$ is a solution of the LCP. Otherwise go to Step 1.

It follows from the description of the steps of the algorithm that each iteration essentially requires a solution of a system with this basis matrix. The LU decomposition of the matrix is used in order to solve this system. Since this matrix B differs from the basis matrix of the previous iteration in exactly one element, then the LU decomposition of B is updated instead of computed from stack. There exist several ways of performing this updating, depending on the sparsity and dimension of the matrix B. Furthermore, the so-called reinversions are performed periodically to get more accurate solutions and reduce fill-in in the large-scale case. We recommend [48, 99] for two implementations of Lemke's algorithm for large-scale LCPs.

It can be shown [21] that the algorithm possesses finite termination if all the basic solutions are nondegenerate. Linear programming techniques for handling degeneracy can be incorporated in the algorithm to prevent cycling [21]. The algorithm may terminate in a solution or in a ray. This last form of termination may have no meaning at all. For instance, if $M \in NSD_+$, then Lemke's algorithm always terminates in a ray [49]. However, there are some cases where such a form of termination cannot occur or only occurs when the feasible set of the LCP is empty. These cases are related with some of the classes of matrices presented in Section 2. Among many results of this type [21, 78], we can state the following properties:

- If $M \in P$ then termination in ray cannot occur.

- If $M \in Z$ or $M \in RSuf$ then termination in ray implies that the feasible set K is empty.

The proofs of these statements can be found in [21, 78]. So Lemke's algorithm can process the LCP (find a solution or show that no solution exists) for special classes of matrices. In particular, strictly monotone and monotone LCPs can be solved by this method.

It is also important to note that the choice of the vector p is important in the performance of the algorithm. The most common choice for this vector is $p = (1, \cdots, 1)^T$. However, the method is able to start with any basis matrix B by simply choosing $p = B\bar{p}$, where $\bar{p}$ is a positive vector. See also [98] for an extension of this algorithm that is able to start with an arbitrary basic or nonbasic solution. Finally, we note that Lemke's algorithm can be extended to the BLCP and process this problem at least when M is a P or a PSD matrix. We recommend [51, 96] for two of these extensions.

3.3. PRINCIPAL PIVOTING ALGORITHMS

These procedures use in each iteration complementary basic solutions and terminate when feasibility is reached or an indication that this is not possible to achieve. A

complementary basic solution for the LCP can be found by the following procedure:
Let F and T be two subsets of $\{1, \cdots, n\}$ such that $F \cup T = \{1, \cdots, n\}$ and $F \cap T = \emptyset$.
Then

$$z = \begin{bmatrix} \bar{z}_F \\ 0 \end{bmatrix} \quad \text{and} \quad w = \begin{bmatrix} 0 \\ \bar{w}_T \end{bmatrix}$$

where

$$M_{FF}\bar{z}_F = -q_F \tag{15}$$

$$\bar{w}_T = q_T + M_{TF}\bar{z}_F \tag{16}$$

We note that M_{FF} must be nonsingular to accomplish this goal.

Suppose that a complementary basic solution is given and consider the so-called
set of infeasibilities

$$H = \{i \in F : \bar{z}_i < 0\} \cup \{i \in T : \bar{w}_i < 0\} \tag{17}$$

The number of elements $|H|$ of this set is called the infeasibility count of the complementary basic solution. If $|H| = 0$ then $z = (\bar{z}_F, 0)$ is a solution of the LCP.
Otherwise we set

$$F = F - (F \cap H_1) \cup (T \cap H_1) \tag{18}$$

$$T = \{1, \cdots, n\} - F$$

where H_1 is a set that is related with H. These algorithms differ in the way that H_1
is chosen in each iteration. Single and block principal pivoting algorithms should be
distinguished. In the first case the set H_1 has always one element, while $|H_1|$ may be
larger than one in some iterations of a block algorithm. Next, we discuss the most
important single principal pivoting algorithms for the LCP. The simpler procedure
of this form simply consists of setting $H_1 = \{\min\{i \in H\}\}$ in each iteration. This
algorithm is due to Murty [80] and can then be presented as follows.

Murty's algorithm

Let $F = \emptyset$ and $T = \{1, \cdots, n\}$.

Step 1. Compute $\bar{z}_F$ and $\bar{w}_T$ by (15) and (16) respectively.

Step 2. Find the set H by (17). If $H = \emptyset$ stop: $z = (\bar{z}_F, 0)$ is a solution of the
LCP.

Step 3. Let $s = \min\{i \in H\}$ and set

$$F = \begin{cases} F - \{s\} & \text{if } s \in F \\ F \cup \{s\} & \text{if } s \in T \end{cases}$$

and $T = \{1, \cdots, n\} - F$. Go to Step 1.

It can be shown [78] that the algorithm works and possesses finite convergence for strictly monotone LCPs ($M \in P$). In particular, the algorithm can be applied to strictly convex LCPs ($M \in PD$). It follows from the description of the algorithm that the main computational effort in each iteration relies on the computation of the vectors $\bar{z}_F$ and $\bar{w}_T$. Any implementation of the algorithm should consider the storage of the matrix M and of the decomposition of the matrix M_{FF} that is required to find $\bar{z}_F$. It is also important to stress that symmetry can be fully exploited in this implementation. In this case the LDL^T decomposition of M_{FF} is used to compute the vector $\bar{z}_F$. As in Lemke's method, the decomposition is not computed from stack, but instead is updated from that of the previous iteration. There are efficient procedures to perform this task that can exploit the sparsity of the matrices M_{FF} [15, 54]. In the unsymmetric case, the LU decomposition of M_{FF} is used to find the vectors $\bar{z}_F$. This LU decomposition can also be updated in each iteration, but the procedure may require some form of partial or threshold pivoting [26] for stability reasons. We recommend [54] and [94] for implementations of Murty's method in the symmetric and unsymmetric cases respectively.

It is also important to note that Murty's method can start with any initial set F. As we see later, this has some important implications on the solution of strictly monotone LCPs by block pivoting techniques. Finally, Murty's algorithm can be extended to the solution of a BLCP with a P matrix [52].

Another important single pivoting algorithm is due to Keller [58]. In this procedure the basic variables z_i, $i \in F$ are forced to maintain nonnegative values, which means that $H \cap F = \emptyset$ during the whole procedure. In each iteration the value of the driving variable z_s is increased from zero. If this variable is not blocked, then the algorithm terminates in a ray. Otherwise this variable is blocked either by its complementary w_s or by a basic variable z_r that becomes negative. In the first case the variable z_s is interchanged with w_s. In the second case z_r becomes nonbasic by changing with its complementary. Furthermore z_s stays the driving variable in the next iteration. Since the set F is finite, there is a finite number of iterations in which z_s maintains its status of driving variable. After this cycle of iterations either z_s becomes basic or the termination in ray occurs. The procedure is then repeated.

The steps of the algorithm can be presented in the following form

Keller's algorithm

Let $F = \emptyset$ and $T = \{1, \cdots, n\}$.

Step 1. If $H = \emptyset$ stop: $z = (\bar{z}_F, 0)$ is a solution of the LCP. Otherwise, let s be the index such that

$$\bar{w}_s = \min\{\bar{w}_i < 0 : i \in T\}$$

Step 2. Find $v \in \mathbb{R}^{|F|}$ and $\bar{m}_{ss}$ by

$$M_{FF}v = -M_{Fs}$$
$$\bar{m}_{ss} = m_{ss} + M_{sF}v$$

Step 3. Compute

$$\theta_1 = \begin{cases} -\frac{\bar{w}_s}{\bar{m}_{ss}} & \text{if } \bar{m}_{ss} > 0 \\ +\infty & \text{if } \bar{m}_{ss} \leq 0 \end{cases}$$

$$\theta_2 = \begin{cases} \min\{-\frac{\bar{z}_i}{v_i} : v_i < 0\} = -\frac{\bar{z}_r}{v_r} \\ +\infty & \text{if } v_i \geq 0 \end{cases}$$

and $\theta = \min\{\theta_1, \theta_2\}$.

Step 4. · If $\theta = +\infty$ stop: termination in ray.
· If $\theta = \theta_1$ set $F = F \cup \{s\}$ and $T = T - \{s\}$. Update $\bar{z}_F$ and $\bar{w}_T$ by $\bar{z}_s = \theta$, $\bar{z}_i = \bar{z}_i + \theta v_i : i \in F - \{s\}$, $\bar{w}_T = q_T + M_{TF}\bar{z}_F$. Go to Step 1.
· If $\theta = \theta_2$ set $F = F - \{s\}$ and $T = T \cup \{s\}$. Update $\bar{z}_F$ and $\bar{w}_s$ by $\bar{z}_F = \bar{z}_F + \theta v$, $\bar{w}_s = q_s + M_{sF}\bar{z}_F$. Go to Step 2.

It follows from the description of their steps that the algorithm can be implemented in a way similar to Murty's method. As before, we recommend [54] and [94] for two implementations of this method when M is symmetric and unsymmetric respectively. The algorithm possesses finite convergence when M is an unsymmetric P matrix [21] or symmetric [58]. In the latter case, the procedure is equivalent to the Fletcher and Jackson active-set method [30].

It is easy to see that if M is a (symmetric or an unsymmetric) P matrix, then the termination in ray cannot occur and the algorithm finds the unique solution of the LCP. If M is a symmetric matrix, then this type of termination means that the quadratic function $f(z) = q^T z + \frac{1}{2}z^T M z$ is unbounded on the nonnegative orthant. In general no conclusion can be stated about the existence of a solution to the LCP, whence the algorithm is not able to process the LCP for a general symmetric matrix. However, if M is symmetric PSD then the LCP is infeasible if this form of termination occurs.

Extensive computational experience presented in [54, 94] has shown that Keller's algorithm is in general more efficient than Murty and Lemke methods for the solution of medium and large scale LCPs with symmetric PD and PSD matrices. However, Lemke's algorithm is competitive with Keller's method for unsymmetric P matrices.

Keller's algorithm can start with any set F such that $\bar{z}_F \geq 0$. This is a disadvantage to the remainning two algorithms in which this nonnegative requirement is not present. As in the other algorithms, Keller's method can be extended to deal with BLCPs in which M is a symmetric matrix. Furthermore this extension also works for an unsymmetric P matrix. However, finite convergence has not been established so far.

There exist some other single pivoting algorithms that deserve to be mentioned in this survey. These are the cases of Cottle's principal pivoting algorithm [19], Graves' method [38] and the so-called Criss-Cross algorithm [24]. We do not concentrate in these algorithms, since they are in general less efficient than Keller's method for symmetric LCPs [54] and perform worse than Lemke's method for unsymmetric LCPs [94]. Another principal pivoting algorithm has been developed by Judice and Faustino [49] and can process a LCP with a NSD_+ matrix. Furthermore this algorithm has been shown to be strongly polynomial and its convergence has implied that NSD_+ is a subclass of the Q_0 matrices [49].

As a final conclusion of this survey on single pivoting algorithms, we recommend the use of Keller's method for the solution of symmetric LCPs and Lemke's algorithm for unsymmetric LCPs. If termination in ray occurs in these algorithms, then in general no conclusion can be stated about the existence of a solution to the LCP. This type of termination cannot occur for strictly monotone LCPs and means that the LCP is infeasible if M is a $RSuf$ matrix. We note that a monotone LCP falls in this category, since any PSD matrix is $RSuf$ and $PSD = RSuf$ for symmetric matrices.

In single pivoting algorithms the set of basic variables z_i is modified in one element per iteration. So these procedures may take too many iterations in the solution of large-scale LCPs in which the initial and final sets of basic variables z_i are too different. By realizing this fact, Judice and Pires [53] have introduced a block principal pivoting algorithm, which allows modifications of many indices of the set F in each iteration. According to a suggestion presented in [62], the set F is modified by the formula (18) with $H_1 = H$. As discussed in [78], we cannot guarantee finite convergence for such a scheme. However, this procedure can be combined with Murty's method in such a way that block modifications (18) with $H_1 = H$ are in general performed and Murty's iterations ($H_1 = \{\min\{i \in H\}\}$) are only included for assuring finite convergence. The switch from one form of iterations to the other is done using the infeasibility count, that is, the number of elements of the set of infeasibilities H given by (17). The steps of the resulting algorithm are presented below.

Block algorithm

Let $F = \emptyset$, $T = \{1, \cdots, n\}$, $p > 0$, $ninf = n + 1$ and $nit = 1$.

Step 1. Compute $\bar{z}_F$ and $\bar{w}_T$ by (15) and (16) respectively and the infeasibility set by (17). Let $|H|$ be the number of elements of H.

  · If $|H| = 0$, stop: $z = (\bar{z}_F, 0)$ is the solution of the LCP.
  · If $ninf > |H|$, set $ninf = |H|$ and $nit = 1$. Go to Step 2.

· If $ninf \leq |H|$ and $nit \leq p$, go to Step 2
(if $nit = 1$ set $\bar{F} = F$ and $\bar{H} = H$).

· If $ninf \leq |H|$ and $nit = p + 1$, go to Step 3 with $F = \bar{F}$ and $H = \bar{H}$.

Step 2. Set $F = F - (F \cap H) \cup (T \cap H)$, $T = \{1, \cdots, n\} - F$, $nit = nit + 1$. Go to Step 1.

Step 3. Let $s = \min\{i \in H\}$ and set

$$F = \left\{ \begin{array}{ll} F - \{s\} & \text{if } s \in F \\ F \cup \{s\} & \text{if } s \in T \end{array} \right.$$

and $T = \{1, \cdots, n\} - F$. Go to Step 1.

It follows from the description of the steps of the algorithm that the integer p plays an important role on the efficiency of the algorithm. This value represents the maximum of block iterations that are allowed to be performed without an improvement of the infeasibility count. It is obvious that this value should be small. However, too small values for p may lead to the use of Murty's method too often with an increase on the number of iterations. Extensive computational experience presented in [29] has shown that $p = 10$ is usually a good choice in practice. An implementation of this procedure for stricly monotone LCPs with symmetric matrices is also described in this paper.

The block algorithm possesses finite termination for strictly monotone LCPs ($M \in P$) independently of the positive value of p. If M is a NSM matrix and $p = \infty$ then the algorithm is strongly polynomial [12]. So $p = \infty$ should be used in this case, which means that Murty's method never comes into operation. Furthermore $F \cap H = \emptyset$ in each iteration [12] and the inequality $\leq$ may replace $<$ in the definition of the set H [84]. Computational experience described in [29] has shown that this modification usually leads to a great reduction of the number of iterations when the dimension of the LCP is quite large. This study has also shown that the block algorithm is usually quite efficient for the solution of large-scale strictly monotone LCPs. However, we have found some hard LCPs for which the algorithm performs poorly [29]. Another feature of the algorithm is its ability to start with any set F. This has been exploited in [29] to develop a hybrid algorithm that combines this block method with an interior-point algorithm. This hybrid scheme will be briefly described in the next subsection.

The block algorithm assumes that M_{FF} is nonsingular for each set F to work. This has prevented its use for the solution of the LCP when M is not a P matrix. However, the algorithm can be applied with $p = \infty$ to Z matrices. In fact, in this case the singularity of M_{FF} means that the LCP is infeasible [12, 84]. Finally, the block algorithm can be extended to the solution of a BLCP with a P matrix. This procedure has been described in [53], where it is also shown its great efficiency for large-scale

BLCPs with P matrices. A strongly polynomial extension of the block algorithm for a BLCP with a NSM matrix has also been discussed in [55, 84]. Computational experience presented in [55] has shown that this algorithm is quite efficient for the solution of large-scale BLCPs with this type of matrices.

3.4. INTERIOR-POINT ALGORITHMS

The LCP can be restated as a system of nonlinear equations in nonnegative variables of the form

$$w = q + Mz \tag{19}$$
$$ZWe = 0 \tag{20}$$
$$z \geq 0, \quad w \geq 0 \tag{21}$$

where Z and W are diagonal matrices of order n whose diagonal elements are components of z and w respectively and e is an n-vector of ones. Interior-point algorithms for the LCP attain to find a solution of the system of equations (19) and (20) by maintaining the positivity of the variables z_i and w_i. A modified Newton algorithm has been designed for this purpose and is fully explained in [60]. This procedure unifies some other interior-point methods that have been previously developed for the LCP. Next, we give a brief description of this algorithm. Suppose that we apply Newton's method for the system (19)-(20). If one of the complementarity equations $z_i w_i = 0$ is verified at the early stages, then the algorithm cannot move from this point. To avoid this type of boundary problems, the iterands are forced to follow the so-called central path. This simply consists of finding in each iteration k of the algorithm the Newton's direction associated with the following system

$$Mz - w = q$$
$$ZWe = \mu_k e \tag{22}$$

where μ_k is the central parameter given by

$$\mu_k = \frac{(z^k)^T w^k}{\delta n} \tag{23}$$

with (z^k, w^k) the current point and δ a positive real number satisfying $1 < \delta \leq n$. As the algorithm progresses, the complementarity gap $(z^k)^T w^k$ tends to reduce to zero and an approximate solution of the LCP is found when $(z^k)^T w^k$ is sufficiently small.

The Jacobian $\nabla H_{\mu_k}(z, w)$ of the mapping $H_{\mu_k}(z, w)$ associated with the system (22) at the point (z^k, w^k) is given by

$$\nabla H_{\mu_k}(z^k, w^k) = \begin{bmatrix} M & -I \\ W_k & Z_k \end{bmatrix}$$

where, as before, W_k and Z_k are diagonal matrices whose diagonal elements are the components of the vectors w^k and z^k respectively. By using this expression of $\nabla H_{\mu_k}(z^k, w^k)$, it is easy to see that the Newton's direction (u, v) is given by

$$(M + Z_k^{-1}W_k)u = -(q + Mz^k) + \mu_k Z_k^{-1}e \qquad (24)$$
$$v = (\mu_k Z_k^{-1} - W_k)e - Z_k^{-1}W_k u \qquad (25)$$

After computing this direction, we set $(z^{k+1}, w^{k+1}) = (z^k, w^k) + \theta(u, v)$, where θ is chosen so that $(z^{k+1}, w^{k+1}) > 0$. Hence this value should satisfy

$$\theta = \gamma \min\{\min\{-\frac{z_i^k}{u_i} : u_i < 0\}, \min\{-\frac{w_i^k}{v_i} : v_i < 0\}\} \qquad (26)$$

with $0 < \gamma < 1$ (γ is usually set equal to 0.99995 in practice). After finding the new point (z^{k+1}, w^{k+1}) a new iteration is performed ($k = k+1$). The algorithm terminates when both the equations (19) and (20) are approximately satisfied, that is, if (z^k, w^k) verifies

$$(z^k)^T w^k < \varepsilon_1 \quad and \quad \| q + Mz^k - w^k \| < \varepsilon_2 \qquad (27)$$

for a certain norm and positive tolerances ε_1 and ε_2.

The steps of the modified Newton's method are presented below.

Modified Newton's algorithm

Let $k = 0$, $(z^0, w^0) > 0$ and ε_1, ε_2 be two positive tolerances.

Step 1. If (27) is satisfied, stop: (z^k, w^k) is an approximate solution of the LCP.

Step 2. Compute
  · The central parameter μ_k by (23).
  · The direction (u, v) by (24)-(25).
  · The stepsize θ by (26).

Step 3. Set $(z^{k+1}, w^{k+1}) = (z^k, w^k) + \theta(u, v)$, $k = k + 1$ and go to Step 1.

It is possible to show that the algorithm works provided $M \in P_0$, since $M + Z_k^{-1}W_k$ is nonsingular for diagonal matrices Z_k and W_k with positive diagonal elements [60]. Furthermore the algorithm possesses global convergence if the LCP has a solution, (z^0, w^0) is an element of the feasible set K and M is a $CSuf$ matrix. In particular, the algorithm can process a LCP with a Sufficient (Suf) matrix. Strictly monotone and monotone LCPs can therefore be solved by this type of methodology. Furthermore the algorithm takes a polynomial number of iterations in the latter case [60].

In general, it is difficult to get an initial feasible point $(z^0, w^0) \in K$ and only positivity is required for the vectors z^0 and w^0. The algorithm still works in this case, but global convergence and polynomiality can only be assured under stronger conditions on the central parameter μ_k and stepsize θ. The study of the convergence of these so-called infeasible interior-point algorithms has been a recent active area of research [10, 74, 100, 104]. In practice there are some strategies for choosing the initial point. We recommend [29] for an implementation of the modified Newton's method in which a discussion of these techniques is also included.

As in linear programming [68, 73], the convergence of the interior-point algorithms can be sped up by the incorporation of a predictor-corrector strategy. In this procedure the Newton's direction $(\bar{u}, \bar{v})$ associated with the system (19-20) is first computed by

$$(M + Z_k^{-1}W_k)\bar{u} = -(q + Mz^k)$$
$$v = -W_k e - Z_k^{-1}W_k u \tag{28}$$

Then the central parameter μ_k is found by

$$\mu_k = \frac{(z^k + \theta\bar{u})^T(w^k + \theta\bar{v})}{\delta n} \tag{29}$$

where $1 < \delta \leq n$ and θ is given by (26) with $u = \bar{u}$ and $v = \bar{v}$. The final direction (u, v) is the Newton's direction associated with the system

$$w - Mz = q$$
$$ZWe = \mu_k e - \bar{U}\bar{v}$$

where $\bar{U}$ is the diagonal matrix whose diagonal elements are the components of the vector $\bar{u}$. It is then easy to see that u and v are given by

$$(M + Z_k^{-1}W_k)u = -(q + Mz^k) + Z_k^{-1}(\mu_k e - \bar{U}\bar{v})$$
$$v = (\mu_k Z_k^{-1} - W_k)e - Z_k^{-1}\bar{U}\bar{v} \tag{30}$$

After finding the direction (u, v), the new point (z^{k+1}, w^{k+1}) is given by $(z^{k+1}, w^{k+1}) = (z^k, w^k) + \theta(u, v)$ and another iteration takes place.

The steps of the predictor-corrector algorithm are presented below.

Predictor-corrector algorithm

Let $k = 0$, $(z^0, w^0) > 0$ and $\varepsilon_1, \varepsilon_2$ be two positive tolerances.

Step 1. If (27) is satisfied stop: (z^k, w^k) is an approximate solution of the LCP.

Step 2. Compute

- The direction $(\bar{u}, \bar{v})$ by (28).
- The stepsize θ by (26) with $(u, v) = (\bar{u}, \bar{v})$.
- The central parameter μ_k by (29).
- The direction (u, v) by (30).
- The stepsize θ by (26).

Step 3. Set $(z^{k+1}, w^{k+1}) = (z^k, w^k) + \theta(u, v)$, $k = k + 1$ and go to Step 1.

The predictor-corrector algorithm possesses global convergence and is polynomial if M is a PSD matrix and the initial point (z^0, w^0) belongs to the feasible set K of the LCP [60]. As before, the second demand is difficult to fulfil in practice. We recommend [29] for an implementation of the predictor-corrector algorithm. In particular, the initialization techniques and the stopping criterium (27) are discussed there. In this implementation the linear systems (28) and (30) are solved by a direct method. However, there are some special structured large-scale LCPs for which this type of implementation is impractical and an iterative solver must be used for the solution of the systems (28) and (30). An interesting example of such a type of LCP comes from the single commodity spatial equilibrium model addressed in [36, 87]. This model leads into a LCP whose matrix M is symmetric singular PSD and is the product of three sparse matrices. The explicit computation of M leads into an almost dense matrix and prohibits the use of a direct solver to be incorporated in the predictor-corrector algorithm. However, it is possible to design a preconditioned conjugate-gradient algorithm [83] to accomplish this goal. We recommend [28] for a description of this type of implementation.

Extensive computational experience with interior-point algorithms for the solution of strictly monotone LCPs with symmetric matrices has shown that the predictor-corrector method is usually superior over the modified Newton's method [29]. A comparison between the predictor-corrector algorithm and the block principal pivoting method for the solution of this type of LCPs has shown that the first procedure is more robust in the sense that it has been able to find an approximate solution for all the test problems. However, the block pivoting method seems to be less dependent on bad scaling and degeneracy of the unique solution of the LCP. Since this latter algorithm can start with any set F, we can combine both the algorithms in such a way that the predictor-corrector algorithm is applied until a certain stage and the block pivoting method is applied with an initial set constructed from the solution found by the predictor-corrector method. This type of hybrid algorithm seems to be the most robust procedure for the solution of strictly monotone LCPs with symmetric matrices. We recommend [29] for a full discussion of this technique and its practical performance. We believe that similar hybrid techniques may be developed for strictly monotone and monotone LCPs with unsymmetric matrices.

Some computational experience with interior-point algorithms in the solution of nonmonotone LCPs has been described in [93, 97] and shows that this type of methodology can process other classes of LCPs. See also [103] for some theoretical results in this context. Finally, the interior-point algorithms discussed in this section can be extended for the solution of the BLCP [10, 60, 92].

3.5. OPTIMIZATION ALGORITHMS

As discussed in Section 2, the LCP can always be reduced into a quadratic programming problem. So it is possible to find a solution of the LCP by solving its associated quadratic program. We start this section by discussion of some approaches for the LCP that exploit these equivalences.

We first consider the case in which M is a symmetric matrix. Then the LCP reduces to the problem of finding a stationary point of the following quadratic program

$$\text{Minimize} \quad f(z) = q^T z + \frac{1}{2} z^T M z \tag{31}$$
$$\text{subject to} \quad z \geq 0$$

If this problem does not contain the nonnegative constraints, then it reduces to the system of equations $Mz = -q$. The partition algorithm [7] for the solution of this system takes the following form

$$Az^k = -(Bz^{k-1} + q), \quad k = 1, 2, \cdots \tag{32}$$

where A and B are matrices such that $M = A + B$ and z^0 is a given vector ($z^0 = 0$ in general). Convergence is assured when M is a symmetric PD matrix and the splitting (A, B) is regular, that is $A - B \in PD$ [7]. An extension of this splitting algorithm for the LCP is obtained by solving the LCP

$$w^k = (q + Bz^{k-1}) + Az^k$$
$$z^k \geq 0, \quad w^k \geq 0$$
$$(z^k)^T w^k = 0$$

in each iteration k, instead of the system (32). It is possible to show that the algorithm still converges under the same hypothesis [21]. Furthermore the requirement $M \in PD$ can be replaced by $M \in PSD$ and there exists a vector z such that $Mz + q > 0$ [21]. We recommend [2, 21, 34, 67, 72] for some convergence results of the splitting algorithm and the use of line-search techniques in this type of procedure.

In an implementation of the splitting algorithm, the $LCP(q + Bz^{k-1}, A)$ should be easily solved. This is the case in which $A = L + \alpha D$, where α is a positive real number smaller that 2 and D, L are the diagonal and the strictly lower triangular part of the matrix M. This leads in the so-called projected Successive OverRelaxation (SOR) algorithm, that is usually stated in the following form.

Projected SOR algorithm

Let $z^0 = 0$, $\alpha \in]0, 2[$, $k = 0$ and ε a tolerance for zero.

repeat

    $k = k + 1$

    for $i = 1$ **to** n **do**

        $w_0 = q_i + \sum_{j<i} m_{ij} z_j^{k+1} + \sum_{j \geq i} m_{ij} z_j^k$

        $z_i^{k+1} = \max\{0, \frac{z_i^k - \alpha w_0}{m_{ii}}\}$

until $w^k = q + M z^k \geq 0$ **and** $(z^k)^T w^k \leq \varepsilon$

We note that the algorithm contains a projection step for the iterands z^k to be nonnegative in each iteration. This requirement is not present in the solution of a system of linear equations and the algorithm reduces to the well-known SOR method in this case. The projected SOR algorithm converges if M is a PD matrix or a PSD matrix such that $Mz + q > 0$ [72]. These two results are consequences of the main convergence theorems for the splitting algorithm.

Due to its simplicity, the projected SOR algorithm has shown to be useful in the solution of special structured large-scale LCPs that arise in some free boundary problems [20]. Block and other forms of this type of algorithm are also efficient in this case [11, 20, 59, 65]. Despite some encouraging results in the solution of these and other strictly convex quadratic programs [66] and in some spatial equilibrium models [40], the projected SOR algorithm is not in general robust for the solution of monotone LCPs. First, the relaxation parameter α has an important effect on the number of iterations of the method and there is no indication about its optimal value in this context. Furthermore, the algorithm is quite sensitive to the ill-conditioning of the matrix M [66]. The benefits and drawbacks of this procedure are well displayed in [28], which contains a computational study of the solution of a monotone LCP that arises in a single commodity spatial equilibrium model. This study also shows that the projected SOR method is not competitive with the predictor-corrector algorithm for the solution of this type of LCP. We believe that similar statements can be done for other spatial equilibrium problems and convex quadratic programs.

Active-set algorithms are certainly the most common techniques for solving quadratic programs. These algorithms simplify if all the constraints are simple bounds on the variables [30] and become equivalent to Keller's single principal pivoting algorithm. As in this latter algorithm, a search direction is computed in each iteration by using a direct solver. It is also possible to perform this task by the preconditioned conjugate-gradient algorithm [82]. As in single principal pivoting algorithms, the set of active constraints is modified in exactly one element per iteration. So active-set algorithms are quite recommended if the dimension of the quadratic problem is relatively small, but tend to become ineficient as the dimension increases.

Projected gradient algorithms have been designed to overcome this slow performance of the active set methods, by replacing many active-set constraints in each iteration. Convergence is usually assured for convex quadratic programs, whence they can be applied for the solution of monotone LCPs. This type of techniques takes the following form

$$z^{k+1} = (z^k + \theta E \nabla f(z^k))^+$$

where $(.)^+$ is the projector operator on the nonnegative orthant defined by

$$(u)_i^+ = \max\{0, u_i\}$$

Furthermore E is a scaling matrix, $\nabla f(z^k)$ is the gradient of f at z^k and θ is a stepsize that is computed by an exact or an inexact Armijo type line-search. In many cases these techniques are combined with active set strategies to achieve global convergence. We recommend [8, 15, 23, 31, 76, 77, 101] for descriptions of the most important algorithms of this category. Computational experience has shown that these algorithms are quite useful in the solution of some special structured LCPs. However, their applicability for a LCP with a ill-conditioned matrix is at least questionable. Further tests are required to make definite claims about the performances of these algorithms.

Penalty techniques have also been recommended for the solution of convex quadratic programs with bounds and can be applied for the solution of a LCP with a symmetric PSD matrix. We recommend [16, 27] for two important algorithms of this type. As is usual in these techniques, the quadratic program is transformed into a nonlinear unconstrained optimization problem, which in turn is solved by a descent method incorporating an inexact line-search or trust region technique. The use of trust region techniques for solving quadratic programs with bounds has also been recommended in some other algorithms that are not based on penalty techniques [17, 18, 32]. Computational experience has indicated that these algorithms seem to be quite suited for the solution of the LCP with a symmetric matrix. In our opinion, a comparison among these techniques with block pivoting and interior-point methods would be quite interesting to draw definite conclusions about the efficiencies of these approaches.

It is important to add that all these techniques are designed for the solution of quadratic programs with bounds of the form

$$\text{Minimize} \qquad f(z) = q^T z + \frac{1}{2} z^T M z$$
$$\text{subject to} \qquad l_i \leq z_i \leq u_i, \quad i = 1, \cdots, n$$

Therefore they can also be applied to the solution of a BLCP with a symmetric matrix.

Consider now the case in which M is an unsymmetric matrix. As stated before, the LCP is equivalent to the following quadratic program

$$\text{Minimize} \qquad f(z) = q^T z + \frac{1}{2} z^T (M + M^T) z \qquad (33)$$
$$\text{subject to} \qquad q + Mz \geq 0, \; z \geq 0$$

Furthermore, if M is a $RSuf$ matrix , then each stationary point of this quadratic program (33) is a solution of the LCP. Hence active-set, reduced-gradient and projected gradient algorithms [13, 35] may be recommended in this case. If M is not $RSuf$, then it is necessary to find a global minimum for (33) to get a solution of the LCP. There exist some techniques for computing global minima of quadratic functions [43, 89]. The application of this type of techniques for the solution of the LCP has been investigated in [90]. This study has shown that the global optimization algorithm can perform its task, but the computational work increases very much with the dimension of the LCP.

Some other optimization problems have also been associated with the LCP. Under certain hypotheses on the matrix M and right-hand side vector q, it is possible to solve a LCP by finding the optimal solution of a linear program of the form

$$\text{Minimize} \qquad p^T z$$
$$\text{subject to} \qquad q + Mz \geq 0, \; z \geq 0$$

where p is a vector depending on M and q [5, 70, 71]. Another reformulation of the LCP has been introduced quite recently [33] and takes the following form

$$\text{Minimize} \qquad \rho \parallel Mz + q - w \parallel^2 + (z^T w)^p \qquad (34)$$
$$\text{subject to} \qquad z \geq 0, \; w \geq 0$$

where $\rho > 0$ and $p > 1$ are given. It is possible to show [33] that if M is a PSD matrix and the LCP is feasible, then a solution of the LCP can be found by computing a stationary point of the nonlinear program (34). Furthermore, it seems that some nonmonotone LCPs can also be processed by this type of approach [33].

The LCP can also be solved by finding global minima of some equivalent optimization problems [4, 6, 95]. However, these approaches are not efficient in general and are not discussed in this paper.

To terminate this section, we mention another approach for the LCP that has become popular in recent times. It is known [69] that a LCP is equivalent to a system of nonlinear equations

$$G(z) = 0 \qquad (35)$$

where G is the mapping from $\mathrm{I\!R}^n$ to $\mathrm{I\!R}^n$ defined by

$$G_i(z) = \min\{z_i, w_i\}, \quad i = 1, \cdots, n$$

As stated in [21, 41, 63], if a complementary basic solution $(\bar{z}, \bar{w})$ is nondegenerate, then G is differentiable at $\bar{z}$ and a block pivoting iteration (18) with $H_1 = H$ reduces to an iteration of Newton's method for the solution of the system (35). This has lead to the design of a damped Newton's algorithm [21, 41] for the solution of strictly monotone LCPs. The method attains to find a solution of the system (35) by using a modified monotone Armijo line-search technique that maintains nondegeneracy throughout the whole procedure. Some limited computational experience presented in [41] indicates that the algorithm holds great promise. Actually, the algorithm reduces to the block pivoting method with $p = \infty$ if the unit stepsize is used in each iteration [41, 63, 62]. The design of nonmonotone [39] and trust region [25] techniques for this system (35) may be an interesting topic for future research.

3.6. ENUMERATIVE ALGORITHM

As discussed in the previous sections, there are many direct and iterative algorithms for the LCP. These procedures are able to find a solution of the LCP or show that none exists when the matrix M possesses some special properties. If no information about the class of the matrix M is known, then the LCP can either be solved by a global optimization technique or by an enumerative method. In this section we describe an algorithm of this type, that has been firstly developed by Al-Khayyal [3] and then improved and implemented by Judice and Faustino [45].

As opposed to direct methods, an enumerative algorithm starts by finding a feasible solution of the LCP. Then, by maintaining feasibility in each iteration, the procedure attempts to find a complementary solution (z, w) satisfying

$$z_i w_i = 0, \quad i = 1, \cdots, n$$

This type of constraints has a combinatorial nature similar to $0 - 1$ integer programming. So, as in this last problem, a binary tree may be exploited to find a complementary solution. This tree has the form displayed in figure , where $i_1, i_2, \cdots$ are integer numbers of $\{1, 2, \cdots, n\}$.

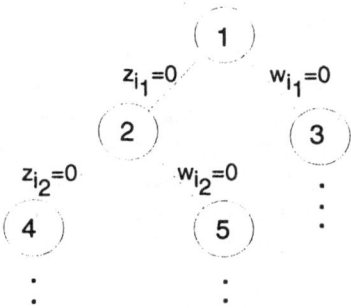

Figure 2: Binary tree for the enumerative method.

As stated before, a feasible solution of the LCP is found in node 1 by solving the linear program

$$\begin{aligned}
\text{Minimize} \quad & z_0 \\
\text{subject to} \quad & Mz + z_0 p - w = -q \\
& z \geq 0, \ w \geq 0, z_0 \geq 0
\end{aligned}$$

where z_0 is an artificial variable and p is a nonnegative vector with $p_i > 0$ for all i such that $q_i < 0$. This is done by the Phase 1 algorithm with a unique artificial variable described in [79]. We can also add in this procedure a basis restricted rule to control the number of pairs of basic complementary variables [50]. Note that this procedure can lead into a solution of the LCP [75], but this rarely occurs.

Each one of the nodes of the binary tree is generated by solving a linear program, which consists of minimizing a variable z_i or w_i subject to the linear constraints of the LCP and some constraints $z_j = 0$ or $w_j = 0$ corresponding to some variables that have been previously fixed in upper level nodes. For instance, to generate the node 4 of the tree of the figure , we have to solve the linear program

$$\begin{aligned}
\text{Minimize} \quad & z_{i_2} \\
\text{subject to} \quad & Mz - w = -q \\
& z \geq 0, \ w \geq 0, z_{i_1} = 0
\end{aligned}$$

This linear program is solved by the Phase II algorithm of the simplex method with a similar basis restricted rule that controls the number of pairs of basic complementary variables. Two cases can then occur:

- If the variable minimized has value equal to zero, then it is fixed at zero in all descent paths of the tree (the variable is said to be starred).

- If the minimum value of the variable is positive, then the branch is pruned and the node is fathomed.

The enumerative algorithm attemps to find a solution of the LCP by generating successive nodes of the tree, according to the process explained above. It contains some heuristic rules to choose the node in each iteration and the pair of complementary basic variables (z_j, w_j) for branching. The algorithm also includes a modified reduced gradient algorithm [3] that is applied in each node and tries to find a local star minimum $(\bar{z}, \bar{w})$ of the function $z_F^T w_F$ on the constraint set K of the LCP, where F is the set of free variables (none of the variables z_i or w_i has been starred in the upper levels of the tree). We note that, by definition, $(\bar{z}, \bar{w})$ is an extreme point of K satisfying

$$\bar{z}_F^T \bar{w}_F \leq z_F^T w_F$$

for all its adjacent extreme points (z, w).

The algorithm only uses basic feasible solutions associated with the constraint set of the LCP. Hence it can be implemented by using a simplex type procedure for small, medium or large-scale LCPs. We recommend [45, 50] for the description and implementation of the enumerative method. It is also important to add that if M is a NSD matrix then the modified reduced gradient algorithm simplifies and reduces to a simplex type method [49]. This makes the enumerative method more efficient for concave LCPs [49]. Computational experience with indefinite and concave medium and large-scale LCPs has shown that the algorithm can find a solution of the LCP in a reasonable amount of time. The computational work increases with the dimension of the LCP, but this increase is not so big as it could be expected [45, 49]. This seems reasonable, since the algorithm searches for only one solution of the LCP and there are some efficient heuristic techniques that help in this context. If the LCP is feasible but does not have a solution, then the search of the tree is usually much longer and the algorithm becomes impractical in this case. In our opinion the study of LCPs which are feasible but have no solutions is an important research area. Some results have already been achieved in this direction [88, 91], but there is still a large amount of work to be done in the future.

4. Existence, complexity and algorithms for the GLCP

4.1. EXISTENCE AND COMPLEXITY OF THE GLCP

As stated in Section 1, given two vectors $q \in \mathbb{R}^n$ and $p \in \mathbb{R}^l$ and matrices $M \in \mathbb{R}^{n \times n}$, $N \in \mathbb{R}^{n \times m}$, $R \in \mathbb{R}^{l \times n}$ and $S \in \mathbb{R}^{l \times m}$ the Generalized Linear Complementarity Problem (GLCP) consists of finding vectors $z \in \mathbb{R}^n$, $y \in \mathbb{R}^m$, $w \in \mathbb{R}^n$ and $v \in \mathbb{R}^l$ such that

$$\begin{aligned} w &= q + Mz + Ny \\ v &= p + Rz + Sy \end{aligned} \tag{36}$$

$$z, w, v, y \geq 0 \tag{37}$$

$$z^T w = 0 \tag{38}$$

We note that the GLCP reduces to a LCP if $l = m = 0$. As in the LCP, (z, w, v, y) is a feasible solution of the GLCP if satisfies its linear contraints (36) and (37). Furthermore (z, w, v, y) is said to be complementary if satisfies (36) and

$$z_i w_i = 0, \quad i = 1, \cdots, n \tag{39}$$

The existence of the variables y may turn the GLCP quite easy in some cases. For instance, if the convex set

$$L = \{y \in \mathbb{R}^m : Ny \geq -q, \ Sy \geq -p, \ y \geq 0\} \tag{40}$$

is nonempty, then any vectors $\bar{y} \in K$ and $z = 0$ constitute a solution of the GLCP. In this section we always assume that this set L is empty. As the LCP, the GLCP may have a unique solution, multiple solutions or no solution. Furthermore we can also associate to the GLCP a quadratic program of the following form

$$\text{Minimize} \qquad z^T w$$
$$\text{subject to} \qquad (z, w, y, v) \in K$$

where K is the set of the linear constraints (36) and (37). This problem can be restated as

$$\text{Minimize} \quad f(z,y) = \begin{bmatrix} q \\ 0 \end{bmatrix}^T \begin{bmatrix} z \\ y \end{bmatrix} + \frac{1}{2} \begin{bmatrix} z \\ y \end{bmatrix}^T \begin{bmatrix} M + M^T & N \\ N^T & 0 \end{bmatrix} \begin{bmatrix} z \\ y \end{bmatrix}$$

$$\text{subject to} \qquad\qquad \begin{bmatrix} M & N \\ R & S \end{bmatrix} \begin{bmatrix} z \\ y \end{bmatrix} \geq \begin{bmatrix} -p \\ -q \end{bmatrix} \qquad (41)$$

$$\begin{bmatrix} z \\ y \end{bmatrix} \geq 0$$

If $\bar{K}$ is the constraint set of this quadratic program, then there are three possible cases:

- $\bar{K} = \emptyset \Rightarrow$ GLCP is infeasible and has no solution.

- $\min_{(z,y)\in\bar{K}} f(z,y) > 0 \Rightarrow$ GLCP is feasible but has no solution.

- $f(\bar{z},\bar{y}) = \min_{(z,y)\in\bar{K}} f(z,y) = 0 \Rightarrow (\bar{z},\bar{y})$ is a solution of the GLCP.

As before, finding a solution of the GLCP by exploring this equivalence turns out to be difficult, since it is in general required to find a global minimum of the quadratic program (41). However, it possible to show [56] that if the matrix

$$U = \begin{bmatrix} M & 0 \\ R & 0 \end{bmatrix} \qquad (42)$$

is $RSuf$, then each stationary point of the quadratic program (41) is a solution of the GLCP. If $R = 0$ then $U \in RSuf$ if and only if $M \in RSuf$. Therefore a GLCP in which M is a $RSuf$ matrix and $R = 0$ can be solved by finding a stationary point of the quadratic program (41). In particular, this occurs when M is a PSD matrix and $R = 0$. Furthermore the GLCP can be solved in polynomial time in this last case [102]. If $R \neq 0$, then a stationary point of the program (41) is not necessarily a solution of the GLCP and it is possible to show that the GLCP is NP-hard even when M is a PSD matrix [56].

4.2. ALGORITHMS FOR THE GLCP

As stated in Section 1, the GLCP has come to our attention in the solution of some global optimization problems by a Sequential LCP (SLCP) technique. Next, we briefly describe this procedure that has been applied to find global minima of linear and linear-quadratic bilevel programs [50, 46, 47], bilinear and concave quadratic programming problems [44]. All these problems can be reformulated in the following form

$$
\begin{array}{ll}
\text{Minimize} & r^T z + s^T y \\
\text{subject to} & w = q + Mz + Ny \\
& v = p \qquad + Sy \\
& z, w, v, y \geq 0 \\
& z^T w = 0
\end{array}
\tag{43}
$$

where M is a *PSD* matrix. The SLCP algorithm attempts to find an ε-global minimum of this problem (43) by solving a sequence of GLCPs of the form

$$
\text{GLCP}(\lambda_k): \quad
\begin{aligned}
w &= q + Mz + Ny \\
v &= p \qquad + Sy \\
w_0 &= \lambda_k - r^T z - s^T y \\
z, w, &v, y, w_0 \geq 0 \\
z^T w &= 0
\end{aligned}
\tag{44}
$$

where λ_k is a strictly decreasing sequence defined by

$$
\lambda_{k+1} = \begin{cases} \lambda_k - \nu|\lambda_k| & \text{if } \lambda_k \neq 0 \\ \lambda_k - \nu & \text{if } \lambda_k = 0 \end{cases}
\tag{45}
$$

for a small positive number ν (usually $\nu = 10^{-3}$). The algorithm starts by finding a solution (z^0, y^0) of the constraint set of the program (43). Then λ_0 is computed by

$$
\lambda_0 = r^T z^0 + s^T y^0
$$

In each iteration $k \geq 1$, the $\text{GLCP}(\lambda_k)$ has to be solved. The algorithm terminates when a $\text{GLCP}(\lambda_k)$ without a solution is found. In this case the solution of the $\text{GLCP}(\lambda_{k-1})$ is an ε-global minimum of this problem (43), where $\varepsilon = \nu|\lambda_k|$.

Next, we study in more detail each one of the iterations k of the SLCP algorithm. In the initial iteration ($k = 0$), we have to solve a GLCP in which M is a PSD matrix and $R = 0$. Therefore a solution of this GLCP can be found by computing a stationary point of its associated quadratic program (41). This can be done by a reduced-gradient or an active-set method [35] or by Ye's interior-point algorithm

[102]. Another procedure [56] consists of considering the following LCP

$$
\begin{aligned}
w &= q + Mz + Ny \\
v &= p \qquad\quad + Sy \\
x &= h - N^T z \qquad\quad - S^T u \\
&z, w, v, y, x, u \geq 0 \\
z^T w &= 0, \; y^T x = 0, \; v^T u = 0
\end{aligned}
\tag{46}
$$

where the vector h satisfies

$$
h_i \geq \max_{(z,y)\in \bar{K}} (N_{.i}^T z), \quad i = 1, \cdots, m
\tag{47}
$$

with $\bar{K}$ the constraint set of the quadratic program (41). Since M is a PSD matrix, then the LCP (46) is monotone. Due to its definition, we can show [56] that there are only two possible cases

- LCP is infeasible and the GLCP has no solution.

- LCP has a solution $(\bar{z}, \bar{y}) \Rightarrow (\bar{z}, \bar{y})$ is a solution of the GLCP.

So a solution of the GLCP can be found by applying Lemke or an interior-point method to the monotone LCP (46). In some cases it is also possible to exploit the structure of the matrix M and solve the GLCP by a finite number of linear programs. This actually occurs in the solution of bilinear and linear bilevel programming problems [44, 47].

For each iteration $k \geq 1$, the GLCP(λ_k) is NP-hard, since M is a PSD matrix and $R \neq 0$. So in general an enumerative algorithm should be used to process this GLCP. The enumerative method described in section 3.6 can be easily extended for the solution of the GLCP. Computational experience presented in [50, 44, 49, 46, 47] has shown that the algorithm is able to find in general a solution of the required GLCP(λ_k). If a solution $(\bar{z}, \bar{y})$ of the GLCP(λ_k) is found, then it is in general possible to find another solution (z^k, y^k) of the GLCP(λ_k) such that

$$
r^T z^k + s^T y^k < r^T \bar{z} + s^T \bar{y}
$$

This can be done by a simplex algorithm with a basis restricted rule that prevents the complementarity condition $z^T w = 0$ to be destroyed [47]. As in the initial iteration, we then set $\lambda_k = r^T z^k + s^T y^k$ and this value of λ_k is used in the updating formula (45).

Consider again the GLCP(λ_k). As before, we can associate to this problem a LCP of the form

$$
\begin{aligned}
w &= q + Mz + Ny && + u_0 r \\
v &= p && + Sy \\
x &= h - N^T z && - S^T u + u_0 s \\
w_0 &= \lambda_k - r^T z - s^T y
\end{aligned}
\tag{48}
$$

$$z, w, v, y, x, u, w_0, u_0 \geq 0$$
$$z^T w = 0, \ y^T x = 0, \ v^T u = 0, \ u_0 w_0 = 0$$

where the vector h is defined by (47). As discussed in [56], if this LCP is infeasible, then the GLCP has no solution. Furthermore, by following a proof similar to that in [56], we can show that if $p \geq 0$ and

$$
\min_{y \in \bar{K} - \{0\}} s^T y > 0
$$

then any solution of the LCP (48) with $y \neq 0$ is a solution of the GLCP (44). However, no conclusion can be stated about the solution of the GLCP if the LCP (48) has a solution with $y = 0$. This nice result shows that a GLCP with a PSD matrix and $R \neq 0$ can be solved in polynomial time in some special cases. Unfortunately, it has no important applications in the solution of bilinear, concave quadratic and bilevel programs, since these conditions are not satisfied in the GLCPs(λ_k) associated with the reformulations of these problems. However, we believe that these GLCPs may be solved by a direct or an iterative algorithm that exploits another type of equivalence. This is certainly a quite important area for future research.

As stated before, the SLCP algorithm terminates when a GLCP(λ_k) with no solution is found. As discussed in section 3.6, the enumerative algorithm is in general impractical in this case, as a great amount of tree search is required to assure that no solution exists. So the SLCP algorithm is in general able to find an ε-global minimum but has difficulty in proving that such a point has been found. It is then necessary to get some practical conditions that assure that a feasible LCP has no solution. This will be quite important on the solution of the global optimization problems mentioned before and in $0 - 1$ integer programming, since this latter problem can be restated as a concave quadratic program [57].

5. Conclusions

In this paper we have surveyed the most important algorithms for the LCP. Keller's method is probably the most interesting procedure for the solution of small and medium scale LCPs with symmetric PSD matrices. However, the algorithm may terminate in a ray for a LCP with a NSD or IND symmetric matrix. If M

is an unsymmetric $RSuf$ matrix of low order then Lemke's algorithm is the most recommended choice for the solution of the LCP. There are some other classes of matrices for which Lemke's method can also be successfully applied, but in general the algorithm is unable to process a LCP with a NSD or an IND matrix. If the (symmetric or unsymmetric) matrix M belongs to these two last classes, then an enumerative algorithm should be in general employed for the solution of the LCP.

Interior-point algorithms are quite efficient for solving monotone large-scale LCPs with symmetric matrices. A block principal pivoting also performs quite well in the solution of strictly monotone large-scale LCPs. Some other nonlinear optimization techniques are still useful for the solution of these types of problems. In our opinion, further tests are required to get definite conclusions about the performances of these algorithms. Hybrid algorithms that exploit the benefits of some of these procedures can also be developed. The application of some of these techniques to more general LCPs has been the subject of recent research and will certainly deserve further interest in the future. The study of feasible LCPs with no solution has not received much attention to date, but should be an important area for future research.

Due to its large number of applications, the GLCP is a quite interesting problem that will certainly deserve much interest in the next coming years. The design of direct and iterative algorithms for the GLCP and the study of feasible GLCPs with no solution are important areas for future research. We believe that the results of this investigation will have a great impact on the solution of some global optimization problems and $0 - 1$ integer programs.

Acknowledgment. I am quite grateful to Luís Portugal for his suggestions and assistance in the writing of this paper.

References

[1] B. Ahn. Iterative methods for linear complementarity problems with upper bounds on primary variables. *Mathematical Programming*, 26:295–315, 1986.

[2] B. Ahn. Solution of nonsymmetric linear complementarity problems by iterative methods. *Journal of Optimization Theory and Applications*, 33:175–185, 1981.

[3] F. Al-Khayyal. An implicit enumeration procedure for the general linear complementarity problem. *Mathematical Programming Study*, 31:1–20, 1987.

[4] F. Al-Khayyal. Linear, quadratic and bilinear programming approaches to the linear complementarity problem. *European Journal of Operational Research*, 24:216–227, 1986.

[5] F. Al-Khayyal. On characterizing linear complementarity problems as linear programs. *Optimization*, 20:715–724, 1989.

[6] J. Bard and J. Falk. A separable programming approach to the linear complementarity problem. *Computers and Operations Research*, 9:153–159, 1982.

[7] A. Berman and R. Plemmons. *Nonnegative Matrices in the Mathematical Sciences*. Academic Press, New York, 1979.

[8] D. Bertsekas. Projected Newton methods for optimization problems with simple constraints. *SIAM Journal on Control and Optimization*, 20:222–246, 1982.

[9] W. Bialas and M. Karwan. Two-level linear programming. *Management Science*, 30:1004–1020, 1984.

[10] S. Billups and M. Ferris. *Convergence of interior-point algorithms from arbitrary starting points*. Working Paper, Computer Sciences Department, University of Wisconsin, Madison, Wisconsin, USA, 1993.

[11] A. Brandt and C. Cryer. Multigrid algorithms for the solution of linear complementarity problems arising from free boundary problems. *SIAM Journal on Scientific and Statistical Computing*, 4:655–684, 1983.

[12] R. Chandrasekaran. A special case of the complementary pivot problem. *Opsearch*, 7:263–268, 1970.

[13] Y. Cheng. On the gradient-projection method for solving the nonsymmetric linear complementarity problem. *Journal of Optimization Theory and Applications*, 43:527–541, 1984.

[14] S. Chung. NP-completeness of the linear complementarity problem. *Journal of Optimization Theory and Applications*, 60:393–399, 1989.

[15] T. Coleman and L. Hulbert. A direct active set algorithm for large sparse quadratic programs with bounds. *Mathematical Programming*, 45:373–406, 1989.

[16] T. Coleman and L. Hulbert. A globally and superlinearly convergent algorithm for convex quadratic programs with bounds. *SIAM Journal on Optimization*, 3:298–321, 1993.

[17] T. Coleman and Y. Li. *A reflective Newton method for minimizing a quadratic function subject to bounds on some of the variables*. Technical Report TR 92-1315, Department of Computer Science, Cornell University, Ithaca, New York, USA, 1992.

[18] A. Conn, N. Gould, and Ph. Toint. Global convergence of a class of trust regions algorithm for optimization with simple bounds. *SIAM Journal on Numerical Analysis*, 25:433–460, 1988.

[19] R. Cottle. The principal pivoting method of quadratic programming. In G. Dantzig and A. Veinott Jr., editors, *Mathematics of Decision Sciences*, pages 144–162, American Mathematical Society, Providence, 1968.

[20] R. Cottle, G. Golub, and R. Sacher. On the solution of large, structured linear complementarity problems: the block partitioned case. *Applied Mathematics and Optimization*, 4:347–363, 1978.

[21] R. Cottle, J. Pang, and R. Stone. *The Linear Complementarity Problem*. Academic Press, New York, 1992.

[22] R. Cottle, J. Pang, and V. Venkateswaran. Sufficient matrices and the linear complementarity problem. *Linear Algebra and its Applications*, 114/115:231–249, 1989.

[23] R. Dembo and U. Tulowitzski. *On the minimization of quadratic functions subject to box constraints*. Technical Report, Department of Computer Science, Yale University, New Haven, Connecticut, USA, 1983.

[24] D. Den Hertog, C. Roos, and T. Terlaky. *The linear complementarity problem, sufficient matrices and the criss-cross method*. Working Paper, Delft University of Technology, Delft, Netherlands, 1990.

[25] J. Dennis Jr. and R. Schnabel. *Numerical Methods for Unconstrained Optimization and Nonlinear Equations*. Prentice-Hall, Englewood Cliffs, New York, 1983.

[26] I. Duff, A. Erisman, and J. Reid. *Direct Methods for Sparse Matrices*. Clarendon Press, Oxford, 1986.

[27] F. Facchinei and S. Lucidi. A class of penalty functions for optimization problems with bound constraints. *Optimization*, 26:239–259, 1992.

[28] L. Fernandes, M. Coelho, J. Júdice, and J. Patricio. On the solution of a spatial equilibrium problem. To appear in Investigação Operacional, 1993.

[29] L. Fernandes, J. Júdice, and J. Patricio. *An investigation of interior-point and block pivoting algorithms for large-scale symmetric monotone linear complementarity problems*. Working Paper, Department of Mathematics, University of Coimbra, Coimbra, Portugal, 1993.

[30] R. Fletcher and M. Jackson. Minimization of a quadratic function subject only to upper and lower bounds. *Journal Institute of Mathematics and Applications*, 14:159–174, 1974.

[31] A. Friedlander, J. Martinez, and M. Raydan. *A new method for box constrained convex quadratic minimization problems.* Working Paper, Department of Applied Mathematics, University of Campinas, Campinas, Brazil, 1993.

[32] A. Friedlander, J. Martinez, and S. Santos. A new trust-region algorithm for bound constrained minimization. To appear in Applied Mathematics and Optimization, 1993.

[33] A. Friedlander, J. Martinez, and S. Santos. *Resolution of linear complementarity problems using minimization with simple bounds.* Working Paper, Department of Applied Mathematics, University of Campinas, Campinas, Brazil, 1993.

[34] U. Garcia Palomares. *Extension of SOR methods for solving symmetric linear complementarity and quadratic programming problems.* Working Paper, Departamento de Processos y Sistemas, Universidad Simon Bolivar, Caracas, Venezuela, 1992.

[35] P. Gill, W. Murray, and M. Wright. *Practical Optimization.* Academic Press, New York, 1981.

[36] C. Glassey. A quadratic network optimization model for equilibrium single commodity trade flow. *Mathematical Programming*, 14:98–107, 1978.

[37] R. Glowinski. *Finite elements and variational inequalities.* MRC Technical Report 1885, Mathematics Research Center, University of Wisconsin, Madison, Wisconsin, USA, 1978.

[38] R. Graves. A principal pivoting simplex algorithm for linear and quadratic programming. *Operations Research*, 15:482–494, 1967.

[39] L. Grippo, L. Lampariello, and S. Lucidi. A class of nonmonotone stabilization methods in unconstrained optimization. *Numerische Mathematik*, 59:779–805, 1991.

[40] F. Guder, J. Morris, and S. Yoon. Parallel and serial successive overrelaxation for multicommodity spatial price equilibrium problems. *Transportation Science*, 26:48–58, 1992.

[41] P. Harker and J. Pang. A damped Newton's method for the linear complementarity problem. In E. Allgower and K. Georg, editors, *Computational Solution of Nonlinear Equations, Lecture Notes in Applied Mathematics 26*, pages 265–284, American Mathematical Society, Providence, 1990.

[42] P. Harker and J. Pang. Finite-dimensional variational inequalities and nonlinear complementarity problems: a survey of theory, algorithms and applications. *Mathematical Programming*, 48:161–220, 1990.

[43] R. Horst and H. Tuy. *Global Optimization*. Springer-Verlag, Heildelberg, 1990.

[44] J Júdice and A. Faustino. A computational analysis of LCP methods for bilinear and concave quadratic programming. *Computers and Operations Research*, 18:645–654, 1991.

[45] J. Júdice and A. Faustino. An experimental investigation of enumerative methods for the linear complementarity problem. *Computers and Operations Research*, 15:417–426, 1988.

[46] J. Júdice and A. Faustino. The linear-quadratic bilevel programming problem. To appear in Canadian Journal of Operational Research and Information Processing, 1993.

[47] J. Júdice and A. Faustino. A sequential LCP algorithm for bilevel linear programming. *Annals of Operations Research*, 34:89–106, 1992.

[48] J. Júdice and A. Faustino. Solution of large-scale convex quadratic programs by Lemke's method. In M. Turkman and M. Carvalho, editors, *Proceedings of First Conference on Statistics and Optimization*, pages 681–695, University of Lisbon, Lisbon, 1991.

[49] J. Júdice and A. Faustino. Solution of the concave linear complementarity problem. In C. Floudas and P. Pardalos, editors, *Recent Advances in Global Optimization*, pages 76–101, Princeton University Press, Princeton, 1991.

[50] J. Júdice and A. Faustino. The solution of the linear bilevel programming problem by using the linear complementarity problem. *Investigação Operacional*, 8:77–95, 1988.

[51] J. Júdice, J. Machado, and A. Faustino. An extension of Lemke's method for the solution of a generalized linear complementarity problem. In P. Kall, editor, *Lecture Notes in Control and Information Sciences 180*, pages 221–230, Springer-Verlag, Berlin, 1992.

[52] J. Júdice and M. Pires. A basic set algorithm for a generalized linear complementarity problem. *Journal of Optimization Theory and Applications*, 74:391–412, 1992.

[53] J. Júdice and M. Pires. A block principal pivoting algorithm for large-scale strictly monotone linear complementarity problems. To appear in Computers and Operations Research, 1993.

[54] J. Júdice and M. Pires. Direct methods for convex quadratic programs subject to box constraints. *Investigação Operacional*, 9:23–56, 1989.

[55] J. Júdice and M. Pires. A polynomial method for a generalized linear complementarity problem with a nonsingular M matrix. *IMA Journal of Mathematics Applied in Business and Industry*, 4:211–224, 1992.

[56] J. Júdice and L. Vicente. On the solution of a generalized linear complementarity problem. To appear in Journal of Global Optimization, 1993.

[57] B. Kalantari and J. Rosen. Penalty for zero-one equivalent problem. *Mathematical Programming*, 24:229–232, 1982.

[58] E. Keller. The general quadratic optimization problem. *Mathematical Programming*, 5:311–337, 1973.

[59] M. Kocvara and J. Zowe. *An iterative two-step algorithm for linear complementarity problems*. Working Paper, Mathematical Institute, University of Bayreuth, Bayreuth, Germany, 1993.

[60] M. Kojima, N. Megiddo, T. Noma, and A. Yoshise. *A unified approach to interior-point algorithms for linear complementarity problems. Lecture Notes in Computer Science 538*, Springer-Verlag, Berlin, 1991.

[61] M. Kojima, S. Mizuno, and A. Yoshise. A polynomial time algorithm for a class of linear complementarity problems. *Mathematical Programming*, 44:1–26, 1989.

[62] M. Kostreva. Block pivoting methods for solving the complementarity problem. *Linear Algebra and its Applications*, 21:207–215, 1979.

[63] M. Kostreva. *Direct Algorithms for Complementarity Problems*. PhD thesis, Rensselaer Polytechnique Institute, New York, USA, 1976.

[64] C. Lemke. On complementary pivot theory. In G. Dantzig and A. Veinott, editors, *Mathematics of Decision Sciences*, pages 95–114, American Mathematical Society, Providence, 1968.

[65] Y. Lin and C. Cryer. An alternating direction implicit algorithm for the solution of linear complementarity problems arising from free boundary problems. *Applied Mathematics and Optimization*, 13:1–17, 1985.

[66] Y. Lin and J. Pang. Iterative methods for large convex quadratic programs: a survey. *SIAM Journal on Control and Optimization*, 25:383–411, 1987.

[67] Z. Luo and P. Tseng. On the convergence of a splitting algorithm for the symmetric monotone linear complementarity problem. *SIAM Journal on Control and Optimization*, 29:1037–1060, 1991.

[68] I. Lustig, R. Marsten, and D. Shanno. On implementing Mehrotra's predictor-corrector interior-point method for linear programming. *SIAM Journal on Optimization*, 2:435–449, 1992.

[69] O. Mangasarian. Equivalence of the complementarity problem to a system of nonlinear equations. *SIAM Journal on Applied Mathematics*, 31:89–92, 1976.

[70] O. Mangasarian. Linear complementarity problems solvable by a single linear program. *Mathematical Programming*, 10:263–270, 1976.

[71] O. Mangasarian. Simplified characterizations of linear complementarity problems solvable as linear programs. *Mathematics of Operations Research*, 4:268–273, 1979.

[72] O. Mangasarian. Solution of symmetric linear complementarity problems by iterative methods. *Journal of Optimization Theory and Applications*, 22:465–485, 1977.

[73] S. Mehrotra. On the implementation of a primal-dual interior-point method. *SIAM Journal on Optimization*, 2:575–601, 1992.

[74] S. Mizuno, M. Kojima, and M. Todd. *Infeasible interior-point primal-dual potential-reduction algorithms for linear programming*. Working Paper, School of Operations Research and Industrial Engineering, Cornell University, Ithaca, New York, 1993.

[75] S. Mohan. On the simplex method and a class of complementarity problems. *Linear Algebra and its Applications*, 14:1–9, 1976.

[76] J. Moré and G. Toraldo. Algorithms for bound constrained quadratic programming problems. *Numerische Mathematik*, 55:377–400, 1989.

[77] J. Moré and G. Toraldo. On the solution of large quadratic programming problems with bound constraints. *SIAM Journal on Optimization*, 1:93–113, 1991.

[78] K. Murty. *Linear Complementarity, Linear and Nonlinear Programming*. Heldermann Verlag, Berlin, 1988.

[79] K. Murty. *Linear Programming*. John Wiley & Sons, New York, 1983.

[80] K. Murty. Note on a Bard-type scheme for solving the complementarity problem. *Opsearch*, 11:123–130, 1974.

[81] A. Nagurney. *Network Economics: a Variational Inequality Approach*. Kluwer. London, 1993.

[82] D. O'Leary. A generalized conjugate-gradient algorithm for solving a class of quadratic programming problems. *Linear Algebra and its Applications*, 34:371–399, 1980.

[83] J. Ortega. *Introduction to Parallel and Vector Solution of Linear Systems*. Plenum Press, New York, 1988.

[84] J. Pang. On a class of least-element complementarity problems. *Mathematical Programming*, 16:325–347, 1979.

[85] J. Pang and L. Chandrasekaran. Linear complementarity problems solvable by a polynomial bounded pivoting algorithm. *Mathematical Programming Study*, 25:13–27, 1985.

[86] J. Pang, I. Kaneko, and W. Hallman. On the solution of some (parametric) linear complementarity problems with application to portfolio analysis, structural engineering and graduation. *Mathematical Programming*, 16:325–347, 1979.

[87] J. Pang and S. Lee. A parametric linear complementarity technique for the computation of equilibrium prices in a single commodity spatial model. *Mathematical Programming*, 20:81–102, 1981.

[88] P. Pardalos. Linear complementarity problems solvable by integer programming. *Optimization*, 19:467–476, 1988.

[89] P. Pardalos and J. Rosen. *Global Optimization: Algorithms and Applications*. *Lecture Notes in Computer Science 268*, Springer-Verlag, Berlin, 1987.

[90] P. Pardalos and J. Rosen. Global optimization approach to the linear complementarity problem. *SIAM Journal on Scientific and Statistical Computing*, 9:341–353, 1988.

[91] P. Pardalos and Y. Ye. *The general linear complementarity problem*. Working Paper, Department of Management Sciences, University of Iowa, Iowa, USA, 1991.

[92] P. Pardalos, Y. Ye, and C. Han. An interior-point algorithm for large-scale quadratic problems with box constraints. In *Lecture Notes in Control and Information 144*, pages 413–422, Springer-Verlag, Berlin, 1990.

[93] P. Pardalos, Y. Ye, C. Han, and J. Kaliski. Solution of P_0-matrix linear complementarity problems using a potential reduction algorithm. To appear in SIAM Journal of Matrix Analysis and Applications, 1993.

[94] F. Pires. *Monotone Linear Complementarity Problems*. PhD thesis, University of Algarve, Portugal, 1993. (in Portuguese).

[95] B. Ramarao and C. Shetty. Application of disjunctive programming to the linear complementarity problem. *Naval Research Logistics Quaterly*, 31:589–600, 1984.

[96] R. Sargent. An efficient implementation of the Lemke's algorithm and its extension to deal with upper and lower bounds. *Mathematical Programming Study*, 7:36–54, 1978.

[97] D. Shanno. *Computational experience with logarithmic barrier methods for linear and nonlinear complementarity problems*. Rutcor Research Report RRR 18–93, Rutgers University, New Brunswick, USA, 1993.

[98] D. Talman and L. Van der Heyden. Algorithms for the linear complementarity problem which allow an arbitrary starting point. In B. Eaves, F. Gould, H. Peitgen, and M. Todd, editors, *Homotopy Methods and Global Convergence*, pages 267–286, Plenum Press, New York, 1981.

[99] J. Tomlin. Robust implementation of Lemke's method for the the linear complementarity problem. *Mathematical Programming Study*, 7:55–60, 1978.

[100] S. Wright. *A path following infeasible interior-point algorithm for linear complementarity problems*. Technecal Report MCS-P 334-1192, Mathematics and Computer Science Division, Argonne National Laboratory, Argonne, Illinois, USA, 1992.

[101] E. Yang and J. Tolle. A class of methods for solving large convex quadratic programs subject to box constraints. *Mathematical Programming*, 51:223–228, 1991.

[102] Y. Ye. *A fully polynomial-time approximation algorithm for computing a stationary point of the general linear complementarity problem*. Working Paper 90–10, College of Business Administration, University of Iowa, Iowa, USA, 1990.

[103] Y. Ye and P. Pardalos. A class of linear complementarity problems solvable in polynomial time. *Linear Algebra and its Applications*, 152:3–19, 1991.

[104] Y. Zhang. On the convergence of a class of infeasible interior-point methods for the horizontal linear complementarity problem. To appear in SIAM Journal on Optimization, 1993.

A HOMEWORK EXERCISE - THE "BIG-M" PROBLEM.

R.W.H. SARGENT

Centre for Process Systems Engineering
Imperial College
London, UK

The following problem was set as a light-hearted challenge to participants during the ASI.

The "Big M" Problem

Most algorithms for solving linear or quadratic programmes require an initial feasible point, while many interior-point algorithms require a point on the "central path", which is both primal and dual feasible.

To avoid finding such a point with an independent "Phase I" algorithm, it has become popular to generate a new problem which has the same solution as the original, but for which an arbitrarily chosen initial point has the required properties. The commonest reformulation of this type seems to have been developed independently by several authors [1, 2, 3], and is as follows:

Suppose we consider a convex quadratic programme in standard form:

$$\min_{x}\left\{c^T x + \tfrac{1}{2}x^T Hx \mid Ax = b, x \geq o\right\} \tag{1}$$

where $x \in R^n$, $b \in R^m$, $c \in R^n$, A is an $m \times n$ matrix and H an $n \times n$ non-negative definite matrix, and we assume (for simplicity) that (1) is regular and has a unique solution.

Then the so-called "big-M" extension of this problem is:

$$
\left.
\begin{aligned}
&\textit{Minimize}: && c^T x + \frac{1}{2}x^T Hx + M_P x_{n+1} \\[1.5em]
&\textit{subject to}: && Ax + a \cdot x_{n+1} = b \\[0.8em]
& && M_D - g^T x - x_{n+2} = 0 \\[0.8em]
& && x_i \geq 0, \quad i = 1,2, \ldots(n+2)
\end{aligned}
\right\} \tag{2}
$$

E. Spedicato (ed.), *Algorithms for Continuous Optimization*, 475–479.

with $M_P > 0$, $M_D > 0$.

The Kuhn-Tucker conditions for a solution of problem (2) can be written:

$$
\left.
\begin{aligned}
r_P &= Ax - b = -a \cdot x_{n+1} \\
r_D &= c + Hx - A^T y - z = -g z_{n+2} \\
x_{n+2} &= M_D - g^T x \\
z_{n+1} &= M_P - a^T y \\
x_i &\geq 0, \; z_i \geq 0, \; x_i z_i = \mu, \; i = 1, 2, \dots (n+2)
\end{aligned}
\right\}
\tag{3}
$$

with of course $\mu = 0$. For interior-point methods, a solution of (3) for some $\mu > 0$ is said to be on the "central path", and this of course implies $x_i > 0$, $z_i > 0$, $i = 1, 2, \dots (n+2)$.

Now suppose we have given initial values $\tilde{y}$, $\tilde{x}_i \geq 0$, $\tilde{z}_i \geq 0$, $i = 1, 2, \dots n$. Then we can choose any $\tilde{x}_{n+1} > 0$, $\tilde{z}_{n+2} > 0$ and:

$$
a = -\tilde{x}_{n+1}^{-1} \cdot \tilde{r}_P \qquad , \qquad g = -\tilde{z}^{n+2} \cdot \tilde{r}_D
\tag{4}
$$

where $\tilde{r}_P$, $\tilde{r}_D$ are evaluated with these values.

Then we choose $M_D \geq g^T \tilde{x}$, $M_P \geq a^T \tilde{y}$, and compute $\tilde{x}_{n+2} \geq 0$, $\tilde{z}_{n+1} \geq 0$ from (3).

It follows that these initial values satisfy all the relations in (3) except for the complementarity condition $\tilde{x}_i \tilde{z}_i = \mu$, and hence are both primal and dual feasible for problem (2), as required.

For an interior-point algorithm, we also wish to satisfy the complementarity condition for some $\tilde{\mu} > 0$. Here we can choose any $\tilde{y}$ and $\tilde{x} > 0$, and then:

$$
\tilde{\mu} = \tilde{x}_{n+1} \tilde{z}_{n+1} = \tilde{x}_{n+1}(M_P - a^T \tilde{y}) = M_P \tilde{x}_{n+1} + \tilde{r}_P^T \tilde{y} > 0
$$

$$
\tilde{\mu} = \hat{x}_{n+2} \tilde{z}_{n+2} = \tilde{z}_{n+2}(M_D - g^T \tilde{x}) = M_D \tilde{z}_{n+2} + \tilde{r}_D^T \tilde{x} > 0
$$

Hence we can choose:

$$
\left.
\begin{aligned}
\tilde{\mu} &> \max \; (0, \tilde{r}_D^T \tilde{x}, \tilde{r}_P^T \tilde{y}) \\
\text{and} \quad \tilde{x}_{n+1} &= M_P^{-1} (\tilde{\mu} - \tilde{r}_P^T \tilde{y}) \; , \; \tilde{z}_{n+1} = \tilde{\mu} / \tilde{x}_{n+1} \\
\tilde{z}_{n+2} &= M_D^{-1} (\tilde{\mu} - \tilde{r}_D^T \tilde{x}) \; , \; \tilde{x}_{n+2} = \tilde{\mu} / \tilde{z}_{n+2}
\end{aligned}
\right\}
\tag{5}
$$

Similarly we define $\bar{z}_i = \bar{\mu}/\bar{x}_i$, $i = 1, 2, \ldots n$, and choose a, g as in (4), to satisfy all the relations in (3).

Now suppose that $(\hat{x}, \hat{y}, \hat{z})$ is the Kuhn-Tucker triple for problem (1). Then to satisfy (3) with $\mu = 0$ we require $x_{n+1} = z_{n+2} = 0$, and this will certainly be the case if:

$$\hat{x}_{n+2} = M_D - g^T \hat{x} > 0 , \quad \hat{z}_{n+1} = M_P - a^T \hat{y} > 0. \tag{6}$$

Moreover it follows from the regularity and uniqueness of (1), and the fact that $M_P > 0$, that if (6) holds then this defines the unique solution of (2), so that condition (6) is both necessary and sufficient for equivalence of the two problems.

Of course, we do not know $\hat{x}, \hat{y}$ a priori, but the equivalence holds if M_D and M_P are chosen "sufficiently large".

A Paradox

We can define a transformation of variables as follows:

$$\left.\begin{array}{llll} \bar{x}_{n+1} = M_P x_{n+1} & , & \bar{z}_{n+1} = M_P^{-1} z_{n+1} & , & \bar{a} = M_P^{-1} a , \\ \bar{x}_{n+2} = M_D^{-1} x_{n+2} & , & \bar{z}_{n+2} = M_D z_{n+2} & , & \bar{g} = M_D^{-1} g . \end{array}\right\} \tag{7}$$

Then (3) becomes:

$$\left.\begin{array}{l} r_P = Ax - b = -\bar{a}\, \bar{x}_{n+1} \\ r_D = c + Hx - A^T y - z = -\bar{g}\, \bar{z}_{n+2} \\ \bar{x}_{n+2} = 1 - \bar{g}^T x \geq 0 , \quad \bar{z}_{n+2} \geq 0 , \quad \bar{x}_{n+2} \bar{z}_{n+2} = \mu , \\ \bar{z}_{n+1} = 1 - \bar{a}^T y \geq 0 , \quad \bar{x}_{n+1} \geq 0 , \quad \bar{x}_{n+1} \bar{z}_{n+1} = \mu , \\ x_i \geq 0 \quad , \quad z_i \geq 0 \quad , \quad x_i z_i = \mu \quad , \quad i = 1, 2, \ldots n \quad . \end{array}\right\} \tag{8}$$

These equations no longer involve M_P and M_D (or equivalently $M_P = M_D = 1$), so that the condition that they should be "sufficiently large" for equivalence with problem (1) seems to have disappeared!

How is this paradox resolved?

Solution

There is of course no real paradox.

To ensure $\bar{x}_{n+2} > 0$, $\bar{z}_{n+1} > 0$ for the solution $(\hat{x}, \hat{y}, \hat{z})$, we have from (8) and (4):

$$\left.\begin{array}{l} \bar{g}^T \hat{x} = -\tilde{r}_D^T \hat{x} / \bar{\bar{z}}_{n+2} < 1 \\[2ex] \bar{a}^T \hat{y} = -r_P^T \hat{y} / \bar{\bar{x}}_{b+1} < 1 \end{array}\right\} \tag{9}$$

Hence we require $\bar{\bar{x}}_{n+1}$, $\bar{\bar{z}}_{n+2}$ sufficiently large, which from (7) is clearly equivalent to choosing $\tilde{x}_{n+1}$, $\tilde{z}_{n+2}$ arbitrarily, then requiring M_P, M_D to be sufficiently large.

The same argument applies to (5), but here, to obtain $\bar{\bar{x}}_{n+1}$, $\bar{\bar{z}}_{n+2}$ sufficiently large, we must choose $\bar{\mu}$ sufficiently large.

Further Comment

Several authors (see [4]) have referred to numerical problems arising from choosing M_P and M_D too large, and also point out that a and g are in general dense columns. Lustig [4] therefore proposed using the limiting direction as $M_P = M_D \rightarrow \infty$ in solving the linearized form of (3), and eliminating the additional rows and columns.

However he took the limit holding $\tilde{x}_{n+1}$, $\tilde{z}_{n+2}$ fixed, which from (3) implies that $\tilde{x}_{n+2} \rightarrow \infty$, $\tilde{z}_{n+1} \rightarrow \infty$, so that $x_i z_i \neq \mu$ for $i = n+1$, $n+2$, and the resulting point is **not** on the central path of the extended problem. (In contrast, taking the limit in (5) above for fixed $\bar{\mu}$ gives $\tilde{x}_{n+1} = \tilde{z}_{n+2} = 0$, which of course reduces (2) to the original problem). Nevertheless, Lustig et al. [5] report excellent numerical performance using Lustig's limiting direction, in spite of the loss of theoretical support. However, in a recent publication (published since the ASI was held), Kojima et al. [6] propose a step-length condition which at least ensures finite convergence of this algorithm, and which is probably often satisfied by the step-length rule used by Lustig et al. [5].

Acknowledgements

Thanks are due to several ASI participants, and particularly to Arnold Neumaier, for discussions which led to the above succinct resolution of the apparent paradox.

References

1.　　Megiddo, N., "Pathways to the Optimal Set in Linear Programming", in Megiddo, N. (Ed) "Progress in Mathematical Programming", pp 131-138, (Springer, New York, 1988).

2.　　Kojima, M., S. Mizuno and A. Yoshise, "A Polynomial-time Algorithm for a Class of Linear Complementarity Problems", Mathematical Programming, **44**, pp 1-26, (1989).

3. Monteiro, R.D.C., and I. Adler, "Interior Path-following Primal-dual Algorithms", Mathematical Programming **44**, Part I pp 27-41, Part II pp 43-66, (1989).

4. Lustig, I.J., "Feasibility Issues in a Primal-dual Interior-point Method for Linear Programming", Mathematical Programming, **49**, pp 145-162 (1991).

5. Lustig, I.J., R.E. Marsten and D.F. Shanno, "Computational Experience with a Primal-dual Interior-point Method for Linear Programming", Linear Algebra and its Applications, **152** pp 191-222, (1991).

6. Kojima, M., N. Megiddo, and S. Mizuno, "A Primal-dual Infeasible-interior-point Algorithm for Linear Programming", Mathematical Programming **61**, pp 263-280 (1993).

DETERMINISTIC GLOBAL OPTIMIZATION [1]

Yu.G. EVTUSHENKO,   M. A. POTAPOV
Computing Center
40 Vavilov Str.
117967 Moscow GSP-1
Russia

Abstract: Numerical methods for finding global solutions of nonlinear programming and multi-criterial optimization problems are proposed. The sequential deterministic approach is used which is based on the non-uniform space covering technique as a general framework. The definitions of the ε-solution for the nonlinear programming problem and the multicriterial optimization problems are given. It is shown that if all the functions, which define these problems, satisfy a Lipschitz condition and the feasible set is compact, then ε-solutions can be found in the process of only one covering of the feasible set on a nonuniform net with a finite number of function evaluations. Space covering techniques are applied to solving systems of nonlinear equations and minimax problems.

1. Introduction

The purpose of this paper is to describe a unified sequential space covering technique for finding approximate global solutions of various optimization problems. The global optimization is of great importance to all fields of engineering, technology and sciences. In numerous applications the global optimum or an approximation to the global optimum is required.

Numerical methods for seeking global solutions of multiextremal problems, in spite of their practical importance, have been rather poorly developed. This is, no doubt, due to their exceedingly great complexity. We do not detail all the available approaches to this problem. Instead, we shall concentrate on one very promising direction, which is based on the idea of a non-uniform covering of a feasible set. This approach has turned out to be quite universal and, as we shall show, can be used not only for seeking global extrema of functions but also for nonlinear programming problems (NLP), for nonlinear integer programming, for solving systems of nonlinear equations, sequential minimax problems and, most importantly, for multicriterial optimization problems (MOP). For these problems we introduce the notion of ε-solutions and describe numerical methods for finding these solutions. Sequential deterministic algorithms and practical results have been obtained for Lipschizian optimization, where all functions which define the problem satisfy a Lipschitz condition. If moreover a feasible set is compact then an ε-solution of MOP can be found after only

---

[1]Research supported by the grant N 93-012-450 from Russian Scientific fund

E. Spedicato (ed.), Algorithms for Continuous Optimization, 481–500.
© 1994 *Kluwer Academic Publishers.*

one covering of the feasible set on a nonuniform net. This property of the proposed approach simplifies the solution process radically in comparison with the traditional approaches which are based on the use of scalarization techniques (convolution functions, reference point approximation, etc.). In our algorithm the computing time which is needed for finding the approximate solution of a multicriterial optimization problem is close to the time which is needed for the search of the global minima of a single function on the same feasible set. This approach was proposed in Evtushenko (1985), Evtushenko and Potapov (1984, 1985, 1987), Potapov (1984).

In this paper we briefly present some results in the field of global optimization which were obtained in the Computing Center of Russian Academy of Sciences. Our methods were implemented on a computer and were used in practice. On the basis of the proposed approach we have developed the packages Solvex and Globex, implemented in Fortran and C language. The libraries of algorithms which were included in these packages enable a user to solve the following classes of problems:

1. Unconstrained global and local minimization of a function of several variables.

2. Nonlinear global and local minimization of a function under the equality and inequality constraints.

3. Nonlinear global solution of a multicriterial problem with equality and inequality constraints.

These packages give opportunity to combine the global approach with local methods and this way speeds up the computation considerably. Global nonlinear problems that are solvable in reasonable computer time must be of limited dimension (of order 10 to 20); however, the use of multiprocessors, parallel computing, and distributed processing substantially increases the possibilities of this approach. Our preliminary results of computations on a parallel transputer system are encouraging. The description of the packages for global optimization and the computational experiments will be given in subsequent papers.

2. General concept

We consider the global optimization problem

$$f_* = \text{global} \min_{x \in X} f(x), \tag{1}$$

where $f : R^n \to R^1$ is a continuous real valued objective function and $X \subset R^n$ is a nonempty, compact feasible set.

Since maximization can be transformed into minimization by changing the sign of the objective function, we shall consider only the minimization problem here. Subscripts will be used to distinguish values of quantities at a particular point and superscripts will indicate components of vectors.

As a special case, we consider the situation, where X is a right parallelepiped P with sides parallel to the coordinate axes (a box in the sequel):

$$P = \{x \in R^n : a \leq x \leq b, a \in R^n, b \in R^n\}. \tag{2}$$

Here and below, the vector inequality $q \leq z$, where $q, z \in R^n$, means that the componentwise inequalities $q^i \leq z^i$ hold for all $i = 1, \ldots, n$.

The set of all global minimum points of the function f (the solutions set) and the set of ε-optimal solutions are defined as follows

$$X_* = \{x_* \in X : f_* = f(x_*) \leq f(x), \forall x \in X\}, \tag{3}$$

$$X_*^\varepsilon = \{x_\varepsilon \in X : f(x_\varepsilon) \leq f_* + \varepsilon\}. \tag{4}$$

For the sake of simplicity of the presentation, we assume throughout this paper that global minimizer sets exist. The existence of an optimal solution in (1) is assured by the well-known Theorem of Weierstrass. The sets X_*, X_*^ε are nonempty because of the assumed compactness of X and continuity of $f(x)$.

The global optimal value of f is denoted by $f_* = f(x_*), x_* \in X_*$. Our goal is to find at least one point $x_\varepsilon \in X_*^\varepsilon$ Any value $f(x_\varepsilon)$, where $x_\varepsilon \in X_*^\varepsilon$, is called an ε-optimal value of f on X. Let $N_k = \{x_1, x_2, \ldots, x_k\}$ be a finite set of k points in X. After evaluating the objective function values at these points, we define the record value

$$R_k = \min_{1 \leq i \leq k} [f(x_1), f(x_2), \ldots, f(x_k)] = f(x_r), \tag{5}$$

where $r \in [1 : k]$; any such point x_r is called a record point.

We say that a numerical algorithm solves the problem (1) after k evaluations if a set N_k is such that $R_k \leq f_* + \varepsilon$, or equivalently $x_r \in X_*^\varepsilon$. The algorithm is defined by a rule for constructing such a set N_k.

We introduce the Lebesque level set in X

$$K(l) = \{x \in X : l - \varepsilon \leq f(x), l \in R^1, \varepsilon \in R^1\}. \tag{6}$$

Theorem 1 *Let N_k be a set of k feasible points such that*

$$X \subseteq K(R_k), \tag{7}$$

then any record point $x_r \in N_k$ belongs to X_*^ε.

Proof. The set of solutions satisfies $X_* \subseteq X \subseteq K(R_k)$. Let a point x_* belong to X_*, then according to (6) and (7) we have $x_* \in K(R_k)$ and therefore $f(x_r) = R_k \leq f(x_*) + \varepsilon$. It means that $x_r \in X_*^\varepsilon$. $\square$

If condition (7) holds then we will say that the set X is covered by $K(R_k)$.

It is very difficult to implement this result as a numerical algorithm because the level set $K(R_k)$ can be very complicated; in general it is not compact and it requires a special program to store it in computer memory. Therefore, we have to weaken the statement of this theorem and impose an additional requirement on the function f. We suppose that for any point $z \in X$ and any level value l, where $l \leq f(z)$, it is possible to define set $B(z, l)$ as follows

$$B(z, l) = \{x \in G(z) : l - \varepsilon \leq f(x)\}, \tag{8}$$

where $G(z) \subset R^n$ is a closed bounded nonempty convex neighborhood of the point z such that the Lebesque measure mes $(G(z))$ of a set $G(z)$ is positive and $z \in G(z)$.

Theorem 2 *(Main Theorem) Let N_k be a set of feasible points such that*

$$X \subseteq \cup_{x_j \in N_k} B(x_j, R_k), \tag{9}$$

then any record point $x_r \in N_k$ belongs to X_*^ε.

Proof. It is obvious that

$$B(z, f(z)) \subseteq K(f(z)), \cup_{x_j \in N_k} B(x_j, R_k) \subseteq K(R_k).$$

Therefore from condition (9) follows (7), that proves the theorem.□

If the set X is compact then a finite set N_k, which satisfies the conditions of this Theorem, exists.

The construction of numerical methods is split in two parts: 1) the definition of a set $B(z, l)$, 2) the definition of covering rule. Let us consider the first part. Assume that the function f satisfies a Lipschitz condition on R^n with constant L. It means that for any x and $z \in R^n$, we have

$$\mid f(x) - f(z) \mid \leq L \parallel x - z \parallel. \tag{10}$$

In this case, we can write that

$$B(z, l) = \{x \in R^n : \quad \parallel x - z \parallel \leq r = [\varepsilon + f(z) - l]/L\}, \tag{11}$$

i.e. $B(z, l)$ is a ball of radius r and a center z. If $x \in B(z, l)$ and $l \leq f(z)$, then from (10) we obtain that condition $l - \varepsilon \leq f(x)$ holds. If $f(z) = l$, then the ball $B(z, f(z))$ has minimal radius $\rho = \varepsilon/L$. The smallest edge of a hypercube inscribed into ball $B(z, f(z))$ is $2\rho/\sqrt{n}$.

Suppose the function f is differentiable and for any x and z of the convex compact set X we have

$$\parallel f_x(x) - f_x(z) \parallel \leq M \parallel x - z \parallel,$$

where M is a constant. In this case, $B(z, l)$ can be constructed as a ball centered at $\bar{z}$ with radius $\bar{r}$:

$$B(\bar{z}, l) = \{x \in R^n : \quad \parallel x - \bar{z} \parallel \leq \bar{r}\}, \bar{z} = z - f_x(z)/M,$$

$$\bar{r}^2 M^2 = [\| f_x(z) \|^2 + 2M[f(z) + \varepsilon - R_k]].$$

These formulas were given in Evtushenko (1971, 1985), Other more complicated cases were considered in Evtushenko and Potapov (1984, 1987).

If $f(x)$ satisfies (10) then, evaluating the function on a regular orthogonal grid of points, $2\rho/\sqrt{n}$ apart in each direction, and choosing the smallest function value solves the problem. But this is generally impractical due to the large number of function evaluations that would be required. Many authors proposed various improvements to the grid method. The main idea of nonuniform grid methods is the following: a set of spheres is constructed such that the minimum of the function over the union of the spheres differs by at most ε from the minimum of the function at their centers. When a sufficient number of spheres is utilized (such that X is a subset of their union), the problem is solved. Such an approach is more efficient than the use of a simple grid.

In recent years a rapidly-growing number of deterministic methods has been published for solving various multiextremal global optimization problems. Many of them can be viewed as realization of a basic covering concept. Theorem 2 suggests a constructive approach for solving problem (1). If inclusion (9) holds, i.e. the union of sets $B(x_j, R_k)$ covers the feasible set, then the set N_k solves the problem (1). As a rule we construct sample sequence of feasible points $N_j = \{x_1, x_2, \ldots, x_j\}$ where the function $f(x)$ has been evaluated and such that the set $W = \cup_{x_j \in N_k} B(x_j, R_j)$ covers X. According to (5) $R_j \geq R_k$ for any $1 \leq j \leq k$, hence $B(x_j, R_j) \subseteq B(x_j, R_k)$. If $X \subseteq W$, then condition (9) holds. A family of numerical methods based on such an approach is called a space covering technique. Many optimization methods have been developed on the basis of the covering idea.

The volume of the current covering set $B(z, R_i)$ essentially depends on the current record value R_i and it is greatest for $R_i = f_*$, but the value f_* is usually not known. Hence, to extend this set it is desirable that the current record value be as close as possible to f_*. For this purpose we use the auxiliary procedures of finding a local minimum in the problem (1). If in the computation process we obtain that $f(x_{i+1}) < R_i$ then we use a local search algorithm and if we find a point $\bar{x} \in X$ at which $f(\bar{x}) < f(x_{i+1})$ then we take the quantity $f(\bar{x})$ as a current record value and the vector $\bar{x}$ as a current record point. After this we continue the global covering process. Therefore, the optimization package which we developed for global minimization includes a library of well-known algorithms for local minimization. The coherent utilization of both these techniques substantially accelerates the computation.

All results given in this paper can be extended straightforwardly to integer global optimization problems and mixed-integer programming. In this case all functions which define a problem must satisfy the Lipschitz condition (10) for any feasible vectors x, y whose components are integers. When solving practical problems, we often take the accuracy $\varepsilon = 0.1$ and the Lipschitz constant $L = 10$, therefore the minimal radius of the covering ball is equal to $\rho = \varepsilon/L = 0.01$. Assume that we compute the value of a function f at a point $x \in X$ with integer coordinates, then we can exclude all points which are inside of the hypercube centered at x and having the edge lengths equal to two. Therefore the minimal radius of the covering balls is greater or equal to one. We take the covering set B into account and exclude the union of the hypercube and B. It is possible to consider another common case where only a part of the variables are integers. For this case we use different covering sets in the spaces of integer and continuous variables. All methods presented here can be used for solving nonlinear integer programming problems and integer multicriterial optimization problems. Moreover, the integer assumption accelerates and simplifies greatly the covering process.

Sometimes we can take advantage of special knowledge about the problem and suggest modifications customized to use the additional information. For example, suppose that we have some prior knowledge about the upper bound Θ for optimal value f_*. We define positive numbers $0 < \delta_1 < \delta_2 \le \varepsilon$ such that $\varepsilon \ge \Theta - f_* + \delta_2$, where $f_* \le \Theta$. Let introduce a covering ball

$$H_j = \{x \in R^n : \ \| x - x_j \| \le r_j\}, r_j = (f(x_j) - \Theta - \delta_1)/L.$$

Now we cover the set X by balls H_j. For a sequence of feasible points $x_1, x_2, x_3, \ldots$ we evaluate the objective function and using (5) we determine the record value and the corresponding record point. Suppose that for all $1 \le j < s$ we have $f(x_j) > \Theta + \delta_2, r_j > (\delta_2 - \delta_1)/L > 0$. If x_s is the first point such that $f(x_s) \le \Theta + \delta_2$, then we conclude that $R_s = f(x_s) \le f_* + \varepsilon, x_s \in X_*^\varepsilon$. We say that the global approximate solution of Problem (1) is found after s evaluations and we terminate computations.

Theorem 3 *Let N_k be a set of k feasible points such that $X \subseteq \cup_{j=1}^k H_j$. Assume that $f(z)$ satisfies the Lipschitz condition(10) and X_* is nonempty. Suppose that an upper estimation Θ of f_* is known. Then there exists an index i such that $x_i \in X_*^\varepsilon, f(x_i) \le \Theta + \delta_2$, where $1 \le i \le k$.*

Proof. By contradiction. Assume that $R_k > \Theta + \delta_2$. According to (10) for any point x, which belongs to H_j, we have:

$$f(x) \ge f(x_j) - L \| x - x_j \| \ge f(x_j) - r_j L = \Theta + \delta_1 > f_*.$$

Therefore $f(x) > f_*$ on H_j. Because of arbitrariness of $j, 1 \le j \le k$, the same property holds everywhere on X. It means that X_* is empty. Hence there exists a point $x_i \in N_k$ such that $x_i \in X_*^\varepsilon.\Box$

We can use Theorem 2 and 3 simultaneously taking sequentially the following covering radius

$$\bar{r}_j = [f(x_j) - \min[R_j - \varepsilon, \Theta + \delta_1]]/L.$$

All covering algorithms can easily be modified and still retain their basic form.

We will not describe here all nonuniform covering techniques, instead we only mention some directions: the layerwise covering algorithm (Evtushenko (1971, 1985)), the bisection algorithm (Evtushenko and Ratkin (1987), Ratschek et.al. (1988)), the branch and bound approach (Volkov (1974), Evtushenko and Potapov (1987), Potapov (1984), Horst and Tuy (1989)), the chain covering algorithm (Evtushenko et.al. (1992)). The covering rules are developed mainly for the case where X is a right parallelepiped. In most of these papers the feasible set X is covered by hypercubes inscribed into covering balls B.

Other nondeterministic approaches for global optimization can be found in the recent survey by Betro (1991), and in the books by Horst and Tuy (1990), Torn and Zilinskas (1991).

3. Solution of nonlinear programming problems

The approach described in the preceding section carries over to solving nonlinear programming problems. The feasible set X can be nonconvex and nonsimply connected. Therefore very often, it is difficult to realize algorithmically a covering of X by balls or boxes. It is easier to cover a rather simple set P that contains the set X. For example, if all components of a vector x are bounded then P can be the "box" defined by (2). Suppose the global minimum is sought:

$$f_* = \text{global} \min_{x \in P \cap X} f(x), \tag{12}$$

$$X = \{x \in R^n : \Psi(x) = 0\}, \tag{13}$$

where $\Psi : R^n \to R^1$ and the intersection $P \cap X$ is nonempty.

The scalar function $\Psi(x)$ is equal to zero everywhere on X, and greater than zero outside X. As before, we denote by X_* the global solutions set of problem (12), which is assumed to be nonempty, $X_* \subseteq P \cap X$. It is obvious that $X_* \subseteq P$. We extend the feasible set X by introducing the ε-feasible set X_ε, define the set of approximate global solutions X_*^ε of the problem (12) and two Lebesque level sets in P:

$$X_\varepsilon = \{x \in R^n : \Psi(x) \le \varepsilon\}, \tag{14}$$

$$X_*^\varepsilon = \{x \in P \cap X_\varepsilon : f(x) - \varepsilon \le f_*\}, \tag{15}$$

$$T(\varepsilon) = \{x \in P : 0 < \varepsilon < \Psi(x), \varepsilon \in R^1\},$$

$$K(l) = \{x \in P : l - \varepsilon \le f(x), l \in R^1\}.$$

Let $N_k = \{x_1, \ldots, x_k\}$ be a set of k points from the set P where the functions $f(x)$ and $\Psi(x)$ have been evaluated. The record point x_r and the record value R_k are defined similarly to (5):

$$R_k = \min_{x_i \in N_k \cap X_\varepsilon} f(x_i) = f(x_r).$$

If the intersection $N_k \cap X_\varepsilon$ is empty, then R_k is not defined and the set $K(R_k)$ is assumed to be empty.

Theorem 4 *Assume that the set of global solutions X_* of problem (12) is nonempty. Let N_k be a set of points from P such that*

$$P \subseteq K(R_k) \cup T(\varepsilon), \tag{16}$$

then there exists at least one record point $x_r \in N_k$ which belongs to X_*^ε.

Proof. The intersection $X_* \cap T(\varepsilon)$ is empty because $\Psi(x) > \varepsilon > 0$ for any point $x \in T(\varepsilon)$. Therefore all points from $T(\varepsilon)$ can not belong to the set X_* of global solutions of problem (12). Hence using (16), we obtain $X_* \subseteq P \subseteq K(R_k)$. It means that $K(R_k)$ is nonempty and the intersection $N_k \cap X_\varepsilon$ is also nonempty. If $x_* \in X_*$ then there exists a record point x_r such that $x_r \in N_k, x_r \in P \cap X_\varepsilon$ and $f(x_r) - \varepsilon \leq f(x_*)$. Taking into account the definition (15), we conclude that $x_r \in X_*^\varepsilon$. □

If functions $f(x)$ and $\Psi(x)$ satisfy the Lipschitz condition with the same constant L, then with each point $x_s \in N_k$ we associate a ball B_{sk} centered in x_s and with radius r_{sk}:

$$B_{sk} = \{x : \| x - x_s \| \leq r_{sk}\}, \quad r_{sk} = \max[\hat{r}_{sk}, \tilde{r}_{sk}],$$

$$\hat{r}_{sk} = (f(x_s) - R_k + \varepsilon)/L, \quad \tilde{r}_{sk} = (\Psi(x_s) - \delta)_+/L, \tag{17}$$

where $a_+ = \max[a, 0], 0 < \delta < \varepsilon$.

Theorem 5 *Assume that the set of global solutions X_* of problem (12) is nonempty. Suppose that the functions f and Ψ satisfy the Lipschitz condition (10) and the set N_k of the points from P is such that $P \subseteq \cup_{i=1}^k B_{ik}$. Then any record point x_r belongs to X_*^ε.*

Proof. The solution set satisfies $X_* \subseteq P \subseteq \cup_{i=1}^k B_{ik}$. Consider a point $x_* \in X_*$. Then there exists at least one covering ball B_{sk} such that $x_* \in B_{sk}$. Hence, according to the definition of a ball B_{sk}, we have $\| x_* - x_s \| \leq r_{sk}$. We prove that $r_{sk} > 0$. Suppose that the radius of this ball is equal to $\tilde{r}_{sk}$, then we have

$$\| x_* - x_s \| \leq \tilde{r}_{sk} = (\Psi(x_s) - \delta)_+/L.$$

If $\Psi(x_s) \geq \delta$, then $\tilde{r}_{sk} = (\Psi(x_s) - \delta)/L$ and, taking into account the Lipschitz condition, we obtain

$$\Psi(x_*) \geq \Psi(x_s) - L \| x_* - x_s \| \geq \delta > 0.$$

The above inequality is impossible because $x_* \in X_*$ and $\Psi(x_*) = 0$.

If $\Psi(x_s) < \delta$, then $\tilde{r}_{sk} = 0$ and $\Psi(x_s) < \varepsilon$. Therefore $x_s \in X_\varepsilon$ and there exist x_r and R_k such that

$$\hat{r}_{sk} = (f(x_s) - R_k + \varepsilon)/L \geq \varepsilon/L > \tilde{r}_{sk} = 0.$$

This contradicts the definition (17) of r_{sk}. Hence we have $r_{sk} = \hat{r}_{sk} = (f(x_s) - R_k + \varepsilon)/L$. Using the Lipschitz condition (10), we obtain

$$f(x_*) \geq f(x_s) - L \parallel x_* - x_s \parallel \geq R_k - \varepsilon = f(x_r) - \varepsilon.$$

Taking into account the definition of X_*^ε, we conclude that $x_r \in X_*^\varepsilon$. $\square$

If a current point $x_s \in N_k$, where $s \leq k$, is such that $x_s \in X_\varepsilon$ then $\Psi(x_s) \leq \varepsilon$, $f(x_s) \geq R_s \geq R_k$ and the radius of a covering ball satisfies $r_{sk} \geq \varepsilon/L$. If $x_s \bar{\in} X_\varepsilon$ then $\Psi(x_s) > \varepsilon$ and $r_{sk} \geq (\varepsilon - \delta)/L$. Consequently, if the set P is compact, then a finite set N_k, which satisfies the conditions of Theorem 5, exists.

As a rule in a practical computation a right parallelepiped (2) plays the role of the set P, which is used in the statement (12). Due to it all covering methods mentioned in previous section can be used for solving (12). During the computational process the sequence R_i monotonically decreases, therefore $B_{ii} \subset B_{ik}$ for any $1 \leq i \leq k$. Problem (12) will be solved, if we find the sequence N_k such that the union of balls B_{ik}, or $B_{ii}, 1 \leq i \leq k$ covers the set P. Such a finite set exists.

Consider the particular case of problem (1) where

$$f_* = \text{global} \min_{x \in P} f(x). \tag{18}$$

It is worthwhile to compare problem (18) of finding the minimum of $f(x)$ on P with the problem (12) under the additional constraint $x \in X$. At first glance, it seems paradoxical (although it is true), that finding the global solution of problem (12) is simpler than solving problem (18). The constraint $x \in X$ provides an additional possibility to increase the radii of the covering balls on $P \backslash X$. Hence, the additional constraints merely simplify the problem of finding global solutions. If we know some properties of the problem we should add them to the definition of the set X. We illustrate this idea using a simple version of problem (18). Suppose that $f(x)$ is differentiable on a box P, then the necessary conditions of the minimum can be written in the form

$$\varphi^i(x) = (x^i - a^i)(b^i - x^i)\frac{\partial}{\partial x^i}f(x) = 0, \quad 1 \leq i \leq n.$$

We introduce the feasible set as follows

$$X = \{x \in R^n : \sum_{i=1}^{n}(\varphi^i(x))^2 = 0\}.$$

Now we solve the problem (12) instead of (18) and simplify the covering process in this way.

The function $\Psi(x)$ which defines the feasible set can be found on the basis of penalty functions. Consider the case where the feasible set is defined by equality and inequality type constraints

$$X = \{x \in R^n : h(x) \leq 0, \quad g(x) = 0\}, \quad h : R^n \to R^c, \quad g : R^n \to R^m. \tag{19}$$

Introduce the vector-valued function $h_+(x) = [h_+^1(x), \ldots, h_+^c(x)], h_+^i = \max[0, h^i]$. Let $\| z \|_p$ denote a Hölder vector norm of a vector $z \in R^s$:

$$\| z \|_p = (\sum_{i=1}^s | z^i |^p)^{1/p}, \quad 1 \leq p \leq \infty.$$

If $p = 1, 2, \infty$, then we have the Manhatten, Euclidean and Chebyshev norms respectively:

$$\| z \|_1 = \sum_{i=1}^s | z^i |, \quad \| z \|_2 = (\sum_{i=1}^s (z^i)^2)^{1/2}, \quad \| z \|_\infty = \max_{1 \leq i \leq s} | z^i | .$$

Suppose that each component of h and g satisfies the Lipschitz condition with constant L:

$$| h^i(x) - h^i(y) | \leq L \| x - y \|, \quad | g^j(x) - g^j(y) | \leq L \| x - y \| .$$

Then using Theorem 1.5.2 from Evtushenko (1985) , it is easy to show that $h_+(x)$ satisfies a Lipschitz condition with the same constant L. If we use the well-knowm inequality

$$| \| a \| - \| b \| | \leq \| a - b \|,$$

then we obtain that the function

$$\Psi(x) = \left\| \begin{array}{c} h_+(x) \\ g(x) \end{array} \right\|_p = [\sum_{i=1}^c (h_+^i(x))^p + \sum_{j=1}^m | g^j(x) |^p]^{1/p}, \tag{20}$$

also satisfies the Lipschitz condition with constant L.

The great advantage of the proposed approach lies in the fact that the constraints need not be dealt with separately and that the classical and modern local numerical methods can be used as auxiliary procedures to improve the record values and in this way to accelerate the computations.

4. Numerical solution of global multicriterial minimization problem

The multicriterial minimization problem has numerous applications in diverse fields of science and technology and plays a key role in many computer - based decision support systems. Various complex engineering problems and practical design require multicriterial optimization. The nonuniform covering technique developed above can be extended for the solution of multicriterial minimization problems.

In minimizing a number of objective functions $F^1(x), F^2(x), \ldots, F^m(x)$ it can not be expected in general that all of the objective functions attain their minimum value simultaneously. The objectives usually conflict with each other in that any improvement of one objective can be achieved only at the expense of another. For such multiobjective optimization the so-called Pareto optimality is introduced.

Define the vector-valued function $F^\top(x) = [F^1(x), F^2(x), \ldots, F^m(x)]$, $F : R^n \to R^m$. Let $Y = F(X)$ be the image of X under the continuous mapping $F(x)$. The problem of global multicriterial minimization of the vector-valued function $F(x)$ on an admissible set X is denoted by

$$\text{global } \min_{x \in X} F(x). \tag{21}$$

The set X_* of solutions of this problem is defined as follows:

$$X_* = \{x \in X : \text{ if it exists } w \in X \text{ such that}$$

$$F(w) \le F(x), \text{ then } F(w) = F(x)\}. \tag{22}$$

To solve problem (21) means to find the set X_*. In papers on multicriterial optimization, X_* is usually called the set of effective solutions, and its image $Y_* = F(X_*)$ is called the Pareto set. The sets X_* and Y_* are assumed to be nonempty.

In many practical problems x is a vector of decisions and $y = F(x)$ is a criterion vector or outcome of the decisions. Therefore we say that x belongs to the decision space R^n and that y belongs to the criteria space R^m. We can rewrite the definition (22) in the criteria space:

$$Y_* = \{y \in Y : \text{ if it exists } q \in Y \text{ such that } q \le y, \text{ then } q = y\}. \tag{23}$$

We say that a vector y_1 is better than or preferred to y_2 if $y_1 \le y_2$ and $y_1 \ne y_2$. From definition (23) it follows that a point y_1 belongs to the Pareto set, if there are no points in Y, which are better than the point y_1. In the same way we can compare the points in decision space. We say that $x_1 \in X$ is better than $x_2 \in X$ if $F(x_1) \le F(x_2)$, $F(x_1) \ne F(x_2)$. In the last case we can say also that x_2 is worse than x_1.

A Pareto-optimal solution is efficient in the sense that none of the multiple objective functions can be downgraded without any other being upgraded. Any satisfactory design for the multiple objectives must be a Pareto optimal solution.

The structure of X_* turns out to be very complicated even for the simplest problems. It often happens that this set is not convex and not simply connected, and every attempt to describe it with the help of approximation formulas is extremely difficult to realize. Therefore, in Evtushenko and Potapov (1984, 1987) and Potapov (1984) we defined the new concept of an ε-optimal solution and gave a rule for finding it.

Definition 1 *A set $A \subseteq X$ is called an ε-optimal solution of problem (21) if*

> *1. for each point $x_* \in X_*$ there exists a point $z \in A$ such that $F(z) - \varepsilon e \le F(x_*)$, e means the m-dimensional vector of ones,*

    2. the set A does not contain two distinct points x and z such that $F(x) \leq F(z)$.

Numerical methods for finding global extrema of functions of several variables can be used for constructing the ε-optimal solution of problem (21). Let $N_k = \{x_1, \ldots, x_k\}$ be a set of k points in X. We shall define a sequence of sets $A_k \subseteq N_k$ as k increases, while trying in the final analysis to find ε-optimal solutions.

    RULE 1 FOR CONSTRUCTING A_k. The set A_1 consists of the single point $x_1 \in N_1 \subseteq X$. Suppose that N_k, N_{k+1}, and A_k are known. We compute the vector $F(x_{k+1})$ at a point $x_{k+1} \in N_{k+1} \subseteq X$. Three cases are possible:

1. If it turns out that among the elements $x_i \in A_k$ there are some such that $F(x_{k+1}) \leq F(x_i), F(x_{k+1}) \neq F(x_i)$, then they are all removed from A_k, the point x_{k+1} is included in A_k and this new set is denoted by A_{k+1}.

2. If it turns out that there exists at least one element $x_i \in A_k$ such that $F(x_i) \leq F(x_{k+1})$, then x_{k+1} is not included in A_k, and the set A_k is denoted by A_{k+1}.

3. If the conditions of the two preceding cases do not hold, then the point x_{k+1} is included in the set A_k, which is denoted now by A_{k+1}.

In the first case we exclude from A_k all points which are worse than the new point x_{k+1}. In the second case the new point x_i is worse than at least one point from A_k, therefore this point is not included in the set A_k. In the last case the point x_{k+1} is equivalent (or equally preferred) to all points which belong to the set A_k.

The definition of the Lebesque set (6) is replaced now by

$$K(l) = \{x \in X : l - \varepsilon e \leq F(x), l \in R^m, e \in R^m, \varepsilon \in R^1\}. \qquad (24)$$

Theorem 6 *Let the finite set A_k of admissible points be such that*

$$X \subseteq \cup_{x_j \in A_k} K(F(x_j)). \qquad (25)$$

Then the set A_k determined by the first rule for constructing A_k forms an ε-optimal solution of the multicriterial problem (21).

    Proof. From the covering condition (25) it follows that the set X_* is also covered, i.e. for any point $x_* \in X_*$ there exists a point $x_s \in A_k$ such that $x_* \in K(F(x_s))$. Then from definition (24) we obtain $F(x_s) - \varepsilon e \leq F(x_*)$. Because of arbitrariness of the point $x_* \in X_*$ we conclude that the set A_k forms an ε-optimal solution set of problem (21).$\square$

    In a manner similar to the second section we suppose that for any point $z \in X$ and level vector $l \in R^m$ where $l \leq F(z)$, it is possible to define the set

$$B(z, l) = \{x \in G(z) : l - \varepsilon e \leq F(x)\}.$$

From the inclusion $B(z, l) \subseteq K(F(z))$ it follows that the set A_k is an ε-optimal solution if

$$X \subseteq \cup_{x_s \in A_k} B(x_s, F(x_s)).$$

We assume now that each component of the vector-valued function $F(x)$ satisfies a Lipschitz condition on X with one and the same constant L. Therefore, for any x and z in X, we have the vector condition

$$F(z) - eL \parallel x - z \parallel \leq F(x).$$

In this case we can use the following covering balls

$$B(x_j, F(x_i)) = \{x \in R^n : \ \parallel x - x_j \parallel \leq r_{jk}^i\},$$

where $r_{jk}^i = [\varepsilon + h_{jk}^i]/L$, the index i is a function of indexes j, k and it is found as a solution of the following maximin problem

$$h_{jk}^i = \max_{x_c \in A_k} \min_{s \in [1:m]} (F^s(x_j) - F^s(x_c)) = \min_{s \in [1:m]} (F^s(x_j) - F^s(x_i)), x_i \in A_k. \qquad (26)$$

The inequality $h_{jk}^i > 0$ holds if there exists at least one point $x_c \in A_k$ such that $F(x_c) < F(x_j)$, otherwise $h_{jk}^i = 0$ and the radius of the covering ball is minimal, i.e. it is equal to $\rho = \varepsilon/L$.

As in the third section, we can take into account constraint restrictions. Suppose that problem (21) is replaced by the following problem

$$\text{global} \min_{x \in P \cap X} F(x). \qquad (27)$$

Let X_* denote the set of global solutions of this problem. This set is defined by (22) with additional requirements: $x \in P \cap X, w \in P \cap X$.

We extend the admissible set by introducing the set $Z_\varepsilon = P \cap X_\varepsilon$, where X_ε is given by (15). The definition of an ε-optimal solution carries over to the case of problem (27) with the following change: instead of the condition $A \subset X$ it is required that $A \subset Z_\varepsilon$. The rule for determining A_k is also changed. Suppose that N_k is a set of k points belonging to P.

RULE 2 FOR CONSTRUCTING A_k. Assume that N_k, N_{k+1}, and A_k are known (A_k may be empty). At the point $x_{k+1} \in N_{k+1}$ it is checked whether $x_{k+1} \in Z_\varepsilon$. If not, then x_{k+1} is not included in A_k, and A_k is then denoted by A_{k+1}; otherwise, the same arguments as in the construction of A_k are carried out with a check of the three cases which were described above.

Denote

$$\bar{B}(x_j, F(x_i)) = \{x \in R^n : \parallel x - x_j \parallel \leq \bar{r}_{jk}^i\},$$

$$\bar{r}_{jk}^i = (1/L) \max[\varepsilon + h_{jk}^i, \Psi(x_j) - \delta],$$

where $0 < \delta < \varepsilon$ and i, h_{jk}^i are given by (26), the feasible set X is defined by (13).

Theorem 7 *Suppose that the set of global solutions X_* of the multicriterial problem (21) is nonempty. Assume that the vector-valued function F and the function Ψ satisfy a Lipschitz condition on P. Let the set N_k of points in P be such that*

$$P \subseteq \cup_{j=1}^{k} \bar{B}(x_j, F(x_i)). \qquad (28)$$

Then the set A_k constructed by the second rule forms an ε-optimal solution of the multicriterial problem (27).

The proof is very similar to that of Theorem 5 and therefore is omitted. Any radius $\bar{r}_{jk}$ cannot be less that the quantity $(\varepsilon - \delta)/L > 0$. Therefore finite sets N_k and A_k satisfying the conditions of Theorem 7 exist in the case where the set P is compact. Theorem 7 is very interesting: it provides a simple rule for finding global ε-solution of a multicriterial problem. A set N_k satisfying condition (28) can be constructed by using diverse variants of the nonuniform covering method which were developed for finding a global extremum of a function of several variables. As before, local methods for improving the current ε-optimal solution N_k can be used. Some variants of local iterative methods for solving multicriterial problem are described in Zhadan (1988).

For an approximate solution of the problem, it suffices to implement a covering of an admissible set P by a nonuniform net. This is an essential advantage of such an approach in comparison with the well-known scalarization schemes. We mention for example the reference point approximation, the method of successive concessions, the method of inequality and other traditional methods which require, for their realization, a multiple global minimization of auxiliary functions on the feasible set X (see, for example Jahn, Merkel (1992)).

The set A_k obtained by computer calculations is transmitted to a user (a designer solving the multicriterial problem). The designer may wish to examine some or all the alternatives, make his tradeoffs analysis, make his judgment, or develop his preference before making rational decision. As the final solution, the user chooses a concrete point from the set A_k, starting from the specifics of the problem, or from some additional considerations not reflected in the statement of the problem (21).

The main result of this section is that, for the constructive solution of multicriterial optimization problems, it is possible to use the non-uniform covering technique developed in research on global extrema of functions of several variables. The approach presented here is developed in Evtushenko (1985), Evtushenko and Potapov (1984, 1987), Potapov (1984).

Another approach to the solution of problem (27) can be obtained if we consider the function $\Psi(x)$ as an additional component of an outcome vector, which also should be minimized. We introduce extended multicriterial problem

$$\text{global } \min_{x \in P} \bar{F}(x), \qquad (29)$$

where
$$\bar{F}^{\mathsf{T}}(x) = [F^1(x), \ldots, F^{m+1}(x)], \, F^{m+1}(x) = \Psi(x).$$
For this problem the set of optimal solutions P_* is defined by the condition

$$P_* = \{x \in P : \text{ if it exists } w \in P \text{ such that } \bar{F}(w) \le \bar{F}(x), \text{ then } \bar{F}(w) = \bar{F}(x)\}.$$

For the solution of problem (27) we can use the method which we used for solving problem (21), where $X = P$. Now the sets N_k and A_k consist of points from the set P. If A_k is an ε-optimal solution of problem (29), then for each point $x_* \in P_*$ there exists a point $x \in A_k$ such that

$$F(x) - \varepsilon e \le F(x_*), \quad \Psi(x) - \varepsilon \le \Psi(x_*).$$

The solutions set X_* of problem (27) belongs to the solutions set P_* of problem (29) because the latter includes the points x_* such that $\Psi(x_*) > 0$. Now the set of ε-optimal solutions can be found in similar way as for problem (21). A third variant of solution of problem (27) can be constructed by analogy with the approach described in Evtushenko (1985).

The multicriterial approach can be used for solving the nonlinear programming problem. In this case we introduce a bicriterial minimization problem. Let the vector-valued function $F(x)$ consist of two components: $F^{\mathsf{T}}(x) = [f(x), \Psi(x)]$. Instead of the original problem (12) with feasible set X defined by (13) we define problem

$$\text{global } \min_{w \in P} F(x).$$

The Pareto set for this problem coincides with the sensitivity function of problem (12). Using the approach described in this section, we obtain an ε-approximation of the sensitivity function.

5. Solution of a system of nonlinear equalities and inequalities

In this section we confine ourselves to the problem of finding a feasible point which belongs to the set X given by (19). To solve this problem approximately it suffices to find at least one point from the set

$$X_\varepsilon = \{x \in P : \quad \Psi(x) \le \varepsilon\},$$

where $\Psi(x)$ is defined by(20).

We assume that further computations for sharpening the solution will involve local methods. When X_* is empty, the algorithm should guarantee that the assertion concerning the absence of approximate solutions be true. The problem of finding a point $x_* \in X_*$ is equivalent to the minimization of $\Psi(x)$ on P. Define

$$\Psi_* = \min_{x \in P} \Psi(x). \tag{30}$$

If X_* is nonempty, then $\Psi_* = 0$; otherwise, $\Psi_* > 0$. Suppose that the mappings $h(x), g(x)$ satisfy a Lipschitz condition on P with constant L.

Now for finding an approximate solution of problem (30) we can use Theorem 5. For the sequence of points $\{x_k\}$ from P we use (5) to determine the record point x_r, we record the value R_k and define a covering ball B_{sk} with radius r_{sk}:

$$B_{sk} = \{x \in R^n : \| x - x_s \| \le r_{sk}\},$$

$$r_{sk} = [\Psi(x_s) - \min[R_k - \varepsilon, \delta]]/L, \quad 0 < \delta < \varepsilon.$$

The stopping rule of covering procedures consists only in verification of inequality $R_i \le \varepsilon, i = 1, 2, \ldots, k$. If for some s it is true, then $x_s \in X_\varepsilon$ and an approximate solution x_s is found, otherwise we have to cover all set P and find a minimal value of $\Psi(x)$ on P with accuracy ε.

Theorem 8 *Suppose that $\Psi(x)$ satisfies a Lipschitz condition and the set N_k of the points from P is such that $P \subseteq \cup_{i=1}^k B_{ik}$. Then*

1. *if X_* is nonempty then $x_r \in X_\varepsilon$,*

2. *if $R_k > \varepsilon$ then X_* is empty.*

The proof follows from the above observations and is therefore omitted.

If we know that X_* is nonempty then we can cover P by balls H_j which we defined in the second section and set $\Theta = f_* = 0, \delta_2 = \varepsilon, 0 < \delta_1 < \varepsilon$, then $r_j = [\Psi(x_j) - \delta_1]/L$.

6. Solution of minimax problems

Let $f(x, y)$ be a continuous function of $x \in X \subset R^n, y \in Y \subset R^m$. We consider the minimax problem

$$f_* = \min_{x \in X} \max_{y \in Y} f(x, y). \tag{31}$$

Here we have internal maximization and external minimization problems. We can rewrite Problem (31) in the following equivalent way

$$f_* = \min_{x \in X} \varphi(x), \tag{32}$$

where $\varphi(x) = f(x, y), y \in W(x), W(x) = \arg\max_{y \in Y} f(x, y)$, i.e. $W(x)$ is the set of all solutions of the internal problem.

Denote $z^\top = [x^\top, y^\top] \in \Omega = X \times Y$ and $\bar{f}(z) = f(x, y)$. By Theorem 1.5.2 from Evtushenko (1985), if $\bar{f}(z)$ satisfies a Lipschitz condition on $\Omega = X \times Y$ with constant L, then the function $\varphi(x)$, defined by (32), also satisfies a Lipschitz condition with the same constant L. This property opens broad possibilities to use the method of finding global extrema of multiextremal function for the solution of minimax problems. The same method can be used sequentially for solving internal as well as external problems. The subprograms of local search usually are chosen differently, since the functions $f(x, y)$ are often differentiable and their local maximization is carried out using properties of smoothness of f in y. The function φ is only directionally differentiable, and it has to be locally minimized by other methods.

Comparing the problem (31) with the problem of finding the global extremum of f in z on Ω, we can conclude that (31) has an important advantage. Indeed, let the current record value of $\varphi(x_r)$ be known. If at some other point $x_s \in X$ we have to find the value of $\varphi(x_s)$, the process of maximization of f in y can be stopped as soon as at least one point $y_1 \in Y$ has been found such that $f(x_s, y_1) \geq \varphi(x_r)$ since in this case a priori $\varphi(x_s) \geq \varphi(x_r)$ and the knowlege of the exact value $\varphi(x_s)$ will not improve the current record value $\varphi(x_r)$. This property makes it possible in a number of cases to terminate the process of solving the internal problem.

Theoretically, this approach makes it possible to solve sequential minimax problems and opens the door to solving the discrete approximation of differential games. For example the simplest Isaacs dolichobrachistochrone game problem was solved in Evtushenko (1985, see pages 463, 464). Full details of the first layerwise variant of the covering algorithm and codes in ALGOL-60 are given in Evtushenko (1972).

7. Computational results

In this section we present some computational experiments using the nonuniform covering techniques. The branch and bound algorithm was applied to a number of standart test functions. This algorithm requires an upper bound L_* for the Lipschitz constant L. At the very begining we run the algorithm with some constants L_i which are much smaller than L_*. Incorrect (diminished) Lipschitz constants are used in order to find good record points and use them in subsequent computations with bigger Lipschitz constants. If we take $L_2 > L_1 \geq L_*$ and run the algorithm, taking $L = L_1$ and $L = L_2$ then the difference between record values must be less than ε. This condition is necessary but not suffisient for the Lipschitz constant L_1 to be greater or equal to the true value.

Everywhere for the local search we use the Brent modification of Powell method (Brent (1973)). The accuracy ε is fixed, we set $\varepsilon = 0.1$. In Tables 1 - 3 we give the Lipschitz constants L_i, the record values R_k, the record points x_r and k - the number of function evaluations.

The following simple examples illustrate the space covering approach.

Example 1. Griewank function (Torn (1989))

$$f(x) = (x^1)^2/200 + (x^2)^2/200 - \cos(x^1\sqrt{2})\cos(x^2\sqrt{2}) + 1,$$

$$x \in R^2, \quad -2 \leq x^i \leq 3, \quad x_* = [0, 0], \quad f(x_*) = 0, \quad x_0 = [3, 3], \quad f(x_0) = 0.8851.$$

Table 1. Results for Griewank function

L	R_k	x_k^1	x_k^2	k
0.1	0.1349	-2.0000	-2.0000	100
0.2	0.0491	2.2104	2.2104	96
0.4	0.0491	2.2104	2.2104	26
0.8	0.0000	0.0000	0.0000	117
1.6	0.0000	0.0000	0.0000	362

Example 2. Goldstein-Price function (Goldstein (1971))

$$f(x) = (1 + (x^1 + x^2 + 1))^2(19 - 14x^1 + 3(x^1)^2 + 3(x^2)^2 - 14x^2 + 6x^1x^2))(30+$$

$$(2x^1 - 3x^2)^2(18 - 32x^1 + 12(x^1)^2 + 48x^2 - 36x^1x^2 + 27(x^2)^2)$$

$$x \in R^2, \quad -2 \le x^1 \le 3, \quad -3 \le x^2 \le 2,$$

$$x_* = [0, -1], \quad f(x_*) = 3, \quad x_0 = [0, 1], \quad f(x_0) = 28611$$

Table 2. Results for Goldstein-Price function

L	R_k	x_k^1	x_k^2	k
100	3.0000	0.0000	-0.9999	114
180	3.0000	0.0000	-0.9999	4963

Example 3. Hartman function (Torn (1989))

$$f(x) = -\sum_{j=1}^{4} C^j e^{-[a_1^j(x^1-b_1^j)^2+a_2^j(x^2-b_2^j)^2+a_3^j(x^3-b_3^j)^2]}$$

where $a_1 = [3, 0.1, 3, 0.1], a_2 = [10, 10, 10, 10], a_3 = [30, 35, 30, 35]$,

$$b_1 = [0.3689, 0.4699, 0.1091, 0.03815],$$

$$b_2 = [0.117, 0.4387, 0.8732, 0.5743],$$

$$b_3 = [0.2673, 0.747, 0.5547, 0.8828],$$

$$c = [1, 1.2, 3, 3.2]$$

$$x \in R^3, \quad -2 \le x^i \le 2, \quad i = 1, 2, 3,$$

$$x_* = [0.1146, 0.5556, 0.8526], \quad f(x_*) = -3.8628,$$

$$x_0 = [2, 2, -2], \quad f(x_0) = 0.$$

Table 3. Results for Hartman function

L	R_k	x_k^1	x_k^2	x_k^3	k
0.1	-3.8628	0.1146	0.5556	0.8526	221
2	-3.8628	0.1146	0.5556	0.8526	6
4	-3.8628	0.1146	0.5556	0.85	1615

We also solved Branin and so - called Camel problems. Starting from various initial points, we found the global solution using the local method from the very begining of computation. Therefore these examples were not very interesting for illustration of global covering technique.

8. Conclusion

The nonuniform covering technique has given rise to numerous theoretical results and effective computational procedures for solving various global optimization problems. Recent developments indicate that these results can be generalized and extended significantly for parallel computations.

References

[1] Betro B. (1991), *Bayesian Methods in Global Optimization*, J. of Global Optimization 1, 1-14.

[2] Brent R. (1973), *Algorithms for minimization without derivatives*, Prentice-Hall, Englewood Cliffs.

[3] Evtushenko Yu. (1971), *Numerical methods for finding global extrema (case of non-uniform mesh)*, U.S.S.R. Comput. Math. and Math. Phys 11, 1390-1403 (Zh. Vychisl. Mat. i Mat. Fiz., 11, 1390-1403).

[4] Evtushenko Yu. (1972), *A numerical method for finding best guaranteed estimates*, U.S.S.R. Comput. Math. and Math. Phys 12, 109-128 (Zh. Vychisl. Mat. i Mat. Fiz., 12, 109-128).

[5] Evtushenko Yu. (1974), *Methods for finding the global extremums in operations research*, Proceedings of Comp. Center of U.S.S.R. Academy of Sciences 4, Moscow, 39-68, (in Russian).

[6] Evtushenko Yu. (1985), *Numerical optimization techniques*. Optimization Software Inc, Publication Division, New-York.

[7] Evtushenko Yu. and Potapov M. (1984), *Global search and methods for solution of operational control problems*, Moscow, VNIPOU, 128-152, (in Russian).

[8] Evtushenko Yu. and Potapov M.(1985), *Nondifferentiable Optimization: Motivation and Applications* (Laxenburg, 1984), Lecture Notes in Economics and Math. Systems 255, 97-102, Springer-Verlag, Berlin.

[9] Evtushenko Yu. and Potapov M.(1987), *Methods of numerical solutions of multicriterion problems*. Soviet Math. Dokl.34, 420-423.

[10] Evtushenko Yu. and Potapov M.(1987), *Numerical solution of multicriterion problem*. Cybernetics and Computational Techniques 3, Moscow, 209-218, (in Russian).

[11] Evtushenko Yu., Potapov M., Korotkich V. (1992), *Numerical methods for global optimization*, in Floudas , C.A. and Pardalos, P.M. (eds.), Recent Advances in Global Optimization, Princeton University Press.

[12] Evtushenko Yu. and Ratkin V. (1987), *Bisection method for global optimization*. Izvestija Akademii Nauk AN USSR, Tehnicheskaya Kibernetika, 119-128, (in Russian).

[13] Goldstein A., Price J. (1971), *On descent from local minima*, Mathematics of Computation, 25, p. 569-574.

[14] Horst R. and Tuy H. (1990), *Global Optimization*, Deterministic Approaches, Springer-Verlag, Berlin.

[15] Jahn I. and Merkel A. (1992), *Reference Point Approximation Method for the Solution of Bicriterial Nonlinear Optimization Problems*, J. of Optimizaton Theory and Application 73, 87-104.

[16] Potapov M. (1984), *Non-uniform covering methods and their use in solving global optimization problems in a dialogue mode*, author's summary of candidate's dissertation, Moscow, (in Russian).

[17] Ratschek H. and Rokne J. (1988), *New computer methods for global optimization*. Ellis Horwood Limited, Chichester.

[18] Torn, A. and Zilinskas, A. (1989), *Global Optimization*, Lecture Notes in Computer Science 350, Springer Verlag, Berlin.

[19] Volkov E. (1974), *Approximate and exact solutions of systems of nonlinear equation*. Proceedings of Mathematical Institute of USSR Academy of Sciences 131, Moscow, (in Russian).

[20] Zhadan V. (1988), *An augmented Lagrange function method for multicriterion optimization problems*, U.S.S.R. Comput. Math. and Math. Phys 28, 1-11 (Zh. Vychisl. Mat. i Mat. Fiz., 28, 1603-1618).

ON AUTOMATIC DIFFERENTIATION AND CONTINUOUS OPTIMIZATION

L. C. W. DIXON
Emeritus Professor
School of Information Sciences
University of Hertfordshire
College Lane
Hatfield
Hertfordshire   AL10 9AB
U.K.

ABSTRACT

The calculation of the gradient vector is an essential requirement of most efficient unconstrained optimization codes; as is the calculation of the Jacobian matrix in constrained optimization. Obtaining analytic expressions for the gradient and Jacobian of simple test functions can often be easy. However, in contrast, correctly differentiating the objective function of an industrial problem when its calculation requires hundreds of lines of computer code is not trivial. Neither is the correct programming of the resulting formulae. The two tasks often require man months of effort.

To avoid this task many researchers used to (and some still do) resort to approximating derivatives by numerical difference formulae. This is both exceedingly expensive of computer time and is prone to truncation error if the step size is chosen inappropriately.

More recently symbolic differentiation codes and preprocessing packages have become available; these, given a listing of the code which calculates a function, produce additional code that calculates its derivatives.

Automatic differentiation works very differently in that it uses facilities available in Pascal SC, C++, and ADA to implement arithmetic on user defined entities:- doublets, triplets and graph-nodes. In this way derivatives can be obtained accurately and efficiently as soon as the function has been coded.

Two versions of automatic differentiation are in use, forward and reverse, and both of these can be implemented in full and sparse form. The relative advantages and disadvantages of all four will be discussed.

The availability of cheap derivatives and the fact that the gradient of any function can theoretically be computed in less than three times the cost of a function evaluation, (in spaces of any dimension) radically alters the relative cost of many aspects of optimization algorithm designs and some of these will also be discussed.

E. Spedicato (ed.), Algorithms for Continuous Optimization, 501–512.
© 1994 *Kluwer Academic Publishers.*

1.   Introduction

Most efficient codes for locating the minimum of an unconstrained function F of n variables x_i assume that the user will provide accurate code for the gradient vector ∇F as well as for the function F. When F is a simple test function of a few variables this is frequently a not too difficult task. In contrast if F is the objective function in an industrial problem, then its calculation may well require hundreds of lines of code. When this is so the manual task of differentiating the coded formulae and correctly programming the result is not trivial and the combined task has been known to require man months of effort. There is of course no guarantee that the resulting program will be free from errors.

In constrained optimisation the equivalent efficient codes also require the Jacobian matrix of the constraints. This is essentially the same task though it does present some additional difficulties, which will be discussed later in the paper.

As the task of analytically differentiating complex objective functions F is so time consuming in man hours, research workers have sought alternatives. For many years it was traditional to use numerical difference formulae, which approximated the gradient component

$$g_i = \nabla F_i = \partial F / \partial x_i$$

by $$g_i(x) = \left\{ F(x + h\hat{a}_i) - F(x) \right\} / h \qquad\qquad (1)$$

or $$g_i(x) = \left\{ F(x + h\hat{a}_i) - F(x - h\hat{a}_i) \right\} / 2h. \qquad\qquad (2)$$

In these formulae $\hat{a}_i$ is the unit vector along the ith axis, and h is a user chosen step. These formulae have three disadvantages:

(i)   if h is chosen too small accuracy is lost due to rounding error in subtracting two very similar numbers.

(ii)   if h is chosen too big accuracy is lost due to truncation errors in the expansion of the Taylor series. Equation (2) is an order more accurate than equation (1) in this regard.

(iii)   computer cost. If $F(x)$ contains M arithmetic operations, equation (1) implies ∇F would cost nM operations while equation (2) would imply $2nM$ operations.

As a safeguard many researchers having coded the gradient analytically would attempt to check it approximately using (1) or (2). If the results disagreed it was difficult to be sure whether the descrepancy was due to the incorrect choice of h in the numerical approximation or an error in the gradient code.

It was this need to be able to obtain reliable accurate values of the derivative that encouraged three separate lines of related research: automatic differentiation, symbolic differentiation and preprocessing packages.

An early investigation was due to Wengert (1964). His basic idea was to decompose the evaluation of complicated functions of many variables into a sequence of simpler evaluations of special functions of one or two variables. The functions of two variables are typically addition, subtraction and multiplication. The functions of one variable are reciprocation, raise to power, exponential, logarithm, trigonometric, hyperbolic.

The second list can be extended to include any other functions of one variable whose analytic derivatives are known.

A "Wengert list" of an objective function consists of a series of steps:

$$\left.\begin{array}{l} \text{Given } x_i \qquad i = 1...n. \\[2em] \text{For} \quad i = n+1,..., \ n+M \\[2em] \qquad x_i = f_i(x_j, \ x_k) \\[2em] F = x_{n+M} \end{array}\right\} \tag{3}$$

In (3) j, k both exist and are the two variables involved if f_i is addition, multiplication or subtraction. Only one of j, k exists for the other operators. For a computable code j and k must be less than i.

Given the Wengert List a simple way of obtaining the derivative $\partial F / \partial x_L$ $(L < n)$ is:

$$\left.\begin{array}{l} \textit{Given} \ \ x_i \quad i = 1 \ ... \ n \\[2em] \textit{set} \quad u_i = 0 \quad i = 1 \ ... \ n \quad i \neq L \\[2em] \textit{and} \quad u_L = 1 \\[2em] \textit{For} \quad i = n+1, \ ..., \ n+M \\[2em] \qquad x_i = f_i(x_j, x_k) \\[2em] \qquad u_i = (\partial f_i / \partial x_j) \ u_j + (\partial f_i / \partial x_k) u_k. \end{array}\right\} \tag{4}$$

$F = x_{n+M}$

$\partial F / \partial x_L = u_{n+M}$

This only requires the user to provide expressions for the derivatives of the special functions, all of which are well known.

Much of the development of automatic differentiation was driven by the limitations of the Fortran-77 language, which is the most common language for numerical applications. Given an objective function code written in standard Fortran, the first task would be to produce the Wengert List (3). This is a preprocessing task. It is now frequently done by special computer codes but in the early days the preprocessor was frequently a programmer. It was then necessary to replace each call to a special function by a subroutine call such as:

CALL ADD (1.0, 0.0, X, 1.0, F1, DF1)

CALL TIMES (X, 1.0, F1, DF1, F, DF)

which not only performed (3) but also (4). So ADD (A, DA, B, DB, C, DC) and TIMES (A, DA, B, DB, C, DC) are assumed to be subroutines which capture rules for evaluating and differentiating sums and products.

Kalabar's (1983) initial Feed algorithm (6) was based on this approach and could be extended to any order higher derivatives (Testfatsion 1991). The results reported in Dixon and Price (1989) were also obtained in this way.

There are now many preprocessors which take a standard Fortran-77 code, effectively produce the Wengert List and thence the gradient by (4) or one of the more sophisticated methods to be discussed later. Soulie (1992) attempted to compare some of those available in 1990 but reports disappointing results in terms of ease of use for an external user.

The output of a symbolic differentiation package is a code that would calculate the gradient when given the numerical values of x_i. Essentially the requirement is to produce the code of (4). Although there is no theoretical reason why this should not be produced reliably, Griewank (1989) reports very disappointing results for the codes available at that time.

2.   Forward Automatic Differentiation

Automatic differentiation is much easier in languages that enable the user to define their own data types and to overload operators. These facilities are now available in a number of programming languages (Pascal-SC, C++, Ada and Fortran-90). The pioneer of this development was L. B. Rall (1981, 1990). Essentially if we introduce a data type, DOUBLETS, which consists of $n+1$ numbers $U = (u, \ \partial u / \partial x_i \quad i = 1 \ \dots \ n)$, then we can overload the meaning of the operators +, -, * and the meaning of the special functions: exponential, log, etc. so that expressions like

$$W = U + V$$
$$W = U * V$$
$$W = \exp (U)$$

give the correct values of the resultant doublet W. This facility allows a user of these languages to employ automatic differentiation while writing code in a natural form. The need to obtain the Wengert List disappears.

This has been implemented in Ada at Hatfield initially by Maany (1989). Simple calculations show that the upper bound for arithmetic operations is $3nM$, the 3 coming from multiplication; while an upper bound on the store is nS where S is the store needed to calculate F. However the use of overloaded operators does involve considerable overheads. There is a switch within the overloaded operators in our packages that can turn the derivative calculations off and in Ada the CPU time for calculating the function value alone but within the overloading structure is approximately 5 times the cost of an ordinary function evaluation. This is because the cost of passing parameters and creating variables of type Doublet still occurs even when the evaluation of the gradient part is suppressed. We therefore have a situation in which the arithmetic costs no longer dominate CPU time (Parkhurst (1990)).

Once the simplicity of using Doublets was appreciated the extension to TRIPLETs was natural and automatic. A TRIPLET consists of the numerical values of a function, its gradient and Hessian.

$$
U = \left(u, \ \partial u / \partial x_i \quad i = 1 \ \dots \ n, \ \partial^2 u / \partial x_i \partial x_j \quad \begin{matrix} i = 1 \ \dots \ n \\ j < i. \end{matrix} \right).
$$

The necessary algebra for the overloading of operators is again simple to program.

However at this point it became obvious that using full matrices within the Triplets was very inefficient. Both Doublet and Triplet codes were therefore rewritten using sparse matrix and vector algebra. Full details are given in (Maany (1989)) where the following results are reported for the CPU time required to differentiate the 50 dimensional Helmholtz function (for details see Griewank (1989)):

	Full Doublets	Sparse Doublets	Full Triplets	Sparse Triplets
Function only	1.36	0.44	60.29	0.44
Function and Gradient	9.24	3.42	68.68	3.52
Function, Gradient and Hessian	-	-	476.36	20.69

The CPU time for calculating the function only within the full Triplet package rises dramatically as although no derivative calculations are being made the calculation still allocates the space needed by the full second derivative matrix.

The method can be easily extended to the calculation of Jacobian matrices. Parkhurst (1990) reports the use of the sparse-doublets package to calculate sparse-Jacobian matrices in up to 15,000 dimensions.

Potential users should note that as Ada does not specify the garbage collector it is necessary to include one efficiently within the packages as otherwise storage problems will be encountered. This is discussed by Maany (1989) and Bartholomew-Biggs et al (1993).

In concluding this section we note that the sparse doublet and sparse triplet codes in Ada enable normal code to be written for the function F and accurate values of ∇F and $\nabla^2 F$ to be obtained reliably by the computer. The major hope for automatic differentiation is therefore achieved.

3.   Reverse Automatic Differentiation

My interest in the method now known as reverse automatic differentiation commenced upon reading Griewank's seminal paper (Griewank (1989)). Historical surveys, e.g. Iri (1992), have located numerous previous discoveries of the method, but all such earlier works were virtually ignored until rediscovered after Griewanks paper was published.

The main theoretical result is that if a function that can be calculated by the Wengert List (3) then its gradient can be calculated in less than $3M$ steps, independent of the dimension n.

The method is simply:

$$
\left.
\begin{array}{l}
\text{Given } x_i \quad i = 1, \, ..., \, n + M \\[2mm]
\text{Set} \quad \bar{x}_i = 0 \quad i = 1, \, ..., \, n + M - 1 \\[2mm]
\bar{x}_{n+M} = 1. \\[2mm]
\text{For} \quad i = n + M, \, ..., \, n + 1 \\[2mm]
\bar{x}_j = \bar{x}_j + \partial f_i / \partial x_j \cdot \bar{x}_i \\[2mm]
\bar{x}_k = \bar{x}_k + \partial f_i / \partial x_k \cdot \bar{x}_i \\[2mm]
\partial F / \partial x_i = \bar{x}_i \quad i = 1 \, ... \, n
\end{array}
\right\}
\tag{5}
$$

The quantities $\bar{x}$ used in this reverse pass are often termed costate or adjoint variables.

This amazing theorem implies, first, that all comparisons of optimisation methods that used the concept of effective function evaluations (E.F.E.'s) now need to be reconsidered. An E.F.E. was conventionally defined as

$$EFE = N_F + nN_g$$

where N_F is the number of function evaluations and N_g the number of gradient evaluations which was logical if the difference approximation (1) was used. Essentially this definition should be replaced by

$$EFE = N_F + 3N_g,$$

and this has a significant effect on comparisons of the relative efficiency of algorithms (Dixon 1991).

Given a Wengert List the reverse calculation (5) is easy to program by hand and gives a very effective way of calculating gradients without the overheads associated with automatic differentiation. A conventional program can be treated the same way, as if we allow the functions f_i in (3) to be completely general functions of all the variables x_j, $j < i$, then in (5) at step i we simply update all the adjoint variables $\bar{x}_j$ that appear in the formula for f_i by:

$$\bar{x}_j = \bar{x}_j + \partial f_i / \partial x_j \bar{x}_i \text{ all } j < i.$$

This fast $3M$ calculation is obtained at the expense of computer store. Since (5) involves retracing the complete computation in reverse order all the M steps must be stored. This is in marked contrast with the forward method where the information can be discarded as soon as the step is complete. A method for alleviating this difficulty is given in Griewank (1992).

My colleague Bruce Christianson wrote an Ada package which constructed the computational graph using the overloaded operator approach, which now has an interface completely compatible with our Ada optimisation codes (Christianson (1992)). As with the sparse DOUBLET and sparse TRIPLET codes the overheads dominate the arithmetic and lead to a factor greater than 3; he reported that it was nearer to 20 for his code.

In Bartholomew-Biggs, Bartholomew-Biggs and Christianson (1993) results are given for a full series of tests on the Watson function ($n = 20$) using a variable metric algorithm. Each type of automatic differentiation took 47 iterations and generated identical points. The relative costs of the various approaches are of interest:

Analytic derivatives	20.1 secs
Numerical approximations	162.7 secs
Full doublets	46.6 secs
Sparse doublets	54.2 secs
Reverse A.D.	37.4 secs

The sparse doublet result is poor due to the very few sparse operations within the Watson function. More benefit from using sparsity was gained when optimising the extended Rosenbrock function ($n = 20$) and then the relative costs were:

Analytic derivatives	20.7 secs
Numerical approximations	55.4 secs.
Full Doublets	36.6 secs.
Sparse Doublets	28.5 secs.
Reverse A.D.	23.9 secs.

Both the Watson function and the extended Rosenbrock function are least squares functions so it is possible to minimize them by a Gauss-Newton algorithm. The comparable times are:

Analytic derivatives	8.5 secs.	111.2 secs
Numeric approximation	14.5 secs	117.6 secs
Full Doublets	11.0 secs.	121.0 secs.
Sparse Doublet	11.3 secs.	119.2 secs
Reverse A.D.	29.5 secs.	126.5 secs.

The times for the Watson function are given first.

We note that the Gauss-Newton method is efficient on the Watson function compared to the variable metric method but not on the Rosenbrock function in 20 dimensions.

The Gauss-Newton method requires the calculation of the Jacobian matrix and for the above results the reverse automatic differentiation code was run independently, through the complete graph, for each subfunction, i.e. 31 times, which accounts for the large CPU time. In the next section we consider how to improve this.

4.   Jacobian Calculations

When F is a sum of subfunctions S_k, then it would be possible to evaluate each subfunction as an independent task requiring M_k operations. Then using sparse doublets the derivative would be obtained in less than $3n_k M_k$ operations where n_k is the number of variables effecting S_k. In contrast reverse automatic differentiation would obtain its gradient in less than $3M_k$ operations and so an upper bound to the cost could be $\sum_k 3M_k$ operations.

For many functions it can be much more efficient to utilise the fact that some strands in the calculation are common to more than one subfunction, for example if the subfunctions are all endpoints of a spacecraft trajectory calculation then the complete simulation is a common strand to all the subfunctions (see Christianson & Dixon (1993)). In these circumstances, the optimum number of operations M will be considerably less than $\sum_k M_k$ and if no structure in the code is used at all then the reverse A.D. approach will need $3mM$ operations where m is the number of subfunctions since it would reverse through the whole code for each subfunction.

In Dixon and Christianson (1992) a vectorised form of reverse automatic differentiation is proposed in which the reverse traversal of the graph is performed for all costates simultaneously and when this was implemented in sparse format the CPU time for the Watson function reduced from 29.5 secs. to 12.5 secs. Similarly on the Rosenbrock function ($n = 20$) the time reduced from 126.5 secs. to 115.5 secs.

These results do indicate that further work is needed into the optimal way to automatically compute sparse Jacobians.

Intuitively sparse doublets will be most successful when the majority of nodes in the computational graph only depend on a few independent variables, whilst sparse reverse A.D. will be most successful when the majority of nodes only effect a few subfunctions.

Dixon (1990), Griewank and Reese (1992) and Dixon and Christianson (1992), discuss such topics in more detail and reach different conclusions. Further research is still needed in this area.

5.   Large Scale Optimisation

As n increases the variable metric approach to unconstrained optimisation, becomes less and less efficient. For many years the conjugate gradient algorithm was used for large scale problems, but more recently the Truncated Newton method of Dembo et al (1985) has been shown to be more efficient by many researchers, e.g. Dixon & Price (1989).

The Truncated Newton method can be written so that it either requires the user to provide F, ∇F and $\nabla^2 F$ at each outer iteration, or alternatively F, ∇F at each outer iteration and $\nabla^2 Fp$ at each inner iteration. The first version is ideally suited to be combined with sparse TRIPLET differentiation. The algorithm was described in Dixon, Mohseninia & Maany (1989a) and results given on functions of up to 3,000 dimensions in Dixon, Mohseninia & Maany (1989b). The second method requires much less store as ∇F, and $\nabla^2 Fp$ are both vectors but the method was generally more expensive of CPU time. Christianson (1992) has however shown that by extending reverse automatic differentiation $\nabla^2 Fp$ can be calculated in less than $6M$ operations. It can therefore be expected that a truncated Newton code based on this extension of reverse automatic differentiation will be even more efficient.

Further research is needed in this area to design codes for large problems that interface effectively with automatic differentiation codes.

In Dixon (1992) the concept of a task matrix was introduced as an alternative to the Wengert List or the computational graph. The task matrix of the operation of obtaining the gradient of a function by applying reverse automatic differentiation to the Wengert List is a sparse matrix with at most 3 non zero elements in each row. This concept of the task matrix can be extended to the calculation of the Newton Step which is shown to be equivalent to the solution of a very sparse matrix in less than $2M + n$ variables. It was shown that more efficient ways of solving this system of equations do exist, than the standard method of forming the Hessian matrix and then solving the Newton equation. This confirmed Griewanks (1989) hypothesis that this must be possible. Further research is still required in this area.

6.   Applications

Readers may be interested in reading more details of applications solved using the N.O.C. Ada automatic differentiation packages.

Dr. Bartholomew-Biggs obtained a Royal Aeronautical Society prize for his paper on an application of SQP and Ada to the structural optimisation of aircraft wings, Bartholomew-Biggs (1989).

Full details of the use of Doublets in a code written for the European Space Operations Centre, Darmstadt, are reported in Bartholomew-Biggs, Bartholomew-Biggs & Parkhurst (1993). More detailed results including the use of sparse Doublets and Reverse A.D. are given in Bartholomew-Biggs, Bartholomew-Biggs & Christianson (1992).

7.   Conclusions

(1) For languages that allow user defined data types and the overloading of operators, DOUBLET and SPARSE DOUBLET codes allow the accurate computation of gradient vectors ∇F. SPARSE-TRIPLET codes allow the accurate computation of Hessian Matrices $\nabla^2 F$.

(2) At the expense of considerable additional storage, reverse automatic differentiation codes allow function gradients ∇F to be calculated in less than qM operations, where $q \approx 3$ for arithmetic operations, and $5 < q < 20$ in practice in Ada. The Hessian Vector product $\nabla^2 F_* p$ can be calculated at about twice this cost. Both are independent of dimension.

(3) If a Jacobian matrix is very sparse and every row contains a small number of variables, then sparse-Doublets can be successfully applied to very large systems to obtain accurate values of J.

Otherwise the relative performance of various automatic differentiation approaches when calculating Jacobian matrices is still an ongoing research topic, but many codes provide accurate results.

(4) The very different (and lower) operation counts now available for the calculation of F, ∇F, $\nabla^2 F$ and $\nabla^2 F p$ from those assumed historically, imply that the relative efficiency of many codes and parts of codes need to be re-examined.

8. References

[1] Bartholomew-Biggs, M. C. (1989) "An application of SQP and Ada to the Structural Optimisation of Aircraft Wings". Journal of the Royal Aeronautical Society. 1989 issue pp.344-350.

[2] Bartholomew-Biggs, M. C., Bartholomew-Biggs, L., and Christianson, B., (1992) "Numerical Experience of Automatic Differentiation and Optimisation in Ada". University of Hertfordshire/ESOC Working Paper.

[3] Bartholomew-Biggs, M. C., Bartholomew-Biggs, L. and Christianson, B. (1993) "Optimisation and Automatic Differentiation in Ada: Some Practical Experience", University of Hertfordshire Technical Report, NOC TR274.

[4] Bartholomew-Biggs, M. C., Bartholomew-Biggs, L. and Parkhurst, S. (1993) "Simplified Ascent-Trajectory Optimisation using Ada Software", ESA Journal, Vol.17, pp.43-56.

[5] Christianson, B., "Automatic Hessians by Reverse Accumulation", (1992), IMAJNA, Vol.12, pp.135-150.

[6] Christianson, B., (1992), "Reverse Accumulation and Accurate Rounding Error Estimates for Taylor Series Coefficients", Optimisation, Methods and Software, Vol.1, pp.81-94.

[7] Dembo, R., & Steihaug, T., (1985) "Truncated Newton Method for Large Scale Optimisation", Mathematical Programming, Vol.26 pp.190-212..

[8] Dixon, L. C. W., (1990) "The solution of partially separable linear equations on a parallel processing system", in Spedicato, E., "Computer Algorithms for Solving Linear Algebraic Equations", Springer Verlag, pp.299-338.

[9] Dixon, L. C. W., (1991) "On the Impact of Automatic Differentiation on the Relative Performance of Parallel Truncated Newton and Variable Metric Algorithms", SIAM Journal of Optimisation, Vol.1, pp.475-486.

[10] Dixon, L. C. W., (1992) "Use of Automatic Differentiation for Calculating Hessians and Newton Steps", in Griewank, A. & Corliss, G. F., "Automatic Differentiation of Algorithms", SIAM, pp114-125.

[11] Dixon, L. C. W. & Christianson, D. B., (1992) "Reverse Accumulation of Jacobians in Optimal Control". University of Hertfordshire Technical Report, NOC TR265.

[12] Dixon, L. C. W., Maany, Z., & Mohseninia, M., (1990), "Automatic Differentiation of Large Sparse Systems", Journal of Economic Dynamics and Control, Vol.14, pp.299-311.

[13] Dixon, L. C. W., Maany, Z., & Mohseninia, M., (1989), "Experience Using the Truncated Newton Method for Large Scale Optimization", in Bromley, K., "High Speed Computing", SPIE, Vol.1058, pp,94-102.

[14] Dixon, L. C. W. & Price, R. C., (1989), "The Truncated Newton Method for Sparse Unconstrained Optimisation Using Automatic Differentiation", JOTA, Vol.60, pp.261-275.

[15] Griewank, A., (1989), "On Automatic Differentiation", in Iri, M., and Tanake, K., "Mathematical Programming: Recent developments and Applications", Kluwer Academic Publishers, pp.83-108.

[16] Griewank, A., (1992), "Achieving Logarithmic Growth of Temporal and Spatial Complexity in Reverse Automatic Differentiation", in Optimisation Methods and Software, Vol.1, pp.35-54.

[17] Griewank, A., & Reese, S., (1992), "On the Calculation of Jacobian Matrices by the Markowitz Rule", in Griewank, A. & Corliss, G. F., Automatic Differentiation of Algorithms", SIAM; pp.126-135.

[18] Iri, M., (1992), "History of Automatic Differentiation and Rounding Error Estimation", in Griewank, A. & Corliss, G. F., "Automatic Differentiation of Algorithms", SIAM,: pp.3-16.

[19] Kalaba, R., Tesfatsion, L. & Wang, J-L., (1983), "A Finite Algorithm for the Exact Evaluation of Higher Order Partial Derivatives of Functions of Many Variables", Journal of Mathematical Analysis and Applications, Vol.12, pp181-191.

[20] Maany, Z. A., (1989), "Ada Automatic Differentiation Packages", Hatfield Polytechnic Technical Report, NOC TR 209.

[21] Parkhurst, S., (1990), "The Evaluation of Exact Numerical Jacobians using Automatic Differentiation", Hatfield Polytechnic Technical Report, NOC, TR224.

[22] Rall, L. B., (1981), "Automatic Differentiation: Techniques & Applications", Lecture Notes in Computer Science, Vol.120, Springer Verlag.

[23] Rall, L. B., (1990), "Differentiation Arithmetics", in Ullrich, C., "Computer Arithmetic and Self Validating Numerical Methods", Academic Press, pp.73-90.

[24] Soulié, E., (1992), "Users Experience with Fortran Compilers for Least Squares Problems", in Griewank, A. & Corliss, G. F. "Automatic Differentiation of Algorithms", SIAM, pp.297-306.

[25] Tesfatsion, L., (1992), "Automatic Evaluation of Higher Order Partial Derivatives for Nonlocal Sensitivity Analysis", in Griewank, A. & Corliss, G. F., "Automatic Differentiation of Algorithms", SIAM, pp.157-166.

[26] Wengert, R. E., (1964), "A Simple Automatic Derivative Evaluation Program", Comm. A.C.M., Vol.7, pp.463-464.

NEURAL NETWORKS AND UNCONSTRAINED OPTIMIZATION

L. C. W. Dixon
Emeritus Professor
School of Information Sciences
University of Hertfordshire
College Lane
Hatfield
Hertfordshire   AL10 9AB
U.K.

ABSTRACT

When performing the unconstrained optimisation of a complicated industrial problem, the main computational time is usually spent in the calculation of the objective function and its derivatives. The calculation of the objective function is usually performed sequentially, so if a parallel processing machine is to be used, then either a number of function evaluations must be calculated in parallel or the sequential calculation of the objective function replaced by a parallel calculation.

The first approach has led to many codes, some of which are efficient. The second approach is very efficient when the objective function has the appropriate, partially separable structure. In this paper a third possibility is introduced where a neural network is used to represent the objective function calculation. A neural network is essentially a parallel processing computer. It is well known that a neural network with a single hidden layer can reproduce accurately any continuous function on a compact set.

The task of calculating the sequential objective function during the optimization is then replaced by a learning task of assigning the correct parameters to the neural network. This learning task is itself a least squares, unconstrained optimization problem. In the neural network field this is often solved by back propagation. For the neural network structure it is possible to apply reverse automatic differentiation effectively to obtain the gradient and Hessian of the learning function and results then show that the truncated Newton method is literally thousands of times faster than back propagation.

This study compares the performance of radial basis function networks with the usual sigmoid networks, but concludes by showing that nets based on the "sinc" function are even better for approximating continuous functions. While a previously published study was restricted to data points on regular grids, this will be extended in this paper to more random distributions.

E. Spedicato (ed.), Algorithms for Continuous Optimization, 513–530.
© 1994 *Kluwer Academic Publishers.*

1. Introduction

In this paper we are considering the optimisation of a class of industrial problems. Industrial problems are usually constrained but for the purpose of this paper the constraints will be ignored or assumed to be combined into a penalty or barrier function. We will therefore be concerned with the unconstrained optimisation of a single function F. Usually in industrial problems this function will depend on two types of variables, the optimisation variables $\underline{x}$, $\underline{x} \in R^n$, and descriptive variables $\underline{d}$, $\underline{d} \in R^m$. When both $\underline{x}$ and $\underline{d}$ are given, the evaluation of $F(\underline{x}, \underline{d})$ will frequently be calculated by a lengthy sequential computation involving M operations. When $\underline{d}$ is fixed, it is required to find the optimal values of $\underline{x}$, denoted by $\underline{x}^*$. Often this optimisation will need to be performed for many different values of $\underline{d}$.

For most optimisation algorithms each iteration involves at least one function and gradient calculation requiring at least cM operations, where $c = n+1$ if traditional methods are used and $c = 3$ if hand crafted reverse differentiation is used. Each iteration also involves an overhead which depending on the optimisation algorithm is proportional to n, n^2 or n^3. For very large values of M the function/gradient calculation dominates.

This is clearly the case when optimising the fuel used on a satellite mission (Bartholomew-Biggs et al (1987)). The calculation of the fuel used involves the solution of a set of simultaneous dynamic equations. The optimisation variables, originally the thrust magnitudes and directions, can be transformed via the adjoint control transformation into six variables, which are augmented by the times at which the engine is turned on and off. Typically there are about 14 optimisation variables. The descriptive variables $\underline{d}$ will include the desired destination, the initial position, the initial mass, the engine efficiency and maximum thrust etc. Some of these can occasionally be treated as optimisation variables, some obviously can not.

The availability of parallel processing machines suggested that larger industrial problems should be solvable conveniently using such devices. As the M operations in the calculation of the objective function dominates, most benefit can be expected if the parallel processing is applied to this calculation. Though the additional benefits of performing the overhead calculations in parallel must not be ignored.

If the objective function calculation $F(x)$ is partially separable in the sense that it naturally divides into P parallel operations, so that on P parallel processors the number of operations on each is M/P, then we have an ideal situation for parallel optimisation. When in addition each parallel calculation only depends on a small number of the n variables, then the overheads are also easy to perform in parallel.

As early as 1983, Dixon, Ducksbury and Patel were applying this principle to solve large problems on the ICL/DAP. This was an SIMD parallel processor with 4096 processors. We solved a series of problems that were formed by converting the solution of sets of partial differential equations into equivalent optimisation problems. The conversion involved the use of finite elements and the objective function became an integral which could be calculated on an SIMD machine by treating each element in parallel. In this way heat conduction problems in 16384 unknowns could be solved faster on the DAP, than problems in 2304 variables on our then sequential machine a DEC1091. The concept of speed up was meaningless as the larger problem could not have been stored on the sequential machine. Results on the nonlinear Burgher equation demonstrated that when the number of processors P = no. of finite elements = (gridsize)^2 then the sequential function/gradient calculation is proportional to the number of finite elements, so that the parallel CPU time remains constant and the number of iterations is approximately linear in the gridsize. The combined result therefore gave

$$\text{sequential CPU } \alpha \text{ (grid size)}^3$$
$$\text{parallel CPU } \alpha \text{ (grid size)}.$$

Two snapshot results; with 4111 unknowns the sequential conjugate gradient code took over one hour the parallel code 34 seconds; whilst with 12036 unknowns the parallel conjugate gradient code required 50.76 seconds and a parallel truncated Newton code 13.21 seconds. These results led us to conclude that when an objective function possessed a suitable structure then the parallel truncated Newton method was very effective. In 1989 these results were confirmed for the Sequent architecture by Dixon & Mohseninia.

For the above approach to be used the function calculation does have to divide naturally into distinct parallel streams. Many calculations do not. The other extreme approach is to calculate the function value at P separate points x in parallel and to redesign the code to utilise this amount of information. Dennis & Torcson (1991) produced a "direct" code for this purpose that uses function evaluations only, whilst Sofer & Nash (1991) produced a modified truncated Newton code based on this principle.

In between these two extreme groups of functions are those for which the concept of a computational graph was originally designed. In a computational graph each operation is fed by its inputs and produces its outputs, so that the inherent parallelism is obvious when viewed. Sometimes however this can be frustrating as the implication is simply that the available parallelism cannot be utilised on the parallel processor available for your use.

In this paper I am advocating an alternative namely that we replace the calculation of $F(\underline{x},\underline{d})$ by a neural network. A neural network is a highly parallel device. Hecht-Nielsen, Cybenko and others demonstrated in 1988 that a neural network with a single hidden layer can reproduce any continuous function to arbitrary accuracy on a compact set, and so we will only be concerned with neural nets having a single hidden layer. This layer will contain N nodes, each with a bias parameter b_i and connected to the input variables $\underline{t}^T = (\underline{x}^T, \underline{d}^T)$ by a matrix of weights W_{ij}. These then pass through an activation function $a(\ )$ and the output is combined using a second set of weights c_i, i.e. given the input $\underline{t}$ the network produces

$$O(\underline{t}) = c_O + \sum_{i=1}^{N} c_i a\left(\sum_j W_{ij}t_j + b_i \right)$$

Essentially the theory states that for any accuracy e there is a value of N for which network parameters $\underline{Z}^T = (c_O, c_i, W_{ij}, b_i)$ exist so that

$$|O(t) - F(t)| < e \qquad \text{all } t \in S.$$

See Cybenko (1988).

The theory was derived for sigmoid activation functions but applies to a wider class. Unfortunately there is no indication how N varies with e.

If we knew the values of N and $\underline{z}$ then by optimising $O(t)$ with respect to x to obtain x_O we would have a point where

$$|F(x_O) - F(x^*)| < 2e$$

which would be acceptable for most practical purposes.

We have therefore replaced a lengthy sequential calculation of $F(x)$ by the highly parallel calculation of $O(x)$. Be warned however that if $O(x)$ is in fact calculated sequentially we may not benefit at all.

The gradient
$$\frac{\partial O}{\partial t_j} = \sum_{i=1}^{N} c_i W_{ij} a'\left(\sum_k W_{ik}t_k + b_i \right)$$

and the Hessian
$$\frac{\partial^2 O}{\partial t_j \partial t_k} = \sum_{i=1}^{N} c_i W_{ij} W_{ik} a''\left(\sum_l W_{il}t_l + b_i \right)$$

are both easy to calculate in parallel on the network, though this does require additional network storage so we can use our favourite optimisation code to optimise $O(t)$.

This leaves us with a major task namely the determination of an appropriate network i.e. values of N, $\underline{z}$: this is known as the neural network learning problem.

<u>Conclusion</u>

(i) There may be considerable advantages in replacing a lengthy sequential calculation of an objective function by a neural network; especially when many different optimisation calculations of related systems have to be undertaken.

(2) The optimisation of the objective function calculated via a neural network can be undertaken by standard optimisation algorithms; analytic formulae for the gradient and Hessian of the function on the network are available.

2. The Neural Network Learning Problem

2.1 The Data Requirement (Dixon & Mills (1992))

Given the theoretical result that a function $O(t)$ exists such that

$$|O(t) - F(t)| < e \qquad \text{all } t \in s$$

the next question is how to determine its parameters. By the nature of $F(t)$ we are only going to be able to obtain its values at a finite number of points t^k, $k = 1 \dots K$ and the method will only be useful if K is less than the total number of function evaluations needed to complete all of the likely performance optimisation calculations.

Given the K values of $F(t^{(k)})$ there are of course an infinite number of functions $G(t)$ that satisfy

$$G(t^{(k)}) = F(t^{(k)}) \qquad k = 1 \dots K$$

and this information alone does not enable us to bound

$$|G(t) - F(t)|$$

at all, at any other point. To do so we need to make additional assumptions about the functions $F(t)$ and $G(t)$.

If we make the simple assumption that both functional derivatives have a Lipshitz bound L

i.e.

$$\left.\begin{array}{c} \|\nabla F(t)\| < L \\[2ex] \|\nabla G(t)\| < L \end{array}\right\} \qquad \text{all } t \in S$$

and the maximum distance of any point $t \in S$ from a member of $t^{(k)}$ is d then we may easily obtain the bound

$$|G(t) - F(t)| < 2Ld.$$

This is therefore the accuracy to which the data determine $F(t)$ and it is therefore necessary that K is large enough to ensure

$$d < \frac{e}{2L}$$

and this is placing all the uncertainty in $F(t)$ and none in $O(t)$ which is unreasonable. The value of K implied by this Lipshitz assumption is usually too large to be practical.

In contrast if we assume K is the appropriate number to enable a complete polynomial of order V to be determined by the K function values and now let L be a bound on the $(V+1)th$ derivative then the data imply

$$|G(t) - F(t)| < \frac{2L}{(V+1)!} d^{V+1}$$

and we only require

$$d^{V+1} < \frac{(V+1)!e}{2L}$$

which is much more reasonable.

If in addition the function $O(t)$ only approximates $F(t)$ at $t^{(k)}$, i.e.

$$\left| O(t^{(k)}) - F(t^{(k)}) \right| < \delta \quad k = 1 \dots K$$

then the Lipschitz assumption leads to the bound

$$|O(t) - F(t)| < \delta + 2Ld$$

and $\qquad\qquad d < (e - \delta)/2L.$

A similar calculation should be undertaken for the complete polynomial assumption. This is left to the reader.

Conclusion

Attempting to fit networks to functions with insufficient data pairs must lead to inaccurate networks.

2.2 The Learning Process (Sigmoid Functions)

The usual method for determining the parameters z is to express the difference between $O(t^{(k)})$ and $F(t^{(k)})$ as a sum of squares

$$E(z) = \sum_{k=1}^{K} \left[O(t^{(k)}, z) - F(t^{(k)}) \right]^2 = \sum_{k=1}^{K} s_k^2$$

and to minimise this as a function of the network parameters z. In the neural network fraternity the back propagation algorithm is popular. This appears in two forms, in the first a fixed step is taken along the direction ∇s_k^2 after each data value is analysed, in the second a fixed step is taken in the direction ∇E at the end of the cycle. Neither of these suggestions would appeal to any member of the optimisation fraternity and it is not surprising that the first method frequently fails and the second takes thousands of iterations to converge.

It is however only fair to note that in back propagation the derivatives have always been calculated by what is now termed reverse automatic differentiation well before this phrase was invented. R.A.D. is very simple on a neural network as the network is its own computational graph. Essentially we pass once forward through the net to calculate s_k, and once backwards to obtain ∇s_k. For any direction p a simple extension gives $\nabla s_k^T p$ and we may traverse the net again to obtain $\nabla^2 s_k p$. More details are given in Christianson (1991) and Dixon (1991). The complete cycle enables ∇E and $\nabla^2 E p$ to be obtained so either the conjugate gradient algorithm or the structured Truncated Newton method can be used cheaply.

Mills (1992) reports the relative costs of calculating these on a simulated neural network to be

N	n	E	$(E, \nabla E)$	$(E, \nabla E, \nabla^2 E p)$
20	80	25.08	35.21	55.64
40	160	44.09	70.19	111.13
60	240	63.13	105.00	166.49

thus emphasing once more the relative efficiency of reverse automatic differentiation in large dimensions.

He also reports the relative performance of four algorithms on four simple problems of matching one variable functions $F(z)$.

$F(z)$	N	Truncated Gauss Newton		Truncated Newton		Variable Metric		Back Propagation	
		cpu	E	cpu	E	cpu	E	cpu	E
z^2	3	9.58	10-6	15.11	10-6	48.42	10-6	211	10-3
z^3	3	10.41	10-6	20.7	10-6	110.6	10-6	211	10-3
$\sin(z)$	8	87.96	10-7	378.4	10-6	2372	10-7	1024	4.7
$\text{sinc}(10z)$	8	131.4	10-6	376.5	10-6	1297	10-6	1024	1.1

It will be noted that the three optimisation codes solved all the problems with reasonable accuracy, but that back propagation fails to solve any. The figures quoted are for the average of ten runs, and the algorithms were enclosed in a multistart global optimisation package. E denotes the average final value of $E(z)$: cpu, the average cpu time in seconds on a transputer.

Although the truncated Newton algorithms performance was satisfactory in that it solved all problems relatively quickly without the storage requirement of the Gauss Newton or Variable Metric methods, we were disappointed by its behaviour near the solution. Under most conditions a Newton method converges superlinearly, but on our network problems the convergence rate is sublinear. Analysis confirmed that the matrix $\nabla^2 E$ is singular at x^* for nearly every network. A similar observation had been made in Saarinen, Bromley & Cybenko (1991).

It is easy to see why this is likely to be true. Suppose the network necessary to represent $F(t)$ to accuracy e contains N^* nodes, and the network being optimised has $N > N^*$ nodes, then three things can happen. First the additional c_i may theoretically be set to zero and then the associated W_{ij} and b_i have no effect on E so $\nabla^2 E$ is singular. Second an optimal c_i may be split into two or more nodes summing to the correct value, again as long as these c_i sum to the correct value E is unaltered so $\nabla^2 E$ is singular. Third the additional c_i may be very small so that they contribute much less than e to E. Such terms will lead to eigenvalues of $\nabla^2 E$ that are effectively zero.

If we wish to obtain a unique neural network to represent a given function it is essential to ensure that N is not set too big, as otherwise optimisation should and does produce different singular optimal networks from different starting points. Though this would not matter when we used $O(x)$ to approximate $F(x)$ in a subsequent study, it is disconcerting to say the least.

The observant reader will have noticed that a constant term c_O has been included in the expression for $O(t)$. This has been included as otherwise sigmoid functions cannot fit a constant without leading to a singular Hessian.

Given that this constant term is included we may note that the standard activation function

$$a(y) = 1 / (1 + \exp(-y))$$

is equivalent to

$$\tfrac{1}{2} + \tfrac{1}{2}\tanh(y / 2)$$

Merging the $\tfrac{1}{2}$ into c_O we note that tanh is an odd function. So every global minimum with c_i, W_{ij}, b_i is repeated with $-c_i, -W_{ij}, -b_i$ (using tanh). Hecht-Nielsen (1992) terms this "sign-flip" and recommends including the constraints $b_i \geq 0$ to avoid this. ($c_i > 0$ would be an appropriate alternative with tanh).

Similarly if we interchange any pair i_1, i_2 the output is unaffected, so we should introduce constraints that prevent such interchanges.

Again either $b_i \geq b_{i+1}$ if $b_i \geq 0$ had been chosen to prevent sign flip

or $c_i \geq c_{i+1}$ if $c_i \geq 0$ had been chosen

would be appropriate.

These are simple mechanisms to rule out most interchanges. There will often, of course, be cases where two values of c_i have the same value. As Hecht-Nielsen pointed out these constraints cut down the search region considerably, they do not however explain the zero eigenvalues as none of these changes are local.

This considerable reduction in the search space may well tie in with an intriguing result of Barron (1992) who has shown that in one sense the number of nodes needed to obtain a given accuracy on "similar" functions is independent of the dimension of the problem. This result obviously depends on the details of the definition of the word "similar".

The fact that networks that are too large lead to singular and non unique networks, has led many people to attempt to evolve networks. This is a very difficult problem with sigmoid or tanh nets.

Let us assume an evolution structure similar to that given by Barron (1992)

$$O^{(V+1)}(t) = \alpha O^{(V)}(t) + (1-\alpha)c_{V+1}a\left(\sum_j W_{V+1,j}t_j + b_{V+1}\right)$$

However if we start with $O^{(O)}(t) = c_O$ and obtain the average value of F and then introduce one tanh function. If we also assume t is scalar, and if $F(t)$ is any even function and the data is evenly distributed, then the best fit to the odd tanh function is $\alpha = 1$, implying no change. Hence the evolution would fail. It is necessary therefore to introduce at least two nodes at each iteration to ensure we can cope with both odd and even functions. Assume now we are in two dimensions, and the function is radially symmetric about a point where $F = c_O$ and the data is radially symmetric; now every direction W is equivalent and along any direction the data is even. Again $\alpha = 1$ and no change would result.

Conclusions

(1) Using Truncated Newton is far better than using back propagation to solve the learning problem.
(2) Overlarge networks are theoretically singular.
(3) All networks have many global minima unless restrictive constraints are included.
(4) Evolving sigmoid or tanh networks is inherently difficult.

2.3 Radial Basis Functions

As an alternative to sigmoid functions considerable interest has been expressed in the properties of radial basis functions.

Six alternatives, at least, have been investigated in the literature

(1)	Cubic	$a(v) = v^3$
(2)	Thin plate	$a(v) = v^2 \ln(v)$
(3)	Gaussian	$a(v) = \exp(-v^2/\sigma^2)$
(4)	Multi quadric	$a(v) = (y^2 + b^2)^{\frac{1}{2}}$
(5)	Inverse Multiquadric	$a(v) = (y^2 + b^2)^{-\frac{1}{2}}$
(6)	Linear	$a(v) = v$

In these radial basis functions v is usually the distance from a centre $\bar{c}_i$ and constant and linear terms are often included so $O(t)$ becomes

$$O(t) = \alpha + \beta^T t + \sum_{i=1}^{N} \mu_i a\left(\|t - c_i\|_2\right)$$

The centres c_i are usually taken as given, indeed $c_i = t^{(i)}, N = K$ is quite common, the extra degrees of freedom being taken up by constraints

$$\sum \mu_i = \sum \mu_i c_{i,k} = 0 \qquad k = 1 \dots n$$

The learning variables $z = (\alpha, \beta, \mu)$ now appear linearly in $O(t)$ so the nonlinear learning problem is replaced by a linear problem. To obtain the appropriate radial basis function network it is necessary to solve a set of linear equations

$$Az = b.$$

The analysis of the nonsingularity of A and of the distribution of positive and negative eigenvalues is due to Micchelli (1986).

If we wish to avoid the problem of storing A and solve the set of equations by a conjugate gradient like iterative scheme, which is implied by the network concept, then it is simpler if A is symmetric and positive definite. This is true for the Gaussian where

$$A_{ij} = \exp\left(-\left\|c_i - c_j\right\|_2^2 / \sigma^2\right)$$

Experiments reported in Dixon & Mills (1993) confirmed reports in the literature that as the number of data points is increased in a given range, then the condition number of A increases and the magnitude of $z, \|z\|_\infty$, increases. As the magnitude of z increases then the network is obtaining $O(t)$ from small differences of large numbers, which is known to be undesirable. The increase in condition number implies that these large numbers will themselves be less accurate.

For fairly simple functions e.g. $F(t) = t^2$ on the interval $-1 \le t \le 1$ with $K = 80$ condition numbers of 10^6 were usual, 10^{14} occurred and values of $\|z\|_\infty$ of 10 were common and 10^5 occurred. As the important question is not whether we can calculate a network but whether it is robust and stable we were not happy.

To overcome the problems Gustafson, Little & Loomis (1992) noted that with a Gaussian net the condition number can be controlled by the choice of σ, the parameter in the Gaussian formula.

In our approach Dixon & Mills (1993) we choose a subset of the data, solve the system with the value of σ that makes it well conditioned then take a finer set of data and form the difference between this data and the previous approximation. A new value of σ is then chosen to make the larger set of equations well conditioned and a new approximation determined. This is continued until as new data is introduced all the differences become less than e.

Because all the equations solved are well conditioned (the maximum condition number was 12), all the parameters are theoretically bounded, the maximum

$\|z\|_\infty$ was 3. We also noted that as the evolution continued the values of $\|z\|_\infty$ steadily decrease.

A sample evolution would be that for sinc($10z$)

Iteration No.	No. of Nodes	Condition No.	σ	$\|z\|_\infty$	$\|F - O\|_\infty$
1	3	5	1.2	1.98	.95
2	5	7	0.56	1.82	.46
3	9	9	0.28	0.64	.013
4	17	11	0.14	0.016	.0002
5	33	11	0.07	0.0002	5×10^{-5}
6	65	11.5	0.035	8×10^{-5}	3×10^{-5}

In the above table $\|F - O\|_\infty$ is calculated over a much finer data set (consisting of 10,000 values).

Conclusion

(1) Radial basis function networks become ill conditioned and numerically unstable as close data points are included.
(2) It is possible to evolve Gaussian Radial Basis function nets that are numerically stable and represent the function well even when the data sets contain close pairs of data.

2.4 Sinc Function Nets

2.4.1 Uniformly distributed data

In his thesis Mills (1992) proposed the use of the sinc function as an activation function for regularly spaced data with interval h:

$$\text{sinc}(y) = \sin(\pi y) / \pi y$$

and $$O(t) = \sum_{j=1}^{K} F(t^{(j)}) \prod_{i=1}^{n} \text{sinc}(t_i - t_i^{(j)}) / h.$$

The method does not require any optimisation or any linear equation solving and has been shown theoretically to converge in the sense that

$$|O(t) - F(t)| \to 0$$

as $K \to \infty$. If the grid is finite then $O(t) \to 0$ outside the range of the gridpoints.

The net works very well and is very fast compared with the methods discussed in sections (2.2) and (2.3), however the measure

$$\|O(t) - F(t)\|_\infty$$

decreased disappointingly slowly with K, being roughly proportional to $1/K$.

Again evolving a net gives far better results. If we start with a small value of K i.e. large h and bring in the midpoints at each iteration we obtain much better values of $\|O(t) - F(t)\|_\infty$

	Unevolved $K = 80$	Evolved $K = 65$
t^2	.1476	.0014
e^t	.3811	.0033
$\sin(2\pi t)$	.0149	.0061
$\text{sinc}(10t)$	.0096	.0005

The main cost in the evolution process lies in forming the data values for the next step of the evolution

$$F^{(k+1)}(t^{(j)}) = F^{(k)}(t^{(j)}) - O^{(k)}(t^{(j)}).$$

It would be wrong to give the impression that all the problems have been solved. The performance of all the methods is disappointing if the objective function takes on a wide set of values and has nonconvex contours. As an example of poor performance, the results obtained when fitting Rosenbrocks function in the range $-1 \le x_i \le +1$ are given below.

Method	K	cpu	$\|O - F\|_\infty$
Sigmoid (Truncated Newton)	289	1×10^5	2×10^5
Multi Quadric	169	6265	4.9
Evolved Gaussian	289	1879	10.8
Sinc	1089	0.511	101.8
Evolved Sinc	800	37.13	1.21

In the range considered Rosenbrocks function takes on values between $0 \leq F \leq 404$ and has the notorious banana shape contours. Essentially the sigmoid networks converge to a good local minimum, but it is not numerically stable and does not generalise well between the data points. The multiquadric network is even more numerically unstable ($\|z\|_\infty = 466$) but generalises better. The evolved Gaussian network learns quicker but does not generalise quite as well, whilst the evolved sinc network is very fast and generalises well.

While an error of 1.21 in 404 would be acceptable at the uninteresting part of the function, an error of 1.21 on 0 would obviously not be acceptable at the solution.

It would therefore seem appropriate to scale the function so that more accuracy is obtained near the solution. For this function this is readily achieved by learning the transformed function

$$F(x) = \ln(1 + Ros(x))$$

Conclusions

The evolved sinc function provides a very quick and very accurate approximation to a function given uniformly spaced data points.

2.4.2   Randomly distributed data

One of the limitations of the sinc function is the need for the data points to be uniformly spaced. It is possible to remove this restriction, as shown below, but to do so it is necessary to consider the basic theory of the method.

Reverting for simplicity to a function of one variable $f(t)$, if $f(t)$ is aperiodic with

$$\int_{-\infty}^{\infty}|f(t)|^2\,dt < \infty$$

it has a Fourier transform

$$F(\Omega) = \int_{-\infty}^{\infty} f(t)\exp(-ct\Omega)dt \tag{1}$$

from which $f(t)$ can be recovered by

$$f(t) = \int_{-\infty}^{\infty} F(\Omega)\exp(it\Omega)d\Omega \tag{2}$$

Essentially with uniform discrete data $f(jh)$, $F(\Omega)$ is approximated by

$$F(\Omega) = \sum_j f(jh)\exp(-it\Omega) \tag{3}$$

this approximation is fed back into (2) and analysis shows that it becomes

$$f(t) = \sum_j f(jh)\text{sinc}((t-jh)/h) \tag{4}$$

If the data is not uniformly distributed, (3) is not an appropriate approximation to (1). A more general formula would be needed which introduced weights on the function data.

$$F(\Omega) = \sum_j w_j f(t^{(j)})\exp(-it\Omega) \tag{5}$$

so that (5) is a good approximation to (1).

The forward analysis of this approach to an equivalent statement to (4) has not been completed. However if instead we choose an h and approximate $f(jh)$ in terms of the data

$$f(jh) = \sum_k W_{jk} f(t^{(k)}) \tag{6}$$

using a local approximation of any order, then the analysis leading to (4) still holds and we would obtain

$$f(t) = \sum_j \sum_k W_{jk} f(t^{(k)})\text{sinc}((t-jh)/h) \tag{7}$$

which will be as accurate as the approximation used in (6).

An alternative generalisation has been tried experimentally by Vespucci (1993) who used

$$f(t) = \sum_k \mu_k \text{sinc}(t - t^{(k)}) / \overline{h} \tag{8}$$

this leads to the solution of a set of linear equations

$$f(t^j) = \sum_k \mu_k \text{sinc}(t^{(j)} - t^{(k)}) / \overline{h} \tag{9}$$

for μ_k and her results are similar to those obtained with radial basis functions.

It would be interesting to see how (7) would perform.

Conclusion

This area needs further research.

3. Conclusions

1.1 There may be considerable advantages in replacing a lengthy sequential calculation of an objective function by a neural network, especially when many different optimisation calculations of related systems have to be undertaken.

1.2 The optimisation of the objective function calculated via a neural network can be undertaken by standard optimisation algorithms.

2.1 Attempting to fit neural networks to a function by using insufficient data pairs must lead to inaccurate networks.

2.2.1 Using truncated Newton is far better than using backpropagation to solve the learning problem of sigmoid networks.

2.2.2 Overlarge networks are theoretically singular.

2.2.3 All networks have many global minima unless restrictive constraints are included.

2.2.4 Evolving sigmoid or tanh networks is inherently difficult.

2.3.1 Radial basis function networks become ill conditioned and numerically unstable as close data points are included.

2.3.2 It is possible to evolve Gaussian Radial Basis function nets that are numerically stable and represent the function well over close data sets.

2.4.1 The evolved sinc function network provides a very quick and very accurate approximation to a function given uniformly spaced data points.

2.4.2 Further research is required to determine the equivalent to sinc function nets for randomly spaced data.

Comment

All the results in this paper assume F is a continuous differentiable function; binary $(0,1)$ functions are therefore excluded from these conclusions.

4. References

[1] Barron, A. R. (1992), "Universal approximation bounds for superpositions of a sigmoidal function", Technical Report of University of Illinois, Urbana-Champagne.

[2] Bartholomew-Biggs, M. C., Dixon, L. C. W., Hersom, S. E., Maany, Z. A., Flury, W. and Hechler, M., (1987), "From High Thrust to Low Thrust: Applications of Advanced Optimisation to Mission Analysis", ESA Journal, 1987, pp.1-37.

[3] Christianson, B., (1992), "Automatic Hessians by Reverse Accumulation", IMAJNA Vol.12, pp.135-150.

[4] Cybenko, G., (1988), "Approximation by Superposition of a Sigmoidal Function", in "Mathematics of Control Signals & Systems", Springer-Verlag, pp.303-314.

[5] Dennis, J. E. & Torcson, V., (1991), "Direct Search Methods on Parallel Machines", SIAM Journal of Optimisation, Vol.1, pp.448-74.

[6] Dixon, L. C. W., (1991), "Use of Automatic Differentiation for Calculating Hessians & Newton Steps", in Griewank, A. & Corliss, G. (ed.), "Automatic Differentiation of Algorithms", SIAM, pp.114-125.

[7] Dixon, L. C. W., Ducksbury, P. & Patel, K., (1983), "Experience running Optimisation Algorithms on Parallel Processing Systems", Plenary Session at 11th International Federation of Information Societies Conference on System Modelling and Optimisation.

[8] Dixon, L. C. W. & Mills, D. J., (1992), "Neural Networks and Nonlinear Optimisation 1", Optimisation Methods and Software, Vol.1, pp.141-151.

[9] Dixon, L. C. W. & Mills, D. J., (1992), "Neural Networks for Massively Parallel Optimisation", in "Science of Artificial Neural Networks", SPIE, Vol.1710, pp.229-238.

[10] Dixon, L. C. W. and Mills, D. J., (1992), "A Method for Evolving Bell Shaped Radial Basis Functions that is Numerically Stable and Well Conditioned", TR.268, NOC Technical Report, University of Hertfordshire, UK.

530

[11] Dixon, L. C. W., Mohseninia, M., (1989), "Concurrent Optimisation on the Sequent Balance 8000", TR.232, NOC Technical Report, Hatfield Polytechnic, UK.

[12] Gustafson, S. C., Little, G. R., & Loomis, J. S., (1992), "Generalisation with Guaranteed Bounds on Computational Effort", Report of the University of Drayton, Drayton, Ohio.

[13] Hecht-Nielsen, R., (1988), "Theory of the Back Propagation Neural Network", in Proceedings of International Joint Conference on Neural Networks, pp.593-608.

[14] Hecht-Nielsen, R., (1992), "The Munificence of High Dimensionality", in Alexander, I. and Taylor, J., "Artificial Neural Networks 2" pp.1017-1030.

[15] Micchelli, C. A., (1986), "Interpolation of Scattered Data in Constructive Approximation, Vol.2, pp.11-22.

[16] Mills, D. J., (1992), "The Optimisation of Neural Networks for Approximation", PhD Thesis, University of Hertfordshire.

[17] Saarinen, S., Bramley, R., & Cybenko, G., (1991), "Neural Networks, Back-propagation and Automatic Differentiation", in Griewank, A. & Corliss, G., "Automatic Differentiation of Algorithms", SIAM, pp.31-42.

[18] Sofer, A. & Nash, S. G., (1991), "A General Purpose Parallel Algorithm for Unconstrained Optimisation", SIAM Journal of Optimisation, Vol.4, pp.530-547.

[19] Vespucci, M., (1993), "Multivariate Scattered Data Interpolation via Sinc Functions", University of Bergamo, Italy, preprint.

PARALLEL NONLINEAR OPTIMIZATION: LIMITATIONS, CHALLENGES, AND OPPORTUNITIES

ROBERT B. SCHNABEL
Department of Computer Science
University of Colorado
Boulder, Colorado 80309-0430, U.S.A.

ABSTRACT. The availability and power of parallel computers is having a significant impact on how large-scale problems are solved in all areas of numerical computation, and is likely to have an even larger impact in the future. This paper attempts to give some indication of how the consideration of parallel computation is affecting, and is likely to affect, the field of nonlinear optimization. It does not attempt to survey the research that has been done in parallel nonlinear optimization. Rather it presents a set of examples, mainly from our own research, that is intended to illustrate many of the limitations, opportunities, and challenges inherent in incorporating parallelism into the field of nonlinear optimization. These examples include parallel methods for small to medium size unconstrained optimization problems, parallel methods for large block bordered systems of nonlinear equations, and parallel methods for both small and large-scale global optimization problems. Our overall conclusions are mixed. For generic, small to medium size problems, the consideration of parallelism does not appear to be leading to major algorithmic innovations. For many classes of large-scale problems, however, the consideration of parallelism appears to be creating opportunities for the development of interesting new methods that may be advantageous for parallel and possibly even sequential computation. In addition, a number of large-scale parallel optimization algorithms exhibit irregular, coarse-grain structure, which leads to interesting computer science challenges in their implementation.

1. Introduction

Parallel computation is having a significant impact upon how large scale scientific computation is performed. To solve the large scientific problems of interest to scientists and engineers today, very powerful computers are necessary, and it appears that many if not all of the most powerful scientific computers of the future (as well as at present) will be parallel computers. To scientific computation researchers, one of the most interesting aspects of this transition from sequential or vector computers to parallel computers is that new algorithm development may be required to use parallel machines efficiently.

In this paper, we attempt to give some indication of how the move to parallel computers is affecting, and is likely to affect, the field of nonlinear optimization. Our focus is on whether, and where, the use of parallel computers is leading to interesting new algorithmic approaches for nonlinear optimization. To address this issue, we try to point out some of the key limitations, opportunities, and challenges in developing parallel

E. Spedicato (ed.), Algorithms for Continuous Optimization, 531–559.
© 1994 *Kluwer Academic Publishers.*

algorithms for nonlinear optimization problems. Our overall conclusion is mixed: in some portions for nonlinear optimization, primarily generic methods for small to medium scale problems, the consideration of parallelism does not appear to be leading to major algorithmic changes or challenges, while in other areas, primarily classes of large-scale optimization problems, there appear to be very interesting opportunities and challenges that arise from the consideration of parallelism.

On the parallel architecture side, our focus in this paper is on moderately to massively parallel MIMD computers. This broad class of machines appears to be the main architectural direction that the parallel computation field is pursuing for general purpose, high performance computation. Section 2 gives a brief overview of parallel computation that should be sufficient for the purposes of this paper. We briefly summarize current trends in parallel computer architectures, and the relationship between parallel computers and the needs of nonlinear optimization methods. This section is by no means intended to serve as a tutorial in parallel computation; some references that fulfill this role include [1,18,29].

In nonlinear optimization, our focus in this paper is on continuous, nonlinear problems. These include solving unconstrained and constrained optimization problems, and systems of nonlinear equations. (We assume that the reader of this paper is familiar with the basics of nonlinear optimization as found in [11,14,17].) It is not, however, our intention in this paper to present a survey of parallel algorithm research in these areas. Rather we will present a small number of examples, primarily from our own research, that illustrate the limitations, opportunities, and challenges in parallel nonlinear optimization. Section 3 discusses parallel methods for unconstrained optimization problems. It primarily illustrates the limitations in developing novel generic parallel optimization algorithms for moderate size problems. Section 4 discusses parallel methods for a special class of large scale problems, block-bordered systems of nonlinear equations. This problem class illustrates the possibility for the consideration of parallelism to lead to the development of new methods that are advantageous on sequential as well as parallel computers. Then in Section 5 we describe parallel methods for large global optimization problems. Section 5 illustrates that large-scale parallel optimization algorithms can be particularly interesting and challenging because they are often irregular, coarse-grain algorithms that are challenging both to develop and to implement. Finally, Section 6 briefly summarizes our views of the limitations, opportunities, and challenges in parallel nonlinear optimization.

This paper is based upon lectures given at the NATO Advanced Research Institute on Algorithms for Continuous Optimization at Il Ciocco, Italy in September 1993. I thank the organizers and participants of the institute for the stimulating environment that it provided. The parallel optimization research described in Sections 3-5 that was performed at the University of Colorado is all joint work; the participants in various projects include Richard Byrd, Tom Derby, Cees Dert, Betty Eskow, Humaid Khalfan, Bart Oldenkamp, Jerry Shultz, Andre van der Hoek, Alexander Rinnooy Kan, Chung-Shang Shao, Sharon Smith, and Xiaodong Zhang.

2. Parallel Computation Background, with Relations to Parallel Optimization

The primary motivation for parallel computation is the continuing need for faster and larger computers, in terms of computational speed, memory and input/output capacity. While computer speeds and memory sizes have continued to increase dramatically, many important scientific problems still require orders of magnitude greater speed and/or memory than is currently available. For decades, these increases in speed and memory were attained by building ever faster and larger sequential electronic computers. However, the evolution of these computers has now reached the stage where further improvements are constrained by fundamental physical limits (e.g. the speed of light, the sizes of atoms) and the limitations these place upon basic machine characteristics such as minimal feature sizes, minimal distances between components, and maximal rates of heat dissipation. Also, very powerful processor/memory chips recently have become widely and cheaply available. These two trends, the limits of sequential computation speed and the availability of commodity chips, along with the continuing need for vastly increased computing power, are largely responsible for the greatly increased production of parallel computers in the last five years.

The challenges in developing and effectively utilizing parallel computers come from many sides: hardware, systems and language software, and applications algorithms and software. The discussion of parallel applications algorithms for optimization is the main focus of this paper, in Sections 3-5, and a few comments on parallel systems and language support are included in these sections. This section briefly discusses parallel architectures, concentrating on those that appear to be of greatest interest to nonlinear optimization. To motivate this, we first briefly consider the motivation and needs for parallel optimization.

2.1 POSSIBILITIES FOR PARALLELISM IN OPTIMIZATION

Parallelism is of interest in optimization because many optimization problems are expensive to solve. To determine what sort of parallel computers may be of interest, it is useful to examine where the expense in nonlinear optimization algorithms comes from. There are at least four different possible sources:
1. The nonlinear functions, constraints, and/or derivatives may be expensive to evaluate.
2. The number of variables or constraints, and hence the cost of each iteration aside from function and derivative evaluation, may be large.
3. Many evaluations of the objective function, constraints, or derivatives may be required.
4. Many iterations may be required.

These in turn lead to at least three levels at which one may consider introducing parallelism into an optimization algorithm:
1. Parallelize each evaluation of the objective function, constraints, and/or their derivatives.
2. Parallelize the linear algebra involved in each iteration.

3. Parallelize the optimization process at a high level, either to perform multiple function, constraint, and/or derivative evaluations on multiple processors concurrently, and/or to reduce the total number of iterations required.

All of these are important ways to create parallel optimization methods, and any may be the best way to utilize parallel computation for a given problem. In this paper, however, we are mainly interested in the third possibility, parallelizing the optimization process, since this is the domain of optimization researchers. The second possibility, parallelizing the linear algebra, may be the concern of optimization researchers if the linear algebra is particular to optimization algorithms. The first possibility is likely to be outside the domain of optimization research; for example if the objective function evaluation involves the solution of a system of differential equations, this option would entail creating or using a parallel differential equations solver.

As will be seen in Sections 3-5, parallelizing the optimization process in a nonlinear optimization algorithm is likely to lead to a "coarse-grain" parallel algorithm. By this we mean an algorithm where each processor performs a significant amount of computation in between each point where it communicates or synchronizes with other processors. For example, if each processor performs at least one function evaluation between communication points, and these function evaluations are even moderately expensive, a coarse-grain parallel algorithm results. Such parallel algorithms are generally well suited to MIMD computers, and less well suited to SIMD computers or vector computers. This is one reason why our brief survey of parallel architectures in Section 2.2 concentrates on MIMD computers.

2.2 BRIEF SURVEY OF PARALLEL ARCHITECTURES

The main classes of parallel computers that are currently used for scientific computation are shared and distributed memory MIMD multiprocessors, and SIMD processor arrays. We briefly discuss the key characteristics of these types of computers in this section, and their main advantages and limitations. We omit discussion of some other architectures, including data-flow computers and systolic arrays, that are not currently in wide use for scientific computation.

An MIMD (Multiple Instruction Multiple Data) computer is a computer with multiple processors that can execute different instructions on different data at the same time. This means that such a computer can execute multiple, similar or dissimilar tasks concurrently. All MIMD computers include multiple processors (arithmetic and instruction processing units) and some method of communicating between them. In a shared memory multiprocessor, the processors all share access to a global memory, and communicate by reading and writing data residing in this memory. Generally, the shared memory is large, many million bytes. In addition, the processors generally each have much smaller (typically 10,000-100,000 byte) local memories. In a distributed memory multiprocessor, there is no shared memory. Instead, each processor has a large local memory (typically 1-100 million bytes), and the processors communicate by sending messages to each other over a network that connects them.

Shared memory multiprocessors have two fundamental advantages over distributed memory multiprocessors. First, communication via the shared memory is usually consid- erably faster than communication by messages in a distributed memory multiprocessor. This enables a wider range of parallel algorithms, in particular those with higher communication/computation ratios, to run efficiently. Secondly, shared memory machines generally are easier to program and utilize, since the programmer needs to give less attention to communication and data partitioning. The main limitation of shared memory multiprocessors is that they appear difficult to build with large numbers of pro- cessors. The difficulty is the cost and complexity of providing all the processors with uniform, fast access to the global shared memory. Currently, shared memory multipro- cessors are being built with two to about 32 processors by many vendors including Cray, but only distributed memory architectures are providing computers with hundreds or thousands of processors. (An in-between class, virtual shared memory multiprocessors, is addressed shortly.) One rather different and interesting alternative for providing a shared memory multiprocessor that may scale effectively is the pipelined instruction and memory paradigm that is embodied in the Tera computer currently under development; the description of this architecture is beyond the scope of this paper.

Among distributed memory multiprocessors, one can currently identify at least three subclasses: "pure" distributed memory multiprocessors, networks of computers, and virtual shared memory multiprocessors. In a pure distributed memory multiprocessor, each processor's memory has its own address space, and all communication is by mes- sage passing (at least at the hardware level). The main advantages and disadvantages of pure distributed memory multiprocessors are exactly the reverse of those for shared memory multiprocessors. The main advantage is that pure distributed memory multipro- cessors are relatively easy to construct and to scale to hundreds or thousands of proces- sors; machines of this size have been offered by several vendors including IBM, Intel, Meiko, NCUBE, and Thinking Machines. Typically each processor is powerful, with roughly the capability of a modern scientific workstation. The interconnection network is generally a hypercube or a two-dimensional grid; this will not be of importance in this paper. The first main disadvantage of pure distributed memory multiprocessors is that communication is generally quite slow, with a typical ratio of the time to communicate a number to the time to do one floating point operation being one thousand. This means that parallel algorithms must have fairly coarse granularity to run efficiently on these machines. A second key disadvantage is that these machines are rather difficult to pro- gram, due to the need to explicitly partition data structures among the processors and manage communication. Current research in parallel programming languages, such as the High Performance Fortran and Fortran D projects ([15,36]), is attempting to diminish or remove the latter disadvantage.

Networks of computers can function as distributed memory multiprocessors. All that is needed is some mechanism for communicating between multiple machines on the network, a facility that is provided by most modern operating systems. In comparison to pure distributed memory multiprocessors, using a networks of computers as a multipro- cessor has the advantages that it can utilize existing computational resources, and that the ratio of hardware cost to floating point speed is lower. The primary disadvantage is that communication is even slower than on a pure distributed memory multiprocessor. Generally this means that parallel algorithms must have very coarse granularity (thousands of computations between communication points) to be efficient in this environment.

Virtual shared memory multiprocessors are a recently introduced class of computers that may be considered to reside between shared and pure distributed memory multiprocessors. Physically, they are distributed memory machines: each processor has its own memory and there is no global shared memory. However, there is a single address space for all the memories, and any processor may directly access any memory location in any processor's memory. If the access is not to the local memory, the data is retrieved by the hardware. As expected, the advantages of this architecture lie between those of shared and pure distributed memory machines. Communication costs are expected to be in between the two extremes, and the programmer may not have to manage the placement of data, although there may be advantages to doing so. A key question is the scalability of this class of architectures, i.e. how many processors can be supported effectively. Kendall Square Research has produced such a machine with up to 128 processors; it remains to be seen whether efficient virtual shared memory machines can be built with considerably higher numbers of processors.

Finally, a SIMD (Single Instruction Multiple Data) computer is a computer with multiple processors that all can execute the same instruction on different data at the same time. This means that such a computer can execute a given program segment simultaneously (in lockstep) on multiple processors using multiple data sets, so long as the program segment contains no data dependent branches. One addition to this model that is generally supported is that some processors may do nothing ("mask") rather than execute a given instruction. Architecturally, such computers consist of a control processor and its memory, which generates the sequence of instructions that all processors follow; multiple processors, each with a local memory, that execute these instructions on their data; and a network that connects all the memories and allows communication between them. A number of such computers have been built by several vendors including MassPar and Thinking Machines. Their advantages are that they are relatively easy to build with large number of processors, and to program. Their primary disadvantage is that the SIMD programming model has limited applicability. For example, it would not usually be possible to evaluate a nonlinear function $f(x)$ at a different point x simultaneously on each processor of such a computer, because most nonlinear function evaluation codes include data dependent branches.

2.3 SUMMARY AND RELATION TO OPTIMIZATION

Among the parallel computers available today, the ones that appear suitable for general purpose optimization algorithms are the various types of MIMD machines: shared memory multiprocessors, virtual shared memory multiprocessors, distributed memory multiprocessors, and networks of computers used as multiprocessors. In considering these four classes of machines, it is important to be aware of several differences between them. First, the ratio of communication speed to (floating point) computational speed generally increases monotonically in the order that the classes are listed above. Second, the ease of producing multiprocessors of these types arguably also increases monotonically in the order they are listed. Third, the ease of programming these machines arguably decreases monotonically in the order they are listed above.

Because many parallel optimization algorithms have very coarse granularity, they are often well suited to any type of MIMD computer. On the other hand, the SIMD architecture is not general enough for most parallel optimization algorithms, particularly those involving multiple, concurrent evaluations of an arbitrary nonlinear objective function, although it may be well suited to parallelizing the linear algebraic calculations in optimization algorithms. For these reasons, the remainder of this paper is oriented to parallel computation on MIMD multiprocessors, usually without being specific about the type of MIMD multiprocessor.

3. General Purpose Parallel Methods for Small to Medium Size Problems -- Unconstrained Optimization

This section illustrates some of the opportunities and limitations that arise in creating general purpose parallel algorithms for small to medium size optimization problems. It does so by considering the example of the unconstrained optimization problem

$$\underset{x \in R^n}{\text{minimize}} \ f(x) : R^n \rightarrow R \ .$$

The most commonly used approach for solving unconstrained optimization problems when the number of variables is not too large, say less than 100, is probably the BFGS quasi-Newton method. In this section we will discuss the opportunities and limitations in creating parallel quasi-Newton methods. This section draws extensively on material in [7].

Like all methods for finding local minimizers, quasi-Newton methods are iterative. Given a current iterate x_c, the function and gradient values at this iterate, and a current Hessian approximation A_c, the basic steps of a quasi-Newton method are:

1. Calculate the search direction d_c: Solve $A_c \cdot d_c = -\nabla f(x_c)$ for d_c.
2. Line search: Find a $\lambda > 0$ for which $f(x_c + \lambda d_c) < f(x_c)$.
3. Calculate $\nabla f(x_c + \lambda d_c)$ and decide whether to stop; if not
4. Update the Hessian approximation: $A_+ = A_c +$ rank-two matrix

The BFGS method corresponds to a specific choice of the rank-two matrix in step 4, but this formula is not important to the discussion in this section. Also, there are various implementations of steps 1 and 4 that are mathematically equivalent but involve different methods of storing and updating A_c and solving for d_c. These variants have interesting consequences for parallelism that are discussed later in this section.

The main opportunities for using parallelism in existing or new quasi-Newton methods correspond to the three general uses of parallelism in optimization algorithms that were mentioned in Section 2.1:

1. One can parallelize the individual evaluations of $f(x)$ or $\nabla f(x)$ in steps 2 and 3 above.

2. One can parallelize the linear algebraic calculations in steps 1 and 4 above.

3. One can perform multiple evaluations of $f(x)$ (or $\nabla f(x)$) concurrently, either within the algorithmic framework above, or by devising new algorithms.

The remainder of this section discusses each of these possibilities briefly, and then draws some overall conclusions.

3.1 PARALLELIZING FUNCTION OR DERIVATIVE EVALUATIONS

As mentioned in Section 2.1, parallelizing the individual evaluations of $f(x)$ may be a very effective way to utilize parallel computers in optimization algorithms if the function evaluations are expensive and can be parallelized effectively, but this endeavor is outside the domain of optimization algorithm research. Parallelizing the gradient evaluations is a little more interesting to discuss since there are several possibilities. If the gradient is evaluated analytically then the same comments apply: this may be an effective use of parallelism but is outside the optimization domain. If the gradient is evaluated by finite differences, e.g. by the formula

$$\nabla f(x_c)_i \approx \frac{f(x_c + h_i e_i) - f(x_c)}{h_i} \quad , \quad i = 1, \cdots, n \tag{3.1}$$

where e_i is the i-th unit vector, then a trivial use of parallelism is to perform the n additional, independent function evaluations $f(x_c + h_i e_i)$ concurrently. If n is considerably larger than the number of processors and function evaluation is expensive, this may be all that is necessary to effectively parallelize a quasi-Newton method that uses finite difference gradients. Finally, if the gradient is evaluated by recently developed automatic differentiation techniques ([21]), there are several possibilities. If the forward version of automatic differentiation is used, analogous possibilities exists as for finite difference gradients. If the reverse mode is used, utilizing parallelism appears to be more difficult and is a topic of current research. In summary, parallelizing individual function or gradient evaluations may be an effective way to utilize parallelism for quasi-Newton methods, but it does not involve any change in the optimization algorithm.

3.2 PARALLELIZING LINEAR ALGEBRAIC CALCULATIONS

Next we turn to the issue of utilizing parallelism during the linear algebraic calculations of a quasi-Newton method. This may be desirable if the number of variables is reasonably large. While this topic also may appear to be outside the domain of optimization research, it is not entirely, because there are several ways of performing the algebraic calculations in steps 1 and 4 that are mathematically equivalent but have interesting differences and tradeoffs with respect to their operation counts, numerical properties, and parallelizability. For example, when the BFGS update (or any other rank-two update that preserves symmetry and positive definiteness) is used, the four main possibilities are:

1. Update A_c to A_+ by adding a rank-two matrix to A_c; each calculation of d_c

requires a Cholesky factorization of A_c and two triangular solves.

2. Update a Cholesky factorization of A_c to a Cholesky factorization of A_+ by adding a rank-one matrix to the Cholesky factor of A_c and then reducing this updated matrix back to lower triangular form by a sequence of $2n-2$ Givens' rotations; each calculation of d_c then requires two triangular solves.

3. Update $(A_c)^{-1}$ to $(A_+)^{-1}$ by adding a rank-two matrix to $(A_c)^{-1}$; each calculation of d_c requires a matrix-vector multiplication.

4. Update the (non-triangular) factorization $B_c B_c^T$ of $(A_c)^{-1}$ to a (non-triangular) factorization $B_+ B_+^T$ of $(A_+)^{-1}$ by adding a rank-one matrix to B_c to obtain B_+; each calculation of d_c then requires two matrix-vector multiplications.

Options #1 and #2 maintain approximations of the Hessian, while options #3 and #4 maintain the inverses of these approximations. Options #2 and #4 keep the approximations in factored form, while options #1 and #3 keep them in unfactored form. For a detailed review of these possibilities, see [11] or [7]. Here we just briefly review their comparison and then discuss the interesting issue from the viewpoint of this paper, which is the impact that the consideration of parallelism has had upon the choice between these options. This topic is also covered in more detail in [7].

Of the four options given above, the most straightforward, #1, requires $O(n^3)$ operations, while the remainder require $O(n^2)$ operations. Thus #1 is not generally used in practice. Of the remainder, #3 is the cheapest, but #2 has been used in most production codes, because it implicitly retains positive definiteness of the Hessian approximation by maintaining a Cholesky factorization of it. This in turn guarantees that all the search directions d_c are descent directions, which is very important to optimization algorithms. It had long been feared that the matrices generated by option #3 might lose positive definiteness due to finite precision error, and then possibly generate non-descent directions, and for this reason it has not often been used in codes.

However the consideration of parallelism introduces a conflicting factor into the comparison between options #2 and #3, because option #3 is much more conducive to parallelism than option #2. Option #3 requires only matrix-vector operations (a rank-two update and a matrix-vector multiplication), which can be parallelized very nicely. On the other hand, option #2 is based upon Givens' rotations and triangular solves, which require sequences of vector-vector operations on vectors ranging from length 2 to n and parallelize very poorly.

These reasons have motivated several researchers to examine whether there really is a difference in the numerical performance of BFGS methods based on option #2 versus option #3. Tests by [7,19,27] have found negligible differences in the iterates produced, over broad sets of problems. Given the considerable advantage of option #3 with respect to parallel implementation, it therefore appears that option #3 is preferable to option #2, at least on parallel computers. This is an interesting example where the consideration of parallelism has led to the re-examination of a part of a basic optimization algorithm, with interesting conclusions.

Finally, it should be noted that option #4 has rarely been considered, but is closely related to a method proposed by Han [20] for use in parallel computation. Option #4 has some attractive properties: like option #3, it parallelizes very well, and like option #2, it implicitly maintains positive definiteness by keeping a factorization of the matrix. However, it is more expensive than option #3. Thus, as long as option #3 does not have numerical problems, which appears to be the case, it would seem to be the preferable option for parallel (and probably also sequential) computation. Otherwise, option #4 would seem to be an attractive choice for parallel computation.

In summary, the consideration of parallelism has led to an interesting re-examination of the implementation of the linear algebraic steps in quasi-Newton methods. Note, however, that these options for parallelism do not involve any change in the basic optimization method, only in the details of its implementation.

3.3 PERFORMING MULTIPLE FUNCTION OR DERIVATIVE EVALUATIONS CONCURRENTLY

Last we turn to the third possibility for parallelizing a quasi-Newton method, the utilization of multiple, concurrent function or derivative evaluations. We only will consider the case when each function evaluation is performed by one processor, and the gradient is evaluated by finite differences (equation 3.1), but the points that this discussion makes are more general.

To motivate this discussion, consider the overall pattern of function and derivative evaluations in a standard quasi-Newton algorithm. Each line search requires one or more function evaluations, and is followed by one gradient evaluation. Extensive computational experience shows that the average number of function evaluations per line search is less than 1.5. Thus a typical sequence of function and gradient evaluations for three iterations of an optimization algorithm might be

$$f(x_{1,1}), \ f(x_{1,2}), \ \nabla f(x_{1,2}), \ f(x_{2,1}), \ \nabla f(x_{2,1}), \ f(x_{3,1}), \ \nabla f(x_{3,1}) \ . \tag{3.2}$$

Here $x_{i,j}$ denotes the j-th point tried in the line search at the i-th iteration. (In some line search algorithms, the gradient is evaluated at a small percentage of unsuccessful points as well, but this does not affect our conclusions and so for simplicity we ignore this possibility.)

As stated previously, if the gradients in (3.2) are evaluated by finite differences, the gradient evaluation can make excellent utilization of parallelism since it involves n independent function evaluations that can be performed concurrently. However, if we assume that each function evaluation is performed by just one processor, and we use a standard line search, the function evaluations in (3.2) present a problem for parallelism. This is because in a standard line search, the selection of the next candidate point $x_{i,j+1}$ depends upon the value of $f(x_{i,j})$, and thus the evaluation of $f(x_{i,j})$ must be concluded before the evaluation of $f(x_{i,j+1})$ is begun. Thus, if we implement a standard line search method on a parallel computer and don't parallelize function evaluations, then while one processor performs a function evaluation, all the others processors will be idle.

This observation has led to the suggestion of developing new parallel line search algorithms that evaluate the objective function at multiple points concurrently. For example if there are p processors, one could pick p points along the search direction, evaluate $f(x)$ at all of them (one per processor), and if any results in a decreased function value, use the point with the lowest function value as the next iterate. This parallel line search suggestion fully utilizes the multiple processors during the line search's function evaluations. Note also that it is the first suggestion we have discussed that actually changes the optimization method, i.e. the iterates produced, instead of just parallelizing it or possibly changing it due to finite precision sffects.

But the property of fully utilizing all processors does not necessarily make a parallel line search algorithm desirable. To assess whether it is, one needs to consider two issues. First, one needs to ask whether the extra evaluations are useful. That is, is the parallel line search algorithm faster on a parallel machine than a straightforward implementation of a standard line search method on the same machine? The answer to this question is probably yes for most problems. All that is needed is that the extra function evaluations in the line searches lead to some decrease in the total number of iterations required to reach the minimizer, and limited experiences indicates that this is usually so.

Secondly, one needs to consider whether there are better alternatives. In this case, an alternative is to use "speculative gradient evaluation". By this, we mean that when a function evaluation is performed in the line search, the remaining $p-1$ processors are used to evaluate $p-1$ components of the finite difference gradient at this new point. If the function value at the new point is too high and the point is rejected (as with $x_{1,1}$ above), this work is wasted unless some new way is found to use these gradient components at the rejected point. If however the function value is acceptable, as it will be roughly 60-80% of the time, then the speculative gradient evaluation has enabled us to do work on otherwise idle processors that we would otherwise have to do next. Thus, it has saved us time in the parallel implementation. To illustrate this, define a concurrent function evaluation step as one where some number between 1 and p of the processors do function evaluations while the remaining processors are idle. Then for the sequence (3.2), if $n = 25$ and $p = 16$, the number of concurrent function evaluation steps is 10 for the straightforward implementation (1 for each function evaluation and 2 for each gradient evaluation) and 7 for the speculative gradient evaluation version (1 for each combined function/partial-gradient evaluation and 1 to evaluate the remaining 10 components for each gradient).

So, to assess parallel line search algorithms, one needs to compare them with parallel speculative gradient evaluation algorithms. This can in part be done analytically. The costs of the parallel line search and speculative gradient evaluation algorithm in concurrent function evaluation steps are easy to express. (We are continuing to assume that gradient evaluation is by finite differences.) For the parallel line search algorithm, assuming the best case where an acceptable lower point is always found among the p candidates, the number of concurrent function evaluation steps is

$$\left(1+\left\lceil\frac{n}{p}\right\rceil\right)*It_p$$

where It_p is the number of iterations required by the parallel line search method. For the speculative gradient evaluation method, the number of concurrent function evaluation steps is

$$\left(\delta+\left\lceil\frac{n+1}{p}\right\rceil\right)*It_s$$

where It_s is the number of iterations required by the standard sequential line search method, and $(1+\delta)$ is the average number of function evaluations per line search. Thus the parallel line search will be superior if

$$\frac{It_p}{It_s} < \frac{\delta+\left\lceil\frac{n+1}{p}\right\rceil}{1+\left\lceil\frac{n}{p}\right\rceil}. \tag{3.3}$$

For the example above ($n = 25, p = 16$), if $\delta = 0.3$, this would require $(It_p)/(It_s) < 0.77$. As another example, if $n = 100$, $p = 16$, and $\delta = 0.3$, the parallel line search algorithm would be superior if $(It_p)/(It_s) < 0.91$. If $p \le n+1$, the parallel line search algorithm is superior if $(It_p)/(It_s) < (\delta+1)/2$.

Unfortunately, researchers who have proposed parallel line searches do not seem to have considered the comparison of their methods to speculative gradient methods, or to have examined whether their methods satisfy (3.3) for various problem sets and values of p. Our intuition is that it will be difficult to create parallel line search algorithms that satisfy (3.3), especially if n/p is not too large. The main reason for this hypothesis is that asymptotically, the first point tried in the standard line search at each iteration is essentially the best point along this line, and so the extra function evaluations are of little or no help. Thus, we suspect that the fairly mundane option of parallel speculative gradient evaluation may be preferable to new parallel line search algorithms, under the assumptions that function evaluations are sequential and gradient evaluations are made by finite differences. Clearly, further research is needed to assess this. The main points of this subsection are that it may be difficult to construct new parallel algorithms for generic, small to moderate sized problems that are superior to intelligent parallel implementations of standard methods, and that one must consider the alternatives when assessing new methods.

3.4 CONCLUSIONS AND DISCUSSION

The discussions in this section illustrate the limitations in the development of parallel algorithms for small to medium sized, generic optimization problems. Our opinion is that at least so far, the consideration of parallelism has not led to the development of exciting new generic optimization methods for small to medium size unconstrained optimization problems. Instead, the capabilities of parallel computers have best been utilized by parallelizing existing sequential algorithms in ways that do not affect the algorithms at the optimization level. These include parallelizing the individual function evaluations, parallelizing the linear algebraic operations, and using speculative derivative evaluations. We consider it likely that similar comments will be true for algorithms for small to medium size constrained optimization problems: the use of parallelism may not lead to the discovery and use of truly novel optimization algorithms, although one may be able to utilize parallel computation effectively by simply parallelizing standard sequential algorithms.

The above statements are not meant to imply that there hasn't been interesting research on new generic parallel algorithms for small to medium sized unconstrained optimization problems. Besides the use of parallelism in line searches, some interesting work has included new quasi-Newton methods for parallel computation ([38]) and the use of partial Hessian information ([8]). So far, however, these new methods do not appear to have had a large practical impact. There has also been interesting research on new derivative-free methods for unconstrained optimization that are especially well suited to parallel computation ([12]). This research may well have an important practical impact for the special problem class for which it is intended, namely problems with very small numbers of variables.

On the other hand, once one considers large-scale problems, there are many opportunities for the development of interesting new parallel optimization algorithms, including possibilities that superior sequential methods may be discovered through this process. The remainder of this paper illustrates this in the contexts of nonlinear equations and global optimization. There are also examples of interesting new parallel algorithms for large-scale unconstrained optimization, and we mention one very briefly. Nash and Sofer [25,26] have developed new block truncated Newton methods that are well suited to parallelism because they utilize p gradient values at each step, as opposed to one gradient value per step in the standard truncated Newton method. The new methods are significantly different than standard truncated Newton methods: in a standard truncated Newton method, a quadratic model is minimized over a subspace that is expanded by dimension one at each inner iteration, whereas in the block method, the subspace is expanded by dimension p at each inner iteration. This results in considerably different sequences of iterates. Preliminary tests by Nash and Sofer show promising results for this approach.

4. Parallel Methods for Large Problems with Special Structure -- Block Bordered Nonlinear Equations

This section illustrates the opportunities that the consideration of parallelism can provide for the creation of new methods for large-scale optimization problems with special structure. It does this by presenting one example, algorithms for the solution of block bordered systems of nonlinear equations. The material in this section is based upon [39].

Block bordered nonlinear equations are a class of large, sparse nonlinear equations that occur in many applications including VLSI design and structural engineering.

Algebraically, they have the following form. The n variables and equations are grouped into $q+1$ subsets, $x_1, ..., x_{q+1}$ and $f_1, ..., f_{q+1}$, where for each i, x_i and f_i have the same number of components. Using this notation, the nonlinear system of equations has the form

$$f_i(x_i, x_{q+1}) = 0, \qquad i = 1, ..., q, \qquad (4.1a)$$

$$f_{q+1}(x_1, ..., x_{q+1}) = 0. \qquad (4.1b)$$

That is, each set of equations except the last depends only upon an "internal" set of variables x_i and a "linking" set of variables x_{q+1}, while the final "linking" set of equations f_{q+1} involves all the variables. A common situation where this block bordered structure arises is in partitioning equations based upon a physical grid. If the variables correspond to values at the grid points, the equations correspond to relationships between values at neighboring grid points, and one partitions the grid into subregions and introduces artificial linking variables at the boundaries of the partition, a block bordered structure results. In this case, x_i, $i=1, \cdots q$ correspond to the variables within each subregion and x_{q+1} corresponds to the artificial variables.

The Jacobian matrix of (4.1) has the form

$$\begin{bmatrix} A_1 & & & & B_1 \\ & A_2 & & & B_2 \\ & & \cdot & & \cdot \\ & & & \cdot & \cdot \\ & & & A_q & B_q \\ C_1 & C_2 & \cdots & C_q & P \end{bmatrix}. \qquad (4.2)$$

Here the square matrices A_i are the derivatives of f_i with respect to x_i, the square matrix P is the derivative of f_{q+1} with respect to x_{q+1}, and the rectangular matrices B_i and C_i are the derivatives of f_i with respect to x_{q+1} and f_{q+1} with respect to x_i, respectively. The matrix (4.2) is referred to as a "block-bordered" matrix, from whence the problem class gets its name.

A standard way to solve (4.1) is to apply Newton's method with a line search or trust region. This results in solving a linear system of equations involving a matrix of the form (4.2) at each iteration. We will refer to this approach as the "explicit method", for reasons that soon will become clear. An important cost of this method is the factorization of the Jacobian at each iteration. Assuming there is no pivoting between blocks, each factorization involves the factorization of each block A_i, $i = 1, \cdots, q$, along with the modifications of B_i and C_i. These q steps can be performed independently of each other. Finally it involves the formation and factorization of the matrix

$$\hat{P} = P - \sum_{i=1}^{q} C_i A_i^{-1} B_i. \qquad (4.3)$$

The implicit method that is discussed next can be motivated by considering the costs of the factorization of (4.2) on a parallel computer. On a parallel computer, the first q steps of the factorization, factoring the diagonal blocks A_i and modifying B_i and C_i, parallelize almost perfectly assuming the q sets of blocks are partitioned equally among the processors. The final step, assembling and factoring $\hat{P}$, can be performed on one or several processors, but is likely to be a bottleneck in either case. The implicit method attempts to do more of the first type of operation, which parallelizes almost perfectly, and less of the second type.

The practical implicit method is derived from a "pure" implicit approach to solving (4.1). In this approach, each of the sets of internal variables x_i, $i = 1, \cdots, q$, is made into an implicit function $x_i(x_{q+1})$ of the linking variables x_{q+1} through the equation

$$f_i (x_i (x_{q+1}), x_{q+1}) = 0 . \tag{4.4}$$

Then the entire system of equations is reduced to solving

$$G (x_{q+1}) = f_{q+1} (x_1 (x_{q+1}), ..., x_q (x_{q+1}), x_{q+1}) = 0. \tag{4.5}$$

It is interesting to note that the Jacobian matrix of this system is given by (4.3). This is because in both (4.5) and the final step of the factorization of (4.2), the equations (4.1a) have been used to eliminate the variables x_i, $i = 1, \cdots, q$. The difference is that in the implicit approach, these variables have been eliminated at the nonlinear rather than at the linear level.

If one applied Newton's method directly to (4.5), one would need to solve exactly for each $x_i(x_{q+1})$ for each new value of x_{q+1}. This would not be efficient. Practical variants of the implicit method come from making an approximation to this process that finds each $x_i(x_{q+1})$ much more approximately at each step. A framework for this approach is given in Algorithm 4.1 below.

Algorithm 4.1 -- Framework of an Implicit Method for Block Bordered Nonlinear Equations

At each iteration:

1. Inner iterations:
   For each $i = 1, \cdots, q$, perform some iterations towards solving the nonlinear equations $f_i(x_i, x_{q+1}) = 0$, with x_i as the variable and x_{q+1} fixed.

2. Outer iteration:
   a. Calculate the Newton step for $G(x_{q+1}) = 0$, with each x_i, $i \le i \le q$, fixed at the value from step 1.
   b. Perform a line search over all the variables $x_1, \cdots, x_{q+1}$, possibly incorporating a correction to the steps in each of the sets of variables x_i, $i = 1, \cdots q$ first.

Zhang, Byrd, and Schnabel [39] investigated a number of variants of the practical implicit method given by Algorithm 4.1, and their comparison to the explicit method. They found that if one performs one "inner" iteration on each x_i at step 1, and the outer iteration with no correction to the x_i's, then the work per iteration is almost identical to the explicit method, but the performance is inferior. In particular, the convergence is only 2-step quadratic. If, however, one also adds the correction

$$x_i = x_i - A_i^{-1} B_i \, \Delta x_{q+1} \, , \quad i = 1, \cdots, q \tag{4.6}$$

in step 2b, then the method is identical to the explicit method. So far, this just constitutes a different derivation of Newton's method.

The interesting implicit methods arise when one performs more than one inner iteration on each x_i at each iteration of Algorithm 4.1, while retaining the correction (4.6). It turns out to be preferable to use the same matrices A_i and B_i for all the inner iterations of one iteration of Algorithm 4.1. Thus the additional inner iterations are quite inexpensive because no derivative matrices need to be computed or factorized.

These practical implicit methods have several good properties with respect to parallelism. First, by performing extra inner iterations, the fraction of time that the entire algorithm spends communicating between processors on a parallel computer is reduced, because the inner iterations require no communication. Secondly, they have good theoretical properties that are consistent with the efficient use of parallelism. Zhang, Byrd, and Schnabel [39] and Feng and Schnabel [13] show that these methods retain 1-step quadratic convergence per iteration of Algorithm 4.1, that the direction used in step 2b can be guaranteed to be a descent direction on $F(x)$ if the inner iteration is monitored properly, and that this monitoring can be done in a manner that does not require communication between the processors. Feng and Schnabel also show how to address singularity of the diagonal blocks A_i in a way that is amenable to parallelization and consistent with global convergence.

Most interestingly, Zhang, Byrd, and Schnabel show that these implicit methods may have computational advantages over the explicit method on sequential and parallel computers. For example, on a circuit model with 115 variables and equations that has four sets of internal variables consisting of 24, 27, 23, and 27 variables each, and 14 linking variables, the implicit method with three inner iterations per outer iteration was 15% faster than the explicit method on a sequential computer. On a parallel computer (an Intel iPSC hypercube) with 4 processors, this implicit method was 21% faster than an efficient parallel implementation of the explicit method. The advantage on the sequential computer arises because the number of outer iterations is reduced from 20 to 12, which more than compensates for the increase in the cost per iteration from the extra inner iterations. The parallel improvement is even greater because the parallel implicit method also requires less communication per iteration than the parallel explicit method, due to the higher ratio of inner to outer iterations. As a second example, on a related circuit model with twelve diagonal blocks, the implicit method with two inner iterations per outer iteration was 19% faster than the explicit method on a sequential computer, and the parallel implicit method was 28% faster than the parallel explicit method on an Intel hypercube with 12 processors. In this case, the number of outer iterations was reduced from 18 to 12.

While these tests are not comprehensive enough for definitive conclusions, they indicate that the new implicit methods outlined in Algorithm 4.1 may be preferable to the standard explicit method for solving block bordered nonlinear equations, on parallel and sequential computers. What we find most interesting about this is that the consideration of parallelism has led to the development of interesting new methods that may be computationally advantageous, even on sequential computers. In short, this is because the consideration of parallelism led naturally to the consideration of whether there were new ways to utilize a partitioning of the problem, and this led to the development of new methods. The new methods happened to be useful computationally; they also led to interesting challenges for theoretical analysis. We speculate that there will be other large scale optimization problems where the consideration of parallelism, particularly ways to subdivide the problem to accommodate parallelism, will lead to the formulation of interesting new methods. This is one of the interesting opportunities and challenges that parallelism presents to optimization.

5. Coarse Grain Parallel Algorithms for Large-Scale Optimization -- Global Optimization

This section illustrates some of the challenges and opportunities that arise in the design and implementation of coarse-grain parallel algorithms for large-scale optimization. By coarse-grain algorithms we mean algorithms where the basic computational units are of large and possibly irregular size. There are many problems in continuous and discrete optimization that give rise to coarse-grain algorithms, and they present interesting opportunities and challenges for parallel computation. These challenges and opportunities arise both in constructing efficient parallel algorithms and in the computer science issues associated with the implementation of these algorithms.

The global optimization problem is a good example of a problem that leads naturally to coarse-grain algorithms, and is used in this section to illustrate this problem class. In particular, this section is based upon our research in constructing parallel global optimization algorithms for generic global optimization problems with rather small numbers of variables (these are still large and expensive computations), and for problems from molecular chemistry with large numbers of variables. It draws upon material in [2-6, 32-35], as well as recent, as yet unpublished work. This section does not give a comprehensive review of this research; rather it attempts to describe this research at a level sufficient to illustrate the interesting issues that arise in conjunction with constructing coarse-grain parallel optimization algorithms.

The global optimization problem is to find the lowest minimizer of a nonlinear function $f(x)$ that may have multiple local minimizers, in some closed subregion D of R^n. In this section we will assume that the subregion D is just given by upper and lower bounds on each variable. This problem arises in many practical applications, such as the molecular configuration problems discussed later in this section. It can be very difficult to solve, for two basic reasons. First, it is very difficult, if not impossible, to find mathematical approaches that lead to efficient and reliable deterministic algorithms for solving these problems. Secondly, solving global optimization problems accurately appears to require a huge amount of computation in many practical cases. For this reason, large-scale nonlinear global optimization problems were hardly attempted until recently. This means that much of the initial algorithm development for these problems has occurred in the context of parallel computation. Therefore, as opposed to the previous sections, it is sometimes impossible to distinguish clearly between "standard" and parallel algorithms for large-scale global optimization.

A wide variety of approaches to global optimization have been proposed, mostly in the context of solving problems with just a handful (say 2-6) of parameters. We do not survey these in this paper; for comprehensive recent surveys, see [22,31]. Rather, we start by describing one approach, the stochastic global optimization approach of Rinnooy Kan and Timmer [30], that was the starting point for our parallel global optimization methods for problems with small numbers of variables.

The method of Rinnooy Kan and Timmer [30] is basically an intelligent "multi-start" algorithm. In a simple multi-start algorithm, one generates a number of random points x in the feasible domain D, starts a local minimization algorithm from each, and chooses the lowest local minimizer found (within D) as the candidate global minimizer. The main inefficiencies of this approach are that it may find all the local minimizers, and that it may find some of the local minimizers many times.

The approach of Rinnooy Kan and Timmer differs from simple multi-start mainly in that local minimizations are only started from carefully selected subset of the sample points, in a way that reduces or eliminates multiple local minimizations that lead to the same local minimizer. In particular, a local minimization is started from a sample point only if it has the lowest function value among all sample points within some "critical distance" from it. Also, the sampling/start-point-selection/local-minimization process is applied iteratively: for some number of iterations, new sample points are chosen, the critical distance is reduced by a carefully derived formula, start points for local minimizers are chosen, and the local minimizations are performed.

It can be shown that under reasonable assumptions, the method of Rinnooy Kan and Timmer finds all the local minimizers in a finite number of iterations, but that even if one iterates forever, the number of local minimizations remains finite. That is, a main inefficiency of simple multi-start has been eliminated. Computationally, this approach appears to be efficient if the number of local minimizers is fairly small. As will be mentioned later in this section, it does not appear to be an efficient approach to large-scale global optimization by itself, but is an important component of our approach to large-scale problems. For this reason, efficient parallel variants of it are of considerable practical interest.

5.1 A SIMPLE PARALLEL GLOBAL OPTIMIZATION ALGORITHM

It is fairly easy to construct a parallel algorithm that simply "parallelizes" the sequential stochastic method of Rinnooy-Kan and Timmer, and this was done as an initial research project in parallel global optimization ([4]). A straightforward and reasonably efficient way to do this is to partition the feasible region D into p subregions, where p is the number of processors. Then at each iteration, each processor samples and selects start points from its subregion. The latter step may require communication with other processors if a sample point is the lowest in its subregion within the critical distance, but is within the critical distance of other subregions. Finally at each iteration, all the start points for local minimizations are collected and distributed among the processors. It is necessary to collect and distribute start points, rather than having each processor simply perform the local minimizations from all the start points in its subregion, because subregions may have highly varying numbers of start points and the time per local minimization may vary significantly. This parallel algorithm is outlined in Algorithm 5.1.

Algorithm 5.1 has several interesting characteristics that begin to indicate the important parallel computation issues that are the focus of much of this section. First, the

---

Algorithm 5.1 -- Framework of a Simple Parallel Global Optimization Algorithm

Given $f : R^n \rightarrow R$, feasible region D, p processors

Partition D into p subregions

At each iteration:

1. **Sampling :** Each processor generates the coordinates of the new random sample points in its subregion, and evaluates $f(x)$ at each new sample point.

2. **Start Point Selection :** Each processor selects a subset of the sample points in its subregion to be start points for local minimizations. A sample point is selected to be a start point if it has the lowest function value of all sample points within the "critical distance" from it. This may require communication with processors that are responsible for neighboring subregions.

3. **Local Minimizations :** When all processors have completed start point selection, one processor collects all the start points and distributes them to the processors to perform the local minimizations from them. (Distribute one start point per processor; if there are more than p start points, distribute the remaining start points to the processors as they complete their current local minimizations.) When all local minimizations are complete, decide whether to stop, if not, begin the next iteration.

---

pieces of the algorithm that are executed in parallel between communication or synchronization points each involve large amounts of computation: sampling and evaluating $f(x)$ at large numbers of points (typically about 100), or performing one complete local minimization. This exemplifies coarse granularity. Second, due to the coarse granularity and the synchronization before and after step 3 at each iteration, it is possible that some processors may be idle for significant amounts of time. This can occur because each local minimization step (and even each start point selection step) can take widely varying amounts of time, and also because the number of local minimizations may not be a multiple of p. Third, the algorithm puts equal sampling effort into each subregion, regardless of whether or not it appears to be a fruitful region for finding the global minimizer. This is also true of the original sequential method and is not a parallel computation issue, but once one starts thinking in terms of subregions due to the consideration of parallelism, it becomes natural to consider varying this effort.

Byrd, Dert, Rinnooy-Kan and Schnabel [4] report computational experience with Algorithm 5.1 on a parallel computer. Basically, their conclusions are consistent with the above observations. On problems with relatively small numbers of variables, the method makes fairly effective use of small numbers of processors (say 8-32). But the coarse granularity and synchronization lead to idle time on some processors, and it is clear that this effect would be more pronounced with a greater number of processors. Also, it is apparent that greater efficiency in the basic method could be achieved by varying the sampling effort per subregion based upon the problem. The algorithm discussed in Section 5.2 is motivated by these observations.

5.2 AN ADAPTIVE, ASYNCHRONOUS PARALLEL GLOBAL OPTIMIZATION ALGORITHM

The limitations of the simple parallel stochastic global optimization method discussed above lead to the more interesting adaptive, asynchronous parallel approach for small, generic global optimization problems that we discuss next. The framework of such an algorithm is outlined in Algorithm 5.2. In comparison to Algorithm 5.1, there are two main new goals in this approach. The first is to concentrate the sampling and minimization effort in "productive" portions of the feasible domain D, i.e. subregions that are considered most likely to contain low minimizers. This goal is equally valid for sequential or parallel algorithms, but is more natural to achieve in an algorithm where the feasible region is divided into subregions, as it is for parallel computation. It is achieved by the adaptive portion of the algorithm, which consists of dynamically identifying subregions that are deemed more likely to contain low minimizers, and concentrating the sampling effort there. The second goal is to improve the load balancing of the algorithm, that is the distribution to work among processors, to eliminate processor idle time. This goal applies only to parallel implementations. It is achieved by the asynchronous part of the algorithm.

The framework that is chosen to accommodate both the adaptive and asynchronous features of Algorithm 5.2 is one where each sampling/start-point selection step for each subregion, and each local minimization from a new starting point, is a separate task. These tasks are distributed among the processors by some scheduling scheme. The overall control of this process is an important issue that is addressed below.

The adaptive, dynamic, and asynchronous aspects of Algorithm 5.2 make it significantly different than the parallel algorithms discussed earlier in this paper. Another way of stating this is that this algorithm is an irregular, task-oriented parallel algorithm, as opposed to the parallel methods discussed previously in this paper, which are either based upon data parallelism, or upon synchronized stages where at each stage, each processor performs the same task. The characteristics of Algorithm 5.2 lead to interesting algorithmic and parallel computation issues, and indicative of one of the key points of this paper: there appear to be a number of optimization problems that lend themselves to coarse-grain, irregular, task-oriented parallelism, and there are many interesting challenges and opportunities in designing and implementing these methods.

We will discuss the algorithmic and parallel computer implementation issues associated with Algorithm 5.2 only briefly, to illustrate these challenges. On the algorithmic side, one has to decide how to make the adaptive decisions in the algorithm. For example, how does one determine which subregions should receive more, or less, attention? In [32,33] this decision was based upon the percentage of low function values in the subregion relative to the overall domain. The sampling density was modified based upon this percentage, and very productive subregions were divided into smaller subregions while very unproductive subregions were skipped at some iterations. New algorithmic procedures are also necessary due to the irregular subregion and task structure: for example, rather than trying to locate neighboring subregions and obtain information about

Algorithm 5.2 -- Framework of an Adaptive, Parallel Global Optimization Algorithm

Given $f : R^n \rightarrow R$, feasible region D, p processors

Partition D into $q \geq p$ subregions

For each subregion :

1. **Sampling** : Generate the coordinates of the new random sample points in the subregion, and evaluate $f(x)$ at each new sample point.

2. **Start Point Selection** : Select a subset of the sample points to be start points for local minimizations. (A sample point is selected to be a start point if it has the lowest function value of all sample points within the "critical distance" from it; special techniques that do not require communication with other processors are used for sample points near subregion boundaries.)

3. **Adaptive Decisions** : Decide whether to split this subregion into smaller subregions, what the new density of sample points for the subregion(s) should be, and the relative priority of continuing to process this subregion. Then apply this algorithm recursively to each of the new subregions as processors become available and as its priority prescribes. (Generally this will be done after the current local minimizations in step 4.)

4. **Local Minimizations** : As processors become available, perform, on some processor, a local minimization from each start point selected in step 2.

Note : A central process generally controls termination of the entire algorithm and may also perform part or all of the scheduling of the parallel algorithm.

---

their sample points in the start point selection step, a new, self-contained "oversampling" strategy was devised. These examples indicate that the consideration of adaptive, dynamic variants of existing optimization methods, like Algorithm 5.2, leads to the emergence of new algorithmic issues.

On the parallel implementation side, a key issue is how one controls and schedules the entire asynchronous, dynamic parallel algorithm. This is an important and significant research topic by itself. Smith [32] and [34,35] investigate this topic extensively. They consider fully centralized and fully distributed scheduling strategies, and introduce a new partially distributed "centralized mediator" strategy. Through modeling, simulation, and parallel implementation, they show that the centralized mediator approach has considerable advantages in scalability over a fully centralized approach, and is very competitive with a fully distributed approach while being easier to implement. The details of these results are not important to this paper, but this topic illustrates that there are significant computer science challenges associated with implementing irregular, coarse-grained parallel algorithms. No matter what the control mechanism, there are also lower level parallel implementation issues that arise for this type of algorithm. For example, one must determine how to schedule processes to minimize data movement, what the priorities among tasks should be, and how to stop the algorithm without resorting to global synchronization.

Computational results for Algorithm 5.2 on some test problems, in a parallel environment consisting of a network of computer workstations, are given in [6,33]. They show that for problems where the local minimizers are unevenly distributed in the domain, the adaptive features of Algorithm 5.2 can lead to large improvements in efficiency over Algorithm 5.1, on sequential or parallel computers. On the other hand, the gains from the asynchronous, rather than synchronous, parallel implementation are more moderate.

These experiments and several others also illustrate several other, important lessons about these adaptive parallel global optimization algorithms. First, controlling them by a centralized scheduled process is reasonable for small numbers of processors (say 8) but likely to be a bottleneck for larger numbers of processors (say 32 or 64). This supports the need for the scheduling research mentioned above. Second, these algorithms are difficult to program and debug on parallel computers, mainly due to their asynchronous, task-oriented nature. This points to the need for research in systems or language support for asynchronous, task-oriented parallel algorithms.

Third, while the adaptive adjustments in Algorithm 5.2 greatly improve its efficiency in solving some small-scale global optimization problems, we have found that the algorithm is still not close to being an effective method for solving the much larger and more specialized problems that we discuss next. (To the best of our knowledge, this statement is equally true for other general purpose approaches to global optimization that have been developed for small problems.) Instead, as we discuss next, larger global optimization problems appear to require considerably more focused, and possibly more problem-specific, approaches.

5.3 A PARALLEL, LARGE-SCALE GLOBAL OPTIMIZATION ALGORITHM FOR MOLECULAR CONFIGURATION PROBLEMS

To illustrate some of the additional challenges and opportunities that stem from larger scale parallel global optimization, the final portion of this section briefly discusses some aspects of our work in global optimization for molecular configuration problems. The molecular configuration problem is to find the configuration of a chemical molecule or compound that has the minimum potential energy. This is believed usually to correspond to the configuration that the molecule or compound assumes in nature. Many problems whose solution is very important to scientists, including the protein folding problem, are posed in this manner.

There are several aspects that make molecular configuration problems very challenging global optimization problems. First, problems of real interest have hundreds or thousands of parameters, whereas until recently, the global optimization community was developing algorithms for problems with fewer than ten parameters. Second, typical potential energy functions have vast numbers of local minimizers; often the number is believed to be an exponential function of the number of variables. Third, there are many local minimizers whose function values are very close to the global minimizer. Finally, all the local minimizers seem to have small basins of attraction (regions in the domain from which local minimizations lead to them). These last three reasons combine to make it very difficult to find the global minimizer.

When one considers constructing global optimization algorithms for problems with so many parameters, and so many hard-to-find low local minimizers, one must address two main new difficulties in comparison to smaller problems. First, how can one explore (e.g. sample) effectively in such a huge dimensional space? Adding to this challenge is the fact that for molecular configuration problems, the function values of randomly selected configurations are often many orders of magnitude higher than the functions values of good configurations. Thus one must find efficient ways to find relatively low points throughout the domain in order to get an idea of whether a particular region is likely to contain low minimizers. Second, how can one find the lowest local minimizers efficiently? One cannot afford to find all the minimizers or even all the low ones; rather, it appears one must find a way to move from low minimizers to even lower ones efficiently. These issues indicate the need for new algorithmic approaches.

Algorithm 5.3 outlines the approach taken by [2,5] to address these issues for a class of molecular configuration problems. (In particular, this approach applies to molecular clusters; some modifications are needed for chains such as proteins or other polymers.) It is not our intent here to justify the merits of this algorithmic approach, but just to explain it sufficiently to be able to point out some interesting issues that it illustrates for parallel optimization.

The framework used in Algorithm 5.3 bears some relation to the stochastic methods discussed earlier in this section, but there are two very significant differences. First, the algorithm has two phases, one that is used to find an initial set of fairly low local minimizers, and a second that is used to move efficiently from low local minimizers to related, lower local minimizers. The first phase is closely related to the stochastic methods discussed previously, but the second phase, which accounts for the bulk of the computational work of the algorithm in practice, is quite different and essentially deterministic. The second difference will be seen to be particularly pertinent to our discussion of interesting issues for parallelism: a key technique used in the algorithm is the consideration of small dimensional subproblems within the large dimensional problem. In the first phase, this involves sampling steps that sample on only a small subset of the variables at once (step 1b). In the local minimizer improvement phase, it involves small-scale global optimizations in which only a few variables are allowed to vary with the remainder temporarily fixed (step 2b). Note that this means that the small-scale global optimization algorithm discussed above in Algorithm 5.2 can be used as a subalgorithm at Step 2b of the large-scale Algorithm 5.3.

Results of applying the approach outlined in Algorithm 5.3 to molecular configuration problems are given in [2,3,5] as well as in some forthcoming papers. One problem class the algorithm was applied to is Lennard-Jones problems. These problems consist of simple mathematical equations that model pairwise attractive/repulsive forces in clusters of identical, spherically symmetric atoms. They are much-studied in the chemistry community, in part because the forces they involve are a crucial part of the mathematical models of the energy of more complex molecular systems, including proteins and polymers. They are also very difficult global optimization problems, with the property of having many low, hard-to-locate local minimizers that was discussed above. Special purpose algorithms have been developed for this problem class based on the

Algorithm 5.3 - Framework of a Parallel Global Optimization Algorithms for a Class of Molecular Configuration Problems

1. **Coarse Identification of Configurations:** On each processor:

   (a) **Sampling in Full Domain:** Randomly generate the coordinates of sample configurations in the feasible domain, and evaluate the energy at each new sample point.

   (b) **One-Atom/Molecule Sampling Improvement:** For some (low energy) sample points: Select the atom/molecule that contributes the most to the energy function value, randomly sample on new locations for this atom/molecule, replace this atom/molecule in the sample configuration with the new location that gives the lowest energy value, and repeat until energy is below a threshold value.

   (c) **Full-Dimensional Local Minimizations:** Perform a local minimization from a subset of the improved sample points. Save these local minimizers for Step 2a.

2. **Improvement of Local Minimizers:**
   On each group of processors, for some number of iterations:

   (a) **Select a Configuration (and Expand it):** From the list of full-dimensional local minimizers, select a local minimizer to improve [and expand it around its center by a fixed factor, generally 1.25-2], and the atom/molecule that contributes the most (or second most) to the energy of this configuration.

   (b) **One-Atom/Molecule Global Optimization:** Apply a parallel global optimization algorithm (on this group of processors) to the energy of the selected configuration with only the coordinates of the selected atom/molecule as variables.

   (c) **Full-Dimensional Local Minimization:** Apply a local minimization procedure, with all atoms/molecules as variables, to the lowest configurations that resulted from the one-atom/molecule global optimization (one minimization at a time per processor), and merge the new local minimizers into the list of local minimizers.

---

known geometrical structure of the solutions. Based upon these, the global minimizers are believed to be known for all the instances with up to at least 150 atoms (450 variables) ([24,28]). Our algorithm has been tested on all the instances with up to 76 atoms (228 variables), and appears to find the best known solution in all cases (including an improved solution for 72 atoms recently discovered by [10]) and a better solution than previously discovered for 75 atoms). A second problem that the approach of Algorithm 5.3 has been applied to is the conformation of water molecules modeled by a well regarded potential energy function ([9]). In the two cases tried, 20 and 21 molecules (180, 189 variables), the energy values found by Algorithm 5.3 are far lower than those found by a previous method ([23]). All these results have been obtained on parallel computers, mainly 16 or 32 processors of an Intel iPSC/860 hypercube for the Lennard-Jones problems, and 64 processors of the Intel Delta at Caltech for the water problems.

These results indicate that parallel global optimization algorithms like the one outlined in Algorithm 5.3 have promise for solving important scientific problems. For this reason, we conclude this section by discussing some of the interesting parallel computation challenges presented by this type of algorithm.

The new challenges and opportunities in the area of parallel computation that are posed by a method like Algorithm 5.3 come mainly from the multi-level nature of the algorithm. For example in Phase II, it is possible for the algorithm to work on improving multiple configurations simultaneously, with each of these improvement steps involving the use of a small-scale global optimization algorithm. This allows for the use of two or even three levels of parallelism. For instance in our runs on a distributed memory multiprocessor with 64 processors, we generally improve 16 configurations simultaneously, using a group of four processors for each configuration. This means that the second level of parallelism involves using four processors for the small-scale global optimization in step 2b, and doing four large scale local minimizations simultaneously in step 2c. (Our experience indicates that using two levels of parallelism is preferable to using all the parallelism at either one of the levels in this case: at one extreme, improving 64 configurations at once is not effective due to the nature of the search space, while at the other extreme, improving one configuration at once and using 64 processors for the small-scale global optimization is not effective since the small-scale global optimization does not effectively utilize this much parallelism.) If we were using more processors, say 256, we could use a third level of parallelism at step 2c, for instance using four processors for each of the four local minimizations for each of 16 configurations.

We expect that algorithms that utilize multiple levels of parallelism will arise naturally in a number of large-scale optimization contexts, as well as many other areas of numerical computation. There is very limited experience with multi-level parallel algorithms so far in any context, and many interesting research issues arise in implementing them. For example, the issues of control and scheduling that were mentioned above for Algorithm 5.2 are even more difficult for these methods. In our two-level parallel implementation of Algorithm 5.3, it appears preferable to allow the algorithm to be asynchronous at both levels due to the irregular computational costs of the steps, and this leads to more difficult challenges in scheduling and controlling the algorithm than for the single-level asynchronous method discussed in Section 5.2. The issues involved in implementing irregular, multi-level parallel algorithms are an interesting part of the challenge in parallel large-scale optimization.

6. Summary

This paper has attempted to point out limitations, opportunities, and challenges in parallel nonlinear optimization through a set of examples. The examples are by no means exhaustive, and have been confined mainly to our own research. But they illustrate some important points.

First, it appears that so far, the consideration of parallelism hasn't led to many significant algorithmic innovations, or new theoretical challenges, for small to medium size generic optimization problems. This may be because, as is often conjectured, the best basic optimization approaches for small to moderate size problems are inherently sequential and already known. If this is true, it may mean that parallelism will mainly lead to new implementations, as opposed to fundamentally new algorithms, for these problems.

Second, it appears that for many classes of large-scale optimization problems, the consideration of parallelism may lead to the discovery of new algorithms that may be advantageous for parallel and possibly even sequential computation. This was illustrated by block-bordered systems of nonlinear equations, where the consideration of parallelism led to the investigation of new, implicit methods. It was also illustrated by global optimization problems with moderate numbers of variables, where the consideration of parallelism led naturally to the investigation of adaptive methods that dynamically partition the domain and decide which subregions should receive more or less emphasis. In both cases, the technique that led to the investigation of new methods was the partitioning of the problem into subproblems, and this partitioning was motivated by the consideration of parallelism. Related possibilities are likely to exist for other large problems.

Third, it appears that a number of large-scale optimization problems give rise to a coarse-grain, task-oriented, irregular type of parallelism. This was illustrated by the discussion of parallel global optimization; similar characteristics are seen, for example, in many branch and bound algorithms for discrete optimization problems ([37]). These types of algorithms not only are challenging from the point of view of the development of the optimization algorithm, but also pose many new challenges on the computer science side. These include the control and scheduling of the algorithms, and the development of communication and language features that make them easier to program and debug.

As a final comment, this paper has not considered problems where the optimization algorithm and the evaluation of the objective function or the constraints are combined. For example, this has been considered by [16] and others in cases where the evaluation of the objective function involves the solution of a system of partial differential equations. At one extreme, which was assumed in Section 3, the solution of this system of differential equations can be considered an atomic unit by the optimization algorithm. At another extreme, the variables within the differential equation solver can be incorporated into the optimization problem, and the solution of the differential equations considered a set of constraints to the optimization problem. This results in a very large-scale optimization problem whose variables are both the optimization parameters and the variables within the differential equations solver. This problem may have considerable exploitable structure, but may also be very difficult to solve ([16]). There are also intermediate approaches, such as splitting the domain of the differential equations into subdomains and making only the boundary points of these subdomains variables to the optimization algorithm. These approaches offer attractive possibilities for the use of parallelism. To the extent that they prove to be fruitful ways to solve these problems, they present additional interesting challenges and opportunities for optimization algorithms that are in part motivated by the consideration of parallelism.

7. References

(1)   G.S. Almasi and A. Gottlieb, *Highly Parallel Computing*, Benjamin Cummings, Redwood City, CA, 1989.

(2)   R.H. Byrd, T. Derby, E. Eskow, K.P.B. Oldenkamp, and R.B. Schnabel, "A new stochastic/perturbation method for large-scale global optimization and its application to water cluster problems," *Large-Scale Optimization: State of the Art*, W. Hager, D. Hearn, and P. Pardalos, eds., Kluwer Academic Publishers, 1994, pp. 71-84.

(3)   R. Byrd, T. Derby, E. Eskow, K. Oldenkamp, R.B. Schnabel, and C. Triantafillou, "Parallel global optimization methods for molecular configuration problems", *Proceedings of Sixth SIAM Conference of Parallel Processing for Scientific Computation*, SIAM, Philadelphia, 1993, pp. 165-169.

(4)   R.H. Byrd, C.L. Dert, A.H.G. Rinnooy Kan, and R.B. Schnabel, "Concurrent stochastic methods for global optimization", *Mathematical Programming* 46, 1990, pp. 1-30.

(5)   R.H. Byrd, E. Eskow, and R.B. Schnabel, "A new large-scale global optimization method and its application to Lennard-Jones problems", University of Colorado Technical Report CS-CS-630-92, 1992.

(6)   R. H. Byrd, E. Eskow, R. B. Schnabel, and S. L. Smith, "Parallel global optimization: numerical methods, dynamic scheduling methods, and applications to molecular configuration", *Parallel Computation*, B. Ford and A. Fincham, eds., Oxford University Press, 1993, pp. 187-207.

(7)   R.H. Byrd, R.B. Schnabel, and G.A. Shultz, "Parallel quasi-Newton methods for unconstrained optimization," *Mathematical Programming* 42, 1988, pp. 273-306.

(8)   R.H. Byrd, R.B. Schnabel, and G.A. Shultz, "Using parallel function evaluations to improve Hessian approximations for unconstrained optimization," *Annals of Operations Research* 14, 1988, pp. 167-193.

(9)   D.F. Coker and R.O. Watts, "Structure and vibrational spectroscopy of the water dimer using quantum simulation", *J. Phys. Chem.* 91, 1987, pp. 2513-2518.

(10)   T. Coleman, D. Shalloway, and Z. Wu, "A parallel build-up algorithm for global energy minimizations of molecular clusters using effective energy simulated annealing." Technical Report CTC93TR130, Advanced Computing Research Institute, Cornell University, 1993.

(11)   J. E. Dennis and R. B. Schnabel, *Numerical Methods for Unconstrained Optimization and Nonlinear Equations*, Prentice-Hall, Englewood Cliffs, NJ, 1983.

(12)   J.E. Dennis Jr. and V. Torczon, "Direct search methods on parallel computers, *SIAM Journal on Optimization* 1, 1991, pp. 448-474.

(13)   D. Feng and R.B. Schnabel, "Globally convergent parallel algorithms for solving block bordered systems of nonlinear equations", *Optimization Methods and Software* 2, 1993, pp. 269-295.

558

(14) R. Fletcher, *Practical Methods of Optimization,* Second Edition, John Wiley & Sons, New York, 1987.

(15) G. Fox, S. Hiranandani, K. Kennedy, C. Koelbel, U. Kremer, C. Tseng, and M. Wu, "Fortran D language specification", Center for Research on Parallel Computation Technical Report CRPC-TR90079, 1990.

(16) P.D. Frank and G.R. Shubin, "A comparison of optimization-based approaches for a model computational aerodynamics design problem", *Journal of Computational Physics* 98, 1992, pp. 74-89.

(17) P. E. Gill, W. Murray, and M. H. Wright, *Practical Optimization,* Academic Press, London, 1981.

(18) G. Golub and J.M. Ortega, *Scientific Computing -- An Introduction with Parallel Computing,* Academic Press, Boston, 1993.

(19) L. Grandinetti, "Factorization versus nonfactorization in quasi-Newtonian methods for differentiable optimization," Report N5, Dipartimento di Sistemi, Universitá della Calabria, 1978.

(20) S. P. Han, "Optimization by updated conjugate subspaces", *Numerical Analysis: Pitman Research Notes in Mathematics Series 140,* D. F. Griffiths and G. A. Watson, eds., Longman Scientific and Technical, Burnt Mill, England, 1986, pp. 82-97.

(21) A. Griewank, "On automatic differentiation," *Mathematical Programming: Recent Developments and Applications,* M. Iri and K. Tanabe, eds., Kluwer Academic Publishers, Tokyo, 1989, pp. 83-108.

(22) R. Horst and H. Tuy, *Global Optimization, Deterministic Approaches*, Springer Verlag, Berlin, 1992.

(23) X. Long, "Molecular dynamics simulations of clusters -- impure van der Waals and $e^- - (H_2O)_n$ systems", Ph.D. Dissertation, Department of Chemistry, University of California, San Diego, 1992.

(24) R.S. Maier, J.B. Rosen, and G.L. Xue, "A discrete-continuous algorithm for molecular energy minimization", AHPCRC Preprint 92-031, University of Minnesota, 1992.

(25) S.G. Nash and A. Sofer, "Block truncated-Newton methods for parallel optimization," *Mathematical Programming* 45, 1989, pp. 529-546.

(26) S.G. Nash and A. Sofer, "A general-purpose parallel algorithm for unconstrained optimization," *SIAM Journal on Optimization* 1, 1991, pp. 530-547.

(27) J. Nocedal, private communication.

(28) J.A. Northby, "Structure and binding of Lennard-Jones clusters: $13 \leq N \leq 147$", *J. Chem. Phys.* 87, 1987, pp. 6166-6177.

(29) M.J. Quinn, *Parallel Computing -- Theory and Practice,* McGraw Hill, New York, 1994.

(30) A.H.G. Rinnooy Kan and G.T. Timmer, "Stochastic methods for global optimization", *American Journal of Mathematical and Management Sciences* 4, 1984, pp.

7-40.

(31) A.H.G. Rinnooy Kan and G.T. Timmer, "Global Optimization", *Handbooks in Operations Research and Management Science, Volume I : Optimization,* G.L Nemhauser, A.H.J. Rinnooy Kan, and M.J. Todd, eds., North-Holland, 1989, pp. 631-662.

(32) S. Smith, "Adaptive Asynchronous Parallel Algorithms in Distributed Computation," Ph.D. Thesis, University of Colorado at Boulder, 1991.

(33) S. Smith, E. Eskow, and R.B. Schnabel, "Adaptive, asynchronous stochastic global optimization for sequential and parallel computation", *Large Scale Numerical Optimization,* T. Coleman and Y. Li, eds., SIAM, Philadelphia, 1990, pp. 207-227.

(34) S. Smith and R.B. Schnabel, "Centralized and distributed dynamic scheduling for adaptive parallel algorithms", *Unstructured Scientific Computation on Scalable Multiprocessors,* P. Mehrotra, J. Saltz, and R. Voigt, eds., MIT Press, Cambridge, Mass., 1992, pp. 301-321.

(35) S. Smith and R.B. Schnabel, "Dynamic scheduling strategies for an adaptive asynchronous parallel global optimization algorithm," University of Colorado Technical Report CU-CS-625-92, 1992.

(36) G.L. Steele, Jr., "High performance Fortran: status report", in *Proceedings of Workshop on Languages, Compilers, and Run-Time Environments for Distributed Memory Multiprocessors, SIGPLAN Notices,* Jan. 1993, pp. 1-4.

(37) H.W.J.M. Trienekens, "Parallel Branch and Bound Algorithms," Ph.D. Thesis, Erasmus University, Rotterdam, The Netherlands, 1990.

(38) P.J. van Laarhoven, "Parallel variable metric methods for unconstrained optimization," *Mathematical Programming* 33, 1985, pp. 68-81.

(39) X. Zhang, R.H. Byrd, and R.B. Schnabel, "Parallel methods for solving nonlinear block bordered systems of equations," *SIAM Journal on Scientific and Statistical Computing,* 13, 1992, pp. 841-859.

Subject Index